Craters of the Near Side Moon

John Moore

Copyright © John Moore 2014
Other than the use of single pages for publication reviews, further reproductions will require consent from the author.

Image Credits & Acknowledgements

LRO and LROC
Except where otherwise noted, all crater images used in this book are credited to the organisations, institutes and universities involved with the Lunar Reconnaissance Orbiter Cameras (LROCs) and the Lunar Reconnaissance Orbiter (LRO).

Launched to the Moon on 18 June 2009 from Cape Canaveral Air Force Station in Florida, LRO's cameras consist of two narrow angle cameras (NACs) capable of producing 0.5 metre-scale, panochromatic images, and a wide angle camera (WAC) providing images at scale of 100 metres/pixel across seven colour bands - 321, 360, 415, 566, 604, 643 and 689 nm respectively.

LROC is designed to address two main objectives: acquire images to facilitate safety analysis of potential landing sites at the lunar poles, and identify regions of permanent shadow and permanent, or near-permanent, illumination therein/at. Thousands of incredible images of the lunar surface have resulted during these objectives, and several global maps have been produced.

Other instruments onboard LRO also are returning data about the global Moon, its geo-morphology and geologic units, along with other physical properties of the surface.

Behind all this is a team of scientists across all disciplines - from geologists to geographers, engineers to aerospace people - each dedicated to various aspects of research and maintenance of LRO and its cameras. With all this in mind, this book is in recognition of their efforts and work, and, of course, a well-respected nod to the wonder that is LRO and LROC.

For more on LRO and LROC see:
NASA Goddard Space Flight Center: http://lro.gsfc.nasa.gov/
Lunar Reconnaissance Orbiter Camera: http://lroc.sese.asu.edu/index.html

International Astronomical Union (IAU)
Nomenclature, latitude and longitude coordinates (and sizes) for craters given in this book are from the official International Astronomical Union (IAU) Working Group for Planetary System Nomenclature (WGPSN). A sub-task group within the WGPSN is responsible for maintaining and updating the relevant nomenclature and data for the Moon, and so it is with special thanks to them also for their efforts.

For more on IAU see:
Gazetteer of Planetary Nomenclature (IAU): http://planetarynames.wr.usgs.gov/

Craters of the Near Side Moon
ISBN: 978-1497324442

Contents

Contents

	Page
● Crater Image Layout...................	1
● Crater Relationships..................	3 to 6
● Coordinates and Statistics...........	7
● Quadrant (NW)..........................	8
● Quadrant (NE)..........................	10
● Quadrant (SW)..........................	12
● Quadrant (SE)..........................	14
● East-West Limbs & Farside.........	16
● North & South Poles...................	17
● Phases.......................................	18
● Craters.....................................	19 to 681
● Features...................................	683
Catenae................................	684
Dorsa/Dorsum.........................	685
Lacus.....................................	688
Maria.....................................	690
Mons/Montes..........................	692
Palus.....................................	695
Promontorium.........................	696
Rima/Rimae...........................	697
Rupes....................................	703
Sinus.....................................	704
Vallis.....................................	705
● Index..	707

Crater Image Layout

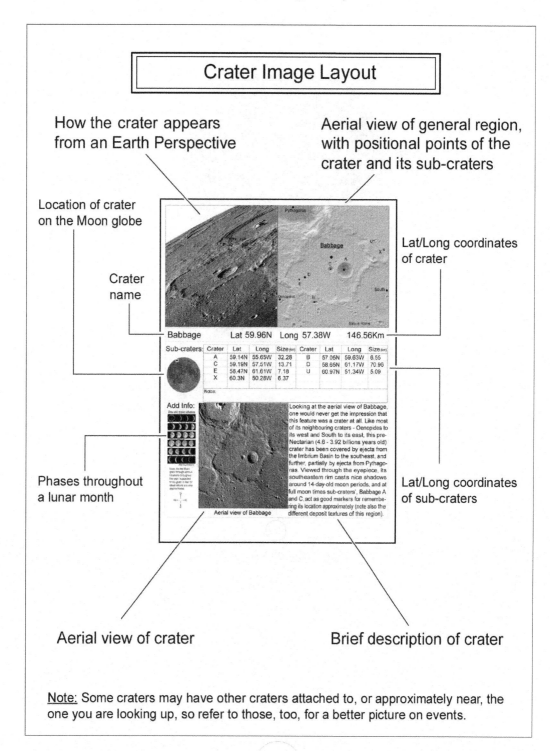

The Near Side of the Moon

Topographic Moon

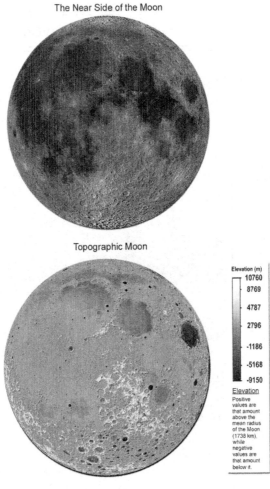

TIME PERIODS	AGE (billions of years)
pre-Nectarian	4.60 - 3.92
Nectarian	3.92 - 3.85
Lower (Early) Imbrian	3.85 - 3.75
Imbrian	3.85 - 3.15
Upper (Late) Imbrian	3.75 - 3.2
Eratosthenian	3.15 - 1.1
Copernican	1.1 - Present

Note: The above ages shown are approximates

Crater Relationships

Craters on the Moon are generally divided in to two main types - simple craters and complex craters. Simple craters are characteristically shaped like a bowl and typically range in sizes of less than 15 km, while Complex craters usually have flat floors, sometimes a central peak (or a ring of peaks), and range in sizes from between 20 to 400 km in diameter. Each type of crater forms differently according to the impactor's size and mass, its make-up, speed and angle of approach, but also make-up of the target rock that the impactor finally strikes. Three stages of a crater's formation are identified: the 'Contact' stage, the 'Excavation' stage, and finally the 'Modification' stage (Fig 1 below).

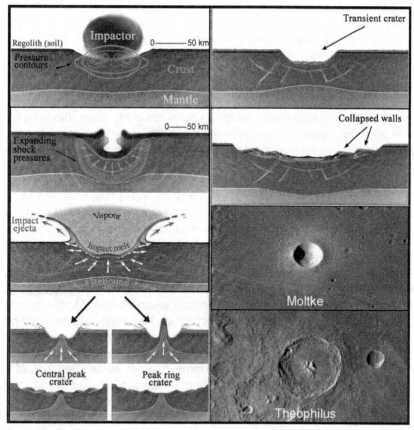

Fig 1. The 'Contact', 'Excavation' and 'Modification' stages of crater formation

Crater Relationships

The Contact stage
When the impactor makes first contact with the surface a series of shock wave energies are created that propagate both into the impactor itself (compressed upwards) and into the target rock (compressed downwards). Travelling sometimes faster than the speed of sound (between 6 and 10 km/s), these shock waves produce highly-confined pressures that affects each particle in the impactor and target rock differently. Propagating deeper and deeper into each, waves of lower pressure immediately behind the highly-confined ones causes a decompression-like effect, which for the impactor vaporises it near-complete, while for the target rock a huge volume of material is excavated outwards at great distance from the initial contact point. The shock effect is so extreme that temperatures in lieu with the pressures produced during the entire event causes partial melting (shock-melted) of the surface, and major fracturing deep down into the unexcavated rock of the forming crater

The Excavation stage
Target rock material isn't only being excavated in every direction from the growing crater, but it also is being directed sideways and downwards. That material going upwards and outwards in every direction results as an ejecta blanket surrounding the crater (a 'continuous' ejecta blanket ~ within one-diameter of the crater), while a 'discontinuous' ejecta blanket forms further away (producing such features like secondary craters, chains, rays etc.,). The deepest and oldest layer material of the target rock will lie closer to the crater (on its outer rim area), while that flung furthest is made up of the topmost, younger layers of the impacted surface. That material going sideways forms a transition zone where some will be excavated up to a particular angle, however, that material going downwards can't escape and so ends up being compressed into the target rock itself. All excavation has ceased when the shock waves end, and cavity-formation of the crater slows down before final modification. The crater, at this stage, is usually referred to as a 'transient' (Fig 1) crater - transient because the crater has yet to undergo further modification.

The Modification stage
Modification of the transient crater is almost immediate after the excavation stage. For the simple type, modification isn't very pronounced as the complex type, where wall material within begins to slump downward along discrete faults concentric to the crater's centre. Terraces form, while rebound of centrally-located material may also occur - depending on the initial impact dynamics and energies produced - leading to an uplifted 'peak' or 'ring-like' series of peaks. A recent *study of these peaks and their mineral make-up suggests an external origin in some - perhaps, of the initial impactor itself. The modification effects can increase the size of a crater by 15 percent and more, with similar, but less dynamic, adjustments continuing for years to millions of years afterwards through such events by seismic and erosion activity.
* *Nature Geoscience Letters* : "Projectile Remnants in Central Peaks of Lunar Impacts" (May 2013).

Crater Relationships

The crater, Moltke (Fig 1), is an excellent example of a Simple crater, while crater Theophilus (Fig 1) shows that of a Complex type. Note both crater types - Simple and Complex - have further classifications, but, in general, it's usually a matter of defining which type is which for simple descriptive purposes.

Fig 2. Crater Schiller - is it the result of a low angle impact

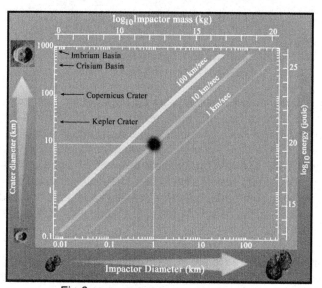

Fig 3. Crater Diameter vs Impactor Diameter

The above graph shows the relationship between the size of an impacting object and the relative size of a crater produced as a result of an impact. While it is generally accepted that the larger an impactor is the larger will be the crater, other factors like the constituent make-up of the impactor and target rock, along with the speed and angle at which the impactor strikes, also plays a role in the final crater size. Craters are almost always round in nature, however, when the incidence angle of an incoming impactor is very low, elongated craters can occur e.g. crater Schiller (Fig 2) may be the result of such an event. In the Fig 3 graphic above, the black dot represents an example of a 1 km diameter impactor striking the surface of the Moon at a speed of 10 km/sec, producing a crater 10 km in diameter.

Crater Relationships

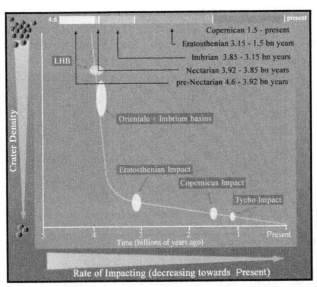

Fig 4. Crater Density vs Rate of Impacting

After the Moon formed over 4.6 billion years ago, a lot of left-over accretionary material impacted the inner planets and their moons. The rate of impacting was especially high during the first 0.8 billion years of lunar history, with a late heavy bombardment (LHB) contributing a series of major craters; affecting again the Moon, Mercury, Venus, Earth and Mars. The rate dramatically dropped off between 3.9 to 3.2 billion years ago, and from 3 billion years to the Present impacting slowed down to a virtual stop. The crater density on the Moon, therefore, acts as a kind of standard rule and useful tool for estimating the relative ages of surfaces and craters on other planets and other moons.

It is said that the older an area is the more craters will be found, so determining the age of geologic units across the lunar surface should simply be a matter of just counting craters. Yes? No! The problem is that while the above method may be of use up to a certain point in determining the relative age, or the geologic unit of an area, the fact is that the absolute age can never really be measured until actual samples are obtained. To-date, only nine sampled sites (Apollo's 11, 12, 14, 15, 16, 17 and Luna's 16, 20 and 24) of the lunar surface have been returned to Earth for analyses. However, while only half have been useful for determining the absolute ages (from 3.8 to 3.2 byo) of specific areas on the Moon (and also estimates for other areas), a huge gap still remains for those years from 3.2 byo down to the Present. Crater counting and samples, therefore, are two areas that need to be researched further in the future for understanding fully the geological history of the Moon.

Coordinates & Statistics

Coordinate System

Longitude coordinates of craters and features in this book mostly have grid designations given within 90 degrees for East and West. However, as some systems represent longitudes that read eastwards from the central meridian point of Long 0.0 degrees through to Long 360.0 degrees, the examples below may help in estimating your conversion.

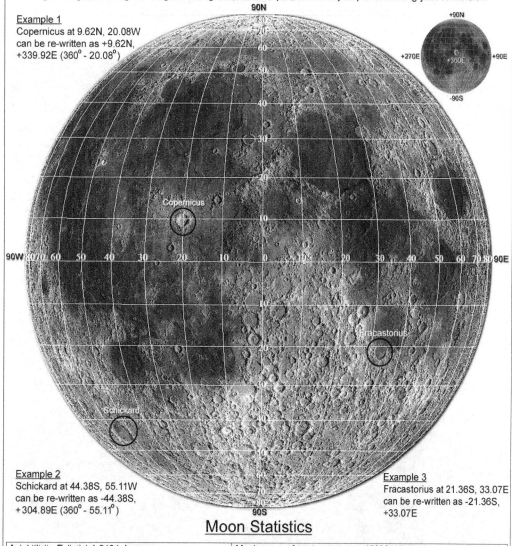

Example 1
Copernicus at 9.62N, 20.08W can be re-written as +9.62N, +339.92E (360° - 20.08°)

Example 2
Schickard at 44.38S, 55.11W can be re-written as -44.38S, +304.89E (360° - 55.11°)

Example 3
Fracastorius at 21.36S, 33.07E can be re-written as -21.36S, +33.07E

Moon Statistics

Axial tilt (to Ecliptic) 1.5424 degrees	Maximum surface temperature 123°Celsius
Equatorial diameter 3,476 km (0.273 of Earth)	Minimum surface temperature -233°Celsius
Equatorial escape velocity 2.38 km/sec	Orbital eccentricity 0.0549 degrees (variable from 0.026 to 0.077)
Equatorial surface gravity 1.62 m/sec^2 (1/6 of Earth)	Orbital inclination 5.1454 degrees (Min 4.98, Max 5.3)
Mass 7.349 x 10^22 kg (1/81 of Earth)	Perigee mean (closest distance) 363,300 km
Mean density 3.34 g/cubic cm (0.6 of Earth)	Polar diameter 3,472 km
Mean distance from Earth 384,401 km	Rotational period (Sidereal month) 27.32166 days
Mean surface temperature (day) 107°Celsius	Surface area 3.793 x 10^7 sq.km (0.074 of Earth)
Mean surface temperature (night) -153°Celsius	Synodic mean period (New Moon to New Moon) 29.53059 days

Quadrant (NW)

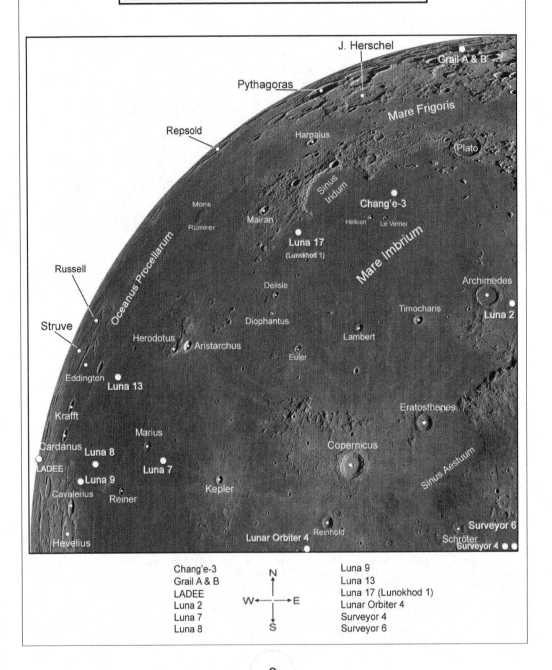

Quadrant (NW)

FEATURE	FEATURE	FEATURE	FEATURE	FEATURE
Anaxagoras	Gruithuisen	Montes Archimedes	Sylvester	
Anaximander	Harding	Montes Carpatus	Timocharis	
Anaximenes	Harpalus	Montes Harbinger	T. Mayer	
Angström	Hedin	Montes Jura	Ulugh Beigh	
Archimedes	Heinrich	Montes Recti	Vallis Schröteri	
Aristarchus	Heis	Montes Spitzbergen	Vasco da Gama	
Babbage	Helicon	Montes Teneriffe	Volta	
Balboa	Herodotus	Mouchez	von Braun	
Bancroft	Hevelius	Murchison	Voskresenkiy	
Beer	Horrebow	Naumann	Wallace	
Bessarion	Hortensius	Nernst	Wollaston	
Bianchini	J. Herschel	Nielsen	Xenophanes	
Birmingham	Kepler	Oc. Procellarum		
Bliss	Kirch	Oenopides		
Bode	Krafft	Olbers		
Bouguer	Krieger	Pallas		
Brayley	Kunowsky	Palus Putredinis		
Brianchon	la Condamine	Pascal		
Briggs	Lambert	Philolaus		
Cardanus	Langley	Piazzi-Smyth		
Carlini	Lavoisier	Plato		
Catena Krafft	le Verrier	Poncelet		
Cavalerius	Lichtenberg	Prinz		
C. Herschel	Louville	Prom. Heraclides		
Cleostratus	Macmillan	Prom. Laplace		
Copernicus	Maestlin	Pythagoras		
Dalton	Mairan	Pytheas		
Dechen	Marco Polo	Regnault		
Delisle	Mare Frigoris	Reiner		
Diophantus	Mare Imbrium	Reiner Gamma		
Dorsa Burnet	Mare Insularum	Reinhold		
Dorsum Arduino	Marius	Repsold		
Dorsum Heim	Markov	Rima Marius		
Dorsum Zirkel	Maupertuis	Robinson		
Draper	Milichius	Russell		
Eddington	Mons Ampere	Schiaparelli		
Einstein	Mons Delisle	Schröter		
Encke	Mons Gruithuisen Delta	Seleucus		
Epigenes	Mons Gruithuisen Gamma	Sharp		
Eratosthenes	Mons Huygens	Sinus Aestuum		
Euler	Mons La Hire	Sinus Iridum		
Fauth	Mons Pico	Sinus Lunicus		
Feuillée	Mons Piton	Sinus Roris		
Fontenelle	Mons Rümker	Sömmering		
Foucault	Mons Vinogradov	South		
Galilaei	Mons Wolff	Spurr		
Gambart	Montes Agricola	Stadius		
Gay-Lussac	Montes Alpes	Stokes		
Glushko	Montes Apenninus	Struve		
Goldschmidt	Montes Archimedes	Suess		

* NOTE: The index above does not cover all features found in this quadrant.

Quadrant (NE)

Northeast Quadrant (NE)

Apollo 11
Apollo 15
Apollo 17
Luna 15
Luna 18
Luna 20

Luna 21 (Lunokhod 2)
Luna 23
Luna 24
Ranger 6
Ranger 8
Surveyor 5

Quadrant (NE)

FEATURE	FEATURE	FEATURE	FEATURE	FEATURE
Agrippa	de la Rue	Joliot	Mitchell	Santos-Dumont
Al-Bakri	Delmotte	Julius Caesar	Moigno	Sarabhai
Aldrin	Dembowski	Kane	Mons Argaeus	Schmidt
Alexander	Democritus	Keldysh	Mons Bradley	Schubert
Alhazen	Deseilligny	Kirchoff	Mons Esam	Schumacher
Appolonius	de Sitter	Lacus Bonitatis	Mons Hadley	Schwabe
Arago	Dionysius	Lacus Doloris	Mons Hadley Delta	Schwarzschild
Aratus	Dorsa Aldrovandi	Lacus Felicitatis	Mons Vitruvius	Scoresby
Archytas	Dorsa Barlow	Lacus Gaudii	Montes Caucasus	Secchi
Ariadaeus	Dorsa Cato	Lacus Hiemalis	Montes Cordillera	Seneca
Aristillus	Dorsa Harker	Lacus Lenitatis	Montes Haemus	Shapley
Aristoteles	Dorsa Lister	Lacus Mortis	Montes Secchi	Sheepshanks
Armstrong	Dorsa Smirnov	Lacus Odii	Montes Taurus	Shuckburgh
Arnold	Dorsa Tetyaev	Lacus Perseverantiae	Neison	Silberschlag
Atlas	Dorsum Buchland	Lacus Somniorum	Neper	Sinas
Autolycus	Dorsum Oppel	Lacus Spei	Newcomb	Sinus Amoris
Auzout	Dorsum von Cotta	Lacus Temporis	Nobili	Sinus Concordiae
Baillaud	Egede	Lamèch	Oersted	Sinus Fidei
Baily	Eimmart	Lamont	Palus Somnii	Sinus Honoris
Banachiewicz	Endymion	Lawrence	Peary	Sinus Medii
Barrow	Esclangon	Lick	Peirce	Sinus Successus
Bernouilli	Euctemon	Linne	Petermann	Sosigenes
Berosus	Eudoxus	Littrow	Picard	Strabo
Berzelius	Fabbroni	Lucian	Plana	Sulpicius Gallus
Bessel	Firmicus	Luther	Plinius	Swift
Blagg	Franck	Lyell	Plutarch	Tacchini
Bobillier	Franklin	Maclear	Posidonius	Tacquet
Boscovich	Galen	Macrobius	Proclus	Taruntius
Brewster	G. Bond	Main	Prom. Agarum	Tebbutt
Burckhardt	Gardner	Manilius	Prom. Agassiz	Theaetetus
Bürg	Gärtner	Manners	Prom. Archerusia	Theophrastus
Byrd	Gauss	Maraldi	Prom. Deville	Tisserand
Calippus	Geminus	Mare Anguis	Prom. Fresnel	Townley
Cassini	Gioja	Mare Crisium	Protagoras	Tralles
Cauchy	Glaisher	Mare Frigoris	Rayleigh	Triesnecker
Cayley	Goddard	Mare Humboldtianum	Rhaeticus	Trouvelot
Cepheus	Godin	Mare Marginis	Riemann	Ukert
Chacornac	Greaves	Mare Serenitatis	Rima Ariadaeus	Urey
Challis	Grove	Mare Smythii	Rima Cauchy	Vallis Alpes
Chevallier	Hahn	Mare Spumans	Rima G. Bond	Virchow
Cleomedes	Hall	Mare Tranquillitatis	Rima Hadley	Vitruvius
C. Mayer	Hansen	Mare Undarum	Rima Hyginus	W. Bond
Collins	Hayn	Mare Vaporum	Rimae Bürg	Watts
Condorcet	Hercules	Maskelyne	Rimae Daniell	Whewell
Conon	Hooke	Mason	Rimae Triesnecker	Williams
Cusanus	Hyginus	Maury	Ritter	Yangel'
da Vinci	Ibn-Yunus	Menelaus	Römer	Yerkes
Daniell	Jansen	Mercurius	Ross	Zähringer
d'Arrest	Jansky	Messala	Rupes Cauchy	Zeno
Debes	Jenkins	Meton	Sabine	

* NOTE: The index above does not cover all features found in this quadrant.

Quadrant (SW)

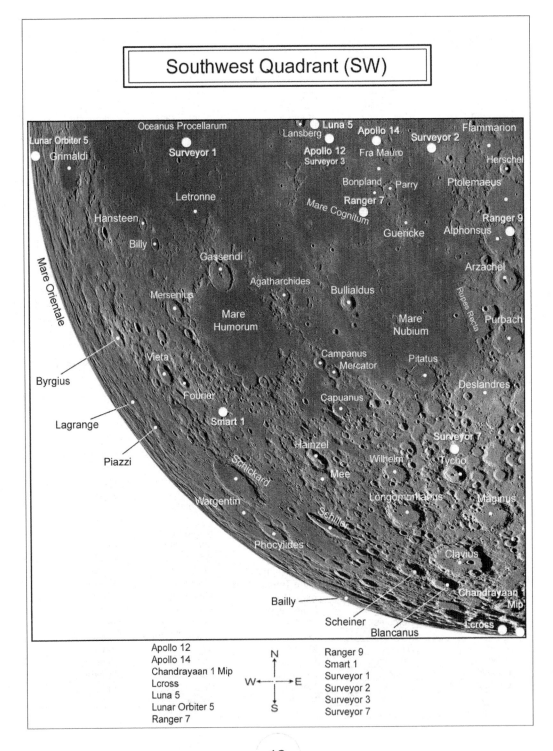

Quadrant (SW)

FEATURE	FEATURE	FEATURE	FEATURE	
Agatharchides	Gould	Mercator	Rutherfurd	
Alpetragius	Grimaldi	Mersenius	Sasserides	
Alphonsus	Gruemberger	Mons Hansteen	Saussure	
Ammonius	Guericke	Montanari	Scheiner	
Arzachel	Haidinger	Montes Cordillera	Schickard	
Bailly	Hainzel	Montes Riphaeus	Schiller	
Bayer	Hansteen	Montes Rook	Schlüter	
Bettinus	Heinsius	Moretus	Segner	
Billy	Hell	Mösting	Shackleton	
Birt	Henry	Nasireddin	Short	
Blancanus	Henry Freres	Nasmyth	Sirsalis	
Boltzmann	Herigonius	Newton	Sporer	
Bonpland	Herschel	Nicholson	Street	
Brown	Hesiodus	Nicollet	Thebit	
Bulliadus	Hippalus	Nöggerath	Tolansky	
Byrgius	Huggins	Oc. Procellarum	Turner	
Cabeus	Inghirami	Opelt	Tycho	
Campanus	Kies	Oppolzer	Vallis Bouvard	
Capuanus	Kircher	Orontius	Vallis Inghirami	
Casatus	Klaproth	Palisa	Vieta	
Catena Davy	König	Palmieri	Vitello	
Cavendish	Kuiper	Palus Epidemiarium	Wargentin	
Cichus	Kundt	Parry	Weigel	
Clausius	Lacroix	Pettit	Weiss	
Clavius	Lacus Aestatis	Phocylides	Wichmann	
Cruger	Lacus Autumni	Piazzi	Wilhelm	
Cysatus	Lacus Excellientiae	Pictet	Wilson	
Damoiseau	Lacus Timoris	Pilâtre	Winthrop	
Darney	Lacus Veris	Pingré	Wolf	
Darwin	Lagalla	Pitatus	Wright	
de Gasparis	Lagrange	Porter	Wurzelbauer	
Deluc	Lalande	Proctor	Yakovkin	
Deslandres	le Gentil	Prom. Kelvin	Zucchius	
de Vico	Lehmann	Prom. Taenarium	Zupus	
Doppelmayer	Lepaute	Ptolemaeus		
Dorsa Ewing	Letronne	Puiseux		
Dorsa Rubey	Lexell	Purbach		
Drebbel	Liebig	Ramsden		
Drygalski	Lippershey	Regiomontanus		
Eichstadt	Loewy	Riccioli		
Elger	Lohrmann	Rima Hesiodus		
Epimenides	Longomontanus	Rimae Hippallus		
Euclides	Lubiniezky	Rimae Mersenius		
Flammarion	Maginus	Rimae Ramsden		
Flamsteed	Mare Cognitum	Rimae Sirsalis		
Fontana	Mare Humorum	Rocca		
Fourier	Mare Nubium	Rost		
Fra Mauro	Mare Orientale	Rupes Kelvin		
Gassendi	Marth	Rupes Liebig		
Gauricus	Mee	Rupes Recta		

* NOTE: The index above does not cover all features found in this quadrant.

Quadrant (SE)

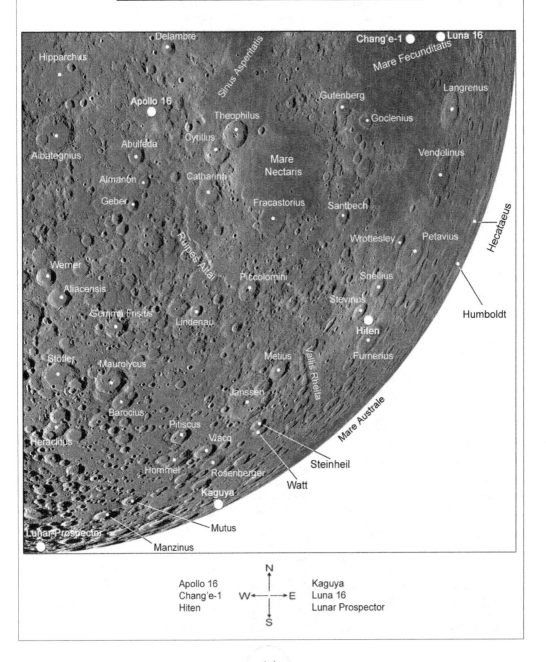

Quadrant (SE)

FEATURE	FEATURE	FEATURE	FEATURE	FEATURE
Abel	Cyrillus	Kaiser	Neumayer	Steinheil
Abenezra	Daguerre	Kant	Nicolai	Stevinus
Abulfeda	Delambre	Kapteyn	Nobile	Stiborius
Acosta	Delaunay	Kästner	Nonius	Stöfler
Adams	Demonax	Kiess	Oken	Tacitus
Airy	Descartes	Kinau	Palitzsch	Tannerus
Albategnius	Dolland	Klein	Parrot	Taylor
Alfraganus	Donati	Krusenstern	Peirescius	Theon-Junior
Aliacensis	Dorsa Geikie	la Caille	Pentland	Theon-Senior
Almanon	Dorsa Mawson	Lade	Petavius	Theophilus
Amundsen	Dove	la Pérouse	Petavius, Rimae	Torricelli
Anděl	Fabricius	Langrenus	Petrov	Vallis Rheita
Ansgarius	Faraday	Lansberg	Phillips	Vallis Snellius
Apianus	Faustini	Lassell	Piccolomini	Van Vleck
Argelander	Faye	Leakey	Pickering	Vega
Asclepi	Fermat	Legendre	Pitiscus	Vendelinus
Atwood	Fernelius	Licetus	Playfair	Vlacq
Azophi	Fracastorius	Lilius	Poisson	Vogel
Baco	Fraunhofer	Lindbergh	Polybius	von-Behring
Balmer	Furnerius	Lindenau	Pons	Walther
Barkla	Gaudibert	Lindsay	Pontanus	Warner
Barnard	Geber	Lockyer	Pontécoulant	Watt
Barocius	Gemma Frisius	Lohse	Rabbi-Levi	Webb
Beaumont	Gibbs	Lubbock	Réaumur	Weierstrass
Behaim	Gilbert	Lyot	Reichenbach	Weinek
Bellot	Goclenius	Maclaurin	Reimarus	Werner
Biela	Goodacre	Maclear	Rheita	Wilkins
Bilharz	Gutenberg	Mädler	Riccius	Wöhler
Biot	Gyldén	Magelhaens	Rimae Goclenius	Wrottesley
Blancanus	Hagecius	Malapert	Rimae Gutenberg	Young
Boguslawsky	Halley	Mallet	Rimae Hypatia	Zach
Bohnenberger	Hamilton	Manzinus	Rimae Janssen	Zagut
Borda	Hanno	Mare Australe	Rimae Petavius	Zöllner
Boussingault	Hase	Mare Fecunditatis	Ritchey	
Breislak	Hecataeus	Mare Nectaris	Rosenberger	
Brenner	Helmholtz	Marinus	Rosse	
Brisbane	Heraclitus	Maurolycus	Rothmann	
Buch	Hind	McClure	Runge	
Burnham	Hipparchus	Messier	Rupes Altai	
Büsching	Holden	Messier A	Sacrobosco	
Capella	Hommel	Metius	Santbech	
Catharina	Horrocks	Moltke	Saunder	
Celsius	Humboldt	Monge	Schomberger	
Censorinus	Hypatia	Mons Penck	Scott	
Clairaut	Ibn-Battuta	Montes Pyrenaeus	Seeliger	
Colombo	Ibn-Rushd	Müller	Shoemaker	
Cook	Ideler	Mutus	Simpelius	
Crozier	Isidorus	Naonobu	Sinus Asperitatis	
Curtius	Jacobi	Neander	Snellius	
Cuvier	Janssen	Nearch	Spallanzani	

* NOTE: The index above does not cover all features found in this quadrant.

East-West Limbs & Farside

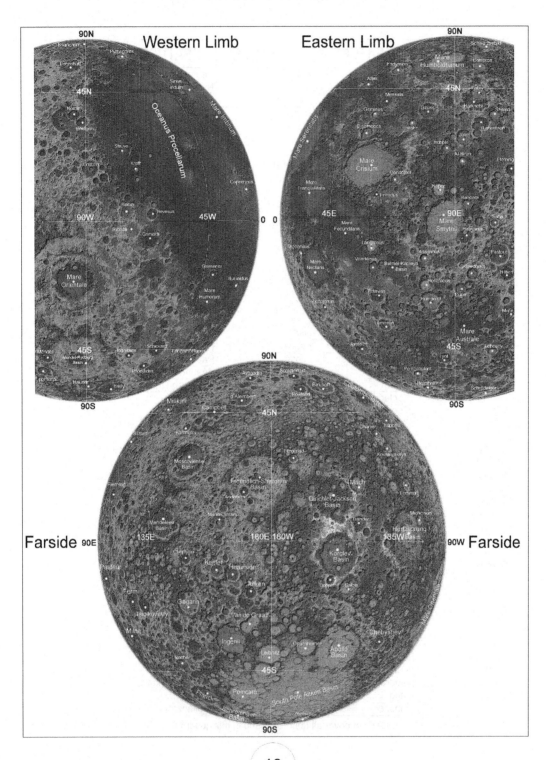

North & South Poles

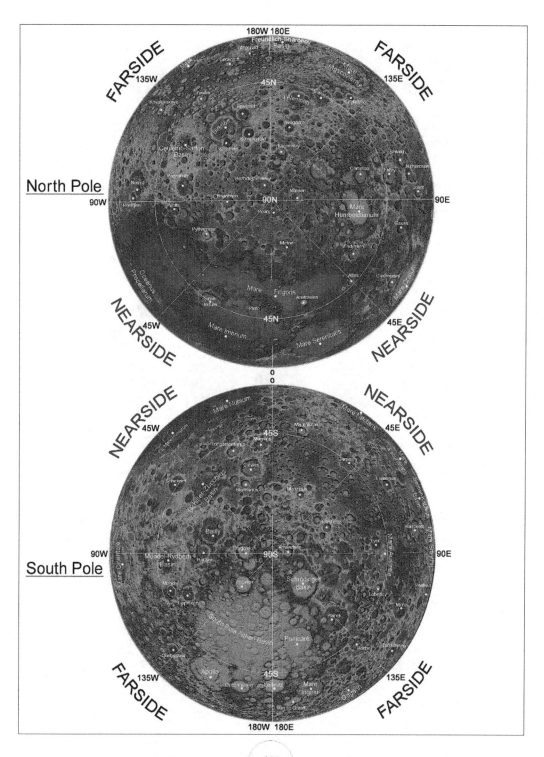

Phases

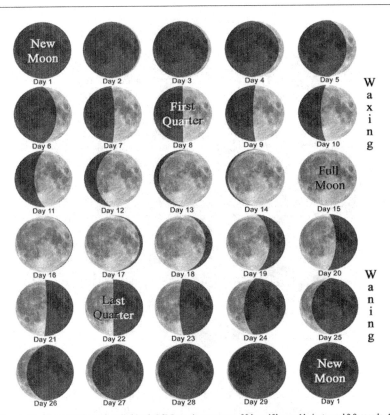

<u>New Moon to New Moon:</u> One Lunar Synodic Month (LSM) equals, on average, 29days, 12hours, 44minutes and 2.9seconds. As the length of the LSM is slightly longer than the Sidereal Month of 27days, 7hours, 43minues and 11.5 seconds (the time it takes the Moon to make one full orbit about the Earth with respect to the stars), and also, because both the Moon and Earth's speed and position are continually changing, the above phase-views fall on different days of the month throughout the year. NB. The above day numbers do not refer to the day-to-day date numbers we normally use in each month, but rather are the numbers of the LSM that relate to what the phase may look like (approximately) on that day.

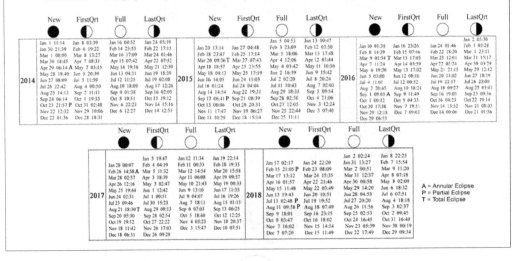

Craters

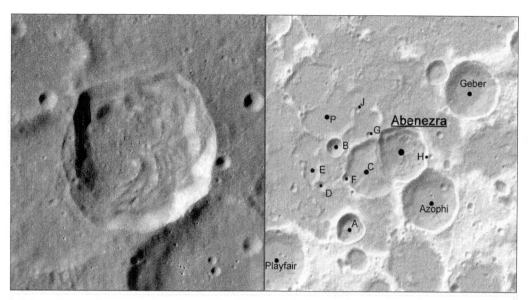

Abenezra   Lat 20.99S   Long 11.89E   43.19 Km

Sub-craters:

Crater	Lat	Long	Size (km)	Crater	Lat	Long	Size (km)
A	22.79S	10.44E	22.19	B	20.82S	10.06E	13.77
C	21.37S	11.09E	43.69	D	21.76S	9.65E	7.25
E	21.46S	9.4E	14.11	F	21.58S	10.34E	6.3
G	20.53S	11.0E	4.92	H	21.09S	12.71E	4.72
J	19.92S	10.69E	4.26	P	20.02S	9.85E	39.27

Notes:

Add Info:
Day-old moon phases

Note: As the Moon goes through various librations throughout the year, suggested times given in text for observations are only approximates.

Forming an easily-recognised triad: Abenezra, sub-crater Abenezra C and Azophi, the three all lie approximately midway between the Nectaris Basin over to the east (~ 650 km away) and the Nubium Basin over to the west (again about ~650 km away). Abenezra is definitely younger than Abenezra C to its southwest as most of its rim overlies nearly half of that crater, however, where the crater meets Azophi to its southeast, the two are vying for attention of who really is the youngest (the relatively sharp-looking rim between the two is what causes the confusion). What produced the unusual palm-tree-like feature of ridges in the eastern part of Abenezra's floor? Would it possibly be due to collapsed material from the eastern inner wall that met low-lying, mountainy (central peaks?) material, which then went on to produce the 'crimple-like' effect seen? Abenezra C's floor also has an unusual feature too that's not so obvious. Viewable during not-so far-off terminator times (easterly or westerly), it appears as a small 'valley' from C's western inner rim wall (starts below Abenezra F), and runs across the floor to where it tapers out before reaching Abenezra's western outer rim. Depth of Abenezra is around ~ 3.7 km approx..

Craters

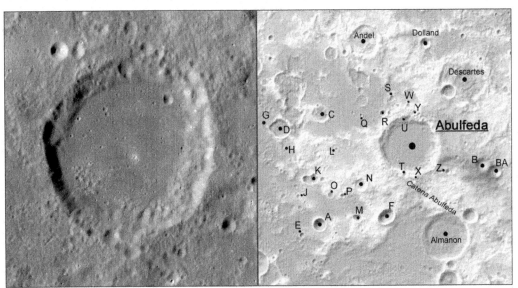

Abulfeda   Lat 13.87S   Long 13.91E   62.23Km

Sub-craters:

Crater	Lat	Long	Size (km)	Crater	Lat	Long	Size (km)
A	16.43S	10.75E	13.26	B	14.51S	16.37E	15.49
BA	14.69S	16.8E	13.99	C	12.79S	10.83E	16.43
D	13.25S	9.46E	19.68	E	16.76S	10.14E	5.55
F	16.19S	13.07E	17.47	G	13.1S	8.94E	6.31
H	13.84S	9.56E	4.22	J	15.49S	10.03E	4.07
K	14.92S	10.58E	9.93	L	14.09S	10.67E	4.95
M	16.23S	12.06E	10.05	N	15.14S	12.2E	13.07
O	15.47S	11.16E	6.6	P	15.5S	11.54E	4.59
Q	12.86S	12.22E	3.51	R	12.79S	12.97E	6.34
S	12.28S	13.3E	5.28	T	14.86S	13.71E	6.16
U	12.97S	13.8E	5.82	W	12.55S	13.85E	5.42
X	15.01S	13.99E	5.8	Y	12.76S	14.05E	4.63
Z	14.73S	15.18E	5.66				

Add Info: Note Catena Abulfeda's alignment - exactly what was the original source of these craters? Are they secondary-related, or are we looking at something exterior - e.g. the break-up of a comet?).

Notes:

Lying not so far away (~ 600 km) from the Nectaris Basin to the east, Abulfeda is of the Nectarian Period (3.92 to 3.85 byo). The crater looks relatively fresh, its floor has probably been filled with ejecta deposits from the above-mentioned basin, and several small craterlets 'dot' this material almost everywhere (note the very fresh, bright signature of one in the centre). But what has happened at the west, inner rim wall? Abulfeda's other walls and terraces all look as we would expect them to be - thick-walled and impacted upon, but why is the west section so edge-like (level of Nectaris's ejecta isn't any higher there). D: ~1.2 km.

Craters

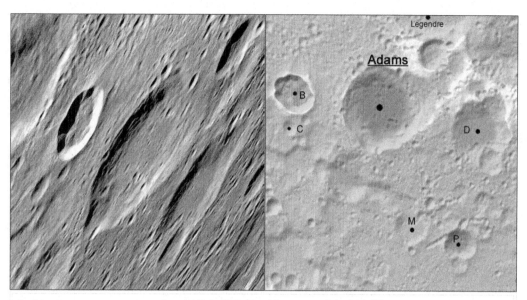

Adams Lat 31.89S Long 68.39E 63.27Km

Sub-craters:

Crater	Lat	Long	Size (km)	Crater	Lat	Long	Size (km)
B	31.54S	65.7E	31.33	C	32.35S	65.54E	11.12
D	32.41S	71.45E	46.26	M	34.73S	69.33E	25.23
P	35.15S	70.91E	25.43				

Notes:

Add Info:

Aerial view of Adams

Nectarian in age (3.92 to 3.85 billions of years old), Adams has a well-worn rim, slumped deposits that almost fill its floor, and numerous small impact craters on nearly every part of it (several or so of the crater chains on its outer southeastern rim may be from crater Petavius - Lower Imbrian in age - to its northwest). The northeast section of Adams's rim adjoins onto what looks like another old crater long lost to geological history (or, is this feature some kind of old oblique signature of Adams?). Bright-walled Adams B, to its west, is well contrasted in sharpness to Adams. Depth ~ 4.5 km.

Craters

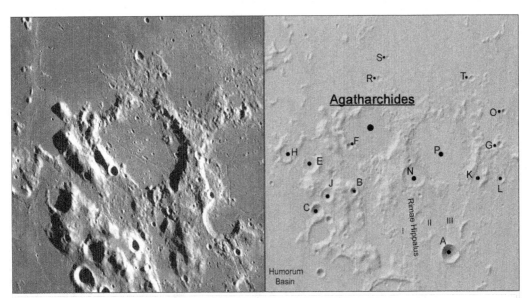

Agatharchides   Lat 19.85S   Long 31.11W   51.98Km

Sub-craters:

Crater	Lat	Long	Size (km)	Crater	Lat	Long	Size (km)
A	23.27S	28.46W	15.68	B	21.52S	31.71W	6.97
C	22.06S	32.94W	11.46	E	20.74S	33.19W	15.27
F	20.26S	31.86W	5.89	G	20.17S	26.84W	5.96
H	20.41S	33.95W	15.97	J	21.62S	32.61W	13.04
K	21.08S	27.53W	13.35	L	21.12S	26.79W	7.99
N	21.09S	29.72W	21.65	O	19.24S	26.71W	4.65
P	20.3S	28.77W	65.98	R	18.36S	30.85W	5.11
S	17.75S	30.59W	3.06	T	18.32S	27.8W	5.62

Notes:

Add Info: One could almost believe from just looking at Agatharchides that it really isn't a crater at all, but simply the approximate shape of one bounded by mountain masses to its east and west. Both its northern and sourthern sectors are all broken up (where they may have been breached?) - the former connecting it to Oceanus Procellarum to its north and the Humorum Basin to its south-west; while the latter to Mare Nubium on its east. The crater interrupts the infamous triad of rilles that is Rimae Hippalus, whose middle rille (designated with a now defunct reference system of Roman numeral, II) is seen to cut across the NE of the floor. At high magnification, another smaller rille to its west can just about be seen (given suitable lighting conditions) to parallel rille II, but it veers off to the northwest part of the rim. The crater does however have its attributes while during low sun terminator times some nice shadows are cast from its mountainy surrounds. Depth of Agatharchides is around 1.2 km approx..

Craters

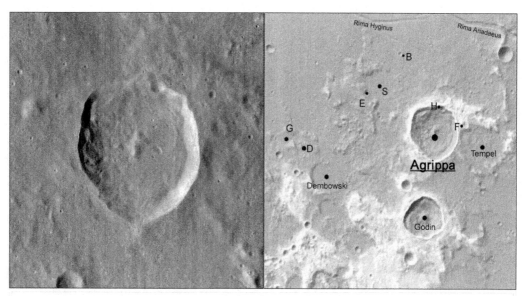

Agrippa	Lat 4.1N	Long 10.47E	43.75Km

Sub-craters:

Crater	Lat	Long	Size (km)	Crater	Lat	Long	Size (km)
B	6.2N	9.46E	4.01	D	3.77N	6.72E	21.36
E	5.16N	8.44E	5.0	F	4.3N	11.37E	5.35
G	3.91N	6.16E	13.24	H	4.75N	10.7E	4.86
S	5.28N	8.89E	31.71				

Notes:

Add Info: Agrippa lies in a region of terrain where the dominant makeup is that of Imbrium ejecta (the Basin's centre is some 1200 km away to the northwest), mixed in with sparse outcrops of original lunar highlands. The crater is Eratosthenian in age - 3.5 to 1.1 billions of years old, it's diameter is some 8 kilometres longer in its NS axis than its EW, it has a central peak ~ 1 kilometre-high, and the crater, generally, has a fresh look to it overall (its sharp rim might suggest it is younger - compare it to crater Godin to its south that is Copernican - 1.1 byo to the present). The crater initially formed on Imbrium's ejecta, but it also clipped the northwestern rim of the very old, barely perceptible crater, Tempel, to its east. Slumping effects only seems to have occurred more at Agrippa's northern, southern and western sectors; giving a rubbly appearance to its floor there in opposite to the more smoother look in the east (the difference shows up nicely during low sun times). Several crater pairs are found all over the Moon, for example, Aristotles and Eudoxes, Aristillus and Autolycus...etc.,, and Agrippa, with Godin to its south, are no different. These pairs may be widely different in features or makeup, but sometimes comparisons between the pair in question serves as a useful exercise in discovery. At full moon times Agrippa's bright walls locate the crater easily, while Godin's rays are seen to cross its floor and the crater's general surrounds. Depth ~ 3.0 km.

Craters

Airy   Lat 18.14S   Long 5.61E   38.9Km

Sub-craters:

Crater	Lat	Long	Size (km)	Crater	Lat	Long	Size (km)
A	17.05S	7.64E	12.19	B	17.67S	8.46E	27.26
C	19.25S	4.77E	30.49	D	18.13S	8.34E	6.61
E	20.75S	7.55E	38.04	F	18.2S	7.26E	4.84
G	18.7S	6.96E	25.32	H	18.74S	5.69E	9.31
J	19.05S	6.08E	3.93	L	20.47S	7.54E	6.02
M	19.15S	7.56E	0.85	N	17.85S	8.18E	7.95
O	16.75S	8.35E	4.53	P	15.88S	8.35E	6.64
R	19.6S	8.78E	6.34	S	17.28S	9.39E	5.31
T	19.27S	9.38E	36.92	V	17.5S	9.14E	4.43
X	18.95S	10.12E	3.87				

Notes:

Add Info: Showing all the characteristic signs of an early-formed impact crater - worn rim all around, smooth central peak, and several smaller craters that impacted nearby - this crater is of the Nectarian Period (3.92 - 3.85 billions years old). Impact melt deposits - possibly from the Imbrium Basin some 1500 km away to its northwest - fill portions of its interior, while all around ejecta from that event covers the region in general. Airy A, to its northeast, is classed into the 'banded' series of craters (that is, they have dark radial bands of material on their inner walls), while east of Airy crater itself - some 60 km away - a bright 'swirl' of material (infomally called "the Airy Swirl") can be seen at high sun times. The sometimes bright peak has a small 3 km-wide crater on its eastern flank.

Craters

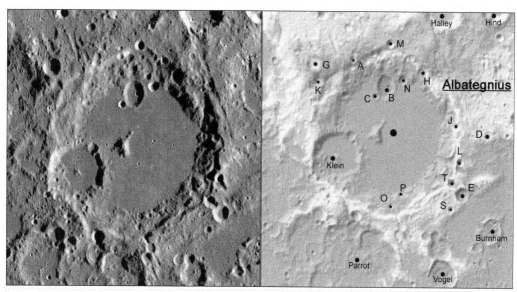

Albategnius   Lat 11.24S   Long 4.01E   130.86Km

Sub-craters:

Crater	Lat	Long	Size (km)	Crater	Lat	Long	Size (km)
A	9.34S	3.03E	6.79	B	10.07S	4.03E	18.98
C	10.33S	3.71E	5.66	D	11.36S	7.14E	8.35
E	12.95S	6.38E	12.67	G	9.46S	1.89E	13.79
H	9.66S	5.16E	11.77	J	11.14S	6.19E	5.97
K	9.9S	2.0E	10.31	L	12.08S	6.31E	7.45
M	8.92S	4.16E	8.07	N	9.89S	4.54E	8.61
O	13.62S	4.43E	4.54	P	13.02S	4.49E	4.01
S	13.35S	6.05E	5.91	T	12.66S	6.06E	9.61

Notes:

Add Info: Not to be confused with the similar-sounding crater, Alpetragius (some 280 km away to its southwest - 16.05S, 4.51W), Albategnius lies in a rough region of the Moon generally known as the Imbrium Sculpture - bombardment-like features of grooves, ridges and secondary cratering produced as a result of formation of the Imbrium Basin (centred some ~ 1400 away to the northwest). These grooves...etc., are radial to Imbrium whose effect caused the whole region to be altered dramatically as blocks of material and ejecta 'sprayed' out in every direction, and at every angle from the central explosive point. The smooth floor of Albategnius - once beleived to be the result of internal volcanic origin - may be of fluidized ejecta from Imbrium. The crater is of the Nectarian Period (3.92 to 3.85 billions of years old), it has a nice central peak (~ 1.8 km high), and crater Klein, which has obviously impacted Albategnius's floor, has its floor level some 1.2 km below that of Albategnius's. At relatively low to high magnifications, the small impact crater (vent?) on Albategnius's peak is easily seen. Depth is around ~ 4 km

Craters

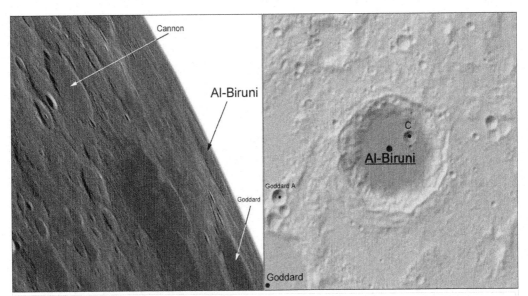

Al-Biruni	Lat 18.07N	Long 92.62E				80.41Km	
Sub-craters:	Crater	Lat	Long	Size (km)	Crater	Lat	Long	Size (km)
	C	18.42N	93.06E	9.53				
	Notes:							

Add Info:

Aerial view of Al-Biruni

Al-Biruni is situated some ~ 170 kilometres away northeastwards from the centre of Mare Marginis. Its depth of over 3.5 kilometres means the impactor tapped into an underlying lava source related to the Mare; leaving its floor filled internally. The albedo of the lava isn't as pronounced (darker) than crater Goddard to the southwest, but then as sub-crater, Al-Biruni C, as well as Goddard A, have dusted lightly the floor, it may explain the contrasting difference between the two. Al-Biruni is younger than Goddard, it shows hint of terraces at the upper parts of its walls, but the lower parts right down to the floor are less defined (more worn).

Craters

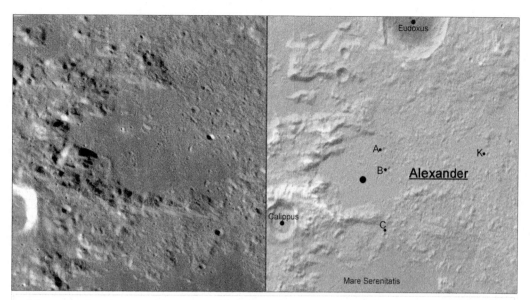

Alexander	Lat 40.25N	Long 13.69E	94.8Km

Sub-craters:

Crater	Lat	Long	Size (km)	Crater	Lat	Long	Size (km)
A	40.78N	14.94E	3.99	B	40.28N	15.14E	3.89
C	38.52N	14.95E	4.23	K	40.52N	19.35E	3.95

Notes:

Add Info: An observer of this 'crater' could easily mistake it as just an enclosure of dark material surrounded by mountains to its west (the Caucasus Mts.,) and rubbly, highland's material to its east. Almost all of its western rim has been altered by Imbrium Basin debris from the west - producing a higher terrain region in contrast to its east. This basin event has also caused the central material in Alexander to take on a lighter hue when one compares it to the darker-looking material found southwards in Mare Tranquillitatis's northern sector. The most eastern part of the rim, where sub-craters Alexander A and B lie, is virtually non-existant - buried or worn? But at terminator times when the Moon is around 8 and 21 days old respectively, the former period shows up the crater's western sector as specularly bright, while the latter period casts prominent shadows across its floor. The floor does appear to be slightly higher on the eastern side (an 8-day-old moon casts a shadow right across its centre in a N-S direction), and at higher sun angles, several small craterlets can be observed - one of which (~ 4 km in diameter) shows bright-ray material. Aerial views of Alexander especially show it to be more rectangular in shape than round; suggesting, perhaps, it to be just a depressed feature, rather than a crater, related to formation of the Serenitatis Basin to its south. Depth of crater is around 0.4 km approx..

Craters

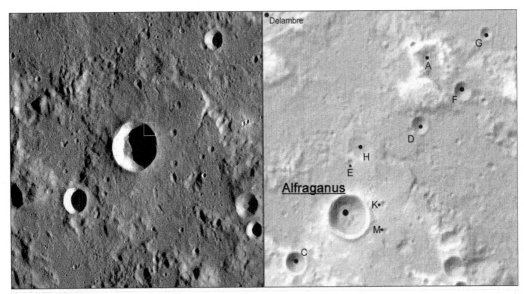

| Alfraganus | Lat 5.42S | Long 18.97E | 20.52Km |

Sub-craters:

Crater	Lat	Long	Size (km)	Crater	Lat	Long	Size (km)
A	3.05S	20.32E	13.87	C	6.13S	18.09E	10.3
D	4.05S	20.14E	8.29	E	4.62S	19.0E	3.85
F	3.54S	20.84E	8.64	G	2.67S	21.23E	5.74
H	4.36S	19.16E	12.0	K	5.25S	19.54E	2.95
M	5.66S	19.56E	3.16				

Notes:

Add Info:

Boulder group on Alfraganus's northeastern rim

Alfraganus lies in a region of smooth terrain possibly related to the Cayley Formation - a form of fluidized, non-melted ejecta produced by the giant impact now known as the Imbrium Basin (centred ~ 1500 km away to the northwest). Alfraganus, therefore, is younger, perhaps, Eratosthenian (3.15 to 1.1 byo) in age. Close-up of the crater's interior show what looks like small slumps occurred at its western wall, which now lie centrally on the floor. Left, shows an extreme close-up view of a boulder group on Alfraganus's northeastern rim (was it at one time a large, single bouder that broke up?).

Craters

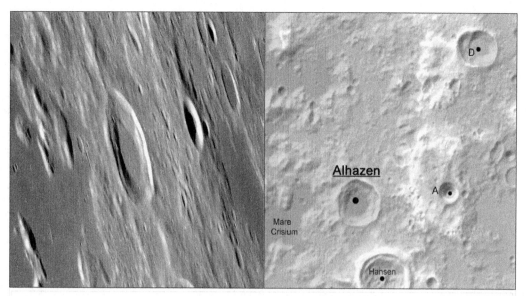

Alhazen	Lat 15.91N	Long 71.83E				34.65Km

Sub-craters:	Crater	Lat	Long	Size (km)	Crater	Lat	Long	Size (km)
	A	16.17N	74.3E	16.14	D	19.68N	75.17E	34.23
	Notes:							

Add Info:

Aerial view of Alhazan

According to the USGS's geological map (Ref: I-948) published in 1977, Alhazen is made up of material that is older than the Orientale Basin and younger than the Imbrium Basin. These geological maps are simply excellent, which border on being artistic, however, while their scientific significance is still of relevance today, our knowledge of the true make-up of the lunar surface still remains a puzzle. Alhazen lies on an outer ring-rim of the Crisium Basin to its west. Its walls have slumped in nearly every sector of the crater, and its floor looks as if it was internally filled with lavas similar to those in the Mare. D: 2.2 km.

Craters

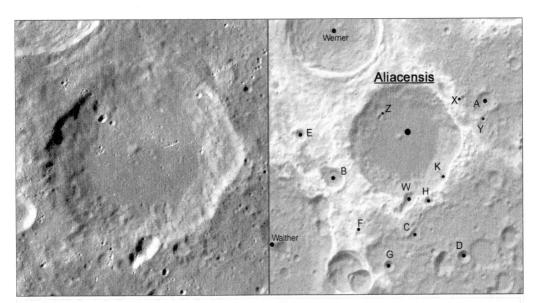

Aliacensis	Lat 30.6S	Long 5.13E				79.65Km

Sub-craters:	Crater	Lat	Long	Size (km)	Crater	Lat	Long	Size (km)
	A	29.71S	7.39E	13.61	B	31.43S	3.2E	15.81
	C	32.7S	5.42E	7.93	D	33.19S	6.82E	9.67
	E	30.42S	2.32E	8.74	F	32.66S	3.84E	4.82
	G	33.39S	4.7E	7.76	H	31.81S	6.01E	5.58
	K	31.43S	6.26E	5.65	W	31.9S	5.32E	10.32
	X	29.64S	6.78E	3.66	Y	30.11S	7.33E	4.7
	Z	30.05S	4.49E	3.83				
Notes:								

Add Info: Aliacensis (Nectarian in age - 3.92 to 3.85 byo) is obviously older than its northwestern neighbour, Werner (of the Eratosthenian Period - 3.15 to 1.1 byo), however, a fine 'age-line' is drawn between its southwestern neighbour, Walther, as the two look very similar in terms of the state of their rims, terraces...etc., (does Walther look younger?). Nearly all of Aliacensis's rim looks worn, and any terracing from its inner walls that once existed has merged its material onto the floor. With an approximate depth of around 3.8 km, and a small central peak that rises to nearly a kilometre in height, the crater has had its fair share of additional ejecta deposits from the Imbrium Basin (some ~ 2000 km away to the northwest), and from younger crater, Tycho, to the southwest (~ 85 km away). The impact of Werner has altered Aliacensis's northern rim - 'pushing' portions of it onto the floor (its lighter deposits can be seen in the northeast part of the floor during full moon times, overlaid, of course, with those from Tycho). The floor has its share of small craterlets (some fresh-looking), and there are about four, very small groups of crater chains in the floor's top half, and outer rim, running in a NE-SW direction (Tycho was the culprit).

Craters

| Almanon | Lat 16.85S | Long 15.14E | 47.76Km |

Sub-craters:

Crater	Lat	Long	Size (km)	Crater	Lat	Long	Size (km)
A	17.79S	15.32E	9.75	B	18.33S	15.26E	23.6
C	16.16S	15.98E	15.62	D	18.6S	15.54E	5.33
E	17.91S	13.63E	5.37	F	15.92S	14.27E	5.23
G	17.93S	14.61E	4.47	H	19.03S	15.22E	6.05
K	15.84S	15.43E	6.46	L	19.0S	16.58E	5.88
P	18.55S	17.0E	19.44	Q	18.08S	17.0E	4.63
R	18.17S	15.88E	3.51				

Notes:

Add Info: Just some 150 km away southwards from where the Apollo 16 lunar module set down (21 April 1972), Almanon lies on a geologic unit known as the Descartes Formation. There also is another companion formation, called the Cayley Formation, that lies just west of the crater (smooth-looking material), and both formations, initially believed to be volcanic in origin, cover some areas of the Nearside. Samples, however, returned by A16 showed the material to be impact in origin, and made up of fluidized ejecta that came from the Imbrium Basin - some 1500 km away to the northwest. The crater itself has a depth of around 2.5 km and looks relatively plain in the eyepiece. Having a smooth, terraced interior with several small craters on the lower half of the floor (ranging from 1 to 2.5 km), two features of note on its floor show up easily under moderate magnification - first: is the small, fresh-looking crater (less than a kilometre across) in the floor's northeast sector; second, a curved (looks like), secondary rim in the floor's western sector ~ some 8 km within Almanon's real rim there.

What would have caused the 'indented' crater rim at Almanon's inner western sector; as it looks very unusual. Does it have something to do with in the dynamics of the crater's formation and the terrain west of the crater?

Craters

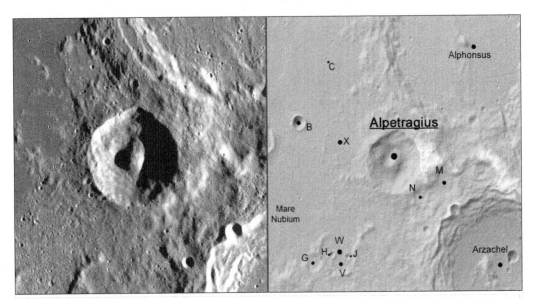

Alpetragius   Lat 16.05S   Long 4.51W   40.02Km

Sub-craters:

Crater	Lat	Long	Size (km)	Crater	Lat	Long	Size (km)
B	15.13S	6.88W	9.73	C	13.75S	6.17W	2.11
G	18.17S	6.56W	12.18	H	18.01S	6.1W	4.43
J	18.05S	5.7W	4.13	M	16.45S	3.27W	23.21
N	16.74S	3.88W	11.27	U	17.71S	5.11W	15.59
V	18.19S	5.83W	15.9	W	17.95S	5.96W	27.68
X	15.61S	5.74W	31.78				

Notes:

Add Info: It's easy to see why astronomer, Gerard Kuiper, referred to the appearance of this crater as an "egg in a nest". The 2-km-high mound of material at its centre fills up about a third of the crater's inner dimensions; and is believed to have been the result of later volcanic activity in the crater afterwards that pushed the mound upwards. While a tiny crater pit, or 'vent', lies on top of the mound, which is extremely hard to observe even at high magnification, the very tip of the mound is well worth looking out for during terminator periods as it 'shines' as a very bright dot (approx., 10-day-old or 23-day-old moons). Alpetragius itself has an average depth of 4 km, and its rim and inner terraced walls look relatively worn; all in an area of ejecta deposits from Imbrium which lies some 1500 km away to its northwest. Alpetragius B, to its west, has been classed into the series of bright-ray craters, however, the effect is best observed during high sun angles as the lighter material is only slightly detected around the 10 km-wide crater.

Craters

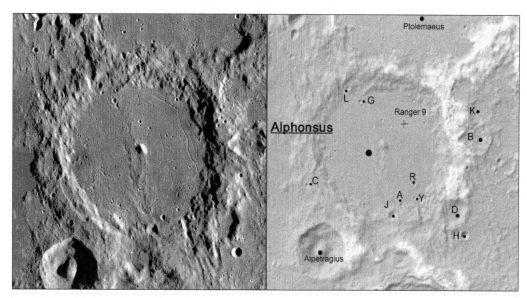

Alphonsus Lat 13.39S Long 2.85W 110.54Km

Sub-craters:

Crater	Lat	Long	Size (km)	Crater	Lat	Long	Size (km)
A	14.87S	2.27W	3.6	B	13.26S	0.2W	22.94
C	14.4S	4.87W	3.37	D	15.05S	0.85W	23.7
G	12.35S	3.39W	3.52	H	15.62S	0.53W	7.01
J	15.14S	2.51W	7.9	K	12.61S	0.11W	20.6
L	12.02S	3.72W	3.75	R	14.39S	1.92W	3.01
X	14.99S	4.46W	4.68	Y	14.71S	1.92W	2.6

Notes:

Add Info: Alphonsus is a wonderfully-situated crater, and forms one of the infamous 'triad' of craters on the Moon - with Ptolemaeus to its north and Arzachel to its south). Aged of the Nectarian - 3.92 to 3.85 billions of years ago, the crater has a central peak that rises to about 2 kilometres, has several fissures predominantly in its eastern sectors, and unusual dark 'spots' known as 'dark-haloed' craters. These features are believed to be volcanic in nature where dark mantled material came from below and 'staining' thereupon the lighter material (possibly fluidized ejecta from formation of the Imbrium Basin, to the northwest) that fills the floor. Alphonsus is famous for reports of Transient Lunar Phenomena (TLP) - changes in the brightness and colour of certain areas of the floor, so these haloed areas are suggested as being hot-spots for the escape of so-called internal lunar gases producing the TLPs (it's a topical area for discussion). Taking in a bigger picture of Alphonsus and surrounds, the immediate striking feature is the striated look that crosses the surface in a NW-SE direction. These 'grooves' are believed to be the result of Imbrium's ejecta 'gouging' out the highest points in the terrain.

Ranger 9 impacted in Alphonsus's floor on 24 Mar 1965 Lat 12.93S, Long 2.4W.

Craters

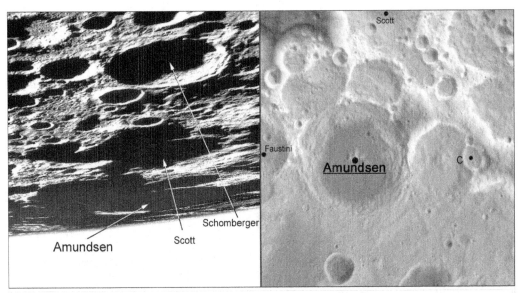

Amundsen	Lat 84.44S	Long 83.07E	103.39Km

Sub-craters:	Crater	Lat	Long	Size (km)	Crater	Lat	Long	Size (km)
	C	80.76S	85.21E	24.22				
	Notes:							

Add Info:

Aerial view of Amundsen

Nine times out of ten attempts at trying to glean any descent detail of Amundsen will surely let you disappointed. The main reason why craters, like Amundsen, on the extreme northern or southern latitudes/longitudes is a problem for viewing is that as sunlight strikes (from the east or west) the 'parallel-wise' axis of the crater (as opposed to the perpendicular-wise aspect for those craters on the eastern/western limbs), the shadows produced by rims and peaks...etc., become lenthier over a lesser period of time. In effect, suitable observations usually occur within a limited window of time only, in lieu, with suitable librations. When the crater is observered during all of the above, its floor, terraces and peaks are just about seen.

Craters

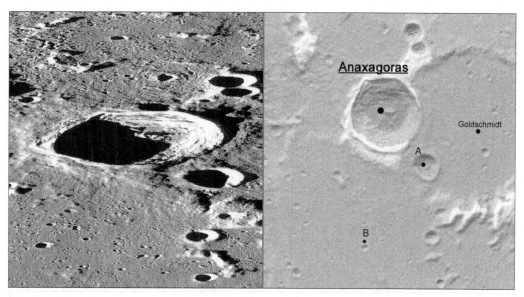

Anaxagoras	Lat 73.48N Long 10.17W			51.9Km			
Sub-craters:	Crater	Lat	Long	Size (km)	Crater	Lat	Long	Size (km)
	A	72.26N	7.02W	20.09	B	70.35N	11.36W	4.95
	Notes:							

Add Info:

Aerial view of Anaxagoras

Like its counter-located opposite, crater Tycho, due south some 3000 km away, Anaxagoras's ray system covers most of its surrounding terrain and features nearby. Copernican in age (1.1 byo to Present), much of its ejecta covers portions of crater Goldschmidt, to its east, and smoothed out nearly all of its own outer rim sector lying to the south. Most of this material is presumed to be made up of Anorthosite, which represents the true crust that initially floated up into a top layer from an ocean of hot magma - believed to have once covered the entire Moon. A pity Anaxagoras lies where it does. Peak ~ 1 km high, depth ~ 3 km.

Craters

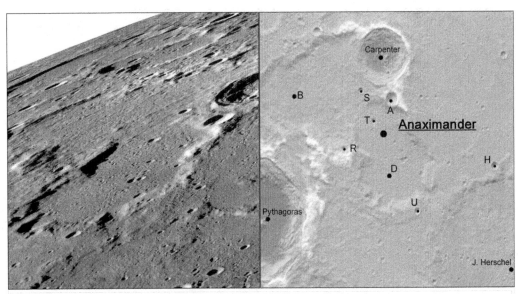

Anaximander	Lat 66.97N	Long 51.44W				68.71Km	

Sub-craters:

Crater	Lat	Long	Size (km)	Crater	Lat	Long	Size (km)
A	68.05N	50.4W	15.24	B	68.08N	61.21W	79.1
D	65.78N	50.68W	96.63	H	65.29N	41.06W	9.03
R	66.32N	55.21W	8.05	S	68.35N	53.6W	7.21
T	67.3N	52.27W	7.01	U	64.14N	48.5W	7.54

Notes:

Add Info: Like most craters in this highland region, Anaximander has been through the ejecta wars from the Imbrium Basin to its South. Pre-Nectarian in age (4.6 - 3.92 billions years), the crater almost becomes lost to larger-sized Anaximander D to its south and the semi-smooth material lying to its northwest (encompassing sub-craters B and S). Both the NW and SE portions of its rim have disappeared, but those remaining portions, which rise to just over 2 km in height, do cast some nice long shadows when the terminator is very close (either side of ~ 13-day and 27-day-old moon periods). Carpenter material overlies the crater.

Aerial view of Anaximander

Craters

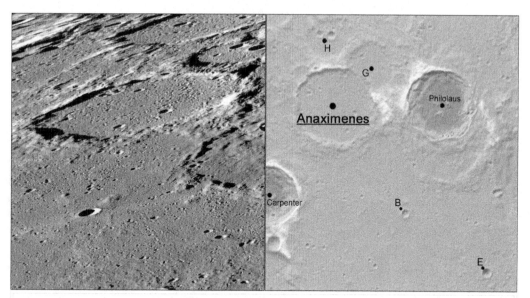

Anaximenes   Lat 72.49N Long 44.98W   81.12Km

Sub-craters:

Crater	Lat	Long	Size (km)	Crater	Lat	Long	Size (km)
B	68.94N	38.06W	8.34	E	66.58N	31.46W	10.0
G	73.75N	40.5W	51.17	H	74.64N	45.7W	42.72

Notes:

Add Info:

Aerial view of Anaximenes

Covered by ejecta from the Imbrium Basin lying to its south, flat-looking Anaximenes borders northwards on an expanse of smooth material which, at times, resembles that of a buried crater three times its size. Pre-Nectarian in age (4.6 - 3.92 billions years old), Anaximenes lies not far from Copernican-aged Philolaus (1.1byo to the present) to its east and sub-crater, Anaximenes G, to its north. Some of the younger, brighter material from Philolaus can be seen on the clump of material that lies between Anaximenes and G during low sun angle (~ 13-day-old moon) times, as well as producing a nice peak-shadow to cross its floor. Depth of Anaximenes is around 0.9 km.

Craters

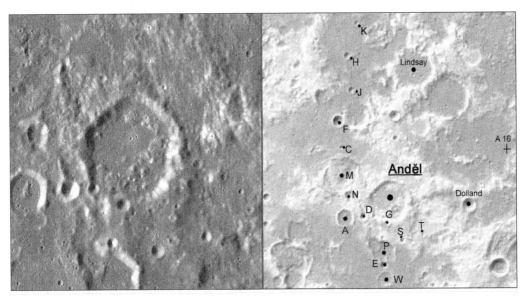

Anděl   Lat 10.41S   Long 12.38E   32.93Km

Sub-craters:

Crater	Lat	Long	Size (km)	Crater	Lat	Long	Size (km)
A	10.81S	11.24E	13.4	C	9.01S	11.15E	3.01
D	10.79S	11.71E	5.98	E	12.02S	12.22E	5.7
F	8.34S	11.08E	9.22	G	10.94S	12.31E	4.12
H	6.67S	11.31E	5.25	J	7.54S	11.41E	5.52
K	5.83S	11.56E	3.69	M	9.75S	11.12E	26.46
N	10.24S	11.34E	8.5	P	11.65S	12.2E	17.56
S	11.4S	12.7E	4.52	T	11.25S	13.24E	3.38
W	12.36S	12.26E	11.19				

Notes:

Add Info: Lying some 100 km away due west from where John Young, Ken Mattingly and Charlie Duke landed on 21 April 1972 in Apollo 16, Anděl, with its worn down rim and relatively flat floor (except for some hilly material in the bottom half), is a crater of the Nectarian period (3.92 - 3.85 billions years old). The general area of the crater's surrounds is well known for the two formations - Cayley and Descartes - that formed as a result of the Imbrium Basin event; whose deposits, initially believed to be volcanic in origin, were latered proved incorrect by the A16 samples returned. Much of the clump of material just outside the crater's eastern rim appears higher than the surrounding terrain, and at termintor times, when the Moon is around 8-days-old, a deep shadow is seen that defines its highest point. At times of full moons, the crater is impossible to find, and only hint of its bright walls is seen. Depth of the crater is around 1.5 km.

Craters

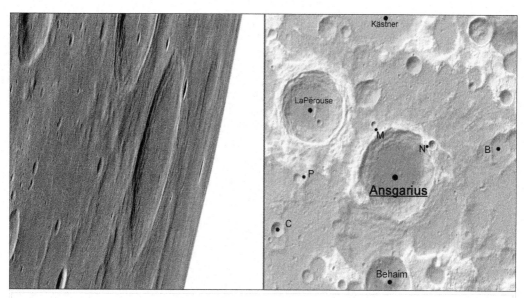

Ansgarius	Lat 12.92S	Long 79.72E				91.42Km

Sub-craters:

Crater	Lat	Long	Size (km)	Crater	Lat	Long	Size (km)
B	11.99S	84.11E	33.1	C	14.79S	74.83E	15.59
M	11.25S	78.79E	7.56	N	11.91S	81.18E	10.23
P	13.06S	75.66E	10.35				

Notes:

Add Info:

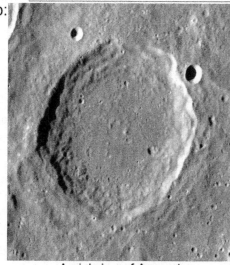

Aerial view of Ansgarius

Relatively fresh-looking for its age ~ 4 billions old, Ansgarius lies just on the main outer ring-rim (~ 750 km in diameter) of Mare Smythii that lies some 300 km to its northeast. Always a challenge to see some detail for this limb-hugging crater, the floor, while flat-appearing, shows a small peak at times around 3-day-old moon periods, and when the longitudinal libration is favourable under suitable lighting conditions. The crater may have formed on two other older, unnamed craters, to its south and southwest respectively - the former nearly twice its diameter, the latter about the same size as Ansgarius. Depth ~ 4 km.

Craters

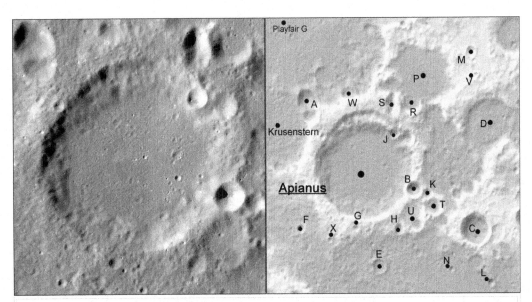

Apianus	Lat 26.96S	Long 7.87E		63.44Km			

Sub-craters:

Crater	Lat	Long	Size (km)	Crater	Lat	Long	Size (km)
A	25.7S	6.52E	13.11	B	27.4S	8.99E	10.21
C	28.13S	10.46E	19.64	D	26.08S	10.66E	33.66
E	28.85S	8.19E	8.39	F	28.14S	6.33E	5.39
G	28.12S	7.64E	4.02	H	28.15S	8.64E	6.57
J	26.33S	8.52E	6.71	K	27.48S	9.31E	6.44
L	29.12S	10.85E	4.67	M	24.78S	10.29E	6.98
N	28.89S	9.91E	3.46	P	25.29S	9.13E	41.59
R	25.79S	8.89E	13.1	S	25.81S	8.47E	23.96
T	27.73S	9.45E	11.52	U	27.91S	8.98E	16.82
V	25.19S	10.4E	3.2	W	25.6S	7.46E	9.68
X	28.34S	7.02E	3.1				

Notes:

Add Info: Lying some 700 km away west of Mare Nectaris with ejecta material overlying from that event, Apianus is pre-Nectarian in age (4.6 to 3.92 billions years old). As a consequence, nearly all the rim and terracing of the crater is worn down, while its floor received ejecta deposits from formation of the Imbrium Basin ~ 1800 kilometres away to the northwest (extreme close-up of the floor shows several small crater chains that 'point' back to Nectaris, but Imbrium's ejecta has covered them everso slightly). Apianus initially formed on another old crater, Playfair G, to its northwest (some of its ejecta can be seen to overlie G's floor there), however, the same isn't so obvious for pre-Nectarian crater, Krusenstern, to its west whose floor is covered by ejecta from crater Werner just 120 km away to its west (but Apianus is slightly younger than Krusenstern). Depth ~ 2.0 km.

Craters

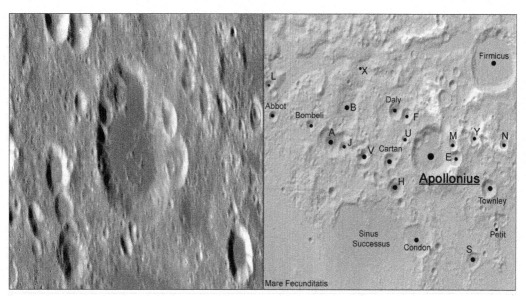

Apollonius   Lat 4.51N   Long 60.96E   50.66Km

Sub-craters:

Crater	Lat	Long	Size (km)	Crater	Lat	Long	Size (km)
A	4.82N	56.85E	24.55	B	5.87N	57.6E	31.17
E	4.34N	61.86E	16.55	F	5.6N	59.94E	14.72
H	3.43N	59.51E	19.04	J	4.63N	57.48E	11.7
L	6.47N	54.57E	10.12	M	4.73N	61.77E	9.54
N	4.75N	63.77E	10.79	S	1.26N	62.58E	16.22
U	4.89N	59.87E	8.44	V	4.39N	58.22E	14.97
X	6.9N	58.18E	28.97	Y	4.88N	62.6E	8.71

Notes:

Add Info: Situated just outside one of the several ripple-like, ring-rims created when Mare Crisium, to its north, formed (the effects of that giant impact undoubtedly is one of the reasons why Apollonius looks the way it does). Fracturing in nearly all of the local rock in this region allowed magma to rise up in between, and fill in all the low spots around (note how this 'ponding' has filled in also craters Firmicus Townley, Mares' Spuman and Undarum...and many more). Apollonius's true floor has thus been covered, and all that is seen is the faint remains of two small craters - the largest being around 10 kilometres in diameter (or, is it just a single small crater close to a half-buried peak of Apollonius?). These lavas also fill-in another crater within Apollonius's northeastern floor, and at extreme close-up, signatures also show the lava just about filled into the smaller craters that lie to its west (on Apollonius's inner rim wall there). Sub-crater's, Apollonius M and E, both deposited ejecta and wall material onto Apollonius's floor. D: 2.8 km.

Craters

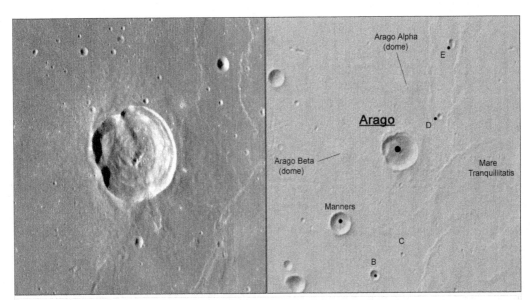

Arago   Lat 6.15N   Long 21.43E   25.51Km

Sub-craters:

Crater	Lat	Long	Size (km)	Crater	Lat	Long	Size (km)
B	3.43N	20.82E	6.9	C	3.89N	21.48E	3.03
D	6.91N	22.39E	3.95	E	8.51N	22.71E	6.28

Notes:

Add Info: Relatively fresh-looking for its Eratosthenian age (3.15 - 1.1 billions years old), Arago lies 'in' and 'on' the mare that is Tranquillitatis (pre-Nec 4.6 - 3.92 byo). Slumping inside the crater has created two very obvious features; firstly, the well-defined terracing - west side is much more extensive than the east side and, secondly, the central peak (or mountainy material) on the floor that connects to the northwestern sector of the rim. It's as if the slumped material here has clumped together through, perhaps, two instances of seperate slumping events - a major western one meeting a minor northern one. In the other, southeastern direction, however, the opposite has occurred where a small depressed cleft has formed (seen at low sun angle periods) that extends onto Arago's ejecta and further into Tranquillitatis's mare. There is what looks like a ridge where it meets this extended ejecta-mare feature, so, perhaps, that had something to do with its formation (note, too, the series of rilles(?) to the west of this feature that fan outwards - they are very unusual). One more feature here at Arago is the two wonderful domes northwards and westwards respectively of the crater. The Alpha dome is slightly more higher (~ 330 metres) than the Beta dome (~ 270 metres), where each shows up nicely during when the terminator isn't that far off - say, around 8-day-old and 21-day-old moons. Depth of Arago ~ 2.5 km.

Arago E
This crater, at extreme close-up, is like a smaller version of crater Messier - hint, that it might be one created by an impactor that came in at a low angle to the surface.

Craters

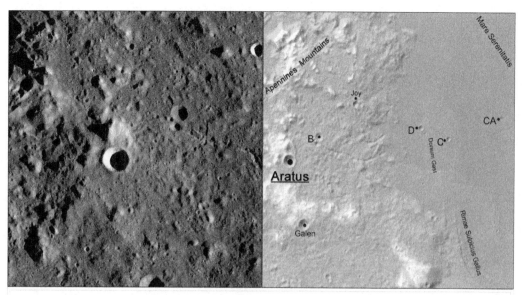

Aratus	Lat 23.58N Long 4.51E	10.23Km

Sub-craters:	Crater	Lat	Long	Size (km)	Crater	Lat	Long	Size (km)
	B	24.18N	5.43E	6.69	C	24.08N	9.46E	3.56
	CA	24.56N	11.18E	2.08	D	24.37N	8.61E	4.05

Notes:

Add Info:

Aerial view of Aratus

Aratus, like its western neighbour, crater Conon, has landed on top of the range of mountains known as the Apennines; associated to formation of the Imbrium Basin off to the northwest. Aratus doesn't look as sharp as Conon (Copernican in age 1.1 byo to the Present), but it probabaly is in that age period anyway. The impactor that produced the crater just missed striking a high peaky mountain over 2 km high to its north. At times of full moon both Conon and Aratus stand out easily against the range, however, while Aratus may be older than Conon, it 'shines' just that little bit ahead of its bigger neighbour. Depth of Aratus ~ 2.0 km.

Craters

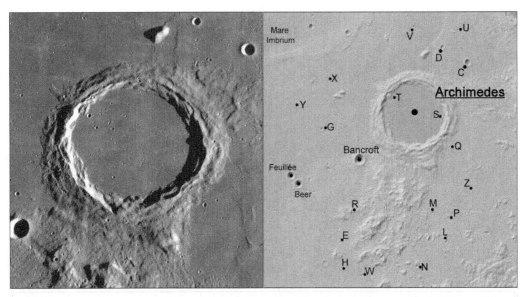

Archimedes   Lat 29.72N Long 3.99W   81.04Km

Sub-craters:

Crater	Lat	Long	Size (km)	Crater	Lat	Long	Size (km)
C	31.63N	1.53W	7.66	D	32.2N	2.69W	4.96
E	25.0N	7.2W	2.56	G	29.14N	8.15W	3.29
H	23.89N	7.02W	3.78	L	25.04N	2.61W	3.21
M	26.12N	3.21W	3.27	N	24.15N	3.89W	3.51
P	25.94N	2.5W	2.63	Q	28.52N	2.43W	2.36
R	26.07N	6.61W	3.5	S	29.56N	2.73W	2.77
T	30.3N	5.03W	2.33	U	32.83N	1.96W	2.74
V	32.98N	4.01W	2.67	W	23.8N	6.25W	3.19
X	31.03N	8.02W	2.15	Y	29.97N	9.5W	2.2
Z	26.88N	1.41W	2.07				
Notes:							

Add Info: Archimedes lies in a region on what is known as the Apennine Bench, that is, a bright patch of mountainy material outside the crater's southern rim believed to be volcanic in origin. Upper Imbrian in age (3.75 - 3.2 billions years old), Archimedes's lava material in the crater's interior probably filled through fissures later on after its formation, as it is slighty lower than the mare material which makes up the Imbrium Basin to its west. Like Plato to its north, Archimedes shows some nice terracing, outer ejecta deposits, small craterlets on its smooth floor (a challenge to see and count), and bright ray material from crater Autolycus to its east (it can be seen at high sun angle times). Apollo 15 landed some 180 km away due southeast. Be sure to take in its easterly neighbours - craters Aristillus and Autolycus in your general view as they make a nice set. Depth ~ 1.6km.

Crater Bancroft to Archimedes's southwest was once called Archimedes A.

Craters

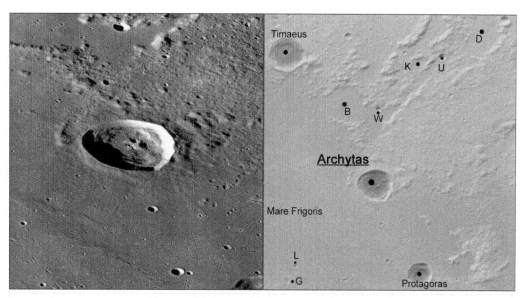

Archytas	Lat 58.87N	Long 4.99E				31.95Km

Sub-craters:

Crater	Lat	Long	Size (km)	Crater	Lat	Long	Size (km)
B	61.44N	3.17E	35.18	D	63.71N	11.9E	44.75
G	55.73N	0.53E	7.21	K	62.66N	7.68E	13.77
L	56.19N	0.9E	4.38	U	62.9N	9.22E	7.39
W	61.29N	5.2E	5.73				

Notes:

Add Info:

Aerial view of Archytas

Kissing mare material in Mare Frigoris to its south and rubbly, highland's material to its northeast, Archytas is a complex crater of the Eratosthenian period (3.15 - 1.1 billons of years old). Archytas is scalloped, angular in shape, and it has a fresh sharp rim all around with inner wall material having slumped into the crater's interior. It has an approximate depth of 3 km and a peak that rises to just over 600 metres. To its northwest, mare material has breached and flooded into sub-crater, Arychytas B at two points, while southwestwards, Archytas G (a concentric crater), will prove a challenge.

Craters

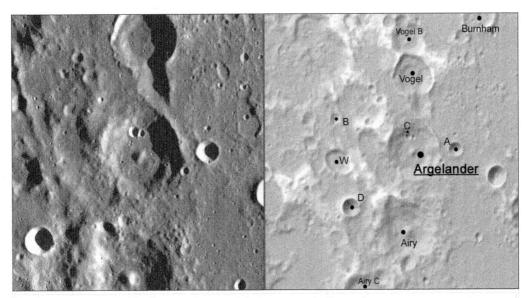

Argelander	Lat 16.55S	Long 5.8E				33.72Km

Sub-craters:

Crater	Lat	Long	Size (km)	Crater	Lat	Long	Size (km)
A	16.54S	6.75E	8.65	B	15.6S	5.1E	5.6
C	16.28S	5.72E	3.87	D	17.64S	4.44E	10.69
W	16.75S	4.18E	18.63				

Notes:

Add Info:

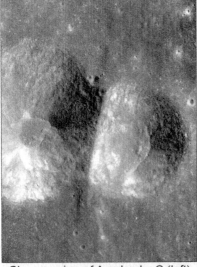

Close-up view of Argelander C (left) and an unnamed crater to its right.

Argelander lies amongst a series of similar sized craters - Airy just below it, Donati and Faye further southwards - each having dimple-like peaks in their centres (a good way of remembering which is Argelander is to note that it is the only one with two small, fresh-looking craters on its floor). These two (left) show a wonderful straight rim-edge where they 'join' - signature that they may have impacted at exactly the same time. Argelander is Nectarian in age (3.92 to 3.85 byo). Ejecta from the Nubium Basin to the west covers the area in general; whose effects has left a series of grooves, ridges and secondary craters spread throughout. The crater isn't as bright as Airy under the same light, and its peak height (~ 650 metres) is some ~ 150 metres lower than it, too. Depth ~ 3.0 km.

Craters

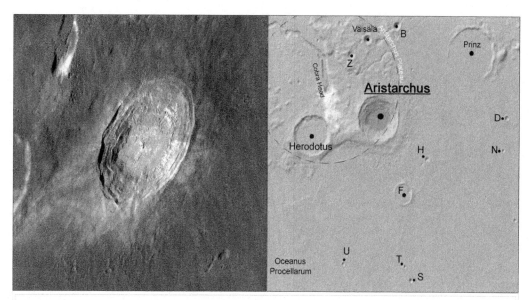

Aristarchus   Lat 23.73N   Long 47.49W   39.99Km

Sub-craters:

Crater	Lat	Long	Size (km)	Crater	Lat	Long	Size (km)
B	26.28N	46.85W	6.95	D	23.73N	42.88W	4.65
F	21.67N	46.57W	17.6	H	22.61N	45.74W	4.43
N	22.83N	43.03W	3.14	S	19.29N	46.28W	3.8
T	19.67N	46.5W	3.33	U	19.73N	48.64W	3.45
Z	25.49N	48.49W	7.74				

Notes:

Add Info:

The plateau was firstly covered with the pyroclastic ash deposits (presumably from a vent at the head of the large rille lying between the two craters), which was later then covered by a dusting of ejecta from Aristarchus. Lavas surrounding the plateau define later flows that covered both the ash and ejecta.

Besides the wonderful bright ray-display that we see from crater Tycho, in the south, Aristarchus towards the west portion of the Moon must be the next best feature that catches the eye. Copernican in age (1.1 byo - Present), this complex ray-crater is the result of deep-down rock material (called Anorthosite) excavated from the moon's original crust, and brought to the surface by the impactor that produced the crater. Lying in and on an extensive deposit known as the 'Aristarchus Plateau', the crater is similar in size to its lava-filled neighbour, Herodotus, to its west, and features terracing, a broad ejecta blanket, hummocky radial grooving, and a sharp crenulated rim all around. From the ray display, one might think that the crater formed as a result of an oblique impact, however, this may not be the case as it lies on a portion of the plateau that is slightly domed. Transient Lunar Phenomenon (TLP) sightings for this area have long been reported upon, but as to what is causing them (e.g. outgassing, impacts...etc.,), if indeed true, is a puzzle that will remain controversial. Two other notes: the Aristarchus Plateau always appears 'brownish to orange' in colour; while the crater is just about visible when in darkness - during 'Earthshine' instances. D: ~ 3 km.

Craters

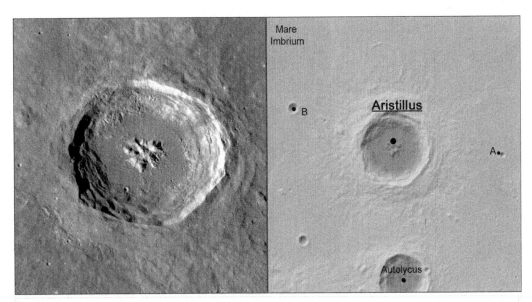

Aristillus	Lat 33.88N	Long 1.21E		54.37Km	

Sub-craters:

Crater	Lat	Long	Size (km)	Crater	Lat	Long	Size (km)
A	33.64N	4.53E	4.44	B	34.8N	1.93W	8.02

Notes:

Add Info: Always a lovely site through the eyepiece at high power, Aristillus has all the charaterisitc features of a young-ish crater (early Eratosthenian 3.15 to 1.1 billions years old). Several well-defined peaks that reach to 1.8 km cover a third of the floor, terracing has occurred all around (except for the NW sector where additional collapse has added material onto the floor) the inner walls, and an outer ejecta blanket surrounds the crater whose effects can be seen through its rays, grooving, and the many small crater chains which cross the mare material of Imbrium. There is an unusual pair of dark streaks (best observed at high sun angles) on the inner terraced walls at the northeast sector of the crater, and these extend for some several kilometres onto the outer rim. One might think that the streaks of the inner section are simply due to fall-back of lighter material exposing dark material underneath, however, explanation for the streaking on the outer section is a problem. Perhaps, they are due to the initial impact and ejecta following that event - a hint, perhaps, of obliqueness (note, how the ejecta is distributed predominantly in the NW and SE where the impactor possibly came in from the NE). With an average depth of 3.5 km, the crater lies some 230 km away due north of the Apollo 15 lander, which set down on the surface on 30th July 1971 (David Scott, James Irwin and Alfred Worden), and where the first ever lunar roving vehicle was driven on the moon's surface.

Craters

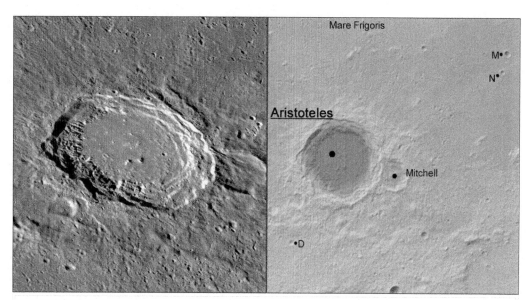

Aristoteles	Lat 50.24N	Long 17.32E			87.57Km		
Sub-craters:	Crater	Lat	Long	Size (km)	Crater	Lat	Long	Size (km)
	D	47.48N	14.71E	5.61	M	53.5N	27.26E	6.98
	N	52.9N	26.84E	5.3				
	Notes:							

Add Info: There are few craters on the Moon that look great in the eyepiece. Craters like Copernicus, Eratosthenes, Plato, Gassendi, Tycho and other large, complex-like craters easily are the favourites, but Aristoteles puts on a great show no matter what time one views it. At low sun angle times, while dark shadows within its western interior walls counter oppositely in brightness of the many terraces on the eastern side, the same can be said for when the easterly side is in shadow and the mountainy, steep-faced scarp on the west side is in light. The outer ejecta blanket that surrounds it, as well as the wonderful ridges and grooving that make up its chains and clusters of secondary craters, signifies that this crater is one of recent event (early Eratosthenian in age - 3.15 - 1.1 billions of years old) in the history of the Moon. At higher sun angles, the crater looks all-round bright, too, with central parts of its floor displaying the several small peaks like white dots. The two main peaks reach a height of around 0.5 km in this 3.5 km depth crater - odd, perhaps, for a crater of this size where one would expect them to be more prominant. Ejecta from the crater covers older Mitchell nearby, however, why does the blanket appear thicker, almost block-like, at Aristoteles's outer northeast and southwest sectors? Would these 'blocks' be some kind of feature of an oblique impact event with large craters like Aristoteles?

Craters

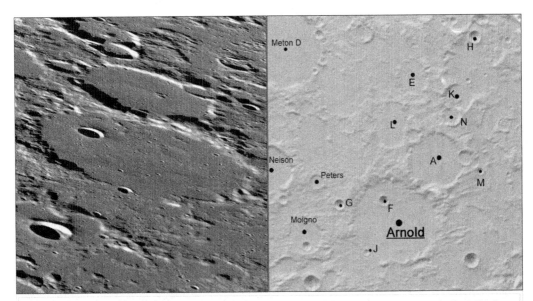

Arnold	Lat 66.98N	Long 35.83E			93.13Km

Sub-craters:

Crater	Lat	Long	Size (km)	Crater	Lat	Long	Size (km)
A	68.77N	39.6E	55.13	E	71.48N	38.1E	31.15
F	67.55N	35.18E	10.41	G	67.39N	31.45E	10.64
H	72.59N	44.98E	12.82	J	65.9N	33.82E	6.17
K	70.71N	42.38E	30.0	L	70.0N	35.77E	34.89
M	68.29N	43.59E	6.77	N	70.15N	41.6E	18.81

Notes:

Add Info:

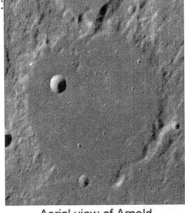

Aerial view of Arnold

Low rims where visible, practically no peak or signs of terracing within, and a smooth floor suggests that Arnold is an old crater - of the pre-Nectarian period (4.6 - 3.92 billions years old). Ejecta deposits from the Imbrium Basin due southwestwards (some 1200 km away) possibly are responsible for the smooth floor; which during low sun angle periods shows not one obvious 'saucer-like' depression but several. Low sun views, particularly easterly ones, show the wonderful 'groove-like' features (Imbrium's ejecta effect) on Arnold's northeast rim, which further extend to as far as crater Archytas ~ 500 km away. Depth of Arnold is ~ 1 km.

Craters

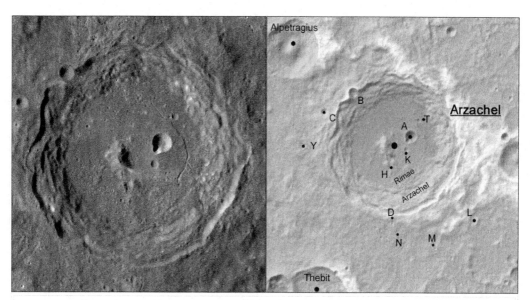

Arzachel Lat 18.26S Long 1.93W 96.99Km

Sub-craters:

Crater	Lat	Long	Size (km)	Crater	Lat	Long	Size (km)
A	18.06S	1.51W	8.69	B	17.06S	2.98W	7.71
C	17.49S	3.71W	5.66	D	20.19S	2.14W	7.46
H	18.69S	2.05W	4.08	K	18.38S	1.62W	3.5
L	19.96S	0.13E	4.41	M	20.65S	0.87W	3.02
N	20.42S	2.27W	2.76	T	17.68S	1.3W	2.96
Y	18.27S	4.26W	3.92				

Notes:

Add Info: Like its compatriots to the north (e.g. Alphonsus, Ptolemaeus), and to its south (e.g. Purbach, Regiomontanus and Walther), Arzachel is a site to behold in the eyepiece. Of the Lower Imbrian Period (3.85 - 3.75 billions years old), it is relatively just younger than those mentioned (all of pre-Nec Period); displaying wonderful terracing all around, a slightly off centre ~ 2.0 km-high peak on its smooth floor, and a faulted rille too obvious to ignore. Several smaller rilles (all encompassed in their 'Rimae Arzachel' designation) can be seen at high magnification, however, the main one is best observed during a ~ 22-day-old moon period where a shadow defines the fault to have down-sloped on its eastern side. This is a floor fractured crater, it is slightly domed towards the centre, has smooth material predominatly in its top half (perhaps, volcanic in origin?), and with an approximate depth of 3.6 km, Arzachel fits into a region where some of the largest crater chains on the nearside of the Moon exist. At full moon times the crater is virtually invisible, and only the slight brightness of its rim and terraced walls show up.

Craters

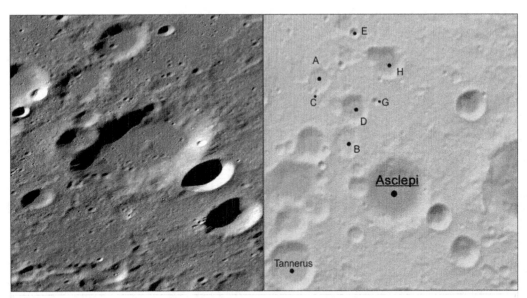

Asclepi	Lat 55.19S	Long 25.52E				40.56Km
Sub-craters:	Crater	Lat	Long	Size (km)	Crater	Lat	Long	Size (km)

Crater	Lat	Long	Size (km)	Crater	Lat	Long	Size (km)
A	53.1S	23.01E	13.43	B	54.24S	23.83E	17.26
C	53.58S	23.55E	9.57	D	53.66S	24.09E	17.86
E	52.26S	24.15E	6.39	G	53.5S	24.8E	4.93
H	52.8S	25.2E	18.22				

Notes:

Add Info:

Aerial view of Asclepi

Situated amongst a region of secondary crater events from both Tycho and the Orientale Basin to its west (some 900 km and 3000 km away respectively), Asclepi is of the Nectarian Period (3.92 - 3.85 billions of years old). Having a depth of around 2.4 km, its flat floor has been covered with plains of material from Mare Imbrium to its northwest (some 2700 kilometres away) that almost has covered its little peak having a height of 0.5 kilometres. Both the small, old-looking crater on the western portion of Asclepi's rim, and a young-looking crater on its SE don't seem to have added any extra material there. Is the set of small craterlets running across the bottom half of the floor secondaries from crater Clavius (~ 650 km to the southwest)?.

Craters

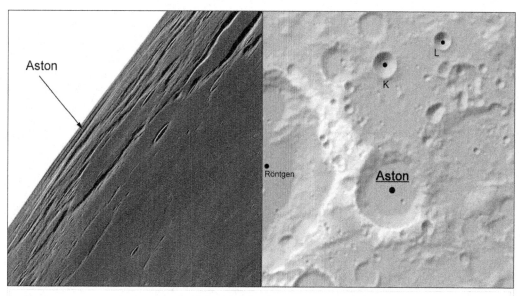

Aston   Lat 32.77N   Long 87.68W   44.48Km

Sub-craters:

Crater	Lat	Long	Size (km)	Crater	Lat	Long	Size (km)
K	35.04N	87.89W	14.46	L	35.47N	86.51W	10.08

Notes:

Add Info:

Librations occur regularly all the time. Many a crater and feature on the extreme limb can be lost to the viewer if planning isn't a part of your observation. So use them well to your advantage, as the results can sometimes be very rewarding.

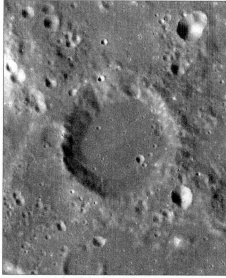

Aerial view of Aston

Aston, from the outset, will prove a challenge to achive any decent view of it. Favourable librations are a bonus, however, even during such occasions, detail of the crater gives only the barest of returns. Its rim has a well worn look to it, most of its ejecta seems to lie at the outer eastern sector, and its floor is some 2 km higher up than the level of material seen to its outer east. Material on the floor looks similar to that found in craters' Röntgen (also seen during good librations) and Nernst to its west, but as to its source - was Imbrium, Orientale, or nearby impacts responsible? Observe just before full moon for best attempts. Depth ~ 1.5 km.

Craters

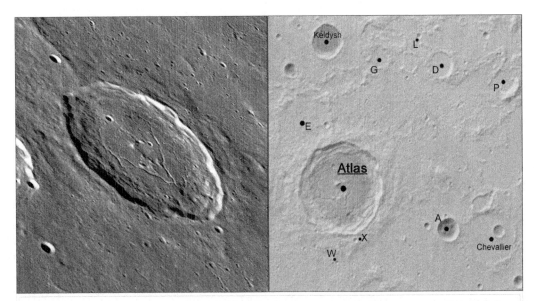

Atlas   Lat 46.74N   Long 44.38E   88.12Km

Sub-craters:

Crater	Lat	Long	Size (km)	Crater	Lat	Long	Size (km)
A	45.34N	49.56E	22.22	D	50.4N	49.65E	25.82
E	48.61N	42.5E	57.98	G	50.73N	46.47E	21.42
L	51.33N	48.62E	5.39	P	49.66N	53.0E	26.9
W	44.43N	44.22E	4.31	X	45.12N	45.1E	5.02

Notes:

Add Info: Together with its neighbour, Hercules, to the west, floor-fractured crater Atlas is never one to disappoint in the eyepiece. Surrounded by a thick ejecta blanket, (except for its northwest sector where it levels out onto the floor of sub-crater, Atlas E), this Early Imbrian-aged crater (3.85 - 3.75 billions years old) has a well-defined rim, a near continuous scarp all around, and slumped terraced material that has become lost to a smooth floor, which is somewhat raised in the centre. With upto five minor peaks, the inner region has a wonderful series of fissures running in every direction, and two very obvious dark spots at its eastern side (both north and south particularly) where patches of pyroclastic deposits have made their way to the surface through volcanic activity. Under high sun angles the peaks - made up of bright Anorthositic crust material - stand out against the darkened spots very easily, and at high magnification the main fissures show smaller, less-obvious ones attached. The crater has a depth of around 3.0 km, while sub-crater, Atlas A, to the east is one of the 'banded' types - where dark to light 'streaking' of ejecta, landslide material etc., appears on a crater's walls.

Banding: The Association of Lunar & Planetary Observers (ALPO) - an amateur group in the USA - maintains a list of craters that show this feature.

Craters

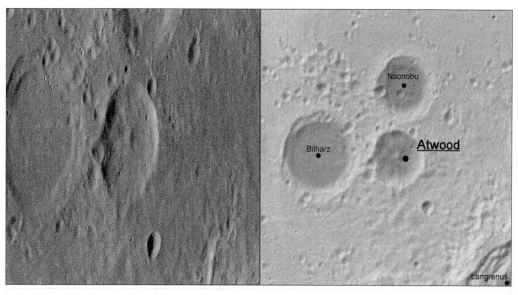

| Atwood | Lat 5.88S | Long 57.78E | 28.64Km |

Sub-craters: No sub-craters

Notes:

Add Info:

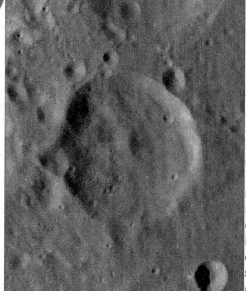

Aerial view of Atwood

There are triplets of craters over all of the Moon. Most are famous for the features seen in each: e.g. Ptolemaeus, Alphonsus and Arzachel, or in Aristillus, Autolycus and Achimedes, however, others, like the triad of Atwood, Bilharz and Naonobu are recognised not for their lack of thereof, but because they look the same (the three blend in seamlessly with the material of Mare Fecunditatis nearby). At full moon times both Bilharz and Naonobu have disappeared, but hint of Atwood's eastern inner rim with its slightly brighter material stands out. The crater has a small peak - seen best during terminator times (around 6 and 17-day-old moons). Depth of Atwood ~ 2.5 km approx..

Craters

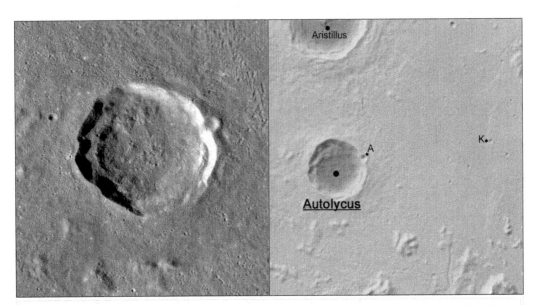

Autolycus	Lat 30.68N	Long 1.49E				38.88Km

Sub-craters:

Crater	Lat	Long	Size (km)	Crater	Lat	Long	Size (km)
A	30.92N	2.17E	4.17	K	31.21N	5.43E	3.02

Notes:

Add Info: Lying within the main ring-rim of the Imbrium Basin (some 1160 km in diameter that, on its eastern side, encompasses mountain regions - Alpes, Caucasus and Apenninus), Autolycus is known more as being paired with its slightly larger 'brother' to the north - that is, the crater Aristillus (55 km in diameter). A comparison of the two together in the eyepiece usually diverts the eye away to look more at 'big-bro' than it, and so the crater isn't one that gets the headline news in terms of any outstanding features. A pity, really, as the crater, at high magnification, shows some nice central hummocky material blending into slumped crater walls all around - particularly in the eastern sector, and a steep scarp that runs along nearly most of its inner rim. The crater looks slightly older than Aristillus (early Eratosthenian Period 3.15 - 1.1 billions years old), and usually mentioned in relation to being the source of the youngest of all rock samples returned throughout the Apollo missions (Apollo 15 landing site lies 120 km away due south-east, and the sample, No. 15405, contained material rich in KREEP - potassium, rare earth elements and phosphorus). Ray material surrounds the crater (best seen at high sun angle times), while sub-crater, Autolycus A, which, in effect, has produced a 'gap' on Autolycus's rim, produces some nice results that show the feature (observe it around 11-day-old and 22-day-old moon periods).

Craters

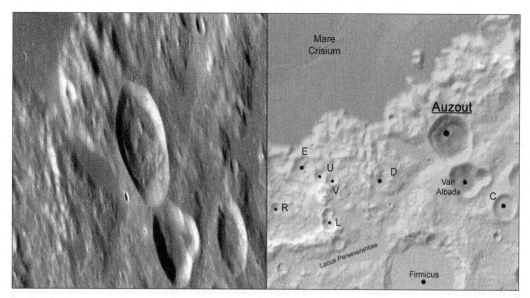

Auzout	Lat 10.21N	Long 64.01E	32.92Km

Sub-craters:

Crater	Lat	Long	Size (km)	Crater	Lat	Long	Size (km)
C	8.79N	65.27E	16.86	D	9.35N	62.43E	11.66
E	9.57N	60.66E	17.36	L	8.34N	61.3E	7.5
R	8.7N	60.04E	7.68	U	9.39N	61.05E	8.15
V	9.34N	61.31E	7.82				

Notes:

Add Info:

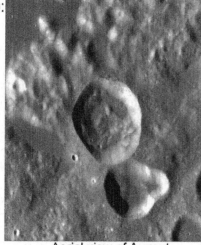

Aerial view of Auzout

Lying just within the main ring-rim (~ 740 km in diameter) of Mare Crisium to its northwest, Auzout, is a crater that formed early on in the Eratosthenian Period (3.15 to 1.1 byo). Having a relatively sharp rim all around, the crater's floor is nearly completely filled with slumped material surrouning a low-level mountain somewhat lost in its 3.8 km depth. The crater is best observed during favourable libration periods, and also around a 4-day-old moon where its shadowed eastern sector cuts into the depressed channel (from Lacus Perseverantiae right upto, and into, the Mare). Luna 24 landed 70 km away due north from this crater - returning 170 grams of samples.

Craters

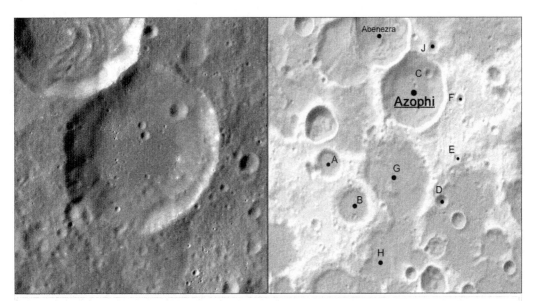

Azophi	Lat 22.19S	Long 12.7E		47.54Km			
Sub-craters:	Crater	Lat	Long	Size (km)	Crater	Lat	Long	Size (km)

Crater	Lat	Long	Size (km)	Crater	Lat	Long	Size (km)
A	24.43S	11.18E	28.3	B	23.58S	10.59E	18.33
C	21.81S	13.06E	5.36	D	24.32S	13.39E	8.49
E	23.48S	13.77E	4.91	F	22.31S	13.84E	5.95
G	23.87S	12.27E	54.65	H	25.58S	11.84E	20.7
J	21.29S	13.17E	7.53				

Notes:

Add Info: Azophi lies some ~ 260 km away due west of the infamous Altai Scarp (a ring-rim nearly 860 km in diameter), and just on the next outer ring-rim (1320 km) - both created by the Mare Nectaris Basin to the east. The crater may be slightly older than its northwestern neighbour, Abenezra (Late Imbrian in age - 3.75 - 3.2 byo), whose impact appears to have imparted portions of the rim onto its smooth floor (hard to see - even at high magnification). However, the bordered section connecting both craters has left an unusually-formed 'sharp' rim between the two. The floor contains material probably of that from the Imbrium Basin event (centred some ~ 1800 kilometres to the northwest), which blended into the already slumped wall material of the crater (note the wonderful long terrace at the crater's southeast inner rim). A range of small craterlets lie on the floor (ranging from ~ 6 km to 1 km in diameter) - each appearing fresher and so younger than the other as they decrease in size (note the particularly bright one in the eastern part of the floor - less than a kilometre in size). Depth ~ 3.7 km.

Craters

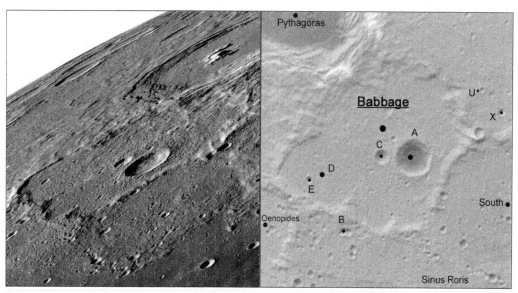

| Babbage | Lat 59.56N | Long 57.38W | 146.56Km |

Sub-craters:

Crater	Lat	Long	Size (km)	Crater	Lat	Long	Size (km)
A	59.14N	55.65W	32.28	B	57.05N	59.63W	6.55
C	59.19N	57.51W	13.71	D	58.65N	61.17W	70.96
E	58.47N	61.61W	7.18	U	60.97N	51.34W	5.09
X	60.3N	50.28W	6.37				

Notes:

Add Info:

Day-old moon phases

Note: As the Moon goes through various librations throughout the year, suggested times given in text for observations are only approximates.

Aerial view of Babbage

Looking at the aerial view of Babbage, one would never get the impression that this feature was a crater at all. Like most of its neighbouring craters - Oenopides to its west and South to its east, this pre-Nectarian (4.6 - 3.92 billions years old) crater has been covered by ejecta from the Imbrium Basin to the southeast, and further, partially by ejecta from Pythagoras. Viewed through the eyepiece, its southeastern rim casts nice shadows around 14-day-old moon periods, and at full moon times sub-craters', Babbage A and C, act as good markers for remembering its location approximately (note also the different deposit textures of this region).

Craters

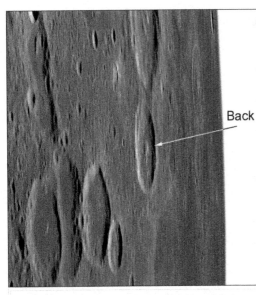
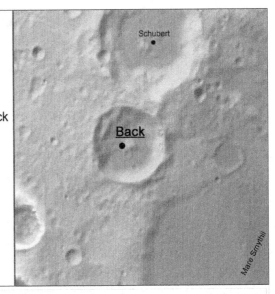

| Back | Lat 1.2N | Long 80.67E | 34.63Km |

Sub-craters: No sub-craters

Notes:

Add Info:

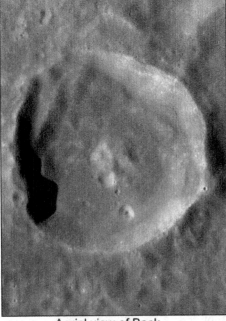

Aerial view of Back

Together with its neighbour, crater Schubert to the north, Back can be quite easy to locate when you remember to use the pair as location markers. Back lies on an outer ring/rim (~ 400 km in diameter) associated to formation of the Smythii Basin to its southeast (part of this ring/rim south of Back drops down sharply in level by some ~ 2 km). Back is probably Nectarian in age (3.92 to 3.85 byo), two main slumps have occurred at its northwestern and eastern walls respectively, it has a small group of peaks - the highest is around 900 metres, and its floor level is about that same amount above the general level of the Mare's floor. The crater has a depth of around 3.5 km approximately. Libration times benefit views.

Craters

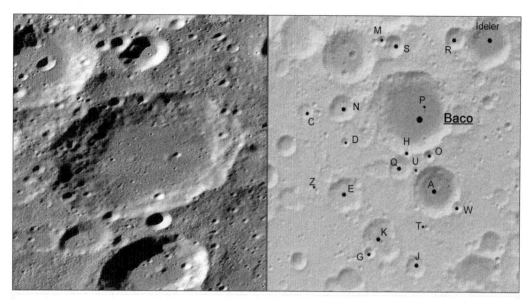

Baco Lat 51.04S Long 19.1E 65.31Km

Sub-craters:

Add Info:

Crater	Lat	Long	Size (km)	Crater	Lat	Long	Size (km)
A	52.99S	20.21E	38.71	B	49.63S	16.58E	41.43
C	50.92S	14.71E	13.21	D	51.73S	16.29E	7.0
E	53.03S	16.17E	27.05	F	50.37S	17.65E	5.79
G	54.47S	17.16E	8.23	H	52.07S	18.95E	6.24
J	54.83S	19.27E	16.64	K	54.08S	17.62E	28.03
L	49.6S	16.75E	6.19	M	49.27S	17.95E	7.22
N	50.9S	16.2E	19.14	O	52.12S	19.88E	9.26
P	51.04S	19.73E	3.15	Q	52.41S	18.62E	17.31
R	49.24S	20.92E	17.43	S	49.41S	18.46E	16.71
T	53.87S	19.77E	4.88	U	52.47S	19.3E	5.8
W	53.39S	21.11E	9.04	Z	53.11S	14.96E	6.53

Notes:

Easily lost, or confusing, to the observer looking through the eye-piece, Baco lies within a region of over-cratering where many are of the same size to it. While relatively fresh-looking for a crater of its age (late Nectarian period - 3.92 to 3.85 billions years old), Baco, looked closely at, shows worn-down rims, smooth terracing, and a levelled floor with numerous tiny craterlets within. Ray material from crater Tycho to its west (some 650 km away) also lies across its floor - best observed at high sun angle periods only and under low magnification as it is difficult to see at times. Most of the region here is said to encompass secondary craters from the Orientale Basin - over 2500 km away to the west.

Craters

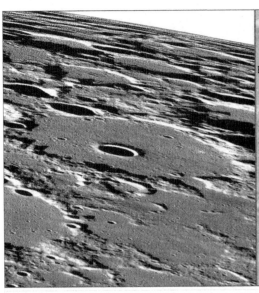
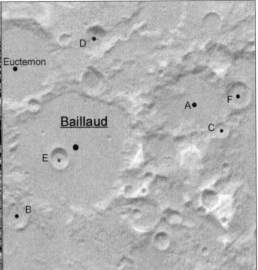

Baillaud Lat 74.61N Long 37.35E 89.44Km

Sub-craters:

Crater	Lat	Long	Size (km)	Crater	Lat	Long	Size (km)
A	75.61N	48.28E	55.61	B	72.93N	32.91E	55.61
C	74.9N	50.91E	11.36	D	73.41N	49.28E	15.42
E	74.33N	35.95E	14.28	F	75.62N	53.02E	18.81

Notes:

Add Info:

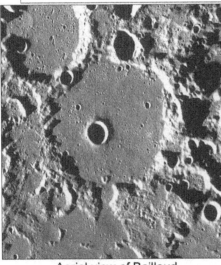

Aerial view of Baillaud

Lying far from the outermost ring/rim (~ 1350 km in diameter) of Nectarian-aged, Humboldtianum Basin (56.92N, 81.54E), pre-Nectarian (4.6 to 3.92 byo) Baillaud looks like it experienced both volcanic and ejecta events. Its floor is very level that shows hint of previous lava-fills (?), followed by ejecta from the Imbrium Basin to the south, which has affected nearly every crater nearby (Meton, Euctemon). A gap in its south rim links into further, lava-filled terrain in this ~ 1.8 km depth crater. Together, with its obvious, younger-looking sub-crater E at centre (where a peak probably lies buried), this crater is one best viewed during favourable librations.

Craters

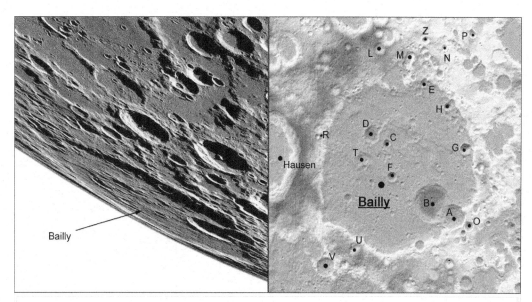

Bailly Lat 66.82S Long 68.9W 300.56Km

Sub-craters:

Crater	Lat	Long	Size (km)	Crater	Lat	Long	Size (km)
A	69.28S	59.57W	42.68	B	68.74S	63.25W	62.21
C	65.79S	70.34W	19.29	D	65.25S	72.38W	26.62
E	62.45S	65.75W	16.42	F	67.46S	69.59W	16.84
G	65.63S	59.47W	18.72	H	63.57S	62.59W	12.93
K	62.73S	76.71W	18.55	L	60.71S	71.13W	21.27
M	61.16S	67.51W	20.19	N	60.53S	63.68W	10.96
O	69.59S	56.92W	18.72	P	59.57S	60.67W	14.38
R	64.66S	79.18W	17.14	T	66.49S	73.83W	19.42
U	71.24S	76.03W	23.97	V	71.91S	81.45W	32.15
Y	61.04S	65.6W	14.26	Z	60.22S	65.86W	13.49

Notes:

Add Info:

Aerial view of Bailly

Not really a crater, but known more for its basin-like features: size (actually, the largest 'crater' on the nearside), no peak, inner ring (seen just about in the aerial view, left), Bailly, on the limb, is one for viewing during favourable latitudinal and longitudinal libration aspects. On such occassions, when the lighting conditions are just right, its southwestern sector displays worn-down terraces (covered by ejecta from younger Hausen nearby), deep shadows in subcrater B, and a slight uplifted look towards its centre. Craters to its north are Orientale in origin.

Craters

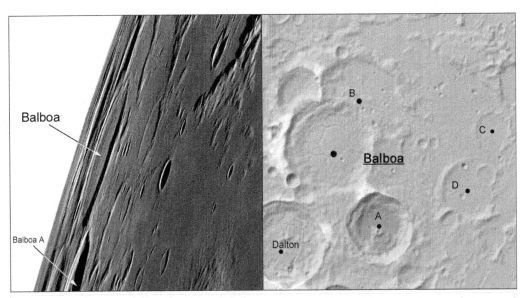

Balboa	Lat 19.24N Long 83.31W	69.19Km

Sub-craters:

Crater	Lat	Long	Size (km)	Crater	Lat	Long	Size (km)
A	17.42N	82.02W	46.66	B	20.49N	82.57W	54.25
C	19.59N	79.17W	26.09	D	18.25N	79.8W	40.37

Notes:

Add Info:

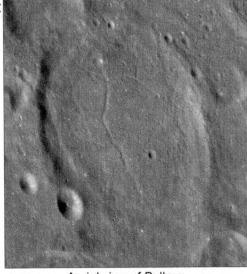

Aerial view of Balboa

Lying on the extreme limb regions of the Nearside, Balboa is a crater that favours a good librational aspect. Put into the fractured floor crater (FFC) category, its slightly raised floor and main fracture running down its centre casts perceptible shadows that defines both features - particularly during 14 to 15-day-old moon periods. With an average depth of 2 km, its appearance reflects its close proximity to the vast lava deposits that is the Oceanus Procellarum to its east, and the Orientale Basin due south - some 1200 km away. Bright ray material from an unnamed, small crater to its west covers its floor.

Craters

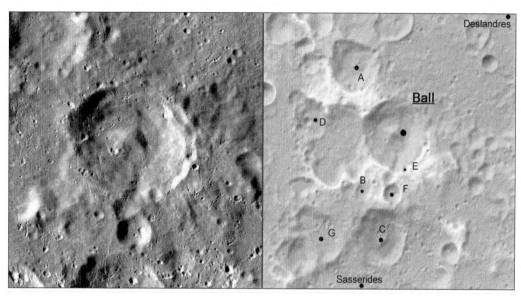

Ball   Lat 35.92S   Long 8.39W   40.31Km

Sub-craters:

Crater	Lat	Long	Size (km)	Crater	Lat	Long	Size (km)
A	34.74S	9.4W	29.23	B	36.92S	9.19W	9.72
C	37.7S	8.79W	31.15	D	35.62S	10.42W	22.45
E	36.53S	8.18W	4.39	F	36.95S	8.49W	11.38
G	37.72S	10.22W	26.73				

Notes:

Add Info: Possibly aged to the Early Imbrian Period (3.85 - 3.75 billions years old), Ball has a well-worn down look to it. Situated just outside the southwestern rim of the Deslandres crater (~ 227 km in diameter - and if a crater at all?), Ball's rim, its terraces and 1.5 km-high peak looks similar to other craters in the region - made up of pre-Nectarian ejecta deposits and secondary craters from Mares' Nubium and Imbrium respectively to its north. Southwards, the influence of the crater Tycho impact - some 235 km away - has thrown bright ejecta material across the crater (hard to see at times), and also, nearer still at just 165 km away, the last Surveyor lander (Surveyor 7) set down on the lunar surface on 10 Jan 1968 at Lat 41.01S, Long 11.01W. As the crater has an approximate depth of 2.8 km, it remains blanketed for some time in shadow when the terminator (during both eastern and western aspects) lies some distance away. However, from these aspects, it appears that the western rim side is slightly lower than its eastern rim side. At times of full moon the crater is very hard to detect (crater Hell ~ 100 km away northwards may at times be mistaken for it).

Craters

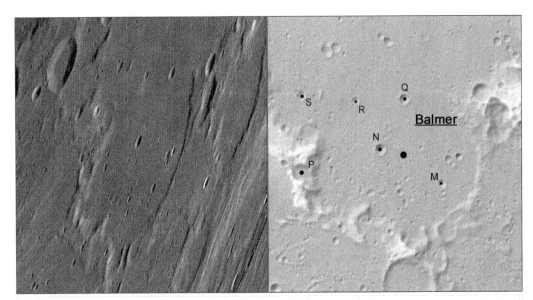

| Balmer | Lat 20.27S | Long 70.22E | | | | 136.3Km |

Sub-craters:

Crater	Lat	Long	Size (km)	Crater	Lat	Long	Size (km)
M	20.76S	71.55E	6.55	N	19.92S	69.82E	8.47
P	20.48S	67.58E	15.25	Q	18.69S	70.5E	7.98
R	18.71S	69.14E	5.72	S	18.57S	67.63E	6.53

Notes:

Add Info:

FeO:
Iron Oxide deposits such as those at, say, regions like Aristarchus, Sulpicius Gallus, Rima Fresnel...and elsewhere on the Moon, can provide invaluable information about the magmatic history and composition of the lunar interior.

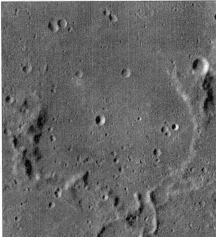

Aerial view of Balmer

Balmer looks almost lost in this region known as the Balmer-Kapteyn Basin - pre-Nectarian (4.6 - 3.92 byo). Having impacted southwards between this two-ringed (~ 260 and 500 km in diameter respectively) Basin, the crater is enhanced in FeO abundances, and suggested as being a site that hides an old, buried mare (a cryptomare). Dark regions east and west within the crater are further evidence of this feature; where dark-haloed craters have brought up such deposits to the surface through impact. With an average depth of 1.96 km, the crater is best observed during favourable libration periods.

Craters

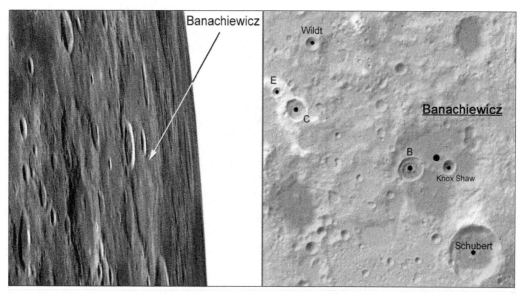

Banachiewicz Lat 5.28N   Long 80.01E   99.09Km

Sub-craters:

Crater	Lat	Long	Size (km)	Crater	Lat	Long	Size (km)
B	5.29N	78.97E	23.02	C	7.02N	75.33E	21.27
E	7.57N	74.75E	8.59				

Notes:

Add Info:

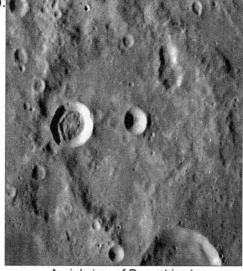

Aerial view of Banachiewicz

Another limb-hugger that demands good longitudinal librations to view any detail within, the crater lies just inside the main ring-rim (~740 in diameter) of Mare Smythii - lying some 270 km away due southeastwards. The crater shows all the signs of its pre-Nectarian age (4.6 - 3.92 byo); of worn rim, no discernable terraces, and no peak, which, if one did exist, may totally have been obliterated by the impact that formed the small crater Knox-Shaw (formerly known as sub-crater, Banachiewicz F). Hard to identify through the eyepiece, Banachiewicz is best found by first locating sub-crater, Banachiewicz B - whose inner rim is almost always bright.

Craters

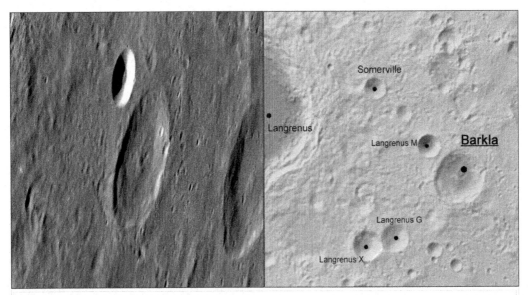

Barkla	Lat 10.67S	Long 67.22E	40.9Km

No sub-craters

Notes:

Add Info:

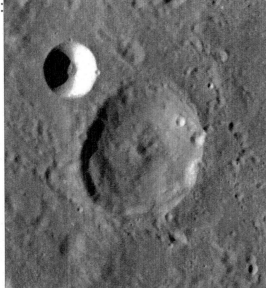

Aerial view of Barkla

Formerly called Langrenus A for [that] crater centred some ~180 km away westwards, Barkla lies within the main ring-rim (~ 500 km in diameter) of the Balmer-Kapteyn Basin; which formed in the pre-Nectarian period (4.6 - 3.92 byo). This 2.9 km-depth crater looks relatively fresh for its age (possibly Late Imbrian - 3.85 to 3.75 byo); showing a small central peak around 900 metres high, and a terrace in the southwest sector only. Ejecta deposits from younger Langrenus nearby covers the crater lightly, and to find it easily in the eyepiece - firstly look out for the bright, inner rim of the small crater, Langrenus M, to its north-west.

Craters

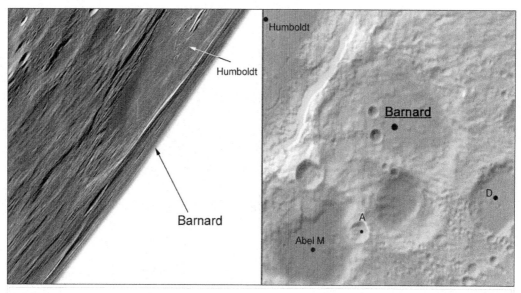

Barnard	Lat 29.79S	Long 85.95E		115.73Km			
Sub-craters:	Crater	Lat	Long	Size (km)	Crater	Lat	Long	Size (km)
	A	32.15S	85.06E	15.58	D	31.34S	89.17E	52.24
	Notes:							

Add Info:

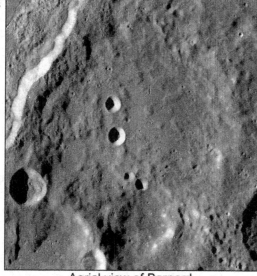

Aerial view of Barnard

Both amateur and professional astronomers would recognise the name Barnard - discoverer of 'Barnard's Star' (the fourth-closest known star to our Solar System whose proper motion was measured by E. E. Barnard). The crater has obviously been affected by impact of nearby, Humboldt (Upper Imbrian in age - 3.75 to 3.2 byo), to its northwest, making it younger, but note also other large craters at its south that have contributed their ejecta on to its floor. It can be quite a challenge to capture any descent views of the entire crater, but the two small, fresh craters on its floor are good signatures. D: ~ 2.7 km.

Craters

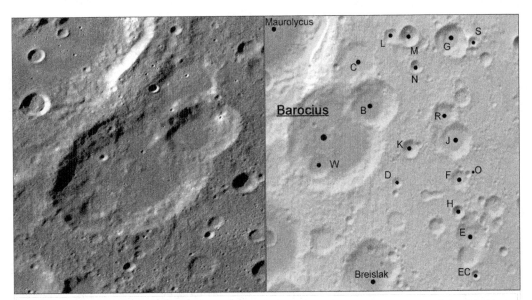

Barocius   Lat 44.98S   Long 16.81E   82.72Km

Sub-craters:

Crater	Lat	Long	Size (km)	Crater	Lat	Long	Size (km)
B	44.13S	18.29E	36.57	C	43.11S	17.48E	35.84
D	46.06S	19.18E	8.69	E	47.24S	22.15E	23.58
EC	48.22S	22.49E	7.67	F	45.92S	21.6E	15.38
G	42.52S	21.03E	27.61	H	46.71S	21.63E	10.56
J	44.97S	21.41E	27.16	K	45.22S	19.63E	13.56
L	42.5S	18.81E	13.03	M	42.45S	19.48E	15.85
N	43.2S	19.76E	10.07	O	45.75S	21.93E	5.36
R	43.91S	21.53E	14.26	S	42.5S	21.82E	8.27
W	45.68S	16.21E	18.9				

Notes:

Add Info: Suggested with origin being from the Serenitatis Basin - over 2000 km away to its north, Barocius's rim, its terraces and floor, have subsequently been covered and pummelled by the ejecta deposits from other major, basin-forming events in the region - e.g. Nectaris to its northeast (~ 900 km away) and Imbrium in the northwest. Several nearby craters of all sizes have also had added their fair share, too - particularly its larger neighbour, Maurolycus, to the northwest, and slightly younger, sub-crater, Barocius B, to its northeast; whose western rim material must have been lost to a void (and now lies on Barocius's floor) as it impacted. The crater does have a small peak that becomes barely perceptible during terminator periods - around a 20-day-old moon (a small shadow is cast), and at high magnification the very smooth-looking material westwards of sub-crater B on Barocius's floor suggests hint of aged impact melt (if it is at all?).

Craters

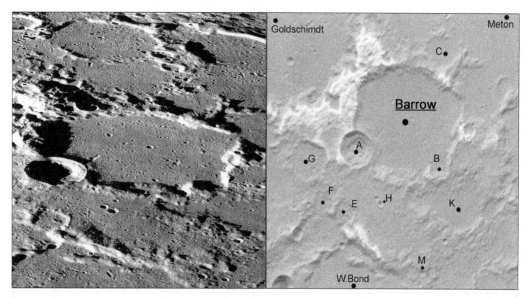

Barrow	Lat 71.28N	Long 7.59E					93.82Km

Sub-craters:

Crater	Lat	Long	Size (km)	Crater	Lat	Long	Size (km)
A	70.56N	3.86E	27.38	B	70.16N	10.53E	16.82
C	73.02N	11.12E	27.99	E	69.0N	3.33E	17.95
F	69.16N	1.81E	18.6	G	70.17N	0.26E	29.36
H	69.28N	6.0E	4.89	K	69.29N	11.74E	45.52
M	67.61N	9.14E	5.86				

Notes:

Add Info:

Aerial view of Barrow

The polygonal shape of Barrow is what first catches the eye as one views it through the eyepiece. Its floor is filled with material from the Imbrium Basin nearby (due southwards), and its worn down, broken rim suggests that this is a crater of age (pre-Nectarian - 4.6 to 3.92 byo, in fact). Bright streaks on the floor is ray material from Anaxagoras crater to its west (some 175 km away), which is also responsible for several hard-to-see crater chains in the same direction. Both its high eastern and western rim casts deep shadows during termintor times, and if observed just before first quarter, a wonderful 'shaft' of light can be seen to stream through the gap in the north-eastern sector of the crater.

Craters

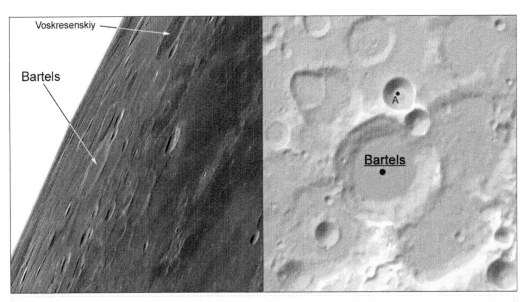

Bartels	Lat 24.51N	Long 89.85W	54.95Km

Sub-craters:	Crater	Lat	Long	Size (km)	Crater	Lat	Long	Size (km)
	A	25.7N	89.6W	16.85				
	Notes:							

Add Info:

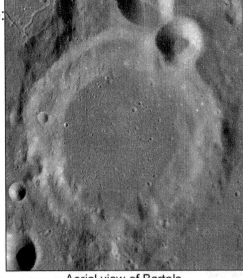

Aerial view of Bartels

If you were to take depth levels of Bartels, crater Voskresenskiy to its northeast, and the mare Oceanus Procellarum over to the east, you would find that Bartels's floor level is roughly a kilometre below the levels of the latter two. Also, you would find that while the other two have relatively dark-filled deposits (lava floodings etc.,), with Bartels it's quite the opposite; where the deposits are brighter. It's obvious, then, that Bartels in its lifetime experienced an additional cover of deposits from another source (are the deposits those of the Orientale Basin to the south, some 1300 km away?). Depth of Bartels ~ 3.0 km.

Craters

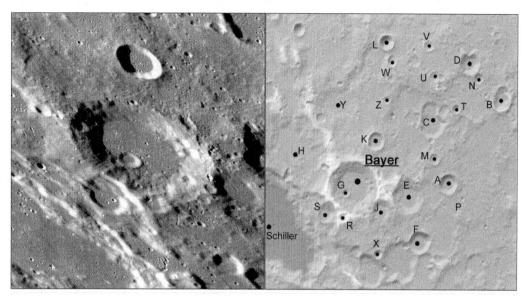

Bayer	Lat 51.62S	Long 35.14W			48.51Km	

Sub-craters:

Crater	Lat	Long	Size (km)	Crater	Lat	Long	Size (km)
A	51.36S	30.44W	17.54	B	48.82S	28.2W	23.09
C	49.77S	31.42W	22.1	D	47.96S	29.84W	19.6
E	51.77S	32.41W	29.35	F	53.03S	31.68W	19.11
G	51.74S	35.49W	7.44	H	53.53S	32.68W	31.54
J	52.26S	33.62W	24.64	K	50.29S	34.06W	15.78
L	47.52S	33.69W	13.35	M	50.69S	31.19W	9.51
N	48.33S	29.37W	8.57	P	51.72S	29.68W	4.25
R	52.42S	35.61W	7.69	S	52.33S	36.45W	15.44
T	49.25S	30.32W	7.98	U	48.4S	31.4W	9.69
V	47.56S	31.76W	9.4	W	48.04S	33.41W	8.83
X	53.42S	33.81W	8.36	Y	49.23S	35.87W	27.59
Z	49.12S	33.61W	6.2				

Notes:

Add Info: Lying just outside the main ring (335 km in diameter) of the Schiller-Zucchius Basin to its west, this crater, like most other craters in the region (e.g. Schiller and Schiller H, Bayer C and E to the east, Rost and Rost A to the south...etc.,), has a worn rim, degraded terraces, and smooth floor. The material that 'fills' (if that is indeed what happened) all of these craters suggest that we are looking at either some kind of ejecta-fill from a major event in the region, or, the result of volcanic-fill into each after they formed. A closer look at Bayer G shows this very nicely whereby, having impacted upon Bayer's southwestern terrace, its small interior is filled with the same material that covers the main crater's floor. D: 2.4 km.

Craters

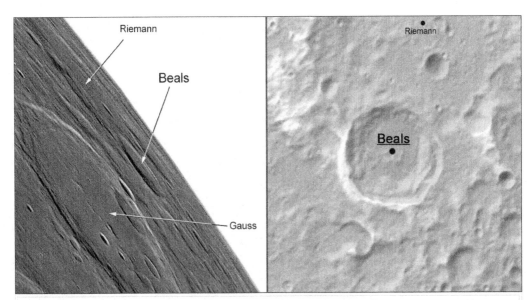

| Beals | Lat 37.11N | Long 86.58E | 52.61Km |

Sub-craters: No sub-craters

Notes:

Add Info:

Aerial view of Beals

See how the impact of Beals on Riemann's southern rim and its outer ejecta, has affected later events, such as, slumping in the crater's southern sectors. Why would this be so? Would it be that Beals's dynamic event, and its impactor, affected more weaker, looser deposits in Riemann's ejecta than that found in its floor; leading to pronounced slumping at Beals's south than at its north? Beals's floor level is some 500 m below the level of Riemann's. It has underdone serious fracturing in one particular direction, which may be a radial signature and feature related somewhat to the Humboldtianum Basin to the north (centred some ~ 600 km away). Depth of Beals ~ 3.0 km.

Craters

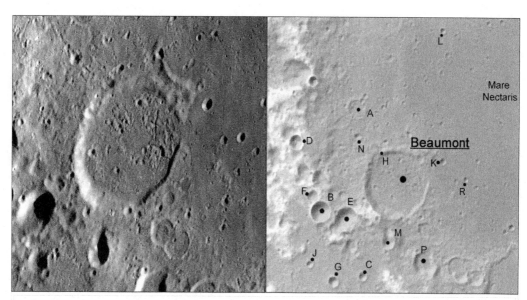

Beaumont Lat 18.08S Long 28.82E 50.69Km

Sub-craters:

Crater	Lat	Long	Size (km)	Crater	Lat	Long	Size (km)
A	16.31S	27.78E	14.32	B	18.71S	26.8E	15.14
C	20.25S	27.93E	6.02	D	17.09S	26.17E	10.49
E	18.89S	27.49E	16.32	F	18.37S	26.59E	8.9
G	20.37S	27.14E	7.73	H	17.08S	28.53E	3.66
J	19.95S	26.5E	4.85	K	17.52S	30.07E	4.95
L	14.46S	30.02E	4.24	M	19.43S	28.63E	10.22
N	16.96S	27.77E	4.9	P	19.92S	29.59E	16.25
R	17.94S	30.72E	4.1				

Notes:

Add Info: Beaumont initially looks to be a smaller version of the floor-flooded crater to its southeast - that of Fracastorius. Indeed, both craters' northeastern rims have a tilted look to them towards Nectaris's centre; where, when lavas filled up the central regions of the basin through several floodings over time, the additional weight produced a pulled-in effect to the outer regions (and thus the tilt effect, too). Beaumont's rim, however, looks more complete than its larger neighbour, and as a small gap exist at its northeastern rim, did the larger extent of lavas from Nectaris breach in here to the crater, or, did volcanic-originated lavas with-in the crater itself breach outwards? Certainly, the inner floor looks very rough; showing a hummocky and fractured apperance, however, adding further to the confusion, the crater also lies close to a northern ridge (unofficially called, Dor-sum Beaumont), and within one of the basin's ring that is ~ 400 km in diameter.

Craters

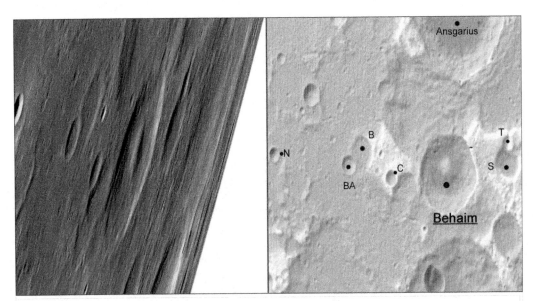

Behaim	Lat 16.61S	Long 79.41E	56.21Km

Sub-craters:

Crater	Lat	Long	Size (km)	Crater	Lat	Long	Size (km)
B	16.1S	76.52E	22.94	BA	16.4S	76.07E	14.3
C	16.71S	77.48E	12.18	N	16.11S	73.55E	9.92
S	16.33S	81.44E	26.68	T	15.9S	81.41E	11.3

Notes:

Add Info:

Aerial view of Behaim

Like all craters lying on the limbs of the Moon, Behaim is best observed during a favourable libration. Some limb craters can be hard to judge at times due to the perspective, however, what gives Behaim away is its broad peak, which, to all intents and purposes, looks overly large for a crater of this size. The rim, like its peak, is well worn down, so this is an old-ish crater; which, according to the USGS geological map by Wilhelms and El Baz (1977 publication), may be older than the Orientale Basin and younger than the Imbrium Basin. Depth of crater is around the 3 km mark, and the peak is approximately 2 km high.

Craters

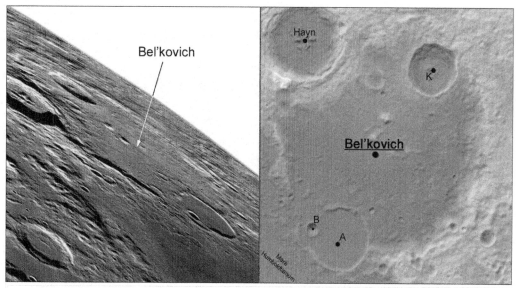

Bel'kovich   Lat 61.53N   Long 90.15E   215.08Km

Sub-craters:

Crater	Lat	Long	Size (km)	Crater	Lat	Long	Size (km)
A	58.65N	87.41E	57.26	B	58.89N	85.71E	12.34
K	63.63N	93.61E	47.03				

Notes:

Add Info:

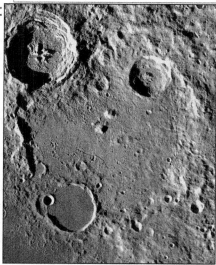

Aerial view of Bel'kovich

At just over 90 degrees longitude, the prospects of observing any details in Bel'kovich is going to be a challenge from the outstart. Libration times will obviously benefit views, but even these may not yield much rewards. The best you may glean from Bel'kovich is of its southwestern rim, its two main peaks (highest is about 1.5 km), and hint of sub crater, Bel'kovich K, which 'peeps' on the limb. The dark mare of Humboldtianum acts as a good marker for locating approximately the crater, with sub-craters', Bel'kovich A and B, further adding to your 'zoning-in' (B more than A). Hayn's ejecta lies on Bel'kovich's floor. The crater's southwestern sector experienced partial flooding (internal?).

Craters

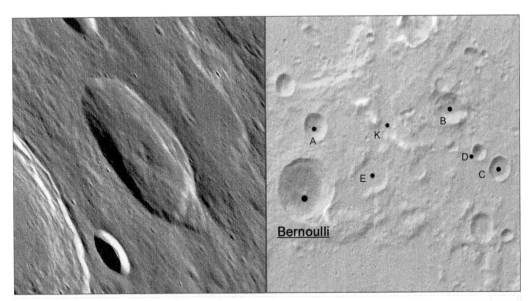

Bernoulli

Bernoulli		Lat 34.93N	Long 60.61E			47.3Km

Sub-craters:

Crater	Lat	Long	Size (km)	Crater	Lat	Long	Size (km)
A	36.41N	60.88E	20.24	B	36.84N	65.51E	23.96
C	35.38N	67.19E	18.0	D	35.79N	66.5E	11.91
E	35.32N	62.86E	26.75	K	36.54N	63.4E	17.68

Notes:

Add Info:

Aerial view of Bernoulli

Lying just outside the ~ 1080 km diameter ring-rim - created during the Crisium Basin impact event to the south, Bernoulli is a crater of the Imbrian Period (3.85 - 3.2 billions years old). Its rim, its 1.4 km-high peak, and its terraces are worn-looking in this deep crater (~ 3.6 km); causing its interior to remain in shadow - even when the terminator (during both E and W aspects) is far from it. Oddly, collapse of the southern sector of the crater's rim seems too extended; hint, perhaps, of the violent effects from later impacts nearby (e.g. younger Germinus to its west), which may have shook its foundations.

Craters

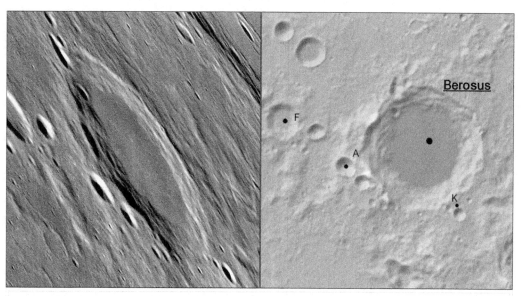

Berosus	Lat 33.5N	Long 69.99E		75.24Km			
Sub-craters:	Crater	Lat	Long	Size (km)	Crater	Lat	Long	Size (km)
	A	33.14N	68.08E	11.31	F	34.0N	66.53E	22.15
	K	32.19N	70.89E	8.17				
	Notes:							

Add Info:

Aerial view of Berosus

Not to be confused with Barocius crater, Nectarian-aged (3.92 - 3.85 byo) Berosus lies just outside the 1080 km diameter ring-rim of Mare Crisium to the south. Most of the regions (N and S) far outside the crater's perimeter resemble that of its smooth interior - all, perhaps, the result of dark mantled material seeping upwards to the surface; which then levelled afterwards. At high magnification, some of the many small craterlets on its floor become obvious, however, these are best hunted for during periods when the sun angles are high e.g. close to full moon. Depth of crater is ~ 4.5 km approx..

Craters

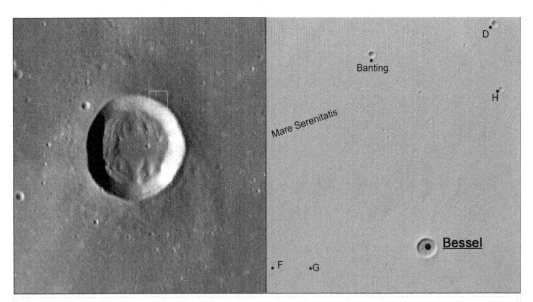

Bessel	Lat 21.73N	Long 17.92E	15.56Km

Sub-craters:

Crater	Lat	Long	Size (km)	Crater	Lat	Long	Size (km)
D	27.34N	19.86E	5.2	F	21.26N	13.84E	0.63
G	21.14N	14.74E	1.11	H	25.69N	20.0E	3.72

Notes:

Add Info:

LROC view of layers in Bessel's northeastern inner wall - exposing possible lava-flows of Mare Serenitatis's past.

Oddly enough, the one thing that makes Bessel stand out against the vast lava deposits covering Mare Serenitatis is the ray system it lies close to. One would initially think that the ray was created by formation of Bessel itself, or, as once believed, sourced from Menelaus to the southwest. The current theory, however, is that Tycho (lying some ~ 2000 km away to the southwest) is the culprit. Judging from the state of Bessel's rim, the crater isn't a fresh one, but more than likely is of a later time - possibly of the late Eratosthenian Period (~3.5 byo). Its inner walls are uniform in thickness, and the material on the outer floor regions are probably due to slumped effects, where huge volumes of the walls collapsed into the floor area. Depth ~ 1.7 km.

Craters

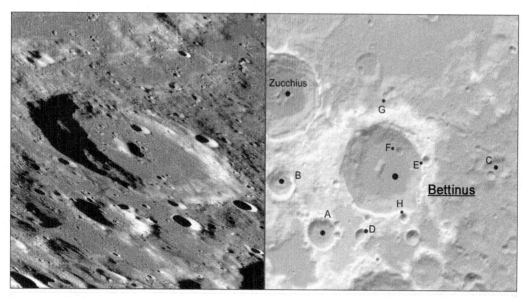

Bettinus Lat 63.4S Long 45.16W 71.78Km

Sub-craters:

Crater	Lat	Long	Size (km)	Crater	Lat	Long	Size (km)
A	64.88S	49.03W	25.67	B	63.54S	51.31W	25.17
C	63.33S	38.09W	22.55	D	65.0S	46.61W	10.46
E	63.21S	42.38W	7.9	F	62.97S	44.02W	6.76
G	61.6S	44.74W	6.52	H	64.62S	43.82W	7.77

Notes:

Add Info:

Aerial view of Bettinus

Lying just outside the 340 km diameter ring-rim of the Schiller-Zucchius Basin to the south, Bettinus is a crater of old lunar times. Well-worn rim and collapsed terraces that overly extend into the floor, the crater may be one of the Late Imbrian to Nectarian Period (3.92 - 3.85 byo). The smooth floor shows characteristic lava (mare?) deposits that settled to a level seamlessly around its central peak (~ 2.5 km high); which all may later have been covered by a thin sheet of fluidized ejecta from creation of the Orientale Basin - some 1600 km away to the north-west. Depth of crater is 3.7 km high approx..

Craters

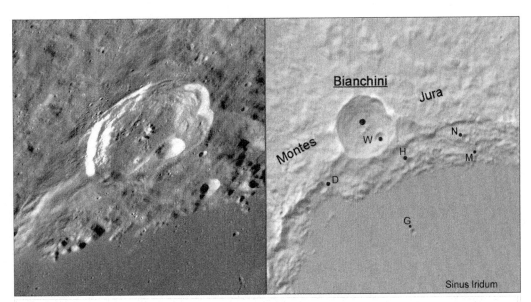

Bianchini	Lat 48.78N	Long 34.37W			37.59Km	

Sub-craters:

Crater	Lat	Long	Size (km)	Crater	Lat	Long	Size (km)
D	47.55N	35.7W	7.06	G	46.7N	32.79W	3.91
H	48.08N	32.86W	6.17	M	48.37N	30.67W	4.24
N	48.55N	31.07W	4.97	W	48.55N	33.81W	8.34

Notes:

Add Info:

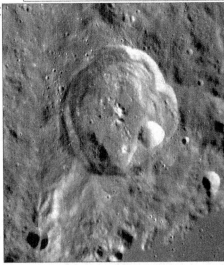

Aerial view of Bianchini

A crater of the Early Imbrian Period (3.75 - 3.2 byo), Bianchini sits on the rim of Sinus Iridum, which in turn sits on the main 1160 km diameter ring-rim of the Imbrium Basin. Maintaining a relatively sharp rim all around, several series of terraces also encompass its inner regions, as well as three to four peaks in the centre. Impact has imparted some of its own ejecta deposits, along with rim deposits from Iridum, onto the mare floor (note, the lighter colour change of this material). The impact also may have affected most of Iridum's eastern rim, as the terraces there seem more extended than in the west - hint, perhaps, of Bianchini's impactor direction.

Craters

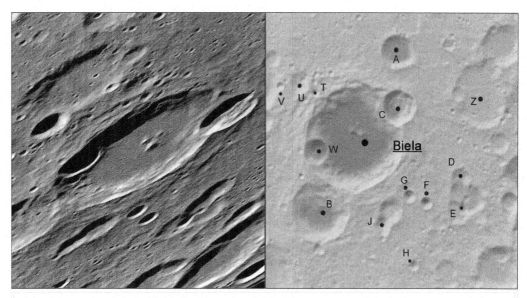

Biela Lat 54.99S Long 51.63E 77.03Km

Sub-craters:

Crater	Lat	Long	Size (km)	Crater	Lat	Long	Size (km)
A	52.94S	53.39E	26.13	B	56.72S	50.03E	44.55
C	54.28S	53.5E	26.89	D	55.89S	56.41E	15.97
E	56.61S	56.58E	8.56	F	56.52S	54.85E	9.4
G	56.38S	54.09E	10.09	H	58.02S	54.4E	7.96
J	57.07S	52.89E	13.95	T	53.92S	49.92E	6.51
U	53.66S	49.31E	16.38	V	53.73S	48.45E	6.4
W	55.31S	49.87E	18.16	Y	54.97S	58.01E	15.71
Z	53.9S	57.0E	50.63				

Notes:

Add Info:

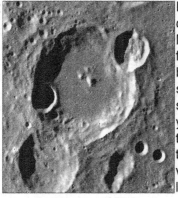

Aerial view of Biela

Lying in a corner of the Moon pock-marked with craters, and chains of craters, that point back to Mares' Nectaris and Crisium, Biela looks quite fresh-looking for its age (Nectarian - 3.92 - 3.85 byo). Both sub-craters, C and W, are probably secondaries from the Imbrium Basin impact - some 3000 km away to its north-west (and so younger), while sub-crater B, which appears to have Biela ejecta deposits within its floor, is therefore older. Like most 'limb-huggers', observations of Biela much depends upon favourable libration periods and near terminator times where shadows play the major role. Depth ~ 5.6 km.

Craters

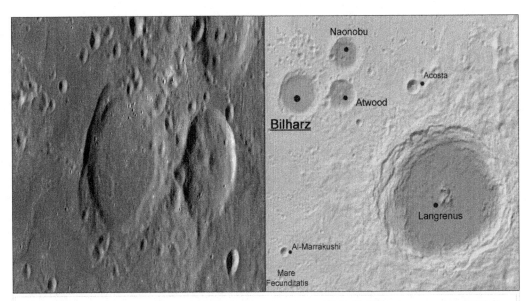

| Bilharz | Lat 5.83S | Long 56.34E | 44.55Km |

No sub-craters

Notes:

Add Info:

Luna 16 landed some 140 km away to the north of this crater on the 20 Sept., 1970 (0.41S, 56.18E).

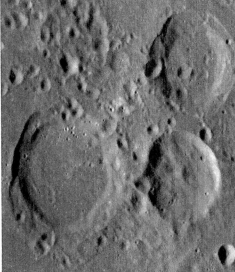

Aerial view of Bilharz (once called Langrenus F), Atwood and Naonobu

Bilharz, Atwood and Naonobu are a crater-triad that formed around the same time - post Crisium to pre-Imbrian Periods. Each looks the same in terms of its aged features - worn rims, well-slumped terraces, and the targets of impact abuse by countless of other craters that formed afterwards. Ejecta from Langrenus to the southeast cover all three, and secondary craters from that event pockmark the surrounding region. Two things to note: (1) the northwest rim of Bilharz looks relatively sharper than the rest - was it because most of Langrenus's ejecta fell short of it? (2) the series of fresh craterlets that cross its northern sector - were they due to an unidentified crater, or the result of a disintegrating meteor or comet?

Craters

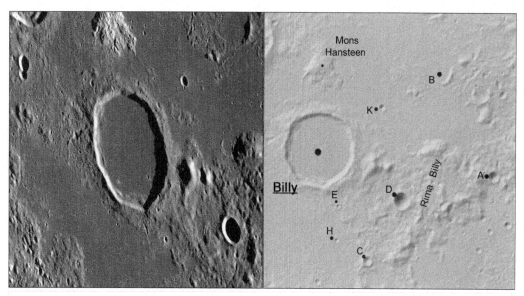

Billy Lat 13.83S Long 50.24W 45.57Km

Sub-craters:

Crater	Lat	Long	Size (km)	Crater	Lat	Long	Size (km)
A	14.34S	46.34W	7.26	B	12.2S	47.67W	20.76
C	16.08S	49.14W	5.59	D	14.86S	48.39W	10.91
E	14.97S	49.81W	2.56	H	15.65S	49.84W	3.26
K	12.97S	48.86W	4.21				

Notes:

Add Info:

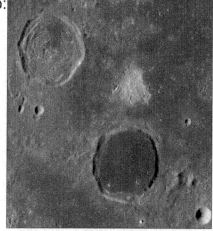

Aerial view of Billy, Hansteen and Mons Hansteen (bright mound)

The lava-filled interior of Billy, to all intents and purposes, looks ever so slightly darker than the vast deposits to its north that make up Oceanus Procellarum (OP). As no breach appears across any part of its rim, and its own outer ejecta deposits are still relatively fresh-looking and intact (not covered with OP lavas), the deposits in the interior must have welled up separately after the crater formed - perhaps, also, of a different consistency. Like Plato to its northeast, Billy has a series of extra-small craterlets that prove a challenge to see, however, the tiny, and bright, impact in the floor's southwest sector is one that contrasts well against the dark deposits and so is easily spotted.

Craters

Birmingham Lat 65.12N   Long 10.7W   89.92Km

Sub-craters:

Crater	Lat	Long	Size (km)	Crater	Lat	Long	Size (km)
B	63.52N	11.24W	7.33	G	64.58N	10.23W	5.19
H	64.46N	10.64W	6.54	K	65.05N	13.17W	5.82

Notes:

Add Info:

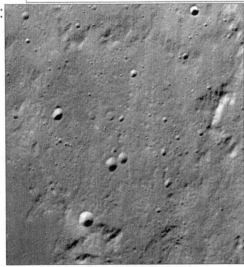

Aerial view of Birmingham

Almost lost amidst the mottled and rubbly-looking material that dominates this region north of Mare Frigoris, Birmingham and its surrounds eastwards, westwards and northwards have been covered by ejecta from the Imbrium Basin event to the south. Its polygonal-shaped rim becomes obvious at terminator periods ~ after first (best) and last quarter moons, where higher sectors cast long shadows across its floor. An easy way to find Birmingham through the eyepiece is to firstly locate the very obvious crater of Plato (some 100 km away to its south), and from there its simply a matter of looking upwards.

Craters

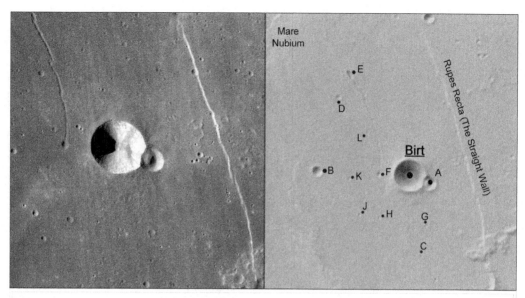

Birt   Lat 22.36S   Long 8.59W   15.81Km

Sub-craters:

Crater	Lat	Long	Size (km)	Crater	Lat	Long	Size (km)
A	22.49S	8.24W	7.16	B	22.29S	10.27W	5.05
C	23.69S	8.37W	1.8	D	21.08S	9.88W	2.63
E	20.72S	9.66W	5.34	F	22.34S	9.15W	2.47
G	23.19S	8.28W	1.84	H	23.07S	9.15W	1.83
J	23.0S	9.5W	1.76	K	22.4S	9.69W	1.76
L	21.66S	9.36W	1.8				

Notes:

Add Info:

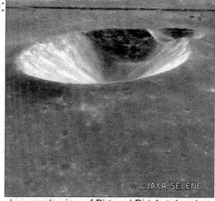

Low angle view of Birt and Birt A, taken by the Japanese Kaguya (Selene) spacecraft.

For such a little crater lying close to one of the most wonderful features on the lunar surface (the Straight Wall), Birt, too, gives some attraction to the viewer, as its bright interior wall 'spotlights' during full moon times. Sub-crater A, has also given the crater some prestige over others of similar size in the region, as under high magnification the lipped feature that joins the two is well worth looking at under any lighting conditions. As can be seen in the Kaguya image to the left, this 'lip' is slightly depressed, and so at certain times light and shadows fill through it forming interesting effects.

Craters

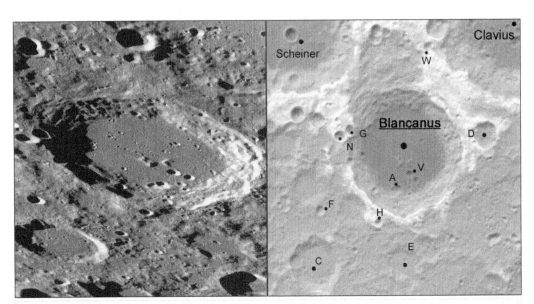

Blancanus	Lat 63.77S	Long 21.63W					105.82Km

Sub-craters:

Crater	Lat	Long	Size (km)	Crater	Lat	Long	Size (km)
A	64.72S	22.15W	6.49	C	66.6S	28.22W	44.16
D	63.22S	16.65W	23.2	E	66.68S	21.83W	31.26
F	65.19S	27.5W	8.82	G	63.13S	25.19W	9.33
H	65.61S	23.59W	6.96	K	60.53S	23.36W	10.87
N	63.26S	25.86W	9.93	V	64.34S	21.43W	6.73
W	60.96S	20.33W	8.49				

Notes:

Add Info:

Not to be confused with similar-sounding crater, Blanchinus, 25.32S, 2.44E.

Aerial view of Blancanus

If you can take your eyes off the wonderful crater - that is Clavius to Blancanus's northeast, which dominates the view, rewards will be gleaned, too. Of Nectarian in age (3.92 - 3.85 byo), the crater, to all intents and purposes, looks like a smaller version of its bigger neighbour without the additional craters; having similar slumped terraces, smooth-ish floor covered in numerous craterlets, and a series of off-centred peaks. At high sun angle periods, the eastern sector of the floor show ray material from Tycho (some ~ 600 km away to its north) crossing it, while at terminator times, deep shadows fill up the crater that are always a treat to view. Depth ~ 6 km.

Craters

Blanchinus Lat 25.32S Long 2.44E 59.9Km

Sub-craters:

Crater	Lat	Long	Size (km)	Crater	Lat	Long	Size (km)
B	25.23S	1.55E	7.22	D	25.03S	4.17E	6.9
K	24.81S	4.99E	8.31	M	25.22S	2.58E	4.11

Notes:

Add Info: The jumble of large-ish-like craters that lie nearly all around Blanchinus sometimes distracts the eye away from this crater, and so it never gets a look-in. Of pre-Nectarian in age (4.6 - 3.92 billions years old), its well-worn appearance is then well justified, however, between additional coverage of ejecta from its younger neighbour, Werner (Eratosthenian - 3.15 - 1.1 byo), to its south, and a small depressed region which extends across to where sub-crater, Blanchinus D, lies, the crater has taken on a somewhat squished look. As most of Werner's ejecta is thicker on the southern sectors of Blanchinus's rim and the depressed region (a buried crater, perhaps?) mentioned, the additional bulk of mountainy material to its north, east and west might also be responsible to the crater's squishy look. At high magnification, three small, fresh-looking crater-chains on the floor (NE, SE and W sectors) all show alignment that run in a NE-SW direction approximately. Each small crater in these chains decreases in size from SW to NE; suggesting the event that produced them came from bottom-left, however, if traced back to find a source crater that might be responsible - none, really, is found of suitable age. Mantled material from the Nectaris Basin (some 900 km to its east) covers the crater also. Depth of crater ~ 1.2 km approx..

Lunar X:
This lighting phenomenon occurs just before First Quarter, as light coming from the east strikes high rim points on the triad of craters' Blanchinus, LaCaille and Purbach.

Craters

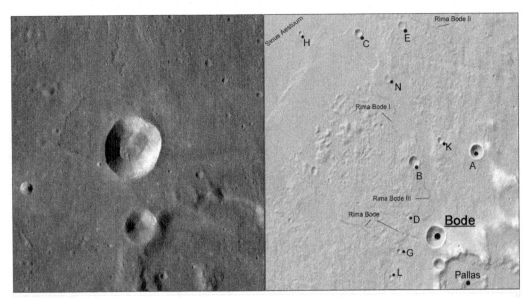

Bode	Lat 6.71N	Long 2.45W				17.8Km
Sub-craters:	Crater	Lat	Long	Size (km)	Crater	Lat	Long	Size (km)

Crater	Lat	Long	Size (km)	Crater	Lat	Long	Size (km)
A	9.0N	1.18W	12.15	B	8.74N	3.08W	9.76
C	12.23N	4.77W	7.05	D	7.27N	3.32W	3.39
E	12.42N	3.45W	6.45	G	6.34N	3.56W	4.21
H	12.21N	6.53W	4.1	K	9.3N	2.28W	5.6
L	5.62N	3.81W	4.32	N	10.97N	3.88W	6.29

Notes:

Add Info:

Note, each individual Bode makes up what is known as the Rimae Bode

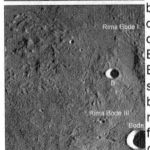

Situated close to one of the largest pyroclastic (ash) deposits on the Moon - see the dark-ish regions north, northwest and west of the crater, Bode has impacted upon an outer ring-rim (~ 1700 km in diameter) of material from the Imbrium Basin event. Having a relatively sharp-looking rim all around, clumps of material from both its north and east inner walls has slumped down onto the floor - produced perhaps by the shaking associated to nearby impacts. With exception to Rima Bode II lying some 200 km away to the crater's north, Bode's I and III are best observed during periods when shadows fill their channels; showing up their meanders better than in full moon times. Rima Bode II, in fact, is marked down by NASA as a 'Region of Interest' for further exploration as it is rich in iron and titanium (resources for exploitation). Depth of crater ~ 3.5 km.

Craters

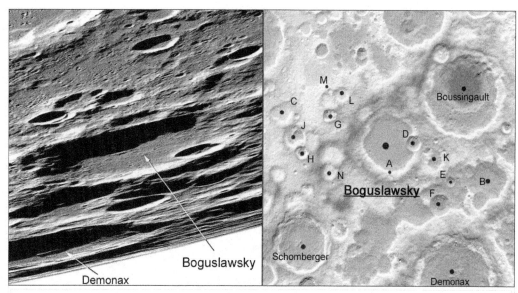

Boguslawsky   Lat 72.9S   Long 43.26E   94.59Km

Sub-craters:

Crater	Lat	Long	Size (km)	Crater	Lat	Long	Size (km)
A	74.41S	43.57E	8.19	B	73.98S	61.12E	63.47
C	70.99S	27.7E	34.48	D	72.86S	47.41E	22.4
E	74.31S	54.33E	14.61	F	75.42S	52.97E	31.05
G	71.48S	34.35E	20.46	H	72.82S	29.06E	21.13
J	72.09S	28.4E	34.73	K	73.47S	50.35E	46.43
L	70.68S	36.47E	21.93	M	70.45S	34.84E	8.09
N	73.86S	32.88E	27.45				

Notes:

Add Info:

Aerial view of Boguslawsky

Lying not far from the 2500 km outer ring-rim of the South Pole Aitken Basin (totally hidden from view for an earth-based observer), this crater is one for the limb-hugger enthusiast. Easily lost amongst the list of similarly-sized craters that lie all around it, the only identifier is sub-crater D on its eastern inner rim, which just after full moon produces shadows that locates the main crater, Boguslawsky. Older than the Nectaris Basin, its rim, terraces and floor with its small, off-central peak has subsequently been covered by ejecta material (west side of rim particularly so) from other impact craters in the region. Depth ~ 4 km.

Craters

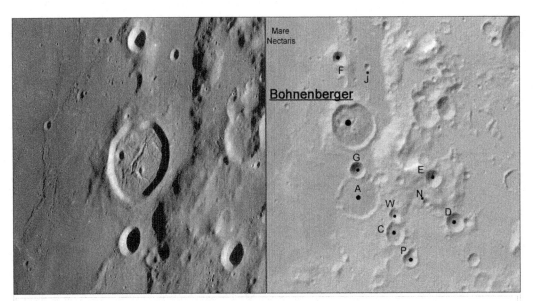

Bohnenberger Lat 16.24S Long 40.06E 31.74Km

Sub-craters:

Crater	Lat	Long	Size (km)	Crater	Lat	Long	Size (km)
A	17.82S	40.17E	31.11	C	18.67S	41.1E	14.11
D	18.39S	42.63E	13.4	E	17.41S	42.07E	11.86
F	14.71S	39.68E	9.34	G	17.23S	40.13E	11.77
J	14.9S	40.39E	4.45	N	17.95S	41.89E	5.73
P	19.25S	41.47E	10.75	W	18.25S	41.1E	10.56

Notes:

Add Info:

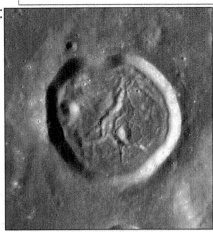

Aerial view of Bohnenberger

At full moon times it's quite easy to see that Bohnernberger lies on brighter rock as opposed to the slightly darker lavas of this Mare Nectaris region. The crater initially formed on ejecta deposits of the basin (it actually lies on the ~ 300 km diameter ring associated to the Basin), whose own ejecta later saw lavas from the Basin lap at its shores (at its west and to its northeast). Volcanic activity in the crater's floor produced uplift events that led to the obvious fractures seen, and it looks like slumping occurred afterwards in the northwest sector that covered over parts there. Depth ~ 2.5 km approx..

Craters

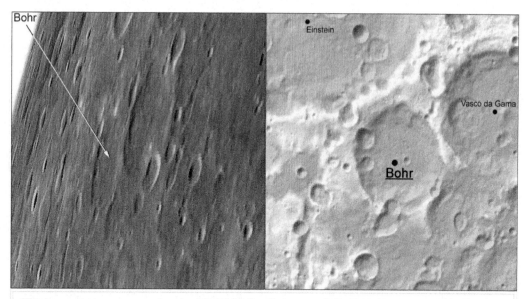

Bohr	Lat 12.71N	Long 86.52W	70.07Km

Sub-craters: No sub-craters

Notes:

Add Info:

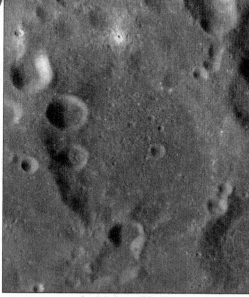

Aerial view of Bohr

Bohr, along with Balboa, Dalton, Vasco da Gama (all to its northeast or thereabouts) and Einstein (to is northwest), makes up a cluster of large craters on the west side of the Moon. All will prove an observer's challenge to see (they are all in the W80's longitude). It's quite hard to see divisions in the rims of Vasco da Gama and Einstein that 'connect' with Bohr, but it's guranteed that all formed early when the Moon was still young. Adding futher to observing any descent detail from Bohr, ejecta deposits from the Orientale Basin (~1000 km to the southwest) has covered the area in general. Good librations are recommended for viewing. Depth ~ 3.5 km.

Craters

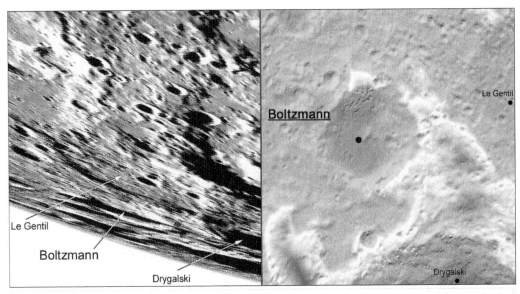

Boltzmann   Lat 74.82S   Long 90.41W   72.31Km

Sub-craters: | No sub-craters
Notes:

Add Info:

Aerial view of Boltzmann

Having a high, westerly longitude designation from the outstart, it's obvious Boltzmann will be a challenge. At best libration times, the only portion of detail that can be gleaned is its high-ish rim (or the mountain) at its west and the small isolated part at its east (the floor is never really seen, nor its small peak either). Good location markers for finding Boltzmann is to look, firstly, for Le Gentil (with its two small craters on its floor), and, secondly, the peak of Drygalski (Boltzmann, then, is just behind Le Gentil, and to the left of Drygalski respectively). Plan well before tryi- to 'catch' Boltzmann. D: ~3 km.

Craters

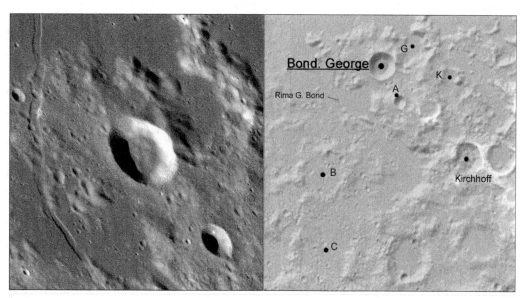

Bond, George   Lat 32.39N Long 36.32E   19.05Km

Sub-craters:

Crater	Lat	Long	Size (km)	Crater	Lat	Long	Size (km)
A	31.58N	36.87E	9.19	B	29.95N	34.67E	32.38
C	28.29N	34.75E	46.65	G	32.74N	37.25E	27.32
K	32.14N	38.33E	12.47				

Notes:

The crater is usually referred to as G. Bond

Add Info: Lying just outside the ~ 920-kilometre diameter ring-rim of Mare Serenitatis to its west, crater George Bond has impacted upon the ejecta from that event. This ejecta, together with additional ejecta deposits from Mare Crisium to the crater's southeast, is essentially the make-up of what are known as the Taurus Mountains, where, in effect, both have interrupted a once-depressed region between the next outer ring-rim ~ 1300 km in diameter of Serenitatis (note the lighter coloured mare material in Lacus Somniorum to G. Bond's northwest that has filled into this possible depression). George Bond, then, is slightly younger than these Basins - pre-Nec., and Nec., periods respectively, so, looking at its relatively sharp-ish rim and smooth floor, its formation and possible source may be secondary in nature from another major, but somewhat later event, nearby - that of the Imbrium Basin to the west. This crater is easily found when one notices the lovely rille of the same namesake, Rima G. Bond, to the crater's west, which probably formed also due to Serenitatis's formation. During full moon the crater's bright walls 'shine'. Sub-crater, G. Bond A, to its southeast shines, too, but try also to detect the small, bright impact crater at A's northeastern wall. D: ~ 2.8 km.

Craters

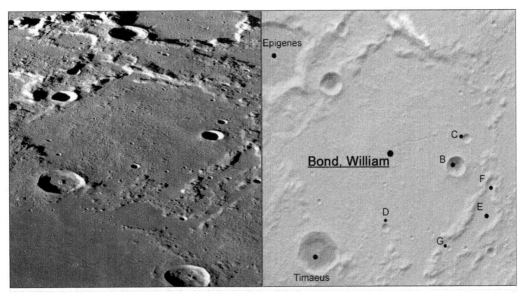

Bond, William Lat 65.41N Long 3.52E   170.53Km

Sub-craters:

Crater	Lat	Long	Size (km)	Crater	Lat	Long	Size (km)
B	65.03N	7.51E	15.24	C	65.69N	8.25E	7.39
D	63.6N	3.21E	6.85	E	63.8N	8.96E	24.95
F	64.45N	9.46E	9.02	G	63.07N	6.86E	3.98

Notes:

Add Info:

The crater is usually referred to as W. Bond

Aerial view of Bond, William

Through the eyepiece, William Bond looks like a square tipped on one of its corners. Covered with gravel-like clumps of projectiled ejecta from the Imbrium Basin event to its southwest and ray material from Anaxagoras to its north, this pre-Nectarian crater (4.6 to 3.92 byo) is one of those 'oldies'. The true rim is very hard to define; particularly in the SE and NW sectors as there seems to be two separate sections with about 20 km-wide valleys in between. The rille at its centre is obviously young and hard to see, however, another rille midway, which runs southeasterly at right angles from it, will prove a challenge for the serious observer. Depth ~ 1.9 km.

Craters

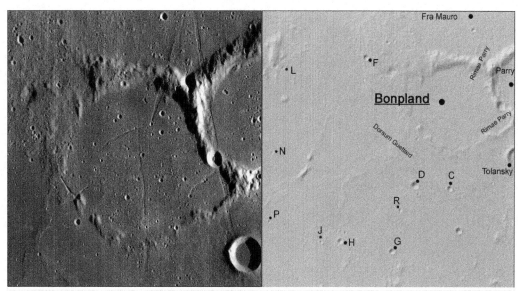

Bonpland Lat 8.38S Long 17.33W 59.25Km

Sub-craters:

Crater	Lat	Long	Size (km)	Crater	Lat	Long	Size (km)
C	10.23S	17.5W	4.22	D	10.16S	18.28W	5.25
F	7.34S	19.37W	4.08	G	11.61S	18.81W	3.7
H	11.37S	19.97W	3.78	J	11.39S	20.4W	2.63
L	7.55S	21.21W	2.84	N	9.41S	21.48W	2.31
P	10.93S	21.56W	1.68	R	10.7S	18.62W	2.48

Notes:

Add Info: Situated in an area of the Moon which got some attention by NASA back in the 70's (Apollo's 12 and 14, as well as Surveyor 3 landed in regions north to northwest, while Ranger 7 crashed some 100 km southwest), Bonpland is a crater whose demise just didn't have a chance in making itself known. With ejecta from the Imbrium Basin event covering most of its north and northwestern rim; lavas having possibly breached up and over its south and southwest rim; and slightly younger nearby, Parry crater, wiping out its remaining eastern rim, it's a wonder that Bonpland exists at all. Adding insult to injury, a series of graben-like features (extension faults) have also cut through several parts of what remains of its rim, while there seems to be some volcanic activities in the area, too (note the darker deposits inside and outside the northern rim where a branch of Rimae Parry lies). Best times to observe Bonpland is when the terminator has just passed some 100 km on either side (e.g. 11 - 13-day-old and 20 - 22-day-old moon periods) of the crater as it starts to disappear under full light conditions. Sub-crater D appears to be low angle impact related. Depth ~ 0.8.

Craters

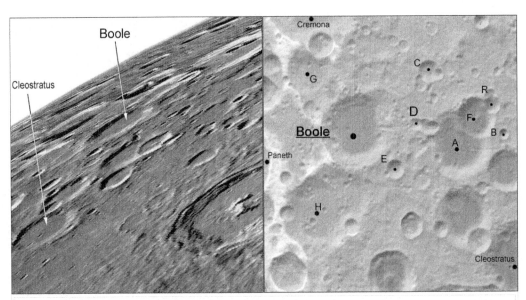

Boole	Lat 63.79N	Long 87.29W		61.34Km			
Sub-craters:	Crater	Lat	Long	Size (km)	Crater	Lat	Long	Size (km)

Crater	Lat	Long	Size (km)	Crater	Lat	Long	Size (km)
A	63.42N	80.57W	55.88	B	63.6N	77.7W	8.96
C	65.37N	82.47W	17.48	D	64.01N	83.38W	10.82
E	62.84N	84.66W	15.95	F	64.1N	79.41W	33.33
G	65.27N	90.68W	40.49	H	61.73N	88.83W	82.4
R	64.39N	78.25W	13.56				

Notes:

Add Info:

Aerial view of Boole

There are quite a few craters of similar size to Boole in front of it (as we view it from our earth perspective) that might lead to confusion. Formed in the Nectarian Period (3.92 to 3.85 billions of years ago), the crater's rim has a relatively smooth look to it, with dozens of small impact craters 'dotting' nearly every part of its inner walls. Like most craters in this region, Imbrium ejecta has 'marked' the crater (note sub-crater D's alignment back to a gash on Boole's northeastern inner rim), while its floor is probably of a fluidized make-up from that same event. What happened at the northeast sector?

Craters

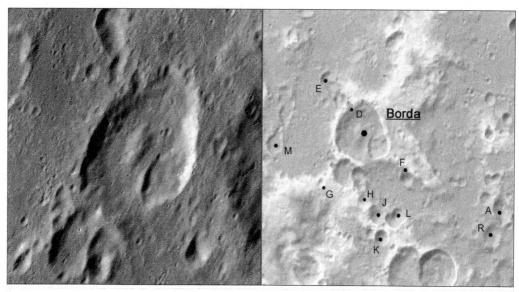

Borda Lat 25.2S Long 46.52E 45.4Km

Sub-craters:

Crater	Lat	Long	Size (km)	Crater	Lat	Long	Size (km)
A	26.87S	50.8E	18.35	D	24.57S	46.12E	5.37
E	24.01S	45.48E	11.95	F	26.19S	47.89E	9.96
G	26.31S	45.31E	6.55	H	26.74S	46.59E	10.84
J	27.02S	46.97E	14.97	K	27.54S	47.14E	12.33
L	27.06S	47.63E	12.65	M	25.43S	43.91E	13.88
R	27.43S	50.53E	14.38				

Notes:

Add Info:

Aerial view of Borda

Lying on the eastern sector of the main ring-rim Nectaris Basin system (~ 860 km in diameter, which shows up much better as the Altai Scarp in the west), Borda is a crater slightly younger than that impact event. Its interior features: terracing and central peak are smooth; numerous impacts, big and small, have struck most of its rim (note the large one in the SE, smaller on its northern rim); while the region, in general, has been subjected to ejecta from Imbrium - some 2500 km away to its north-west, and possible lava flows in the trough-like depression between Nectaris's next inner ring-rim. Depth ~ 3.5 km.

Craters

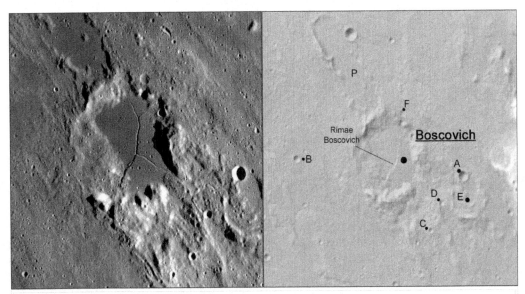

Boscovich	Lat 9.71N	Long 11.01E		41.53Km			
Sub-craters:	Crater	Lat	Long	Size (km)	Crater	Lat	Long	Size (km)
	A	9.46N	12.63E	5.83	B	9.79N	9.22E	4.34
	C	8.47N	11.96E	3.06	D	8.97N	12.19E	3.48
	E	9.03N	12.72E	19.77	F	10.44N	11.37E	4.0
	P	11.47N	10.31E	64.82				
Notes:								

Add Info:

General view of Boscovich region

Situated just outside the main ring-rim formed as a result of the Serenitatis Basin event to the northeast, Boscovich is a crater that has almost been lost due to high modification. Covered, and 'bombed', by ejecta from the Imbrium Basin event to its northwest, filled with dark mare deposits, and then 'cracked' open by interior stress effects (Rima Boscovich), this feature is rather a sculpture than a crater. Note the dark region where it lies (left). Depth ~ 1.8 km approx..

Craters

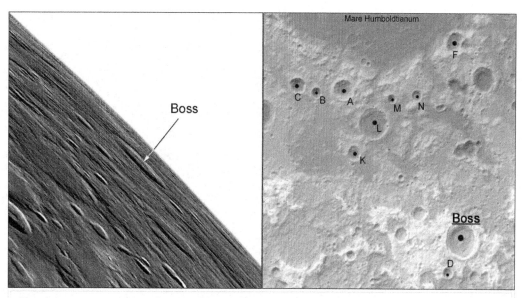

Boss	Lat 45.75N	Long 88.68E	50.2Km

Sub-craters:

Crater	Lat	Long	Size (km)	Crater	Lat	Long	Size (km)
A	52.21N	80.18E	29.19	B	52.02N	78.13E	14.5
C	52.26N	76.67E	22.78	D	44.36N	87.59E	15.59
F	54.17N	89.09E	30.49	K	49.53N	81.15E	20.44
L	50.85N	82.48E	40.95	M	51.9N	83.87E	12.95
N	52.03N	85.8E	15.42				

Notes:

Add Info:

Aerial view of Boss

It's a pity that we can't get a descent view of Boss because of its limb location, as it lies just outside one of the rings/rims (of 650 km in diameter) associated to formaation of the Humboldtianum Basin to the north - centred some ~ 350 km away. The crater and the ring/rim look almost like a mirrored version of crater Piccolomini on the Altai Scarp ring/rim at Mare Nectaris's southwest (have a look at them first as it will give a good idea of Boss's location and surrounds). The crater has to be younger than Nectarian-aged (3.92 to 3.85 byo) Humboldtianum, and like Piccolomini itself, it is probably of the Upper Imbrian (3.75 to 3.2 byo). Depth of Boss ~ 3.7 km approx..

Craters

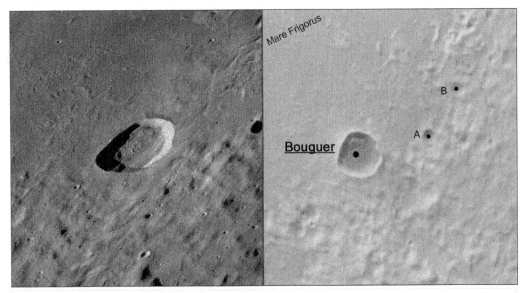

Bouguer	Lat 52.32N	Long 35.82W	22.23Km

Sub-craters:	Crater	Lat	Long	Size (km)	Crater	Lat	Long	Size (km)
	A	52.61N	33.89W	7.37	B	53.36N	33.07W	8.0
	Notes:							

Add Info:

Aerial view of Bouguer

Copernican in age (1.1 billions of years to the Present), Bouguer lies just on the extreme rim remain-deposits from the Sinus Iridum crater to the southeast, which may lie on the main, northern ring-rim of the Imbrium Basin. Having a sharp rim all around, the crater's shape has changed somewhat - due to what looks like a massive landslide of material in the southeast sector that covers a third of the floor there, and a similar, less-dynamic event in the southwest (in shadow). The floor has some smooth areas which hint of impact melt deposits that settled into low spots, while higher areas are possibly clumps of the original target rock. Depth of crater is around the 3.2 km mark.

Craters

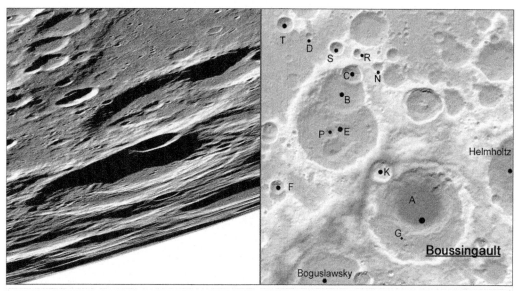

Boussingault   Lat 70.21S   Long 53.73E   127.61Km

Sub-craters:

Crater	Lat	Long	Size (km)	Crater	Lat	Long	Size (km)
A	69.96S	53.87E	75.4	B	65.69S	47.04E	57.72
C	65.15S	48.06E	23.13	D	63.58S	44.81E	8.85
E	67.08S	46.64E	104.09	F	68.97S	39.65E	16.64
G	71.44S	52.51E	4.59	K	68.83S	50.1E	27.22
N	71.37S	61.2E	15.1	P	67.29S	45.59E	13.34
R	64.4S	48.66E	11.92	S	64.14S	46.91E	15.83
T	63.02S	43.06E	19.43				

Notes:

Add Info:

Aerial view of Boussingault

First look at Boussingault's features - worn rim and terraces - suggests to the viewer that this is an old crater. That would be so if not for the big crater at its centre, (sub-crater A), that has totally obliterated any semblance of a possible central peak formed by the initial impact, and whose floor is almost covered by the ejecta from A. This latter impact has thus increased Boussingault's depth (~ 5.7 km), and brought to the surface substantial deep deposits over a wider area. Both craters, however, were later struck by formation of sub-crater K, whose effect has shifted additional material from A's rim and terraces onto A's own floor (floor 2?).

Craters

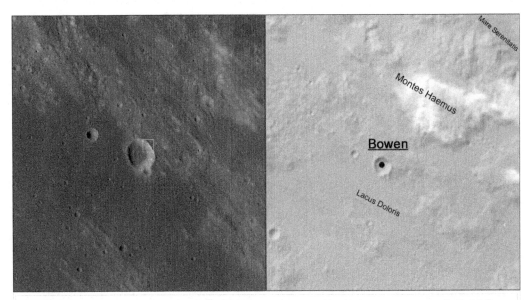

Bowen	Lat 17.63N Long 9.1E	8.09Km

Sub-craters: No sub-craters

Notes:

Add Info:

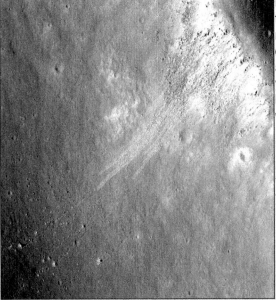

Boulder roll trails (and break-up of boulders) on Bowen's northeastern wall

Bowen formed just outside the main Serenitatis Basin ring/rim (~ 670 km in diameter), and on the edge of a a trough-like zone just beyond it, which connects to Mare Vaporum in the southwest. The ring/rim material is of course ejecta from the Basin, and it has subsequently been 'striated' by ejecta from the Imbrium Basin off to the northwest (centred ~ 800 km away). Bowen itself hasn't been affected, so it is younger than Imbrium, however, it must be older than the lavas in Lacus Doloris that have lapped up to its southwest rim (internal flooding, too?). D: 1.0 km.

Craters

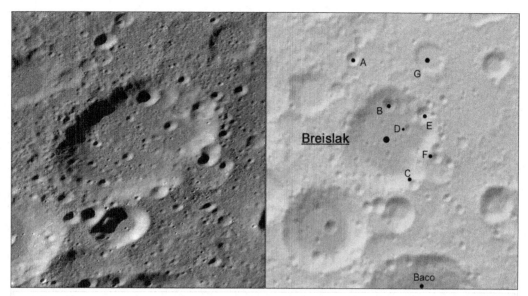

Breislak   Lat 48.31S   Long 18.31E   48.64Km

Sub-craters:

Crater	Lat	Long	Size (km)	Crater	Lat	Long	Size (km)
A	47.01S	17.35E	6.73	B	47.72S	18.21E	7.37
C	48.96S	18.85E	5.26	D	48.09S	18.68E	4.06
E	47.87S	19.08E	7.6	F	48.53S	19.42E	6.26
G	47.0S	19.2E	15.71				

Notes:

Add Info: Easy to get lost in this region covered with similarly-sized craters, Breislak, like most of its neghbouring craters, is probably late Nectarian (3.92 - 3.85 billions years) in age. Most of this region experienced secondary cratering from the Orientale Basin event to the west (some 2500 km away) - created during the Late Imbrian Period (3.75 - 3.2 byo), while a more younger, somewhat modern draping of ejecta from Tycho crater (Copernican - 1.1bn years - Present) - again to the west ~ 650 km away, covered Breislak's floor. Under high magnification, most of the small craters (< 7 km in diameter) on Breislak's rim and floor look relatively worn; which should prove a challenge for the observer who is into counting them (exceeding 40). Several of these craters have impacted both the northeastern and southwestern sectors of Breislak's terraces there, however, an older, and bigger-sized crater (~ 15 km) in the east sector (on which sub-crater F has impacted upon) seems to have imparted additional material onto the floor. This area of the Moon, in general, is usually ignored by the observer, perhaps, for its confusing amount of similarly-sized craters, however, a lot has gone on here that is worth noting and investigating for its history alone. Depth ~ 3.5 km.

Craters

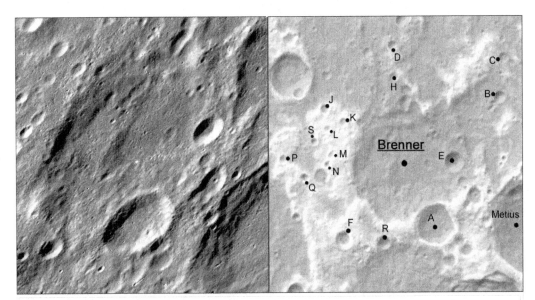

| Brenner | Lat 39.09S | Long 39.11E | 90.01 Km |

Sub-craters:

Crater	Lat	Long	Size (km)	Crater	Lat	Long	Size (km)
A	40.4S	39.95E	31.69	B	37.45S	41.9E	8.5
C	36.56S	41.99E	8.01	D	36.29S	38.65E	7.59
E	38.97S	40.51E	14.26	F	40.72S	37.03E	13.72
H	37.01S	38.67E	7.39	J	37.71S	36.55E	7.74
K	38.01S	37.12E	6.76	L	38.16S	36.61E	5.46
M	38.76S	36.72E	6.54	N	38.97S	36.58E	5.12
P	38.81S	35.21E	6.31	Q	39.26S	35.84E	7.19
R	40.79S	38.27E	10.8	S	38.44S	38.27E	5.77

Notes:

Add Info:

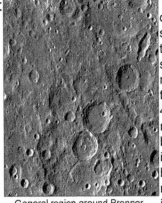

General region around Brenner

Try hard as you might to observe crater Brenner, this crater will always prove a challenge to find. At high sun angle periods it's almost invisible, while during terminator times (~ 8-day-old and 17-day-old moons) shadows from their respective easterly and westerly rims define its location. Of course, the reason for all this is that Brenner is an old crater (pre-Nectarian - 4.6 - 3.93 byo), and so is well worn and well covered by ejecta from numerous impacts (craters and basins) in the region. For example, sub-crater A has imparted its ejecta onto the floor - joining with, perhaps, Brenner's central peak, while a chain of secondary craters radial to Nectaris (see left) cut across its northeast rim. Depth of Brenner ~ 1.8 km.

Craters

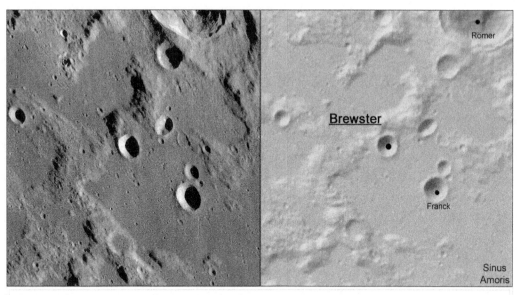

Brewster   Lat 23.27N   Long 34.69E   9.83Km

Sub-craters: No sub-craters

Notes:

Add Info:

On the 11 Dec 1972, Apollo 17 landed some 140 km away to the southwest of this crater (in the Taurus-Littrow valley region 20N, 31E).

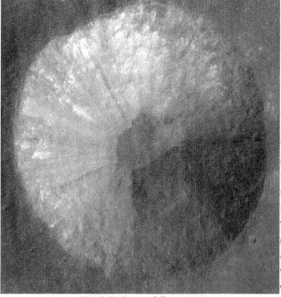

Aerial view of Brewster

You'd hardly notice crater Brewster in the eyepiece, as it hasn't got any real features of note. At full moon times the northern sectors of its inner walls 'shine' a bit more than the other smaller craters nearby (including Franck), while at other times radial aspects of its ejecta are just about seen. Close-up (left), however, shows the crater has a lot more to offer: with wonderful dry debris flows (light & dark); impact melt on its floor; and hint of a slump at the southeast sector. What would it be like to actually walk down into this crater?

Craters

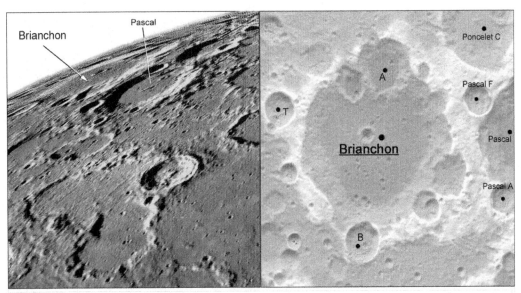

Brianchon	Lat 74.75N	Long 88.36W	137.26Km

Sub-craters:

Crater	Lat	Long	Size (km)	Crater	Lat	Long	Size (km)
A	76.81N	87.29W	49.2	B	72.15N	89.18W	31.9
T	75.28N	99.31W	29.25				

Notes:

Add Info:

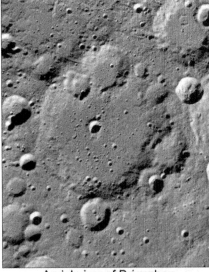

Aerial view of Brianchon

One would expect for such an extreme limb-type crater that Brianchon would pose a challenge for the observer. It's a big crater, but an old one - of the pre-Nectarian Period (4.6 to 3.93 byo), and it only is really observable during favourable librations of the 'nodding', latitudinal kind (that is, the moons poles from our earth-based perspective appears to nod back and forth throughout its orbit). During such occasions (try about a day after full moon), the unnamed ~ 13 km-sized crater at its centre appears like a ring peak, its rim to floor depth of 5 km casts slight shadows that define its rim, and the 'gash' (an ejecta feature possibly Orientale in origin) through Brianchon A is just about seen. Take photos!

Craters

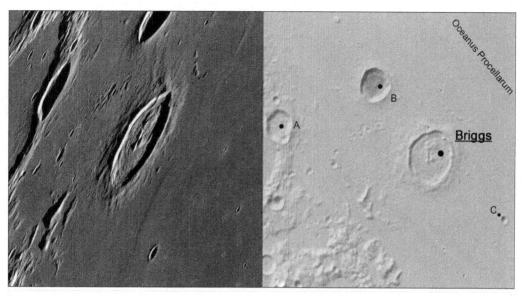

Briggs	Lat 26.45N	Long 69.19W				36.75Km

Sub-craters:

Crater	Lat	Long	Size (km)	Crater	Lat	Long	Size (km)
A	27.12N	73.78W	25.13	B	28.15N	70.91W	24.96
C	25.02N	66.95W	5.5				

Notes:

Add Info:

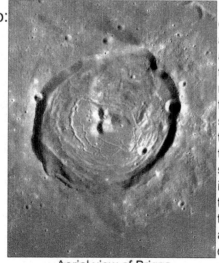

Aerial view of Briggs

It's a pity that Briggs is located where it is on the Moon for us earth viewers, as there are some wonderful features to be seen all around. Most of its floor contains concentric rilles (with one or two radial, too) where magma activity underneath uplifted the interior slightly, which then 'cracked' existing solidified material on top. The crater's central peak doesn't seem to have been affected by all this, however, the northern rim sector has collapsed (perhaps due to other impacts nearby); imparting material onto the floor, and filling into some of the cracks there. This crater is young judging by the appearance of its sharp rim, and has a depth of around 1.2 km. Lavas from Oceanus Procellarum later licked its outer ejecta.

Craters

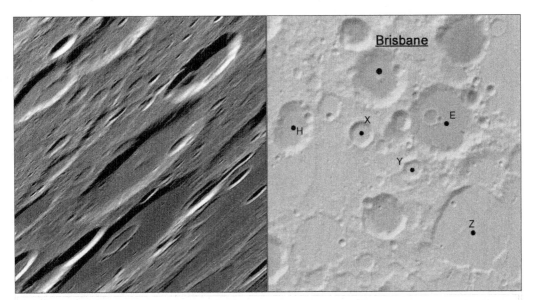

Brisbane	Lat 49.2S	Long 68.76E				44.32Km

Sub-craters:

Crater	Lat	Long	Size (km)	Crater	Lat	Long	Size (km)
E	50.26S	71.41E	57.35	H	50.37S	65.12E	43.7
X	50.58S	67.98E	20.07	Y	51.47S	70.17E	16.11
Z	52.89S	72.82E	64.73				

Notes:

Add Info:

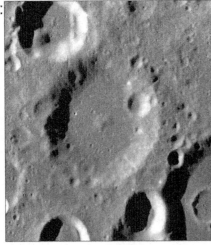

Aerial view of Brisbane

This crater is one that can be a bit hard to find in this limb-region - of Mare Australe's ancient environs - where several other craters look similar to it. Unlike subcrater, Brisbane Z, to the southeast, Brisbane itself has escaped the mare floods of lava that encompassed most of the general region, however, ejecta from other impact events nearby have fallen into its interior - covering what remains of its once prominant peak (note this, too, in the small, unnamed crater on Brisbane's inner eastern wall). Favourable librations are thus a necessity to observe this crater, whose depth reaches around 3 km approx.., while the peak reaches to about ~ 1.5 km.

Craters

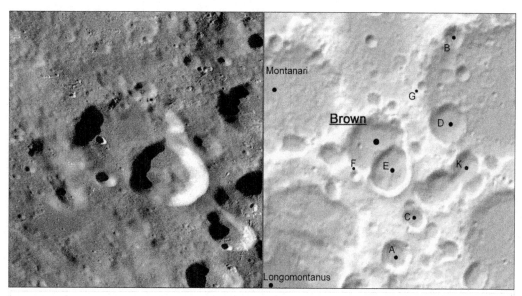

Brown Lat 46.53S Long 17.99W 34.03Km

Sub-craters:

Crater	Lat	Long	Size (km)	Crater	Lat	Long	Size (km)
A	48.15S	17.44W	15.75	B	44.73S	16.17W	11.66
C	47.58S	17.0W	12.33	D	46.07S	16.25W	19.51
E	46.86S	17.63W	23.19	F	46.91S	18.48W	5.5
G	45.51S	16.88W	4.74	K	46.74S	15.75W	14.94

Notes:

Add Info: Lying some ~ 170 km away to Tycho's southwest, Brown is mixed amongst a jumble of smaller craters all of similar size. This region has been subjected to secondary cratering from the huge impact that created Orientale to the crater's west, but also to ejecta-fill from subsequent basins e.g. Humorum to the north-west, and craters like Longomontanus to the southwest. Due to Tycho's initial, low angle impactor that came in from the south, less of the ray material from it - that one would expect to see for such a close encounter - covers Brown at all. However, a wonderful series of tiny crater-chains lying particularly close to its north, northeast and eastern borders are worth noting under high magnification. Much of Brown's rim has undergone a bombardment of sorts; where nearly all its northeastern sectors have been obliterated by small impactors, whose area was later impacted by a relatively bigger one to the southeast that created sub-crater E. This deeper crater's formation has totally wiped out any semblance of Browns rim there; imparting amounts of material onto Brown's floor, which is composed of a smoother material (impact melt, perhaps, with a mixture of light ejecta from elsewhere?). Depth of Brown is around 1.8 km approx.

Not to be confused with crater D. Brown found on the Farside of the Moon - 41.65S, 212.84E, which is named in honour of Columbia Shuttle astronaut, David McDowell Brown.

Craters

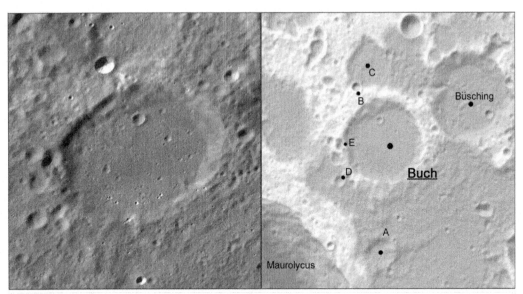

Buch   Lat 38.9S   Long 17.68E   51.31Km

Sub-craters:

Crater	Lat	Long	Size (km)	Crater	Lat	Long	Size (km)
A	41.06S	17.58E	18.47	B	37.89S	16.95E	6.27
C	37.47S	17.25E	27.38	D	39.68S	16.48E	6.9
E	38.98S	16.5E	8.31				

Notes:

Add Info: One look at Buch (pronounced 'Book') immediately says this is an old crater of the pre-Nectarian age (4.6 - 3.92 billions of years old). With its worn-down rims and hard-to-see terraces, its smooth floor - filled with ejecta from the Mare Nectaris Basin event (~ 800 km away to its northeast), and by several other impact craters nearby, for example, Maurolycus to its southwest, this crater is one that is easily missed amongst all the others in the region. The floor of Buch is cover- with many tiny craterlets ranging from 0.5 to 1 km upwards (some old, some fresh); the largest in its northern sector coming in at around 3 km. Ray material from Tycho (~ 700 km away) can just about be seen on the western, exterior flanks of this crater, with possible traces also seen on the southern sector of the floor (note the light-coloured material there, as well as the small fresh craterlets). Additional ray material from sub-crater B (an oblique impact) definitely covers the crater's northern rim and inner terraces there, which requires high magnification to observe (B, in fact, shows up quite nicely under full moon). Neighbouring Büsching crater to its northeast looks similar in nature to Buch, however, as this, too, is a crater of age, both then formed around the same time (Büsching, possibly, first as it looks slightly more worn). Depth of Buch is around ~ 1.5 km.

Craters

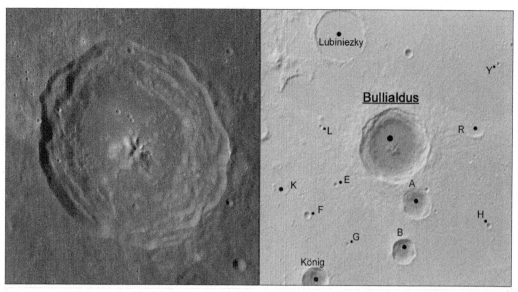

Bulialdus Lat 20.75S Long 22.26W 60.72Km

Sub-craters:

Crater	Lat	Long	Size (km)	Crater	Lat	Long	Size (km)
A	22.21S	21.57W	25.48	B	23.46S	21.96W	21.27
E	21.74S	24.03W	3.91	F	22.51S	24.91W	5.9
G	23.27S	23.7W	3.4	H	22.75S	19.36W	4.38
K	21.78S	25.69W	12.01	L	20.23S	24.46W	3.59
R	20.23S	19.84W	16.82	Y	18.58S	19.18W	3.23

Notes:

Add Info: Wonderful, and typical, example of your average complex crater, Bulialdus has all the features one would expect to find from such events. Just look to similar craters like, say, Piccolomini, Eratosthenes, Theophilus, and you'll see that all have relatively shallow floors, a series of central uplifted peaks, terraced walls, and a crenulated outline to their rims due to further collapse of inner material. This 'collapsing' is due to frictional forces between solid rock fragments in the target rock behaving like a fluid. This 'fluidized' state doesn't mean that the rocks are molten during the process, rather that each rock fragment is separated enough from each other to permit a fluid-like flow of the total. After a certain time, as gravity - the main component of the collapse - produces normal frictional forces on each fragment, the total rock-flow then returns to a static state. This also applies to the central peak formation where gravitational equilibrium is the cause, too, and not elastic properites in rocks once believed to produce a 'rebound' effect. Age of Bulialdus - Eratosthenian Period (~ 3.15 byo) to Late Imbrian (~ 3.2 byo). Depth ~ 3.5 km, while peak height is around ~ 1 km.

The north-western rim of sub-crater, Bulialdus B, shows a small impact crater some ~ 0.5 km in diameter that has just about 'nicked' the rim (is the ray display nearby a consequence of the impact, or is it of Tycho's?).

Craters

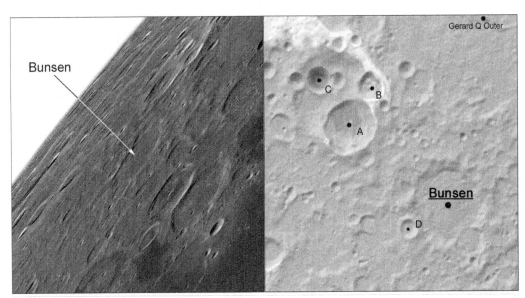

Bunsen	Lat 41.4N	Long 85.46W	55.22Km

Sub-craters:

Crater	Lat	Long	Size (km)	Crater	Lat	Long	Size (km)
A	43.17N	88.71W	40.7	B	44.11N	88.03W	21.49
C	44.21N	89.82W	19.02	D	40.85N	86.71W	13.6

Notes:

Add Info:

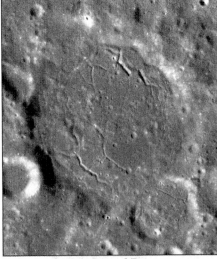

Aerial view of Bunsen

It's obvious from looking at the aerial view of Bunsen that something seriously geologic has happened within its floor. Fractures show up everywhere, particularly on the outer sectors of the floor, where they also appear to have been more structurally violent - some having widths at over two kilometres. The fractures are believed to have formed by magma (under pressure) pooling underneath the crater, but not quite able to reach the surface. In effect, it 'domes' up the floor (Bunsen's central floor, in parts, is some 400 metres above its outer floor sectors), which then leads to the fracturing. The effects, and extent, of this 'magma uplift' can be seen in several other craters nearby.

Craters

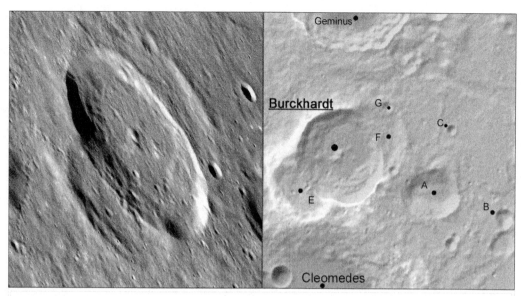

Burckhardt   Lat 31.11N   Long 56.39E   54.36Km

Sub-craters:

Crater	Lat	Long	Size (km)	Crater	Lat	Long	Size (km)
A	30.4N	58.48E	35.08	B	29.93N	60.2E	11.14
C	31.6N	59.0E	6.13	E	30.57N	55.55E	37.46
F	31.35N	57.11E	45.02	G	32.04N	57.49E	7.56

Notes:

Add Info:

Aerial vierw of Burckhardt

Lying northwards - just outside the main ring-rim of the Mare Crisium Basin (~ 750 km in diameter), Burckhardt is very easily found from its two ear-like appendages (sub-craters E + F respectively) on either side. The impactor that created Burckhardt struck the confluence sectors of these old, sub-crater rims, which must have affected material distribution in either direction (two things to note: firstly, see how the south-western inner-wall section of Burckhardt is thicker there than on the opposite side; secondly, ejecta from Burckhardt may have raised the floors of E and F). Whatever the after-effects, all three craters are old - Burckhardt being the youngest, E slightly younger than F.

Craters

| Bürg | Lat 45.07N | Long 28.21E | 41.04Km |

Sub-craters:	Crater	Lat	Long	Size (km)	Crater	Lat	Long	Size (km)
	A	46.87N	33.06E	11.41	B	42.7N	23.46E	5.65
	Notes:							

Add Info:

Bürg region under different lighting conditions

What a wonderful little crater is Bürg. It's got all the true character of a relatively young, complex crater: peaks, slumped terraces, sharp rim, and ejecta evenly distributed all around - except for the western side. That regional side of Lacus Mortis filled with, perhaps, the same lavas of Mare Frigoris, which then subsequently underwent stress and collapse effects that gave rise to the fractures, rilles and ridges we see today.

Craters

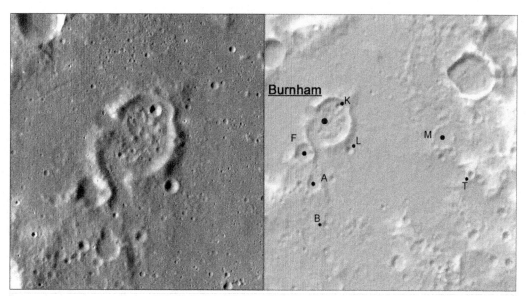

Burnham   Lat 13.92S   Long 7.25E   24.09Km

Sub-craters:

Crater	Lat	Long	Size (km)	Crater	Lat	Long	Size (km)
A	14.8S	7.03E	6.11	B	15.36S	7.22E	3.4
F	14.35S	6.88E	8.42	K	13.69S	7.4E	3.12
L	14.3S	7.57E	3.8	M	14.12S	9.02E	9.32
T	14.63S	9.47E	3.42				

Notes:

Add Info: One's initial reaction to seeing Burnham crater for the first time is - "crater, what crater?". It's got no real structure-of-a-rim to it, hasn't any identifiable hint of a floor that we're used to seeing, and it looks more rather like a mix of isolated blocks that happened to come align together into a so-called 'crater'. All that said, however, Burnham is a crater - odd as it looks! Slightly oval in a NE to SW configuration, the crater has been subjected to the dramatic effects from creation of the Imbrium Basin - some 1500 km away northwestwards. Blocks of material more than likely obliterated Burnham's missing rim sections in the northwest and south, as they plowed through this whole region, while ejecta showered down over the crater - infilling its floor. Any semblance of Burnham's own outer ejecta has thus well been covered, too, leaving only small, recognisable signatures - predominantly exposed at the northeast and southeast sectors of the crater. Pull back your view of this region and you'll instantly see several other grooves that align back towards Imbrium - e.g. at craters Vogel or Parrot to the west, or at the rim of Albategnius to the northwest. This is known as the 'Imbrium Sculpture' - all due to the bombardment from that event. Depth of Burnham ~ 0.8 km.

Craters

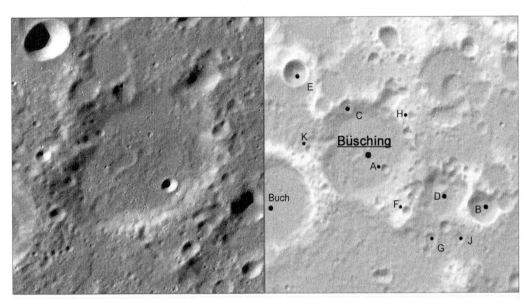

Büsching   Lat 38.04S Long 19.96E   53.49Km

Sub-craters:

Crater	Lat	Long	Size (km)	Crater	Lat	Long	Size (km)
A	38.37S	20.43E	5.19	B	39.01S	22.75E	17.92
C	37.3S	19.53E	7.41	D	38.75S	21.8E	30.55
E	36.71S	18.38E	14.7	F	39.1S	20.95E	5.73
G	39.57S	21.54E	7.3	H	37.46S	21.06E	4.71
J	39.55S	22.21E	6.71	K	38.02S	18.66E	5.07

Notes:

Add Info: Like its neighbour, Buch, to the southwest, Büsching is a crater of the pre-Nectarian Period (4.6 to 3.92 billions of years old). Its worn-down rim and inner walls blend seamlessly together as they meet the floor; whose makeup is that of ejecta-fill from creation of basin events - for example, Nectaris to its northeast (~ 700 km away) and Imbrium to the northwest (~ 2400 km away). The crater may be slightly older than Buch, however, as both have undergone material coverage, as mentioned above, signatures of own ejecta by each, as well as their true floors, lay hidden underneath a possible thin layer of the basin ejecta. Under high magnification, many craterlets can be seen nearly all over the floor (each around a kilometre and less), while several slightly larger depressions ~ 4 to 6 km in diameter hint at signatures of buried craters (that is ghost craters). Bright ray material from sub-crater, Buch B, to the west cover parts of Büsching's western rim, which is best looked for during full moon times before shadows start to develop there. Additionally, ray material from Tycho (some ~ 700 km away to the west) has crossed this region, however, any obvious signature of this is truly hard to find. Depth of Büsching ~ 1.8 km.

Craters

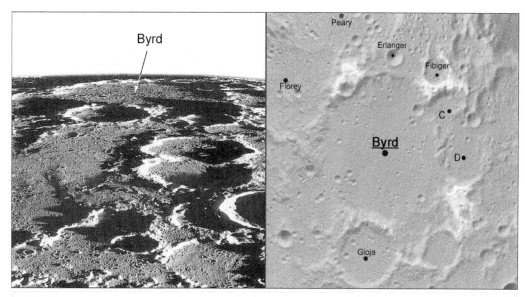

Byrd	Lat 85.43N	Long 10.07E	97.49Km

Sub-craters:

Crater	Lat	Long	Size (km)	Crater	Lat	Long	Size (km)
C	84.35N	28.34E	34.26	D	85.52N	33.41E	24.47
Notes:							

Add Info:

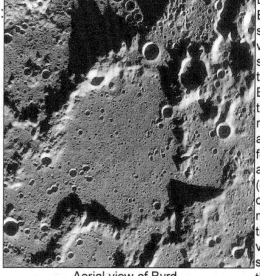

Aerial view of Byrd

Like several other craters nearby Byrd, for example, Meton to its southeast or Mouchez to its southwest, all have been affected in some way or another by the event that was formation of the Imbrium Basin - centred some 1500 km to the south. The only part of Byrd's rim remains at its northwest, as all other parts experienced some form of impact cratering or coverage by ejecta other than Imbrium (exactly what has happened at the crater's eastern sectors is a nightmare to solve visually). Librations times are recommended for best views, while crater Gioja on its southern rim helps in aid for locating Byrd. Depth ~ 1.5 km approx..

Craters

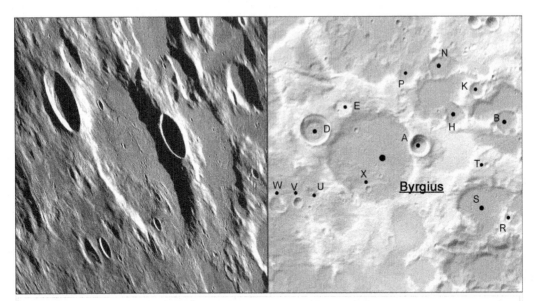

| Byrgius | Lat 24.73S | Long 65.38W | 84.46Km |

Sub-craters:

Crater	Lat	Long	Size (km)	Crater	Lat	Long	Size (km)
A	24.55S	63.81W	18.45	B	23.88S	60.96W	23.03
D	24.08S	67.28W	27.06	E	23.49S	66.36W	17.19
H	23.76S	62.6W	21.33	K	23.01S	61.94W	14.98
N	22.39S	63.16W	22.85	P	22.53S	64.23W	18.68
R	26.5S	60.85W	7.18	S	26.06S	61.77W	45.94
T	25.06S	61.6W	4.74	U	25.84S	67.36W	10.42
V	26.07S	67.9W	8.74	W	26.12S	68.64W	13.64
X	25.74S	65.55W	5.9				

Notes:

Add Info:

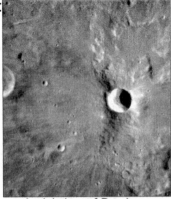

Aerial view of Byrgius

Lying between the two outermost rings-rims of the Orientale Basin (~ 1300 km and 1900 km in diameter respectively, and some ~ 800 km away to the west), Byrgius undoubtedly has been covered by ejecta from that gigantic event. Several smaller impact craters ranging in size from 1 km to 5 km lay strewn across its floor, however, subject to all these is the wonderful ray material from Byrgius A that covers them and Byrgius itself. The rays extend in every direction - some of which reach right out to Orientale, and all are an easy sight through the most smallest of telescopes - especially during full moon times. Depth of Byrgius is ~ 2.0 km.

Craters

Cabeus — Moretus

Cabeus   Lat 85.33S   Long 42.13W   100.58Km

Sub-craters:	Crater	Lat	Long	Size (km)	Crater	Lat	Long	Size (km)
	A	82.08S	40.24W	40.88	B	82.26S	54.48W	59.62

Notes:

Add Info:

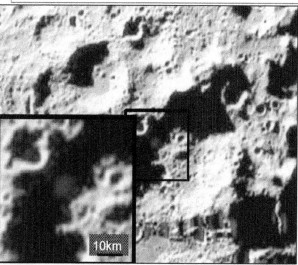

Close-up of a plume (smudge) created by Centaur on impact (recorded by LCROSS before it, too, impacted). Credit: NASA

Cabeus is famous: not for its size, its worn rim, or its awkward location that will require a good libration to view it, but more for it as the crater chosen by NASA to impact an upper stage rocket (called Centaur) onto its innards. Centaur was part of a launch vehicle from the LRO/LCROSS mission that went to the Moon on 18 June 2009 (the stage impacted on 9 October 2009). Why? To see if water trapped as ice existed in the crater's sub-surface soil. It did!

Craters

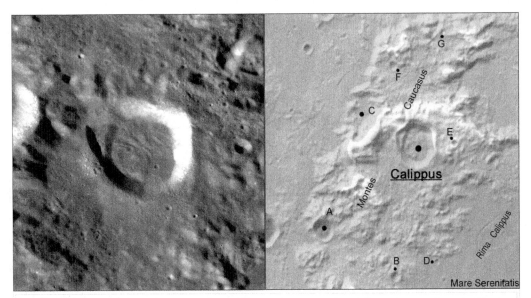

Calippus	Lat 38.92N	Long 10.72E		34.03Km		

Sub-craters:

Crater	Lat	Long	Size (km)	Crater	Lat	Long	Size (km)
A	37.06N	7.87E	15.72	B	36.08N	10.07E	7.21
C	39.43N	9.16E	37.69	D	36.38N	11.31E	3.99
E	38.92N	11.89E	4.82	F	40.58N	10.0E	6.75
G	41.3N	11.55E	3.78				

Notes:

Add Info:
Day-old moon phases

Note: As the Moon goes through various librations throughout the year, suggested times given in text for observations are only approximates.

Calippus lies on a region known as the Caucasus Mountains (a range that is said to be that of Imbrium ejecta, whose concentric configuration doesn't quite 'line up' to that basin's centre). Aged of the Upper Imbrian Period (3.75 to 3.2 billions of years old), the crater's rim and steep inner walls follows an irregular shape all around, while its floor, with no sign of an expected peak, is filled with material taking on a hummocky appearance. On the western side of the rim, a huge clump of the wall looks like it has slided down by about a kilometre or so in a sharp angled direction; creating a well-defined division between it and the floor. This might suggest that Calippus was created through processes from a low, oblique impactor coming in from the west (like we see with crater Proclus), however, this may not be so as no signatures of bright ray material shows up in the easterly surrounds. Sub-crater, Calippus C, has entirely lost all of its west rim, however, its east rim more than enough makes up for the missing portion - coming in at over 5 kilometres high at certain sections in its southern sector. In periods when the terminator is not far off (~ 20 day-old moon), this high peak can keep Calippus covered in shadow for some time, leaving only its bright, eastern rim in light. Depth of crater is around 3 km.

Craters

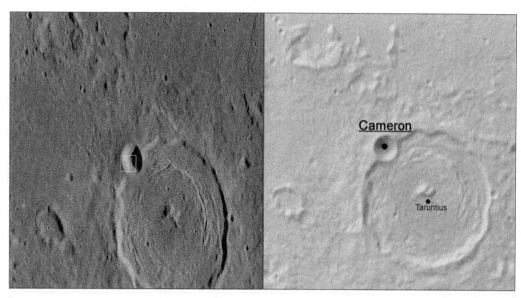

| Cameron | Lat 6.19N | Long 45.93E | 10.91Km |

Sub-craters: No sub-craters

Notes:

Add Info:

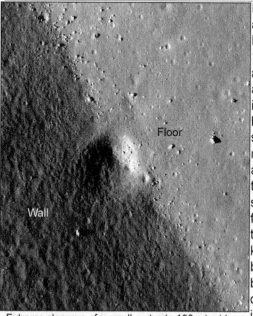

Extreme close-up of a small crater (~ 100 m) midway between Cameron's southwestern wall and floor.

It's obvious that Cameron is of a younger age than Taruntius - Eratosthenian Period - 3.15 to 1.1 byo, however, as its rim isn't as sharp as we would expect for a very young crater, did it form in the same Period? The crater has no distinguishing features to speak of, but around the crater's rim, dark-ish deposits, both north and south, are seen (full moon times shows them up easily) to surround Cameron. Ash deposits from vents may be the cause for those at the south (note how darker they appear near the peaks), but north's deposits are slightly brighter (did Cameron's impactor produce them initially; tapping into deeper deposits?). Depth of Cameron is just over ~ 2.0 km.

Craters

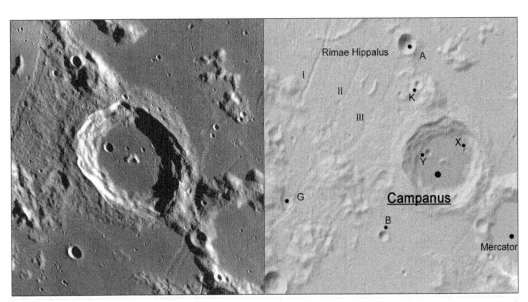

Campanus	Lat 28.04S	Long 27.9W	46.41Km

Sub-craters:

Crater	Lat	Long	Size (km)	Crater	Lat	Long	Size (km)
A	25.99S	28.67W	11.14	B	29.32S	29.24W	5.85
G	28.66S	31.36W	8.22	K	26.7S	28.42W	4.76
X	27.84S	27.41W	3.13	Y	27.88S	28.22W	4.39

Notes:

Add Info: Not to be confused with the other, similar-sounding named crater, Capuanus, to its south, this crater is one of the Early Imbrian Period (3.85 to 3.75 billions of years old). Lying in a lava-filled region where possibly two trough-like depressions from Mare Humorum to the west (Nectarian in age) and Mare Nubium to the east (of pre-Nectarian age) meet, and which subsequently might have given rise to Palus Epidemiarum to the south, the historical and geological sequence of Campanus's creation is one not to be messed with. The crater looks as if it has formed on a sector of ejecta from Nubium's main southwest ring-rim (690 km in diameter), and just kissed on a small piece of remained ejecta that was initially connected to that of Humorum's main southeast ring-rim (425 km in diameter) - now missing due to subsidence (note the Hippalus rilles' region). The crenulated rim of Campanus looks relatively sharp, however, its worn-looking terraces are a giveaway of its age where time has took its toll on them. There are two small peaks in the centre of the floor, however, like both portions of Campanus's own ejecta lying outside the crater's northeast and southwest rim, lavas have licked at its shores, producing a seamless join. Lava in the floor possibly seaped upwards from beneath through the main fracture there, as its colour is slightly darker than the surrounding mare lavas. Depth of crater ~ 2 km.

Craters

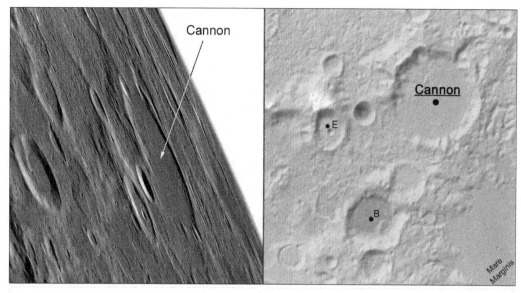

Cannon	Lat 19.88N	Long 81.36E			57.58Km	

Sub-craters:	Crater	Lat	Long	Size (km)	Crater	Lat	Long	Size (km)
	B	17.54N	80.05E	38.78	E	19.27N	78.99E	25.87
	Notes:							

Add Info:

Aerial view of Cannon

Cannon is situated just outside a main ring/rim (~ 357 km in diameter) and trough-zone associated to the Marginis Basin (?) - centred 400 km to it southeast. The terrain here looks very crusty-like, which seems also to reflect in Cannon's walls where horizontal-type terraces aren't seen, but more vertical-running valleys go down on to the floor (they show up in several other nearby craters, too). If that's odd, look at the configuration of terrain off Cannon's northeast - what has happened there? Partial flooding of Cannon's southwestern floor may have occurred at some time; later dusted over by brighter material - but from where? Depth 2.8km.

Craters

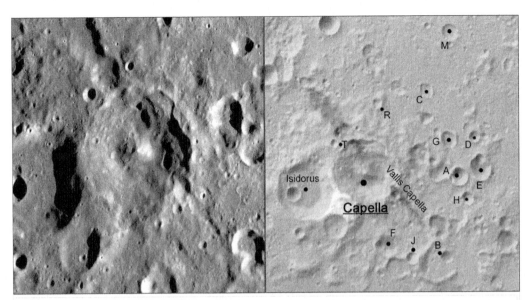

| Capella | Lat 7.65S | Long 34.92E | 48.13Km |

Sub-craters:

Crater	Lat	Long	Size (km)	Crater	Lat	Long	Size (km)
A	7.68S	37.18E	11.76	B	9.47S	36.87E	8.38
C	5.76S	36.39E	11.17	D	6.79S	37.53E	7.96
E	7.53S	37.7E	14.33	F	9.22S	35.42E	13.58
G	6.86S	36.92E	11.02	H	8.17S	37.38E	8.94
J	9.5S	36.06E	8.93	M	4.47S	36.95E	11.8
R	6.05S	35.21E	7.28	T	6.94S	34.18E	5.72

Notes:

Add Info: Lying on a chunk of highland material encircled by lavas in Basins: Fecunditatis to its east; Tranquillitatis to its northwest; and Nectaris to its south, Capella is a crater of age - formed not long after the Nectaris Basin, so it's of the Nectairan Period. It's an odd-looking crater however, in that while the eastern sector of its rim is relatively well-defined, the opposite, western rim has somewhat been lost to the crater, Isidorus, that it partially impacted upon. Is this latter impact event the cause as to why the central peak in Capella looks very broad and fills the floor. Was the westerly-bound direction of 'some' of the ejecta from Capella in one way affected reverse-wise by Isidorus's eastern rim? Or, has it to do with another feature here - the obvious 'gash' that is Vallis Capella - seen wonderfully whenever low sun angle conditions occur. The 'gash' is produced by the merging of a chain of craters that decrease in size - from the NW (~ 12 km in diameter) to the SE (~ 3 km in diameter). The chain 'points' back to Mare Serenitatis as its source, however, as ejecta from the Imbrium Basin covers this entire region, it, too, may be a candidate. D: 3.5 km.

Craters

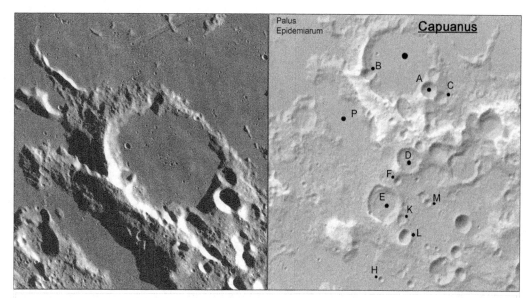

Capuanus   Lat 34.09S   Long 26.73W   59.69Km

Sub-craters:

Crater	Lat	Long	Size (km)	Crater	Lat	Long	Size (km)
A	34.69S	25.7W	12.98	B	34.32S	27.73W	10.79
C	34.88S	25.31W	10.31	D	36.44S	26.32W	20.99
E	37.54S	27.17W	28.37	F	36.97S	26.71W	6.97
H	39.4S	27.27W	4.4	K	37.91S	26.56W	8.15
L	38.34S	26.41W	11.45	M	37.5S	25.68W	6.37
P	35.44S	28.59W	68.52				

Notes:

Add Info: Remember, this is Capuanus we're talking about, and not its close neighbour to the northwest (some 200 km away) known as Campanus - many observers have confused the two when referencing them. Lying in the Palus Epidemiarum region (Palus = 'swamp'), the crater's characteristic feature that makes it easy to find immediately is the three fingers of material connected to its outer northwest. The top two fingers are probably made up of older material related to an outer, ring-rim of Mare Nubium to the northeast, while the third finger may be of younger material from Mare Humorum - perhaps, of its ~ 800 km diameter ring-rim system (note the difference in texture). The crater's interior has a smooth floor containing over a dozen small craterlets all about a kilometre in size, and at low sun-angle times two or three domes (lunar volcanoes) are seen. Note, also how the northeastern sector of Capuanus's rim is lower than the southwest sector - was subsidance in Nubium responsible (like we see with other craters that lie on the edges of mares, e.g. Fracastorius, le Monnier)? Depth ~ 1.6 km.

Craters

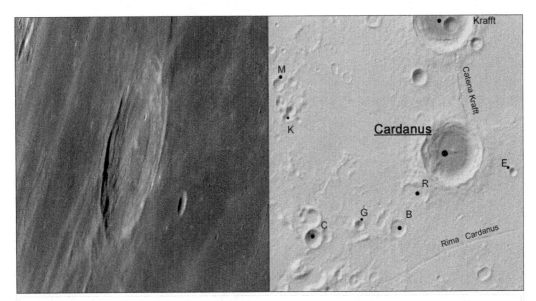

Cardanus Lat 13.27N Long 72.5W 49.57Km

Sub-craters:

Crater	Lat	Long	Size (km)	Crater	Lat	Long	Size (km)
B	11.45N	73.96W	13.18	C	11.18N	76.3W	13.65
E	12.8N	70.8W	6.42	G	11.51N	75.04W	7.61
K	14.12N	77.04W	7.9	M	14.9N	77.31W	8.88
R	12.25N	73.5W	20.99				

Notes:

Add Info:

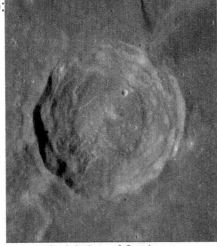

Aerial view of Cardanus

Quite easy to find but not easy to observe because of its limb location, this crater is more recognisable for its tied link with cr-crater Krafft to the north by a chain of craters known as the Catena Krafft. Both are aged of the Late Imbrian (3.75 - 3.2 byo), Cardanus's rim and broad terraces have had plenty of time to degrade - the latter of which edges upto to rilles on the floor where dark deposits extruded (note the extensive deposit meeting the inner eastern terrace). A small, off-central peak rises to about 700 metres in this crater having an approx., depth of ~ 3.5 km, where rays from the crater Glushko (Copernican in age) to its southwest cross all.

Craters

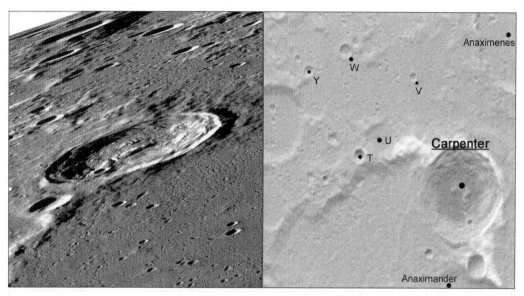

Carpenter	Lat 69.52N	Long 51.23W	59.06Km

Sub-craters:

Crater	Lat	Long	Size (km)	Crater	Lat	Long	Size (km)
T	70.28N	58.69W	9.38	U	70.61N	57.08W	25.08
V	71.91N	54.68W	5.76	W	72.37N	60.24W	9.62
Y	71.86N	63.02W	7.93				

Notes:

Add Info:

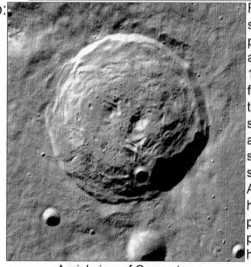

Aerial view of Carpenter

From its fresh-looking appearance - sharp rim and terraces, and central peaks, it's obvious that Carpenter is a young crater (Copernican in age - 1.1 byo to Present). The crater has formed on Imbrium ejecta deposits that cover this region, in general, but signatures of its own ejecta deposits are only obvious where material outside the north, south and eastern rim sectors are all higher then at the west. At this angle, the floor is somewhat hard to define as several clumps dispersed throughout meet the two main peaks - the eastern peak being the highest at around 1 km high and the western at 700 m. Depth ~ 4.2 km.

Craters

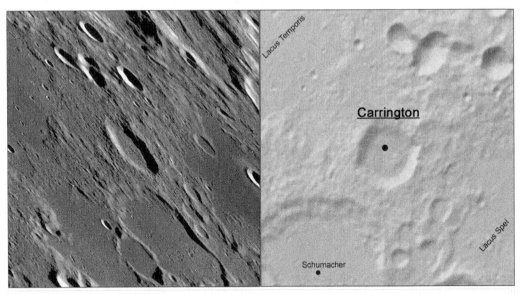

Carrington   Lat 43.97N   Long 62.04E   27.77Km

Sub-craters: No sub-craters

Notes:

Add Info:

Aerial view of Carrington

Carrington has impacted on the outer remnants of ejecta that is related to Nectarian-aged Humboldtianum - centred some 500 km to the northeast. The crater also lies between two lakes: that of Lacus Spei with its overly-dark floor, to southeast, and the slightly lighter-coloured floor of Lacus Temporis, to its northwest. The floors of these lakes are some 500 metres above that of Carrington's floor, and are presumed to have been flooded at some point in their lives. But Carrington's floor doesn't show any signs of such. Why? The large clump on its floor may be an answer - is it a chunk of ejecta plopped on to its centre. If it is that, the question to ask then is: 'where did it come from?'. The depth of Carrington is ~ 2.7 km.

Craters

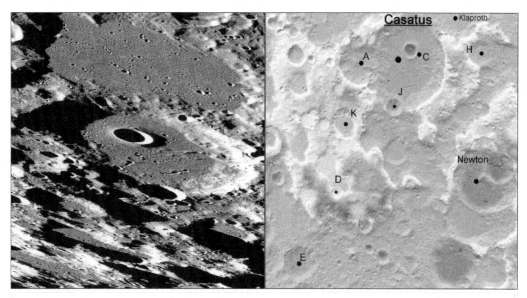

Casatus   Lat 72.7S   Long 30.75W   102.85Km

Sub-craters:

Crater	Lat	Long	Size (km)	Crater	Lat	Long	Size (km)
A	72.68S	37.63W	54.64	C	72.25S	30.29W	17.36
D	76.89S	43.3W	37.66	E	77.25S	53.97W	44.3
H	72.11S	21.4W	35.67	J	74.35S	32.94W	20.45
K	74.8S	40.7W	35.67				

Notes:

Add Info:

Aerial view of Casatus

Casatus is one of those limb-huggers that benefits the observer when viewed through favourable librations. But even at that, it's still a hard crater to view as it lies in a jumble of similarly-sized craters and mountainous terrain where shadows continually play an unhelpful role. The crater's formation has wiped out a portion of crater Klaproth's southwestern rim - imparting ejecta onto its floor, however, Casatus itself hasn't escaped similar circumstances where sub-crater A has done the same type of depositing. Depth of Casatus is just over 5 km, but formation of sub-crater C has dug down even further - reaching older layers.

Craters

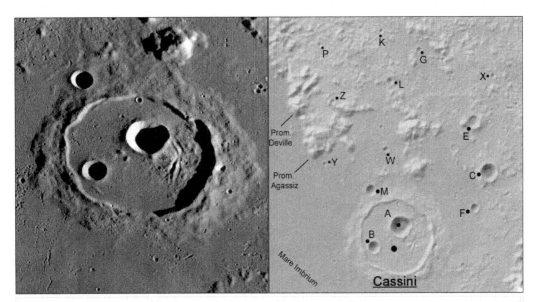

Cassini	Lat 40.25N	Long 4.64E				56.88Km	

Sub-craters:

Crater	Lat	Long	Size (km)	Crater	Lat	Long	Size (km)
A	40.51N	4.78E	16.76	B	40.02N	3.87E	9.42
C	41.76N	7.8E	13.79	E	42.98N	7.34E	9.45
F	40.92N	7.28E	6.72	G	44.73N	5.46E	4.72
K	45.19N	4.09E	3.47	L	43.99N	4.41E	6.14
M	41.38N	3.75E	7.85	P	44.83N	1.86E	3.37
W	42.35N	4.25E	5.46	X	44.06N	8.08E	4.4
Y	41.98N	2.16E	3.55	Z	43.48N	2.38E	4.69

Notes:

Add Info: What a wonderful sight is Cassini under low sun angle views - both its outer and inner features show up very nicely. Aged of the Lower Imbrian - 3.85 - 3.75 byo, the crater lies just within the main ring-rim (1160 km in diameter) system of Imbrium, and it formed most likely on semi-liquified, semi-cooled lava deposits interperspersed with ring-rim material from that major event (note its outer ejecta's texture). Lava levels inside and outside the crater are about the same, and while all are relatively smooth in appearance, the latter outer sectors, particularly in the southwestern regions, are pock-marked with minute crater chains that point back to crater Aristillus. Sub-crater A, at high magnification, and under a high western sun, shows up an additional feature worth viewing - a clump of material inside the crater's eastern sector. It doesn't seem to suggest it is due to slumping of the wall there, as the amount looks too little to have done so. Instead, is the feature somewhat related to elongation of the crater by an oblique impact event or even one of a double impact event? Depth of Cassini is ~ 1.3 km approx..

Craters

Catharina   Lat 17.98S   Long 23.55E   98.77Km

Sub-craters:

Crater	Lat	Long	Size (km)	Crater	Lat	Long	Size (km)
A	20.21S	22.27E	12.97	B	16.96S	24.29E	21.7
C	20.38S	24.31E	27.28	D	16.94S	21.4E	8.36
E	17.16S	21.26E	6.05	F	19.54S	23.06E	6.31
G	17.54S	24.91E	15.94	H	19.31S	25.37E	5.45
J	19.45S	22.18E	4.89	K	20.06S	23.89E	7.19
L	20.96S	24.21E	3.87	M	19.27S	20.73E	5.4
P	17.28S	23.32E	46.8	S	18.92S	23.37E	15.77

Notes:

Add Info: Like its neighbour, Cyrillus (Nectarian in age), to the north, Catharina has all the signs of an aged crater. Its overly-battered rim and levelled terraces have been completely changed by both ancient and fresh impacts, while its slightly raised floor has received debris from Imbrium (some 1800 km away to the northwest) and several nearby impacts. Sub-crater P, had previously extended the crater's shape from circular to pear-like - changing the geology of nearly a third of Catharina's floor, and wiped out any semblance of a peak that may have existed (is there still a hint-of-a-peak at close terminator times?). Under high magnification, the floor and northeastern rim of P also shows up a series of fine striations that look like rilles, however, as they point back towards younger crater, Theophilus (Eratosthenian in age) to the northeast, they are more than likely fine crater chains created by that impact. Catharina lies within the main ring-rim system (~ 860 km in diameter) of the Nectaris Basin, and has a depth of 3 km approx..

Craters

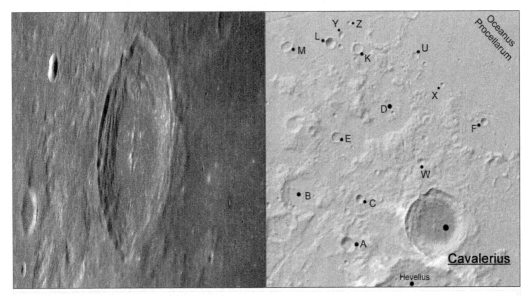

Cavalerius Lat 5.1N Long 66.93W 59.35Km

Sub-craters:

Crater	Lat	Long	Size (km)	Crater	Lat	Long	Size (km)
A	4.5N	69.68W	14.14	B	5.97N	71.12W	39.51
C	5.84N	69.32W	8.19	D	8.66N	68.47W	51.77
E	7.67N	70.05W	9.55	F	8.11N	65.39W	7.18
K	10.29N	69.38W	9.9	L	10.46N	70.25W	9.9
M	10.35N	71.63W	11.79	U	10.08N	67.55W	6.76
W	6.94N	67.4W	7.85	X	9.21N	66.7W	4.33
Y	10.69N	69.91W	6.92	Z	11.05N	69.67W	4.18

Notes:

Add Info:

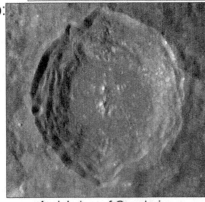

Aerial view of Cavalerius

Of the Eratosthenian Period (3.15 - 1.1 byo), Cavalerius in the eyepiece is easily recognised by its two 'pointy' north and south rims. The features are due to a predominance of collapse at the terraced walls in these areas, however, such collapse has occurred at all sectors of the crater's inner regions. Ejecta deposits from Cavalerius cover older Hevelius (Nec.,) to the south, while eastwards a wonderful series of crater chains are worth noting. Rays radiate from the crater here, too, but are they related to Cavalerius? Depth ~ 3.6 km, main peak height ~ 1 km.

Craters

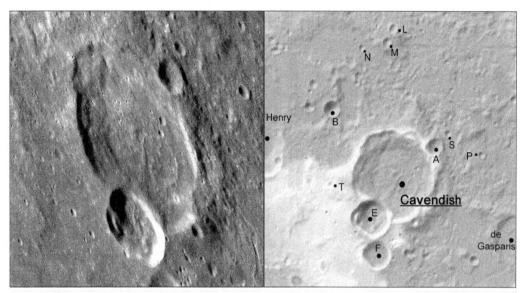

Cavendish	Lat 24.63S	Long 53.78W			52.64Km		
Sub-craters:	Crater	Lat	Long	Size (km)	Crater	Lat	Long	Size (km)
	A	24.02S	52.83W	10.66	B	23.27S	55.19W	10.29
	E	25.43S	54.27W	23.53	F	26.14S	54.16W	17.76
	L	21.71S	53.77W	5.68	M	21.97S	53.91W	5.86
	N	22.07S	54.48W	4.31	P	24.16S	51.68W	4.45
	S	23.76S	52.51W	5.01	T	24.76S	55.22W	4.18
Notes:								

Add Info: Lying on one of the outer ring-rims (~ 800 km in diameter) from Mare Humorum to the east, Cavendish's age must then be younger than that event - formed in the Nectarian Period (3.92 to 3.85 billions of years old). The crater's northern and southern rim sectors have experienced terrace-collapse and impact (sub-crater E) respectively throughout its lifetime, while own ejecta deposits lie even-spread all around its outer rim. The floor of Cavendish is a different matter! Hint of a peak is seen best at low sun angle periods, but also note two, relatively big depressions (each about 15 km in diameter) in the bottom half of the floor. The right-most (east) depression looks, to all intents and purposes, like a buried crater, however, the other may in fact be only chance alignment of ejecta material related to E. A faint rille also crosses the floor from sub-crater E right up to the light-coloured terracing in the north, which more than likely is due to the number of fresh, but small, craterlets that pounded the surface soil. Another fainter rille is seen to cross over the first rille at right angles to it, however, extremely good and suitable conditions, or a very high magnification, is required. Depth ~ 2.5 km.

Craters

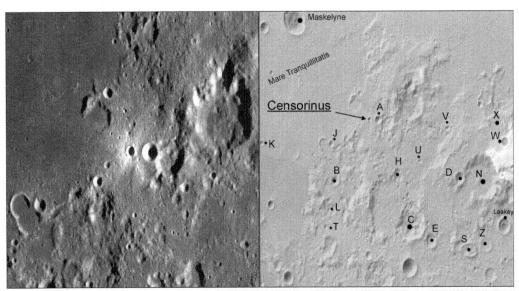

Censorinus Lat 0.42S Long 32.69E 4.1Km

Sub-craters:

Crater	Lat	Long	Size (km)	Crater	Lat	Long	Size (km)
A	0.43S	32.99E	6.76	B	2.02S	31.44E	6.89
C	3.09S	34.15E	27.7	D	1.95S	35.86E	10.3
E	3.58S	34.84E	11.26	H	1.84S	33.66E	8.83
J	1.04S	31.33E	5.27	K	1.02S	28.85E	4.49
L	2.77S	31.23E	3.71	N	1.93S	36.6E	36.31
S	3.82S	36.19E	15.55	T	3.22S	31.16E	3.73
U	1.51S	34.44E	3.44	V	0.65S	35.45E	4.24
W	1.02S	37.49E	8.26	X	0.59S	37.19E	16.9
Z	3.68S	36.8E	11.49				

Notes:

Add Info:

Apollo 11, Ranger 8 and Surveyor 5 landed not so far away to the west of this region.

Close-up of Censorinus

For such a small crater, Censorinus packs a big punch. Best observed at high sun-angle times (or anytime when in light), its bright ejecta has splayed out in every direction - extending as far as to sub-crater K in the west (~ 120 km away) and beyond sub-crater A in the east, where the rays are a bit harder to define. Having formed upon the remains of highland material that now surrounds Mare Tranquillitatis, its bright signature will eventually fade after a billion years - through weathering, cosmic rays continually bombarding the surface, and micrometeorites which further disturbs the lunar soil at a low level. Depth ~ 1 km.

Craters

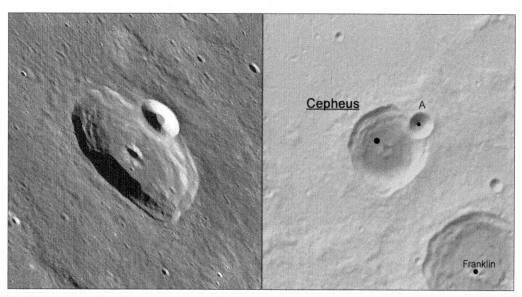

Cepheus	Lat 40.68N	Long 45.78E	39.43Km

Sub-craters:	Crater	Lat	Long	Size (km)	Crater	Lat	Long	Size (km)
	A	41.04N	46.51E	12.47				
	Notes:							

Add Info:

Aerial view of Cepheus

Aged between the Eratosthenian (3.15 - 1.1 byo) and Copernican (1.1 - Present) Periods, Cepheus lies on a hilly clump of material that may once have been associated to a basin (~ 400 km in diameter and centred near Shuckburgh in the east). It has a crenulated rim, wonderful series of terraces, and a central peak whose eastern sector seems to have been struck by an impactor (small crater is ~ 3 km in diam.,). Cepheus's 'ring' accepted a nice little diamond, too, that is, sub-crater A. Ejecta from A lies on the floor in an unusual mound near that crater - was this an oblique impact event? Note the slight ray off to the northwest of A where some bright material lies on the exterior of Cepheus's own ejecta. Depth ~ 5 km approx..

Craters

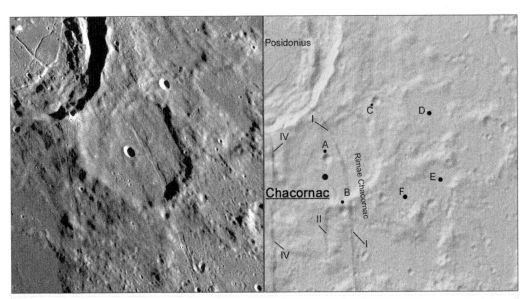

Chacornac	Lat 29.88N	Long 31.67E			50.44Km		

Sub-craters:

Crater	Lat	Long	Size (km)	Crater	Lat	Long	Size (km)
A	29.82N	31.58E	5.24	B	28.94N	31.91E	5.03
C	30.78N	32.58E	4.89	D	30.73N	33.63E	25.13
E	29.45N	33.81E	20.06	F	29.13N	33.05E	21.58

Notes:

Add Info: Chacornac is an odd-looking crater in shape – resembling more a pentagon than the usual circular shape we expect to see from impacts. Lying on material between two ring-rims (620 and 920 km in diameter respectively) of the Mare Serenitatis system, the crater has to be older than its famous neighbour, Posidonius (Upper Imbrian - 3.75 to 3.2 byo), to its northwest, as ejecta from that impact has filled about a third of Chacornac's floor. Besides the small, relatively fresh-looking crater, sub-crater A, in the centre of Chacornac's floor, the only other defining feature that this crater has going for it is the rille (Rima Chacornac I) cutting across its middle. This rille is concentric to Mare Serenitatis, so it may be related to the after-effects of that formation (e.g. sagging in the central Mare), however, as a portion of the rille can be seen to 'cut' into Posidonius's ejecta on Chacornac's floor, its creation and age most likely occurred in the Lower Imbrian Period. Another, less obvious, rille can be seen just to the left (west) of Rima Chacornac I, called Rima Chacornac II, however, good lighting conditions and high magnification is required. Note, too, the small, but bright, unnamed crater (~ 4 km in diameter) on Posidonius's south-eastern rim, which has imparted some of its high-albedo ejecta onto Chacornac. Depth ~ 1.5 km.

Craters

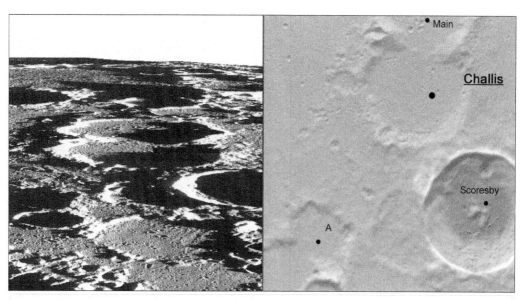

Challis	Lat 79.58N	Long 9.09E		53.21Km			
Sub-craters:	Crater	Lat	Long	Size (km)	Crater	Lat	Long	Size (km)
	A	77.27N	1.98E	31.82				
	Notes:							

Add Info:

Aerial view of Challis

A limb-hugger for sure, this is one crater for the serious astrophotographer to capture under favourable lighting conditions. Like its neighbour, Main, to the north, both are of the Nectarian Period (3.92 - 3.85 byo), both are covered in material from Imbrium to the south, and both are of around the same depth (1.8 - 1.9 km). It looks like Main impacted Challis, so it may be slightly younger - but not by much (perhaps, a double impact event occurring at the same time is the cause for the small cleft of material between the two?). Younger Scoresby (Erastosthenian) to the southeast doesn't seem to have imparted any of its ejecta onto Challis's floor, but is has raised the area between the two craters somewhat.

Craters

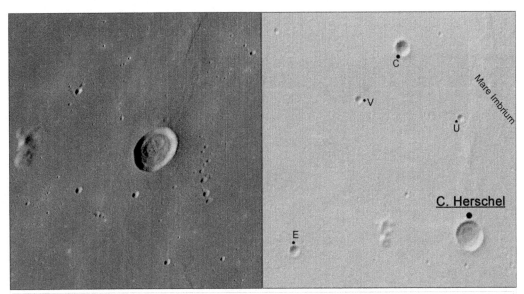

C. Herschel	Lat 34.48N	Long 31.29W					13.7Km

Sub-craters:	Crater	Lat	Long	Size (km)	Crater	Lat	Long	Size (km)
	C	37.23N	32.62W	7.04	E	34.21N	34.69W	5.03
	U	36.2N	31.44W	3.3	V	36.46N	33.49W	3.46

Notes:

Add Info:

Is it possible to spot the bright signature of a small, fresh impact crater ~ 100 m in diam., on the floor's southwest sector?

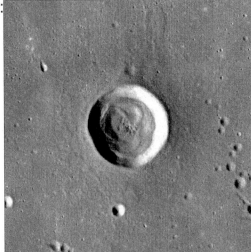

Aerial view of C. Herschel

Confusion can sometimes arise if when finding, or referencing, the three-named Herschel craters on the Moon. There's Caroline Herschel (this one); John Frederick William Herschel (Lat 62.31N, Long 41.86W) - usually referred to as crater J. Herschel; and finally, Wilhelm (William) Herschel (Lat 5.69S, Long 2.09W) - usually known just as crater Herschel. Crater C. Herschel has two distinct features: its mottled display of ejecta at its exterior, and its puddle-like floor - due to what appears to be the result of two sub-levels of terracing around a low central peak. Depth 1.9 km.

Craters

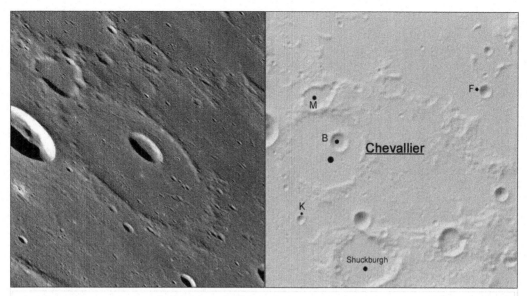

Chevallier   Lat 45.01N   Long 51.57E   51.83Km

Sub-craters:

Crater	Lat	Long	Size (km)	Crater	Lat	Long	Size (km)
B	45.15N	51.96E	12.05	F	46.2N	56.6E	8.53
K	43.52N	50.86E	5.56	M	46.04N	51.16E	16.27
Notes:							

Add Info:

Aerial view of Chevallier

It's quite easy to overlook this crater as most lies underneath a volume of lava deposits related to a possible basin ~ 400 km in diameter and centred just east of Shuckburgh. Take a wider view of this area and note the extended deposits of lava that encompasses Lacus Temporis to the northeast right across to deposits in the southwest near Franklin crater. The exposed rim of Chevallier looks very worn, its northern sector has been impacted upon (sub-crater M), while its southeastern sector has a 'kink' in it - possibly where lavas breached into its centre. Sub-crater B has a shallow, saucer-like shape to it at depth ~ 1.5 km, while Chevallier is at ~ 1 km.

Craters

Cichus		Lat 33.29S	Long 21.18W			39.18Km

Sub-craters:

Crater	Lat	Long	Size (km)	Crater	Lat	Long	Size (km)
A	34.82S	21.52W	19.82	B	33.19S	19.38W	13.84
C	33.57S	21.86W	11.51	F	35.68S	22.52W	7.83
G	35.5S	23.61W	21.55	H	32.81S	22.5W	7.25
J	32.05S	21.37W	13.12	K	36.63S	20.05W	6.2
N	30.58S	21.74W	7.42				

Notes:

Add Info: Cichus is a very strange-looking crater! Its entire rim spills seamlessly into its outer ejecta that extends just too much beyond its expected borders, while the same seems to have happened inside the floor area. The western side of ejecta at its outer rim is higher than its east, which under a low sun shining from that direction causes a thick shadow to end abruptly shorter than one would expect. As this region is near the confluence of two major basins - Nubium and Humorum respectively, the flat lava plain of Palus Epidemiarum where Cichus's ejecta overlies may then be the cause for the sharp fall-off in the shadow seen. The eastern side of the outer rim and ejecta is odd, too, where deposits have fallen on, or covered, a linear feature that forms part of Weiss's western rim. The south-eastern floor of Cichus, or what 'peeps' through as only a slight portion remains exposed, has what looks like a small, partially buried crater similar in size to sub-crater C (which seems to have impact material in its centre?). The western floor of Cichus blends in with possible, slumped material (caused by impact of sub-crater C?) from Cichus. Age - Eratosthenain (3.15 - 1.1 byo). Depth ~ 2.8 km.

Craters

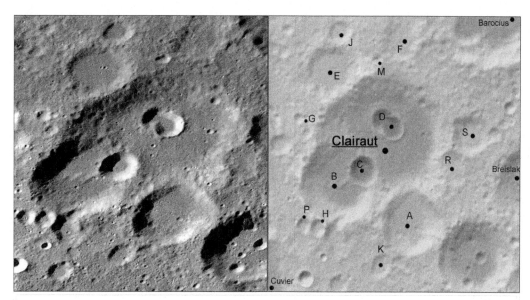

Clairaut   Lat 47.84S   Long 13.86E   76.89Km

Sub-craters:

Crater	Lat	Long	Size (km)	Crater	Lat	Long	Size (km)
A	49.05S	14.75E	34.19	B	48.36S	12.64E	41.45
C	48.17S	13.39E	16.18	D	47.43S	14.19E	12.66
E	46.48S	12.5E	29.14	F	45.78S	14.35E	31.18
G	47.26S	11.64E	6.24	H	49.12S	12.11E	9.21
J	45.81S	12.76E	13.02	K	49.78S	13.91E	12.42
M	46.26S	13.76E	5.83	P	49.09S	11.71E	8.26
R	47.99S	15.84E	13.55	S	47.51S	16.32E	22.04

Notes:

Add Info: We're in 'confusion territory' again - so many craters jumbled here and there, it's quite easy to get lost. Clairaut, in the view above, is not the slightly oval, or pear-like shaped crater that encompasses sub-crater B, but rather whose diameter can be determined from what remains of the northern half of its rim - lying just below sub-crater M, and between sub-craters G and S respectively. Nearly all of its southern half has been entirely lost to impactors nearby; with sub-crater B wiping out any semblence of the rim there, and sub-crater A leaving just a hint of the rim as its own ejecta mixed into it on impact. The floor has subsequently experienced several other impacts through sub-crater C, and D (which appears to be two craters merged into one). The material, most likely, that now fills Clairaut's floor, and the other sub-craters, too, is that from Nectaris to the NE (~ 1100 km away), and Imbrium to the NW (~ 2600 km away), but also of nearby craters, e.g. Maurolycus in the north. Western side of Clairaut is higher. Depth ~ 2.2 km.

Craters

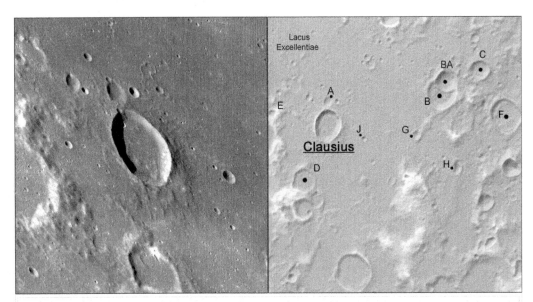

Clausius	Lat 36.9S	Long 43.93W				24.2Km	
Sub-craters:	Crater	Lat	Long	Size (km)	Crater	Lat	Long	Size (km)
	A	36.32S	43.91W	7.57	B	36.06S	40.24W	23.91
	BA	35.75S	40.08W	18.52	C	35.46S	39.05W	14.76
	D	38.22S	44.71W	17.26	E	36.58S	45.73W	6.74
	F	36.55S	38.13W	25.77	G	37.09S	41.01W	7.03
	H	37.87S	39.63W	7.0	J	37.21S	42.73W	4.39

Notes:

Add Info:

SMART-1 impacted closeby in Lacus Excellentiae on the 3 Sept., 2006 at Lat 34.24S Long 46.13W

Aerial view of Clausius

Situated just off-centre in a deposit of lava that is Lacus Excellentiae, Clausius, looks very saucer-like for a crater of its size. Except for a few hilly mounds predominantly in the southwest portion of the floor, the rest looks relatively flat. Two small (~1 km) craters lie close together in the northwest sector - one on the floor whose outline is obviously brighter against the dark lava deposits, and the other that has clipped Clausius's rim. Note the bright 'streaks' of material on the inner rim's southeastern slopes during westerly sun periods, while during low sun angle times note Clausius's outer ejecta deposits. Depth ~ 4.9 km.

Craters

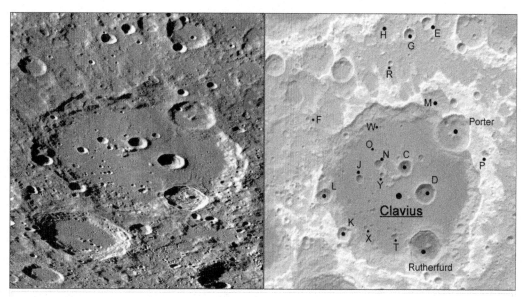

Clavius   Lat 58.62S   Long 14.73W   230.77Km

Sub-craters:

Crater	Lat	Long	Size (km)	Crater	Lat	Long	Size (km)
C	57.72S	14.19W	20.97	D	58.82S	12.43W	27.85
E	51.51S	12.74W	15.75	F	55.44S	22.08W	7.31
G	52.02S	14.0W	16.75	H	51.84S	15.86W	32.56
J	58.17S	18.17W	12.38	K	60.46S	19.86W	19.57
L	58.77S	21.34W	23.39	M	54.76S	11.63W	47.89
N	57.57S	16.5W	12.78	O	56.99S	16.66W	4.41
P	56.93S	7.5W	13.16	R	53.25S	15.42W	7.65
T	60.68S	15.22W	8.12	W	56.0S	16.25W	5.01
X	60.25S	17.81W	6.52	Y	58.02S	16.22W	7.12

Notes:

Add Info:

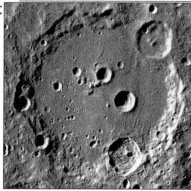

Aerial view of Clavius

One word is all that suits the crater Clavius - 'magnificent'. Even the smallest of telescopes, or binoculars, produces a view of a crater that seems to have it all going for it. Note its crimpled-looking rim all around, its wonderful series of terraces, its floor filled with craters from ~ 30 km downwards, and several other features - rays from Tycho, ejecta from sub-craters, peaks, mounds, crater chains and more...phew! Observe this crater at times other than full moon (very hard to see), and note on how shadows fill its floor during other times. Depth ~ 4.9 km approx..

Craters

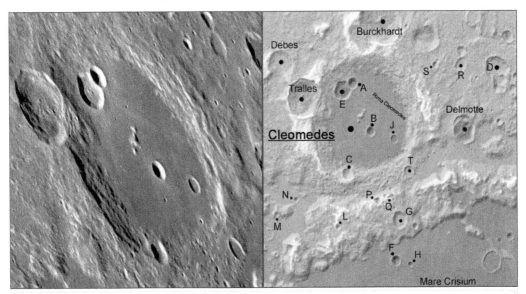

| Cleomedes | Lat 27.6N | Long 55.5E | | | | 130.77Km |

Sub-craters:

Crater	Lat	Long	Size (km)	Crater	Lat	Long	Size (km)
A	28.88N	55.06E	12.57	B	27.15N	55.89E	10.68
C	25.66N	54.83E	13.85	D	29.21N	61.86E	28.94
E	28.61N	54.58E	21.09	F	22.57N	56.99E	11.82
G	24.01N	57.17E	19.35	H	22.43N	57.63E	6.5
J	26.87N	56.96E	10.01	L	23.79N	54.41E	7.08
M	24.19N	51.66E	6.29	N	24.72N	52.52E	6.0
P	24.7N	56.2E	9.49	Q	24.81N	56.69E	4.47
R	29.46N	60.16E	14.75	S	29.46N	58.95E	7.58
T	25.74N	57.61E	12.37				
Notes:							

Add Info: Like its larger neighbour, Mare Crisium, to the south, Cleomedes was created sometime in the Nectarian Period (3.93 to 3.85 billions of years old). The crater just lies in between, and on the edge of, two major ring-rim systems (540 km and 740 km in diameter respectively) from Crisium - essentially, forming a trough, whose effect has led to flooding of the interior. Several small rilles that criss-cross the floor can be seen under reasonable magnification, while at slightly higher mag.,, over seven dark-halo craters (volcanic or impact created) close to the rilles stand out. The peak, which consists of three small sub-peaks, rises to near a kilometre high, and during near teminator times (~ 5-day-old or 16-day-old moons), a small shadow crosses the floor. The rim and inner terraces are well worn, and these, particularly the northwestern sectors, have had their fair share of bombardment (note the wonderful, pudding-like 'clump' of ejecta from Tralles on the floor), and the straightness in the southern terraces. Depth 3.8 km.

Dark haloed craters are believed to form either through volcanoes spewing ash that falls close to their general exterior, or from small impacts that excavate dark material under the surface.

Craters

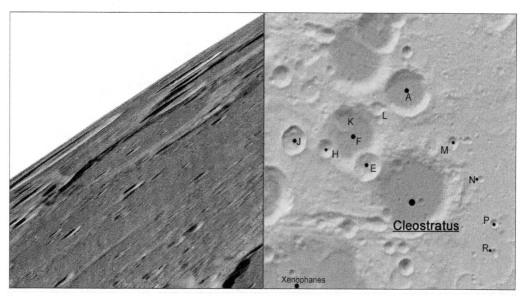

Cleostratus   Lat 60.32N   Long 77.4W   63.23Km

Sub-craters:

Crater	Lat	Long	Size (km)	Crater	Lat	Long	Size (km)
A	62.67N	77.65W	35.61	E	60.93N	79.73W	20.96
F	61.53N	80.58W	49.22	H	61.28N	81.97W	12.67
J	61.39N	83.66W	20.2	K	61.98N	81.08W	16.94
L	62.28N	79.3W	9.85	M	61.5N	75.2W	9.09
N	60.62N	73.77W	5.26	P	59.58N	73.2W	7.13
R	58.89N	73.28W	6.44				

Notes:

Add Info:

Aerial view of Cleostratus

An easy way to locate any difficult feature on the moon's limb is to use a very obvious one, and then work back from it towards your intended target (for Cleostratus, the obvious feature is crater Pythagoras east of it with its wonderful peak). Aged in the pre-Nectarian, Cleostratus lies in a zone where the dominant lavas of Oceanus Procellarum haven't quite reached through a depresssed valley between it and Pythagoras (whose ejecta overlies the lavas underneath). The crater obviously has been subjected to Imbrium's ejecta also, and note how its effect has produced 'gouges' along its eastern rim, and particularly a major one at its southwest. Depth is around 4 km.

Craters

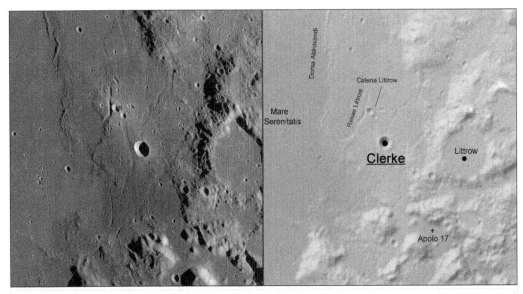

Clerke   Lat 21.68N   Long 29.8E   6.66Km

Sub-craters: No sub-craters

Notes:

Add Info: The impactor responsible for the chain of craters, known as 'Catena Littrow', most likely came in from the west. While the general direction of surface material ejected was eastwards, note the different set of ray patterns that 'point' back to some impacts in the chain.

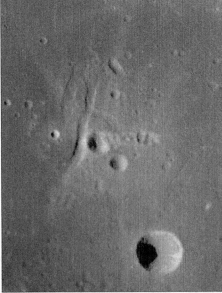

Aerial view of Clerke and Catena Littrow

For such a small crater lying at the eastern-most edge of Mare Serenitatis, Clerke is quite easy to see as its relatively bright inner wall material contrasts nicely, at times, against the dark lavas it lies upon. These lavas are of a younger series than those towards the centre of the Mare (note the colour difference in these from a wider-viewed perspective), which have been subjected to additional stress as the centre of the mare sagged - giving rise to the myriad of rilles and ridges seen also in the region. Clerke's impact has obviously ejected lighter material from below onto the surrounds, but the small chain of craters (Catena Littrow) north of the crater is the real 'duster-er' for the 'light display'. Depth ~ 1.4 km.

Craters

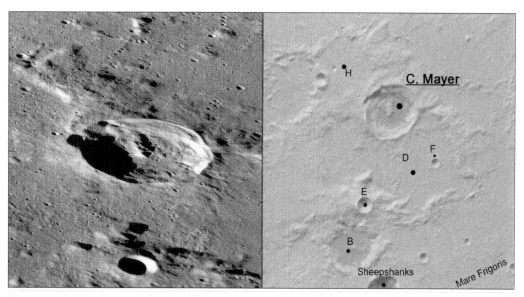

C. Mayer	Lat 63.26N	Long 17.31E			37.54Km

Sub-craters:

Crater	Lat	Long	Size (km)	Crater	Lat	Long	Size (km)
B	60.19N	15.44E	32.7	D	62.2N	18.51E	64.82
E	61.18N	16.04E	11.23	F	62.09N	19.62E	6.65
H	64.21N	14.74E	42.7				

Notes:

Add Info:

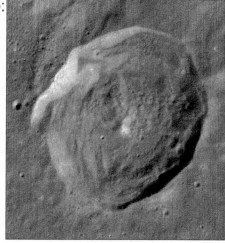

Aerial view of C. Mayer

Named in honour of Czech astronomer, Christian Mayer, this crater is one that tends to be overlooked because of its location. Having a sharp rim all around and freshly-slumped terraces, it lies on Imbrium ejecta that has covered this region of the Moon, and possibly on the northwestern rim of an old, unnamed crater lost to obscurity. The peak (peaks) rises to about 0.8 km high, however, while separated from the terraces to the north, at times during shadowed occassions, it looks like they are connected. The crater's ejecta to the southeast can also be clearly seen during these shadowed times, too. Depth ~ 2.4 km approx..

Craters

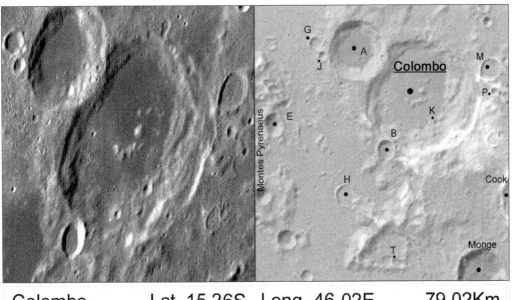

Colombo   Lat 15.26S Long 46.02E   79.02Km

Sub-craters:

Crater	Lat	Long	Size (km)	Crater	Lat	Long	Size (km)
A	14.18S	44.46E	40.78	B	16.41S	45.16E	13.48
E	15.82S	42.38E	14.93	G	14.01S	43.44E	8.95
H	17.45S	44.13E	14.14	J	14.3S	43.62E	5.96
K	15.83S	46.44E	5.0	M	14.64S	47.8E	15.65
P	15.11S	47.9E	5.9	T	18.97S	45.46E	9.96

Notes:

Add Info: Colombo lies in a region where two ring-rim systems from both Nectaris (~ 620km ring) and Fecunditatis (~ 690 km ring) respectively meet. With its worn rim and terraces, its smooth-like central peaks (compare these to the peaks in Theophilus over to the west), this is a crater of old age - possibly of the early Nectarian. Look at its terraces more closely and you'll immediately note that the southeastern sectors are nearly three times the width of those in the opposite direction. What happened to produce such an inequality? The height in topography just outside the southeast rim is higher than in the northwest, so did that extra volume of material play a role afterwards in the formation of the extended width in the terraces? Take a 'bigger-picture' look of the area in which Colombo lies in and you'll see that large pockets of lava plains dominate, as, too, seen in nearly all of the heighbouring craters around. These lavas are the same as those that lie in the crater's floor, so we're probably talking about infilling through fractures - now, obviously, buried. Other features: look at how sub-crater A's ejecta has covered Colombo's terraces there; the odd floor of sub-crater B. Depth: 2.5 km.

Craters

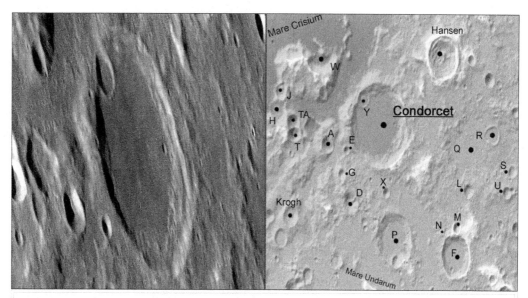

Condorcet   Lat 12.1N   Long 69.58E   74.85Km

Sub-craters:

Crater	Lat	Long	Size (km)	Crater	Lat	Long	Size (km)
A	11.51N	67.29E	15.38	D	9.84N	68.35E	19.8
E	11.35N	68.16E	8.03	F	8.27N	73.09E	39.66
G	10.65N	67.97E	10.31	H	12.45N	65.08E	21.04
J	13.07N	65.2E	16.25	L	10.15N	73.67E	12.57
M	9.07N	73.12E	10.54	N	8.98N	72.61E	5.02
P	8.76N	70.29E	48.4	Q	11.23N	73.36E	35.81
R	11.75N	74.75E	19.89	S	10.67N	75.64E	10.2
T	11.8N	65.84E	14.67	TA	12.16N	65.72E	14.6
U	10.08N	75.4E	10.97	W	13.94N	67.0E	35.83
X	10.12N	69.95E	8.88	Y	12.86N	68.92E	14.19

Notes:

Add Info:

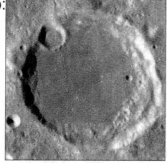

Aerial view of Condorcet

Lying on one of the ring-rim (~ 740 kilometre ring) systems (ejecta) from Mare Crisium, Condorcet is an old crater. Portions of the area in which it lies has experienced some breaching and flooding of lava-flows over millions of years, however, as Condorcet doesn't have any 'break' in its surrounding rim, most likely the lava that fills its floor is due to infill through fractures - now well buried underneath. Condorcet's eastern walls and terraces look overly extended as opposed to its western walls and terraces, and note, too, the change in albedo. D: 2.6 km.

Craters

Conon   Lat 21.66N   Long 1.95E   20.96Km

Sub-craters:

Crater	Lat	Long	Size (km)	Crater	Lat	Long	Size (km)
A	19.68N	4.38E	6.03	W	18.72N	3.07E	4.01
Y	22.34N	1.82E	3.72				

Notes:

Add Info:

Apollo 15 landed some 140 km away to the northeast from this crater - in the Hadley Rille area.

Close-up of Conon

If you can drag your gaze away from the wonderful range of mountains that make up the Apennines, Conon, does, in a way, catch the eye. This Copernican-aged (1.1 byo - Present) crater formed, essentially, on the rim of Imbrium's main 1160 km diameter ring/rim, and at the edge of a valley, whose feature shows up quite nicely just after a first quarter moon. Under high magnification and suitable lighting conditions, the inner rim appears bright, the 'wavey' characteristic to the slumped material in the northern sector of the crater is obvious, but ejecta from the crater is a bit hard to see. Depth ~ 2.9 km approx..

Craters

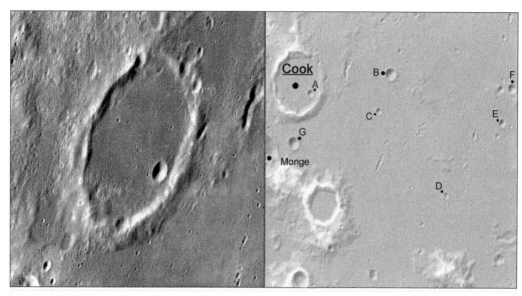

Cook	Lat 17.5S	Long 48.81E	45.16Km

Sub-craters:

Crater	Lat	Long	Size (km)	Crater	Lat	Long	Size (km)
A	17.76S	49.21E	5.83	B	17.3S	51.64E	8.99
C	18.19S	51.28E	4.95	D	20.17S	53.38E	4.75
E	18.43S	55.1E	6.15	F	17.59S	55.37E	7.05
G	18.94S	48.69E	9.9				

Notes:

Add Info:

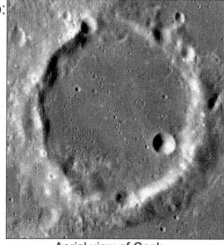

Aerial view of Cook

A 'bigger picture' view of this area in where Cook lies shows that the region experienced flooding of lavas in a low depressed zone. Three separate Basin rings/rims intersect here: the main 860-km one of Nectaris to the west; the 690 km one of Fecunditatis to north; and the 1000 km one of Balmer-Kapteyn in the east - centred west of sub-crater, Kapteyn B. Cook's rim in the northeast is slightly lower, so did a breach there allow lavas to slowly pour into the interior, or, was the infill an internal one (e.g. fractures etc.,)? The bright inner southeast rim wall and sub-crater A show up easily during full moon times. D: 1 km.

Craters

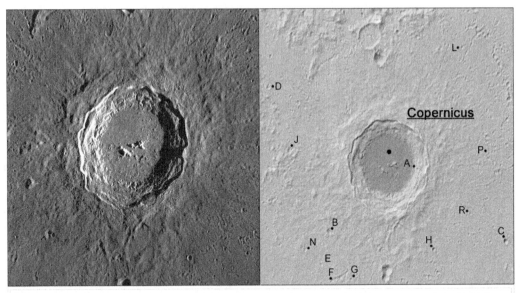

Copernicus	Lat 9.62N	Long 20.08W				96.07Km	

Sub-craters:

Crater	Lat	Long	Size (km)	Crater	Lat	Long	Size (km)
A	9.52N	18.9W	3.22	B	7.5N	22.39W	7.55
C	7.12N	15.44W	5.73	D	12.2N	24.8W	5.09
E	6.4N	22.7W	4.12	F	5.89N	22.24W	3.33
G	5.92N	21.51W	4.14	H	6.89N	18.29W	4.4
J	10.13N	23.94W	5.71	L	13.48N	17.08W	3.89
N	6.91N	23.31W	6.33	P	10.11N	16.06W	4.32
R	8.06N	16.84W	3.57				
Notes:							

Add Info:

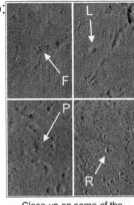

Close-up on some of the 'hard-to-see' sub-craters.

For more on the numerous small secondaries see LAC 58
http://planetarynames.wr.usgs.gov/images/Lunar/lac_58_lo.pdf

Like earthly 'New York', lunar Copernicus crater should have been named twice - it's such a wonderful view in a scope of any size. Its crenulated rim, inner walls and terraces, outer ejecta and rays almost never cease to put on a spectacular show; even under the least favourable of lighting conditions. The peaks in relation to the crater are somewhat disappointing if one were to grade them as like the crater's overall magnificence, but note the floor in which they lie - showing impact melt (particularly in the northwest), and hundreds of blocky clumps and mounds in the remaining sectors. Other features: note the radial lineations (+ impact melt) on/in the NE and SW terraces; secondary cratering - radial and concentric (+ crater chains) on outer sectors. Depth 3.8 km.

Craters

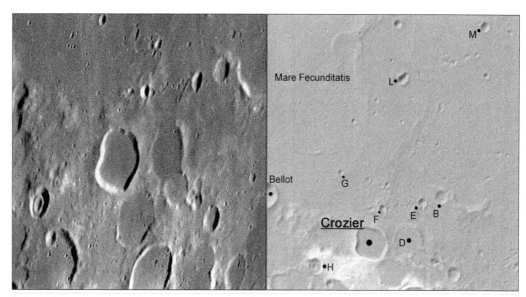

Crozier   Lat 13.56S   Long 50.72E   22.51Km

Sub-craters:

Crater	Lat	Long	Size (km)	Crater	Lat	Long	Size (km)
B	12.55S	52.39E	8.86	D	13.46S	51.59E	19.37
E	12.67S	51.95E	6.03	F	12.79S	51.0E	4.64
G	12.02S	49.99E	4.07	H	14.01S	49.38E	10.52
L	10.0S	51.5E	8.36	M	8.84S	51.36E	5.92

Notes:

Add Info:
Concentric Craters (CCs): Thought to be the result of a normal impact event forming in a layered, lava region (each layer affecting, differently, the dynamics of the crater's material and its rebound properties). For more on CCs, see Leakey.

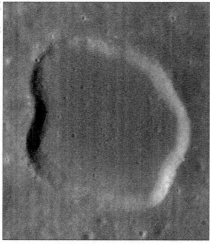

Aerial view of Crozier

Crozier lies on material related to the events and system of rings-rims created through formation of Mare Fecunditatis (to the north). The crater doesn't have any main distinguishing features to speak of - except for its odd shape, however, like most of its neighbouring craters of similar size, nearly all have experienced some flooding of lavas of sorts (note their smooth rims and flat floors). See also how these group of craters align radially back to the Imbrium Basin - a suggestion that they may be of secondary impact origin. Other features: note sub-crater H - a wonderful example of a 'concentirc crater'. Depth of Crozier ~ 1.3 km approx..

Craters

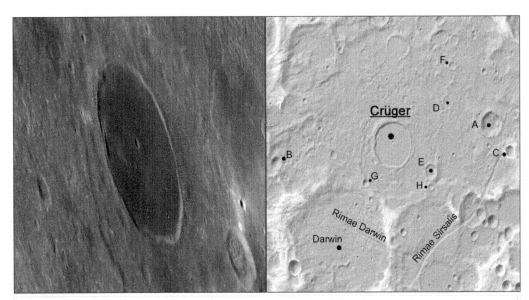

Crüger		Lat 16.68S	Long 66.96W				45.94Km

Sub-craters:

Crater	Lat	Long	Size (km)	Crater	Lat	Long	Size (km)
A	16.01S	62.85W	24.89	B	17.2S	71.78W	12.95
C	16.85S	62.03W	12.12	D	15.33S	64.65W	12.51
E	17.53S	65.34W	14.85	F	14.18S	64.47W	8.47
G	17.89S	68.07W	7.94	H	18.01S	65.35W	6.9

Notes:

Add Info:

Crüger lies centrally in a possible basin - having an approximate diameter of around 330 km (the crust in the central zone is some 30 km thick, or less).

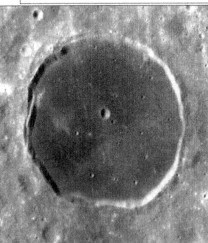

Aerial view of Crüger

Crüger is a crater of the Lower (Early) Imbrian Period - 3.85 to 3.75 billions of years in age. It formed in a depressed region of the Moon (note the other dark patches of lava here); which explains its floor as to why it may contain the same lavas seen in Grimaldi crater to the north, and in a more wider area seen in southwestern Oceanus Procellarum. A closer look at the floor shows lighter coloured material (note the band in the southern sector crossing in an ~ E-W direction and hints of other in a ~ N-S direction). Are these rays from sub-craters Sirsalis F (in the northeast) and Byrgius A (southwards) respectively? Depth ~ 0.6 km.

Craters

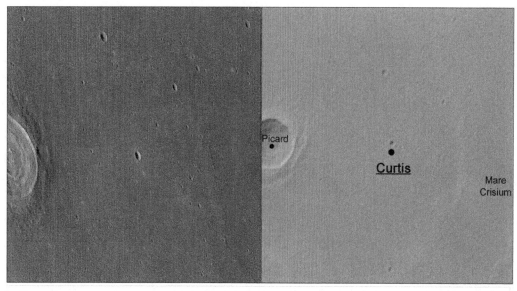

| Curtis | Lat 14.57N | Long 56.79E | 2.88Km |

Sub-craters: No sub-craters

Notes:

Add Info:

Material excavated from the deepest part of a crater usually lies on the near outer rim (therefore is older), while that excavated further away from the rim is younger.

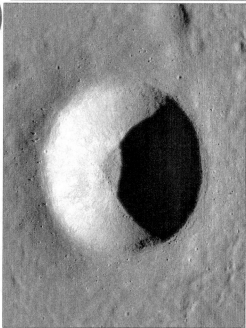

Extreme close-up aerial view of Curtis

Formerly known as Picard Z, craters as small as Curtis are usually, and generally, overlooked by the observer. Even though there may be tens to hundreds in any single view at a time, these 'little guys' can sometimes give other details about the area in which they formed. Examples include: noting how far they are from nearby bigger craters (are they then secondaries from them?), is one 'guy' brighter than the other (so, target deposit-types come into play), are they in groups or clusters (for ageing other craters). Close-ups (left) shows a whole new world (look at the wonderful 'boulder field' on the outer rim). Such boulders are useful for determining the make-up of sub-surface material of the crater's centre.

Craters

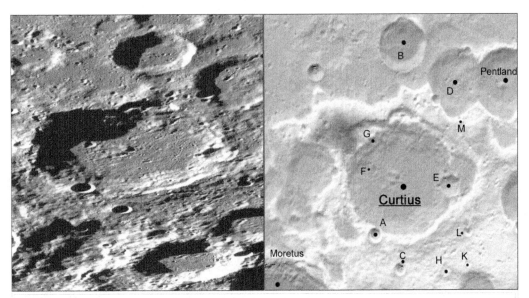

Curtius   Lat 67.08S   Long 4.4E   99.29Km

Sub-craters:

Crater	Lat	Long	Size (km)	Crater	Lat	Long	Size (km)
A	68.5S	2.52E	11.73	B	63.76S	4.63E	40.26
C	69.36S	4.38E	9.94	D	64.7S	8.35E	57.19
E	67.23S	7.98E	16.95	F	66.86S	2.65E	5.54
G	66.0S	2.92E	5.44	H	69.35S	7.99E	9.73
K	69.15S	9.68E	6.58	L	68.35S	9.47E	7.7
M	65.58S	8.43E	5.12				

Notes:

Add Info:

Aerial view of Curtius

Not to be confused with crater Curtis (and found in Mare Crisium - 14.57N, 56.79E), Curtius lies in a heavily-cratered region of the Moon that sometimes is overlooked. It's easy to locate as it lies close to similar-sized crater, Moretus, with its lovely peak, but the two are quite different when it comes to spotting respective features - like state of rims and terraces, as Curtius's are much older and worn. It does have a small peak that really never makes its presence felt during low sun times, but the collapsed wall at its north has been known to cast some nice dark shadows that define the different levels. Note the small, but obvious crater chain on the terraces here, too, and at really high powers another, less smaller chain on the floor's eastern sector can also be seen to parallel the first. Depth ~ 3.75 km.

Craters

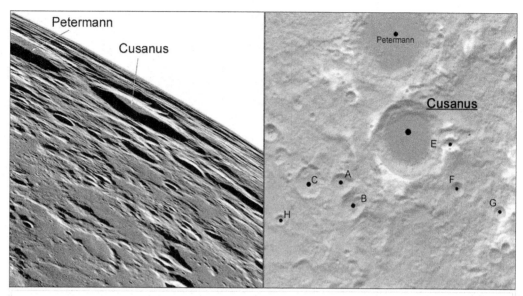

Cusanus   Lat 71.82N   Long 69.4E   60.87Km

Sub-craters:

Crater	Lat	Long	Size (km)	Crater	Lat	Long	Size (km)
A	70.58N	63.96E	16.81	B	70.09N	65.04E	21.13
C	70.49N	61.31E	23.45	E	71.73N	73.12E	10.25
F	70.59N	73.65E	12.01	G	69.93N	77.01E	10.57
H	69.43N	59.59E	8.3				

Notes:

Add Info:

Aerial view of Cusanus

The most distinctive feature about crater Cusanus must be the two 'bite-marks' on its eastern rim. They show up very nicely (just after full moon times), and were probably produced by a 'duo' set of impactors clipping the higher part of Cusanus's rim (note the similarly-sized feature to the east of the duo also - was there originally a 'triad' in this proposed impactor set?). A further puzzle: if the above impact scenario is correct, which direction did they come from, and what was there original source? Striated effects on Cusanus's and Petermann's rims and surrounds 'point' back to the Imbrium Basin in the southwest, but the two 'bites' seem to have a different orientation (would Moscoviense on the farside 27.27N,148.12E be a source?).

Craters

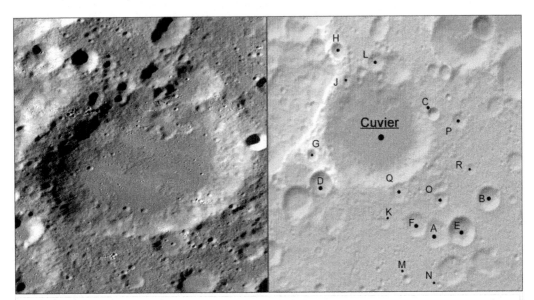

Cuvier Lat 50.29S Long 9.69E 77.3Km

Sub-craters:

Crater	Lat	Long	Size (km)	Crater	Lat	Long	Size (km)
A	52.48S	11.91E	17.86	B	51.71S	13.8E	15.76
C	50.0S	11.74E	8.52	D	51.41S	7.75E	16.32
E	52.4S	12.88E	18.72	F	52.29S	11.17E	15.13
G	50.84S	7.47E	7.54	H	48.69S	8.43E	10.31
J	49.28S	8.74E	5.17	K	52.28S	10.02E	7.07
L	48.92S	9.76E	12.61	M	53.42S	10.85E	5.71
N	53.53S	12.07E	4.09	O	51.78S	12.05E	10.67
P	50.12S	12.66E	10.78	Q	51.64S	10.53E	12.66
R	51.13S	13.15E	6.33				
Notes:							

Add Info: Before you take a closer look at Cuvier, take a broader view of the other craters around it, along with the inter-terrain between them, that all seem to have a consistant 'flatness'. This area experienced secondary impacts from the Orientale Basin (some ~ 2600 km away off to the northwest) event, and undoubtedly ejecta too - that might contribute to the overall flatness. However, as most of the craters here are generally considered of an older age (~ pre-Nectarian), the material make-up in and around them possibly contains ejecta-fill from other major previous events (e.g. Nubium, Humorum, and Imbrium etc.,). Cuvier, as such, doesn't have any major features to speak of - from a general view anyway, however, in a higher magnification view, note the tiny fresh craters (secondaries from Tycho over to the west ~ all less than a kilometre in size) that cross the northern sector of the crater! Rays on floor are from Tycho, too. Depth ~ 3 km.

Craters

Cyrillus	Lat 13.29S	Long 24.07E				98.09Km

Sub-craters:

Crater	Lat	Long	Size (km)	Crater	Lat	Long	Size (km)
A	13.77S	23.12E	13.69	C	12.36S	21.5E	11.75
E	15.9S	25.31E	10.46	F	15.32S	25.49E	43.9
G	15.66S	26.63E	7.76				

Notes:

Add Info: Like Theophilus to the northeast of this crater and Catharina to its south, Cyrillus lies in a region between a two ring/rim system (~ 620 km and 860 km in diameter respectively) created by the Nectaris Basin (eastwards) event. The crater is considered of this age too (Nectairan - 3.92 to 3.85 billions of years old), but subsequent events, for example, impact of younger Theophilus (Eratosthenian in age - 3.15 to 1.1 byo) nearby, has altered the appearance of how we might expect it to look. Volumes of ejecta from Theophilus cover the floor, and look at Cyrillus's peaks, too, which undoubtedly were draped with the same material. The floor has a central uplift that predominantly runs in a NE-SW direction, the southern sector of the rim is lower (looks like a gap), and all (including the rounded peaks) show up nice shadows during low sun angle times that defines each feature. Sub-crater A looks as if it might be of low impact origin - created, perhaps, by an impactor coming from the west, however, did the small crater (~ 1 km) within it have something to do with A's final shape (or, does this sub-crater have its origin in Theophilus as it sort-of points back to it?)? As to sub-crater F, well, it looks a total mess and is very hard to define any semblance of a crater here. Other features: fractures on floor; secondary craters on NW ejecta, and more. Depth 3.5 km.

Apollo 16 landed in the Descartes Highlands at some 270 km away to the northwest of this crater on 21 April 1972.

Craters

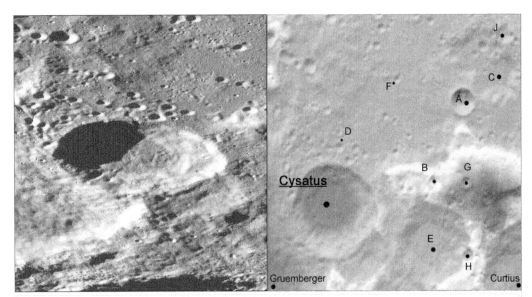

Cysatus	Lat 66.21S	Long 6.34W	47.77Km

Sub-craters:

Crater	Lat	Long	Size (km)	Crater	Lat	Long	Size (km)
A	64.32S	0.85W	13.47	B	65.7S	1.83W	8.19
C	63.89S	0.47E	26.67	D	65.17S	6.08W	5.01
E	66.62S	1.29W	48.59	F	64.03S	3.68W	4.1
G	65.65S	0.43W	6.09	H	66.86S	0.02W	9.04
J	63.15S	0.67E	10.11				

Notes:

Add Info:

Aerial view of Cysatus

If it wasn't for the wonderful crater that is Moretus (with is identifiable peak) to the south of Cysatus, one could easily become lost for finding it. Cysatus has impacted on Gruemberger to the SW (imparting ejecta onto its floor), so it is younger than that pre-Nectarian (4.6 to 3.92 byo) crater, but as its own rim and terraces are worn, too, its age is probably the same. Obviously, this is a crater for observing during favourable libration periods, however, even during those occasions, no outstanding feature from the crater states its presence. The exterior clump of material to the east of Cysatus is slightly higher than the remaining terrain, and ejecta from Moretus overlies, thinly, on the crater. Depth ~ 4 km.

Craters

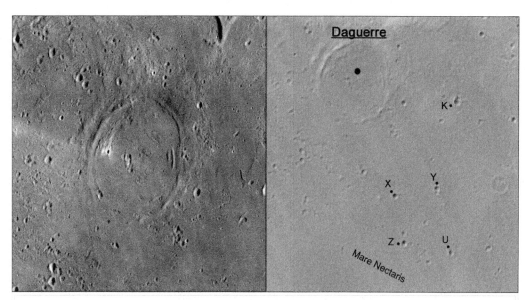

Daguerre

Daguerre   Lat 11.91S   Long 33.61E   45.79Km

Sub-craters:

Crater	Lat	Long	Size (km)	Crater	Lat	Long	Size (km)
K	12.25S	35.78E	5.44	U	15.14S	35.68E	3.16
X	14.07S	34.46E	3.43	Y	13.92S	35.36E	3.08
Z	14.97S	34.67E	3.67				

Notes:

Add Info:

Day-old moon phases

Note: As the Moon goes through various librations throughout the year, suggested times given in text for observations are only approximates.

Close-up views of sub-craters

Daguerre, as a crater of survival, just didn't have a chance! Situated awkwardly just too close to Nectaris's central regions; as mare lavas flooded across that Basin, they more than likely then poured (breached?) in through Daguerre's south-western rim filling up the crater and, perhaps, overflowing outwards to the remaining rim. Two features that obviously stand out in Daguerre are, firstly, the dark deposits at the rim's northeast sector, and, secondly, the additional rim in the west. The first possibly has something to do with volcanic activity and dark pyroclastic ash from a vent (not identified yet), however, the second is a bit of a puzzler. Look closely and you'll see that the additional rim actually encircle's nearly all the outside of Daguerre's (real?) rim. Is this a signature of an older rim from a different crater altogether, or, is it due to odd behaviour from Daguerre's impact? Other features: note the small bright impact crater on Daguerre's western floor, and the odd-aligned ray, too (from crater Madler to the west?). D: 0.1 km.

Craters

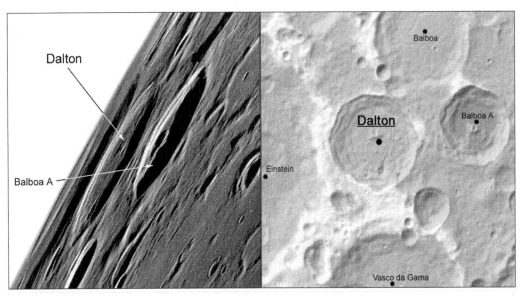

Dalton   Lat 17.07N   Long 84.45W   60.69Km

Sub-craters: No sub-craters

Notes:

Add Info:

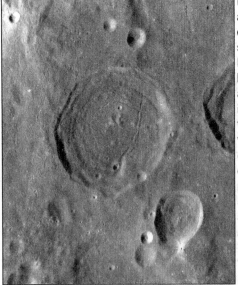

Aerial view of Dalton

Lying not far from the vast lava plains of Oceanus Procellarum to its east, and the Orientale Basin to its south (centred some ~ 1200 km), Dalton has, in a way, connection to both. For the former feature, the crater exhibits fractures (due to possible uplift) on its floor where similar-type lavas welled up to partially cover Dalton's peaks (highest is about 800 metres). While for the latter, its ejecta sparsely fell on Dalton itself, producing striations on the local terrain, and secondary impact craters as well. Dalton's semi-worn appearance suggests it probably is of the Imbrian Period (perhaps, Early- 3.85 to 3.75 byo). Favourable librations are obviously required for viewing, and try also just before Full Moon.

Craters

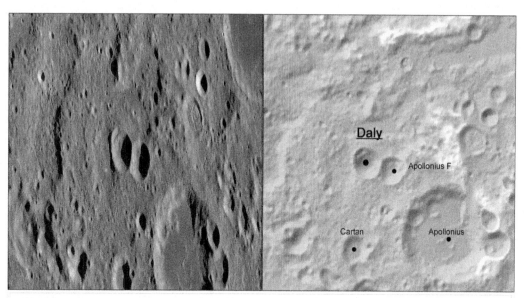

| Daly | Lat 5.74N | Long 59.5E | 14.96Km |

Sub-craters: No sub-craters

Notes:

Add Info:

Aerial view of Daly

The obvious feature of Daly itself is not the crater, but rather how it relates to its fellow companion, Apollonius F, on its eastern shoulder. Usually, when you see such a sharp dividing ridge between craters, the likely cause is that they formed by two impactors simultaneously striking the surface, and, which came in at high angles. There are quite a few dual-like impact craters all over the lunar surface - have a look at one pair due N by W (Apollonius L and an unnamed crater, some 160 km away) from Daly, while another pair lies northeastwards (Condorcet T and TA). The ridge formed also depends on how close together the impactors were on striking the surface, have a look at Plato K and KA to see the difference.

Craters

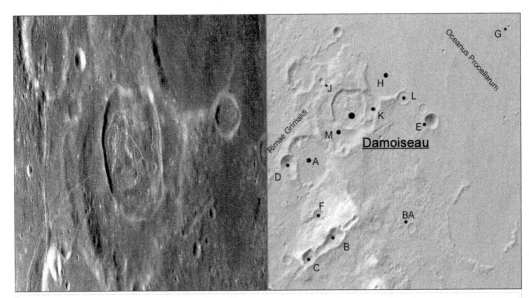

Damoiseau	Lat 4.85S	Long 61.25W	36.66Km

Sub-craters:

Crater	Lat	Long	Size (km)	Crater	Lat	Long	Size (km)
A	6.36S	62.59W	47.29	B	8.51S	61.71W	20.4
BA	8.28S	59.12W	8.67	C	9.14S	62.64W	13.73
D	6.43S	63.29W	16.31	E	5.22S	58.42W	13.63
F	7.89S	62.3W	9.53	G	2.51S	55.76W	3.76
H	3.87S	59.96W	44.89	J	4.07S	62.15W	6.71
K	4.74S	60.47W	18.54	L	4.55S	59.4W	14.87
M	5.08S	61.38W	58.3				
Notes:							

Add Info: Laccoliths form at shallow depths that 'dome' upwards the material lying above them; as magma fills in between deposit layers underneath the region in which they occur.

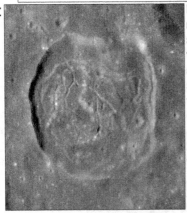

Aerial view of Damoiseau

Look at craters like Encke, Vitello, Bohnenberger, Atlas, and many, many others similar in size to Damoiseau, and immediately you'll see two common features - firstly, they all have a series of rilles, or 'cracks' on their floors and, secondly, all aren't that far from extensive lava deposits like mares etc.,. Why? The answer is that most of these craters experienced some sort of sub-surface, volcanic/magmatic activity underneath them associated to formation of the extensive deposits; producing pressure-related features like uplifting, fracturing, laccoliths, lava-escape. Damoiseau shows some of these effects that are best observed during low sun angle times. Depth ~ 1.25 km.

Craters

Daniell	Lat 35.42N	Long 31.16E	28.2Km

Sub-craters:

Crater	Lat	Long	Size (km)	Crater	Lat	Long	Size (km)
D	37.05N	25.85E	5.81	W	35.93N	31.55E	3.58
X	36.6N	31.81E	4.42				

Notes:

Add Info:

Aerial view of Daniell

Note here in the aerial view of Daniell its shape. The crater isn't actually circular because most of its northern and southern sectors has undergone some slumping of material (a large volume, in fact, in the latter) that now lies on the floor. And what of the floor - its central deposits look completely darker to those in Lacus Somniorum to the north, and more resemble those lavas found westwards of crater Posidonius in Mare Serenitatis. Daniell lies within the main ring/rim (~ 920 km in diameter) of the basin, so it's quite possible that the source of the lava is somewhat related. At high magnification, the slumped deposit appear to cover the rilles, suggesting they are older. Depth 2.0 km.

Craters

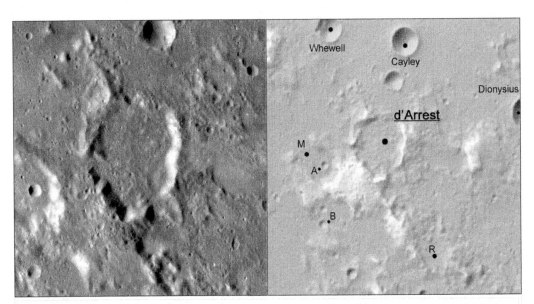

d'Arrest		Lat 2.26N		Long 14.6E			29.67Km

Sub-craters:

Crater	Lat	Long	Size (km)	Crater	Lat	Long	Size (km)
A	1.94N	13.64E	3.75	B	0.96N	13.61E	4.99
M	1.95N	13.5E	25.27	R	0.47N	15.63E	17.37

Notes:

Add Info: Crater d'Arrest how are ya? This is one of those craters that just doesn't look like your normal, average-looking crater. Lying outside one of the rings/rims (the ~950 km diameter one) from Mare Tranquillitatis's to the east, d'Arrest is an impact crater that seems to have formed on a clump, or block, of material -maybe associated to the Mare? Did the clump affect the impact dynamics that led to d'Arrest's odd shape, or, are we just seeing the results of other impact events nearby that caused the subsequent effects like slumping, added ejecta to occur. Whatever the cause, the crater looks a right mess, it has no real rim of its own to speak of as several 'breaks' occur all around, while its floor has been filled with deposits - possibly of Imbrium origin to the northwest. The floor has also received some ray covering - both dark and light - from Dionysius crater to its east (some ~ 100 km away), but these don't show up very good under any favourable lighting conditions. Depth of d'Arrest is around the 1.5 km mark.

Aerial view of d'Arrest (note how Dionysius's rays cross the crater)

Craters

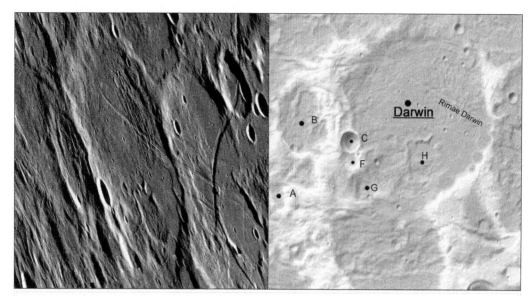

Darwin   Lat 19.93S   Long 69.21W   122.18Km

Sub-craters:

Crater	Lat	Long	Size (km)	Crater	Lat	Long	Size (km)
A	21.77S	73.14W	23.48	B	19.96S	72.29W	53.74
C	20.51S	71.12W	15.3	F	20.97S	71.12W	17.73
G	21.49S	70.87W	16.98	H	20.93S	69.0W	28.83

Notes:

Add Info:

Outer ejecta deposits from Orientale (for example, like in this area near Darwin and other craters nearby) are sometimes referred to as that of the 'Hevelius Formation'

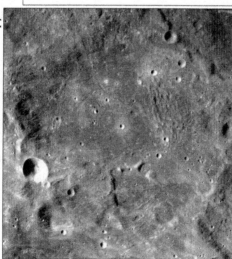

Aerial view of Darwin

Darwin's location, its depth (~ 1.4 km), and its worn-down, old appearance makes this a crater best observed during favourable librations, or, in times when the terminator (around 13 or 27-day-old moon periods) isn't that far from it. The region, in general, has received ejecta from Mare Orientale ~ 650 km away to the SW (note Darwin's NE sector with its dune-like hills), but the rille crossing its floor is probably a consequence of effects related to Oceanus Procellarum to the NE, rather than to Orientale; as it looks more concentric to it than radially to the latter. Ray material from Byrgius A crater, to the SE, cover parts of Darwin's floor.

Craters

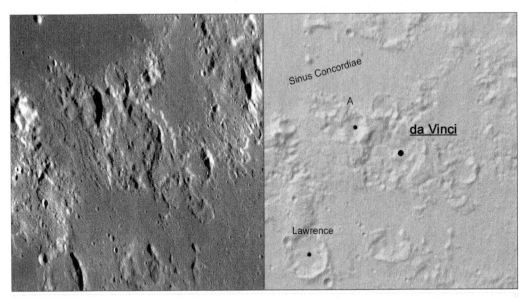

da Vinci	Lat 9.1N	Long 44.95E		37.46Km			
Sub-craters:							
Crater	Lat	Long	Size (km)	Crater	Lat	Long	Size (km)
A	9.66N	44. 2E	15.93				
Notes:							

Add Info: What a messy-looking crater is da Vinci. In fact, the only semblance of it looking like a cater at all is the chunk, or blocky lump of material, that makes up its eastern rim. It's hard to say exactly what the material is in which da Vinci lies; as here we are looking at an intersection region where ring/rim systems meet: from Mare Fecunditatis to the south; Mare Crisium to the northeast; and Mare Tranquillitatis to the west. Add to that, this area has also received additional deposits from Mare Serenitatis and Mare Imbrium - both to the northwest, so it's a wonder that we can 'see' da Vinci amongst all that has happened in its lifetime. The crater itself doesn't have any main features to speak of! At times of low sun when in the east, one can spot the odd level of terraces (?) in its eastern inner sector, and a small central peak also shows up. The sourthern sector of the rim has a small indent to it where lavas from Fecunditatis have filled in, but these haven't quite reached into the crater's interior as some hilly terrain there may have caused 'blocking' of said. Other small patches of these lava-pockets can also be seen at other near extremities of the crater nearly all around, so do try and see how [they] relate to the respective mares. During first quarter to full moon times da Vinci cannot really be discerned, so use the darkish lava inlet of Sinus Concordiae to its north as a location marker for it. Depth ~ 1.1 km approx..

Sinus Concordiae (Bay of Harmony)

Craters

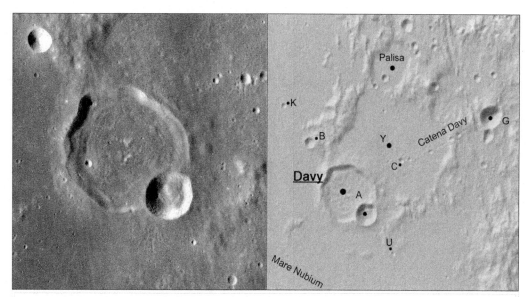

Davy	Lat 11.85S	Long 8.15W	33.94Km

Sub-craters:

Crater	Lat	Long	Size (km)	Crater	Lat	Long	Size (km)
A	12.23S	7.75W	13.21	B	10.89S	8.94W	6.57
C	11.22S	7.0W	2.97	G	10.37S	5.1W	15.36
K	10.19S	9.48W	3.05	U	12.96S	7.17W	2.76
Y	11.05S	7.25W	69.56				
Notes:							

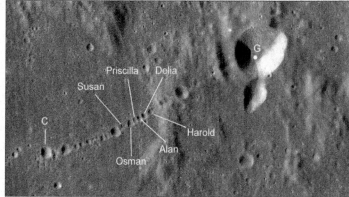

Close-up of the Catena Davy (length about 50 Km)

Add Info: A nice little crater is Davy! Its east and western rim shows signs of collapsed inner walls, where a series of concentric rilles have also developed. Note the three small peaks in the floor during low sun times, and the odd slump in sub-crater, Davy A. The *pièce de résistance*, however, has to be Catena Davy - the result, it is believed, due to break-up of a comet. Depth of Davy ~ 0.9 km approx..

Craters

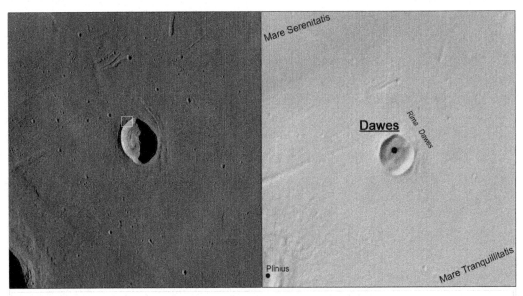

Dawes   Lat 17.21N   Long 26.34E   17.6Km

Sub-craters: No sub-craters

Notes:

Add Info:

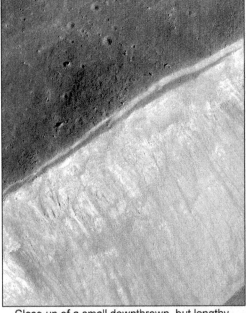

Close-up of a small downthrown, but lengthy (~ 4.0 km), shift of Dawes's northwestern rim

When Dawes is in full light, say, after first quarter to full moon, a slight change in albedo brightness of the dark Tranquillitatis lavas around it show that the crater formed on them. The bright material is oddly dispersed, and it doesn't have the ray-like features that we sometimes can see from young craters, for example, Kepler over to the west. Like Eratosthenian-aged (3.15 to 1.1 byo), Plinius, over to its west, Dawes has a relatively sharp rim, a hint of a small peak, but slumping in its walls (particularly at the north and south sectors) seem as single events for each, as opposed to several on Plinius's walls. Rimae Dawes on the crater's east looks to interupt Dawes's ejecta. Depth of Dawes is around 2.5 km.

Craters

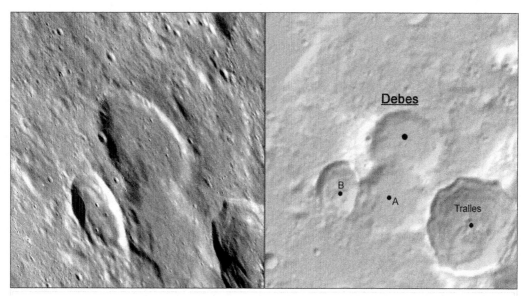

Debes		Lat 29.47N	Long 51.62E				31.92Km
Sub-craters:	Crater	Lat	Long	Size (km)	Crater	Lat	Long	Size (km)
	A	28.71N	51.47E	34.27	B	28.92N	50.53E	19.27
	Notes:						

Add Info: Debes, while lying in a region just between two of Mare Crisium's ring/rim system ~ 740 km and 1080 km ring/rim respectively to the southeast, is a crater of age (older than the Imbrian Period - possibly of Nectarian). No real features stick out as such, except for its floor that show signs of a smooth material (note this too in sub-crater, Debes A). This is also an area where flooding of lavas has taken place (see the extent of these due south and southwest of Debes - at eastwards of Lacus Bonitatis), however, these may not explain Debes's smoothed floor, so the material is more than likely that of ejecta from Imbrium and other impact craters nearby. Still, let not this be a definition! From the state of sub-crater A's barely perceptible rim, it looks older than Debes, however, as it obviously impacted upon Debes it has to be younger. Tralles, younger again, to the southeast, must surely have imparted some of its ejecta onto both Debes and Debes A, however, it isn't very obvious. Why would this be so? Would it be that Tralles was an oblique impact event, where the impactor came in from the west, and thus directed its ejecta eastwards (look at Cleomedes's floor), and so away from both Debes and Debes A? Whatever the reason, Tralles and Cleomedes together should factor into your observing of Debes as they both act as good location markers for finding it (even during full moon times). Depth ~ 1 km.

Craters

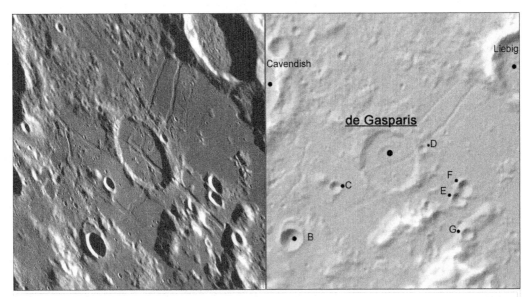

de Gasparis   Lat 25.83S   Long 50.83W   30.9Km

Sub-craters:

Crater	Lat	Long	Size (km)	Crater	Lat	Long	Size (km)
A	26.74S	51.37W	38.41	B	27.05S	52.65W	11.87
C	26.24S	51.82W	6.19	D	25.64S	50.21W	4.43
E	26.43S	49.54W	6.89	F	26.25S	49.44W	7.82
G	26.98S	49.39W	5.79				

Notes:

Add Info:

Have a look also at crater Palmieri to the southeast ~ 120 km away, which has similar features within its floor.

Aerial view of de Gasparis

"All aboard" - for rille junction! Lying on a lower clump of material associated to one of Mare Humorum's ring/rim system (the 570 km in diameter ring/rim), de Gasparis is more recognised for its series of rilles that criss-cross its floor. A closer look at 'junction central' shows it's really just where three rilles (II, III and IV) meet; whose extended arms of each signify some serious fractures in the region (see how far II and III join to another major rille that reaches as far as crater Latronne - 450 km away to the northeast). View the 'junction' during low sun angle times; when shadows accentuate its network best. Depth ~ 1.1 km.

Craters

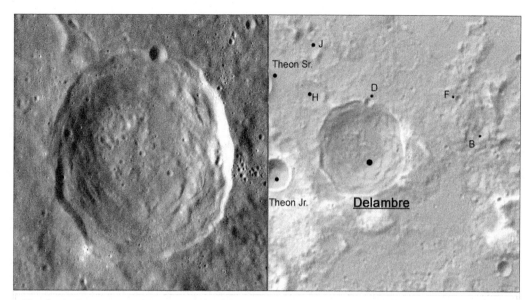

Delambre	Lat 1.94S	Long 17.39E	51.49Km

Sub-craters:

Crater	Lat	Long	Size (km)	Crater	Lat	Long	Size (km)
B	1.64S	19.64E	9.67	D	1.12S	17.54E	5.11
F	1.06S	19.25E	4.25	H	1.0S	16.44E	15.46
J	0.34S	16.75E	12.49				

Notes:

Add Info: Delambre lies in a region of terrain believed to be made up of material from the formation of Mare Tranquillitatis to the northeast. Its location is placed nicely so as to view the crater's features head-on: which shows its rim is relatively sharp all around; the series of collapsed terraces (predominantly in the southern sectors) to be fairly evenly distributed onto the floor - covering nearly a third of it; and smooth areas can be seen in the northeast portion of the floor (is it impact melt material or ejecta from other craters/basins nearby??). The remaining area of the floor looks 'hummocky' as it meets a 'hilly' section, also with the sign of a central peak; all of which wind and weeve across the crater in a NW-SE fashion. The exterior, eastern rim of the crater has a small group of craters - all around a kilometre in size, and worth having a look at so as to determine their origin (would crater, Theon Senior to its northwest be their parent, or are they sourced further?). The crater has a small fresh impact crater on its northern rim, which shows up easily during low sun views, but full moon times it just isn't seen at all. As expected, Delambre is quite hard to find also during full moon, however, both Theon Junior and Theon Senior to its west act as good location markers for approximating its whereabouts. Depth of Delambre is around 3.6 km approx..

Craters

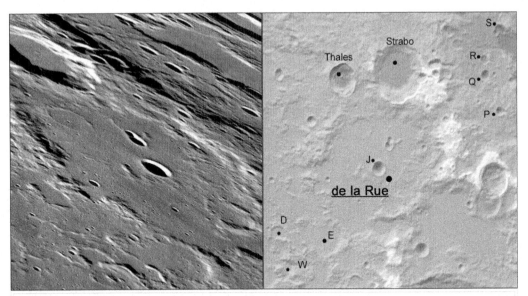

de la Rue   Lat 59.02N   Long 52.84E   135.22Km

Sub-craters:

Crater	Lat	Long	Size (km)	Crater	Lat	Long	Size (km)
D	56.81N	46.34E	17.58	E	56.85N	49.82E	31.92
J	58.99N	52.88E	14.09	P	60.43N	61.58E	9.02
Q	61.6N	60.86E	10.36	R	62.09N	60.91E	9.58
S	62.94N	62.48E	13.26	W	55.78N	47.08E	17.0

Notes:

Add Info:

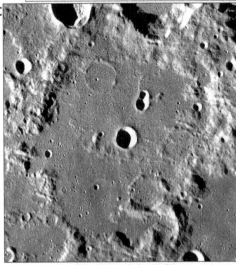

Aerial view of de la Rue

An odd-looking crater, is de la Rue! From an observer's earth-based persective, one can see some shape of sorts to that of a crater, however, as in the aerial view over, its hard to say where the actual 'crater' begins and ends. Something has seriously happened at the southern end; perhaps, a large crater lies buried there by lavas from Mare Frigoris to the west? Look also at de la Rue's northern sectors where impact of crater Strabo has almost 'squared' the crater there. Strabo is of Nectarian - 3.92 to 3.85 byo, so de la Rue must be older. Note other buried craters, too (five?), in the floor. Depth ~ 3 km approx..

Craters

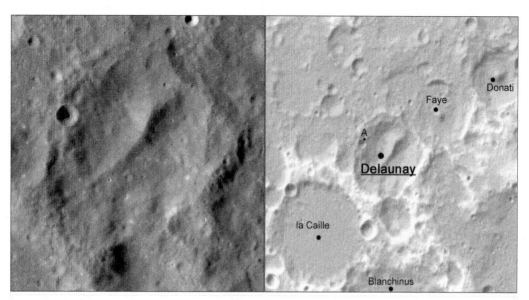

Delaunay	Lat 22.26S	Long 2.62E		44.63Km			
Sub-craters:	Crater	Lat	Long	Size (km)	Crater	Lat	Long	Size (km)
	A	22.01S	2.02E	5.62				
	Notes:							

Add Info: Looking like some kind of 'hoof-print', Delaunay seems, to all intents and purposes, like it got a right battering throughout life. Crater la Caille to the southwest looks like it totally obliterated any semblance of Delaunay's rim and walls there; dumping and mixing a load of ejecta onto and into these. Equally, at the opposite, northeast direction, crater Faye looks like it, too, wiped out a portion of Delaunay's rim and walls, turning the whole area into a mess. The central part of the 'hoof'; which looks like a sharp 'ridge' running in a NE-SW direction, is slightly higher at the Faye end, so is the 'ridge' a product of that crater? But wait! A closer look at Delaunay's battered environs shows it to have several old, 'over-lapped' and 'under-lapped' craters - big and small, so are we just seeing the last remnants of a very old crater peeking through all the 'rubble' generated by these impacts and associated ejecta? Surprisingly enough, Delaunay isn't that hard to find amongst all that has happened in the region in general, but it's best observed when shadows cross its floor. Try and capture, particularly during a low easterly sun time (~ 10 day-old moon), the ridge when light strikes it as it 'splits' the crater in half. Trying to find it during full moon times is useless, however, rays from Tycho to the southwest do appear to cross its approximate location. High end of ridge (peak?) is ~ 2 km, while depth is ~ 2.5 km.

Craters

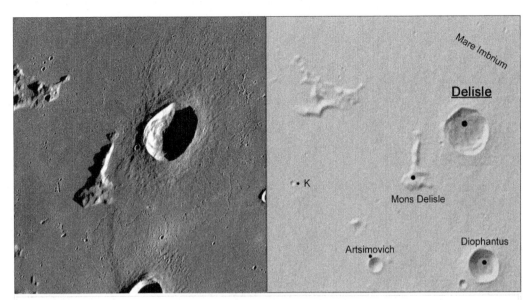

Delisle	Lat 29.98N	Long 34.68W	24.83Km

Sub-craters:

Crater	Lat	Long	Size (km)	Crater	Lat	Long	Size (km)
K	29.0N	38.46W	2.92				

Notes:

Add Info: Delisle lies just within the main ring/rim (1160 km in diameter) system of Mare Imbrium. Aged of the Eratosthenian Period (3.15 to 1.1 billions of years ago), its lofty, sharp rim always puts on a nice shadow show during a low sun angle pass, along with its wonderful display of ejecta all around the outside which always tries to steal the limelight. During higher sun angle periods, the crater's bright floor stands out easily amongst the slightly darker mare deposits of Imbrium, and if a closer inspection of the crater is taken, hints of a small central peak, collapsed terraces all around the inner rim, and a section in the northeast of smooth material can be seen. The presence of Mons Delisle to the west of the crater has in some way affected direction of Delisle's ejecta - acting, perhaps, like a 'buffer zone'; where, if you look on the western side of the mountain, no ejecta is seen. Is this a valid event/reason? But, this is also a region in where a series of flooding by lavas from formation of the Imbrium Basin has occurred over millions of years afterwards, so does the 'missing ejecta' lie underneath layers of thin mare deposits (note the eastern side of the Mons)? If so, this would imply that formation of Delisle occurred within a specific time frame 'close' to the Eratosthenain and Early Imbrian border Periods. Together with crater Diophantus to its south, both form a nice comparison pair in the eyepiece. Depth ~ 2.5 km.

Craters

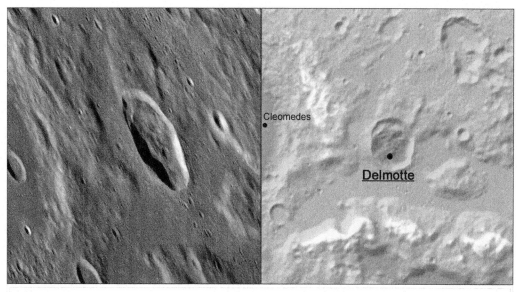

Delmotte	Lat 27.16N Long 60.2E	32.16Km

Sub-craters: No sub-craters

Notes:

Add Info:

Aerial view of Delmotte

Eratosthenian in age (3.15 to 1.1 billions of years old), Delmotte lies just on the edge in a clump of material of possible ejecta from Mare Crisium (?) to the south, and in a low-level region between a two ring/rim system ~ 540 and 740 km in diameter (produced as a result from formation of the Basin). Its rim looks relatively sharp, and no real terracing is seen except for the northwestern sector where possible slumping of material there has flowed (a 'dry flow?') onto the floor. This material covers over half the floor towards the south, but as its volume looks way too much for the slippage seen, it may have joined some other material (perhaps a peak?) already in the centre. Depth - just about over 4 km approx..

Craters

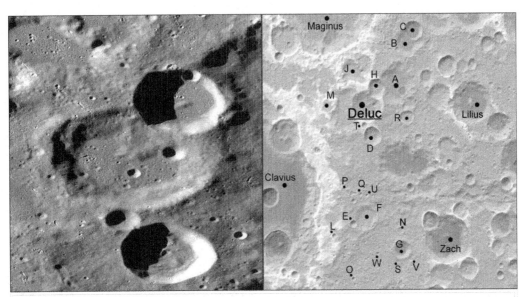

Deluc	Lat 55.02S	Long 2.98W	45.69Km

Sub-craters:

Crater	Lat	Long	Size (km)	Crater	Lat	Long	Size (km)
A	54.07S	0.5W	53.6	B	52.15S	0.27E	32.81
C	51.49S	0.85E	27.87	D	56.38S	2.45W	27.31
E	60.4S	4.39W	11.48	F	60.12S	3.07W	35.84
G	61.55S	0.66E	26.52	H	54.2S	2.13W	27.04
J	53.41S	4.13W	39.64	K	60.9S	6.4W	7.37
M	54.92S	6.32W	20.44	N	60.69S	0.38E	10.2
O	62.93S	4.53W	6.65	P	58.94S	4.85W	7.07
Q	59.05S	3.62W	5.64	R	55.48S	0.58E	21.03
S	62.11S	0.21E	6.35	T	55.89S	3.13W	9.88
U	59.1S	2.97W	5.25	V	61.97S	1.59E	8.76
W	61.76S	1.91W	6.0				

Notes:

Add Info: If you can tear yourself away from observing craters Maginus (pre-Nectarian in age) to the north and Clavius (Nectarian) to the southwest, Deluc, with its worn down rim and barely perceptible terraces, is probably a crater of these periods, too. It lies on a clump of material that almost resembles the northwest portion of a crater (some ~ 200 km in diameter), but not really as this clump is probably just the chance alignment of rims and ejecta by numerous craters nearby (look southeastwards and under low sun-angle times to see this effect). The floor of Deluc is smooth and possibly of Imbrium ejecta in origin, however, influence of Orientale (westwards) also is in this area, so could the material be from that? Features: tiny craterlets on floor and rim; sub-crater H's ejecta. Depth ~ 5 km.

Craters

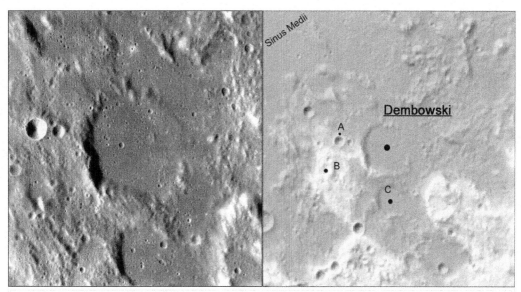

| Dembowski | Lat 2.88N | Long 7.27E | 26.11Km |

Sub-craters:

Crater	Lat	Long	Size (km)	Crater	Lat	Long	Size (km)
A	3.01N	6.46E	5.17	B	2.52N	6.19E	7.39
C	2.05N	7.33E	14.99				

Notes:

Add Info: Dembowski lives in an area of sorts! Firstly, this region is well associated with the Imbrium Basin to the northwest (~ 1000 km away); where, some four billions years ago, this huge impact event shot out in every direction massive chunks of ejecta (solid and fluidized) that became known as the Fra Mauro Formation - named because of the deposit make-up as found through the Apollo 14 mission near Fra Mauro crater. This ejecta oblitered many older craters closer to the focus of impact, filled partially the floors of others further away that survived, and carpeted many terrain types for over 2500 km away. Secondly, lavas from Sinus Medii to the west and from Mare Varorum to the north have flooded the region, in general - breaching, perhaps, Dembowski's eastern rim, and filling its floor. Thirdly, another blanket of material overlies the Fra Mauro Formation and Dembowski - known as the Cayley Formation - named because the smooth-looking deposits east of crater Cayley had similar material characteristics to those explored during the Apollo 16 mission in the Descartes region. Origin of the Cayley deposits are suggested as being those of fluidized ejecta from Imbrium, but possibly, also, of the Orientale Basin to the west. Not much was said about Dembowski here, but do observe it for the few features it has. Depth ~ 1 km approx..

Craters

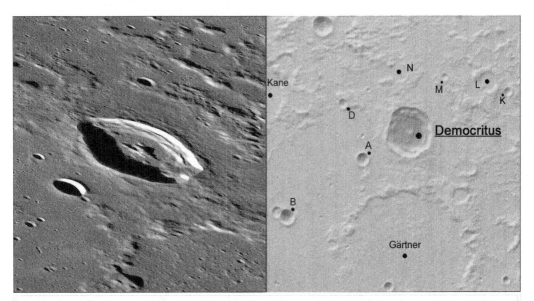

Democritus	Lat 62.31N	Long 34.99E			37.78Km	

Sub-craters:

Crater	Lat	Long	Size (km)	Crater	Lat	Long	Size (km)
A	61.61N	32.47E	9.88	B	60.08N	28.63E	11.72
D	62.95N	31.23E	7.39	K	63.15N	40.81E	6.98
L	63.42N	39.65E	17.39	M	63.61N	37.12E	5.7
N	63.73N	34.4E	15.42				

Notes:

Add Info:

Aerial view of Democritus

Lying in a slightly raised area - just outside Mare Humboldtianum's outermost ring/rim (~ 1350 km in diameter), and on the edge of Mare Frigoris, Democritus, with its sharp-ish rim, its series of terraces, and its central peak, looks odd in sorts when comparing it to its older, worn-looking neighbours all around. Democritus's 4.5-km depth produces nice shadows on its floor and from its peak (note how shadows look from its shallower neighbours), and rays from Thales (Copernican in age) to the east have dusted it thinly - observe the slight change in albedo in the surrounding material.

Craters

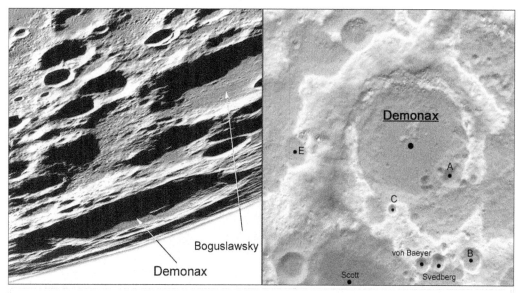

Boguslawsky
Demonax

Demonax

Demonax	Lat 78.09S	Long 59.36E	121.93Km

Sub-craters:

Crater	Lat	Long	Size (km)	Crater	Lat	Long	Size (km)
A	79.35S	65.86E	20.12	B	81.47S	72.05E	23.17
C	80.28S	56.29E	10.37	E	78.24S	43.21E	33.75

Notes:

Add Info:

Aerial view of Demonax

Demonax lies some 350 km away from the lunar South Pole, and just outside the main, 2500 km ring/rim of the South Pole Aitken Basin, which aside from the Procellarum Basin (?) on the Nearside, is the largest on the Farside. Being close to the limb favourable librations and good lighting conditions are required to 'catch' any descent view of Demonax. The crater formed on odd terrain - to its west it gets as high as ~ 7 km while to its southeast it's as low as 3 km - possibly a depresssed zone related to SPA (Demonax may also have impacted upon an old unnamed crater at its southeast). Highest peak on its floor is about 0.8 km. Depth ~ 5 km.

Craters

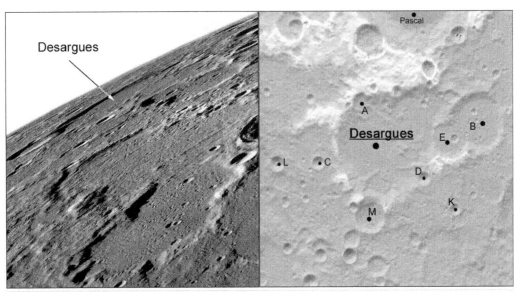

| Desargues | Lat 70.25N | Long 73.42W | 84.85Km |

Sub-craters:

Crater	Lat	Long	Size (km)	Crater	Lat	Long	Size (km)
A	71.34N	75.64W	14.27	B	70.67N	65.35W	48.61
C	69.67N	78.83W	10.87	D	69.37N	70.09W	11.16
E	70.33N	67.66W	31.82	K	68.52N	67.69W	9.23
L	69.49N	82.23W	12.94	M	68.42N	74.42W	27.47

Notes:

Add Info:

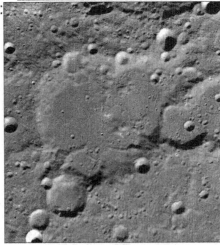

Aerial view of Desargues

Given its high latitude and longitude coordinates, Desargues, from the outstart, is one that will demand favourable librations (a good way to note its location is think of it lying somewhat behind the more obvious craters, Pythagoras and Carpenter). All of Desargues's southeastern sector is gone, where several unnamed impact craters (big and small) make it difficult to determine the crater's actual size. Southeastwards from the crater, ejecta deposits from Imbrium dominate the terrain and interior of craters, but Desargues's interior is smoother. Is it just finer Imbrium material, ejecta from younger Pythagoras, or something else?

Craters

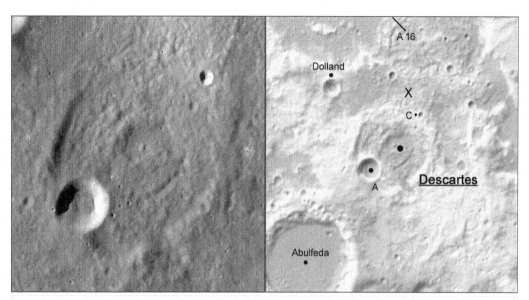

Descartes	Lat 11.74S	Long 15.67E	47.73Km

Sub-craters:	Crater	Lat	Long	Size (km)	Crater	Lat	Long	Size (km)
	A	12.08S	15.19E	14.25	C	11.02S	16.3E	4.33

Notes:

Add Info: Initial observation of Descartes suggest that we are looking at a very old and battered crater - covered, perhaps, by ejecta deposits from the Nectaris Basin to the east. But in a closer view you'll see that these deposits look very odd - particularly the bright material north of Descartes; where it has a bubbly texture to it that extends beyond Dolland and right up to where Apollo 16 landed. This bubbly feature is known as the Descartes Formation, but its make-up has proved controvercial: is it volcanic in origin, or, ejecta from Nectaris. A-16 samples taken, [not] from the formation itself, suggest a highland makeup, so it really is a puzzler that future missions will have to resolve. One thing is known for sure, the bright region has an intense magnetic field, centred approx., on the X above. Other bright regions: Reiner Gamma; within Mare Ingenii (Farside); east of Mare Marginis - all have such magnetic anomalies associated to them. But as to why they exist e.g.: antipodal impacts (that converge global mag., fields); local impacts (that magnetised the local material on impact); electrical fields associated with the mag., fields (that separates brighter dust deposits from darker deposits according to net charges applied)...etc., is still unknown. The crater itself doesn't have any main features, but there is a circular imprint (about 20 km in diameter) in the central region. Is it a buried crater, or a hilly terrain that looks circular? Depth ~ 0.8 km.

Craters

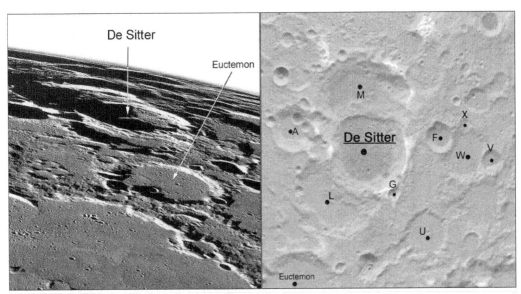

De Sitter Lat 79.81N Long 38.57E 63.79Km

Sub-craters:

Crater	Lat	Long	Size (km)	Crater	Lat	Long	Size (km)
A	80.24N	26.57E	35.98	F	80.03N	49.8E	22.93
G	78.78N	42.0E	9.71	L	78.86N	34.52E	69.4
M	81.11N	37.57E	75.52	U	77.73N	46.05E	35.38
V	79.06N	56.47E	19.24	W	79.51N	53.04E	44.37
X	80.18N	54.12E	11.02				

Notes:

Add Info:

Aerial view of De Sitter

Look at how De Sitter has imparted its ejecta on to sub-crater, De Sitter M, to its north and De Sitter L to its southwest (De Sitter's floor level is about 1.8 km below M and about 1.2 km below L). There is a hint of a small peak (~ 200 m high) on the floor, but the main feature here is the series of fractures within its northern rim and walls. Is the fracturing due to some volcanic activity underneath the floor, or has the weighty volume of De Sitter's ejecta in M's floor pulled some way at De Sitter's sides, and thus leading to the fractures? Depth ~ 3.7 km approx..

Craters

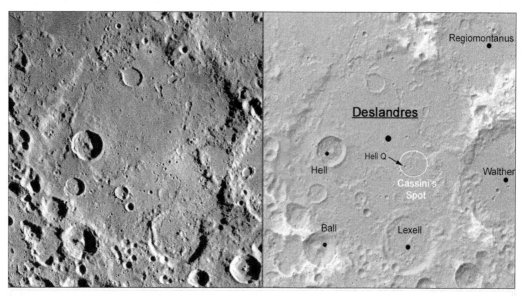

Deslandres Lat 32.55S Long 5.77W 227.02Km

Sub-craters: No sub-craters

Notes:

Add Info:
The bright area north of Lexell may be due to ejecta from Hell Q, which is believed to have been an oblique impact event. Though not officially recognised, the bright area is also known as 'Cassini's Bright Spot' (see above for approximate location).

Of pre-Nectarian in age (4.6 to 3.92 billions of years old), Deslandres joins the elite group of large craters on the Nearside favourably suited for observing. Its rim all around has obviously seen the impact wars: what with Walther wiping out Deslandres eastern rim partially (is the odd clump of rock just west of Walther the crater's original rim?); Lexell taking a 'chip off the old southern block'; while other numerous impacts in the west and southwest has turned the whole area there into a 'messy mass' in history of events. The floor of Deslandres is also a wonder to look at: note the many hints of buried craters (e.g. the large one south-east of Hell, two small ones north of Lexell, and several smaller ones in the top half - resembling crater chains). The floor also has a slight rise to it centrally, hint, perhaps, due to some volcanic uplift below - note the long rille-like feature in the northwest portion of the crater (a fracture-of-consequence?), or, is the rise due to ejecta from basins near and far away (examples: Imbrium, Orientale), or, floo-ding? Certainly, this latter event would explain the smooth-ish, dark-ish material in the northeast, which is also seen in the small crater group west of it and in nearby surrounds. At high sun angle times note the bright rays from the small crater (~ 2 km in diameter) at mid-floor - created by an oblique impactor that pro-bably came in from the west, while at low sun times note how shadows show up features above-mentioned (rilles, ejecta, rise and moat...etc.,) Depth ~ 1.58 km.

Craters

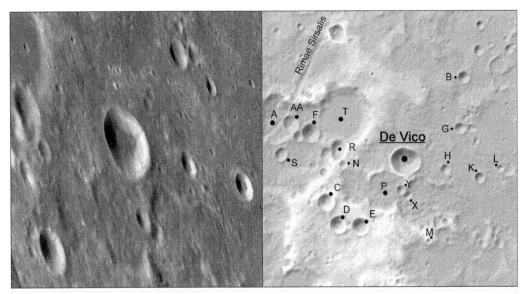

De Vico Lat 19.71S Long 60.32W 22.13Km

Sub-craters:

Crater	Lat	Long	Size (km)	Crater	Lat	Long	Size (km)
A	18.75S	63.62W	34.08	AA	18.84S	63.23W	11.34
B	17.85S	58.84W	8.71	C	20.62S	62.37W	12.59
D	21.12S	62.05W	12.52	E	21.16S	61.5W	12.22
F	19.06S	62.73W	12.52	G	18.98S	58.93W	8.19
H	19.9S	59.3W	8.07	K	20.08S	58.43W	7.49
L	19.87S	57.89W	5.25	M	21.06S	59.29W	5.74
N	19.77S	61.93W	6.44	P	20.4S	60.8W	29.15
R	19.44S	62.03W	13.1	S	19.53S	63.49W	9.63
T	18.69S	61.82W	39.72	X	20.47S	60.26W	6.42
Y	20.36S	60.44W	6.64				

Notes:

Add Info:

Aerial view of De Vico

For such a little crater, De Vico certainly has a lot of sub-craters related after it. From the aerial view shot over, its shape suggest a possible oblique impactor was involved (from the west), but as no hint of rays is seen, perhaps, the extension of its easterly sector may just be due to collapse. But that said, look at De Vico's sub-craters B right up to H and you'll notice that they all look similar - with a concentric attitude to the Orientale Basin towards the west. So, these, and De Vico itself, may just be secondaries from that event (try spot others in a wider view of this area).

Craters

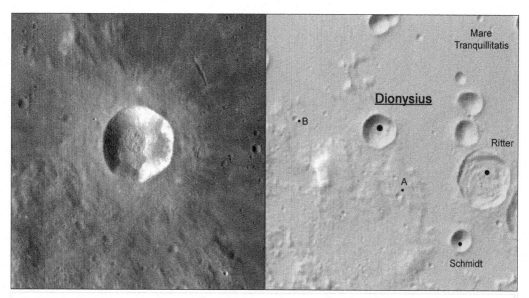

Dionysius	Lat 2.77N	Long 17.29E		17.25Km			
Sub-craters:	Crater	Lat	Long	Size (km)	Crater	Lat	Long	Size (km)
	A	1.68N	17.65E	3.57	B	2.96N	15.81E	3.57

Notes:

Add Info: Pronounced as 'Die-O-Nis-E-Us' or 'Die-O-Nish-Us', the crater, in whatever way you say it, remains a unique one in that it has two types of rays (see*). The first type is the bright ray material lying close to the outer rim all around. Made up of small amounts of Iron and Titanium Dioxide, the majority composes of anorthositic norite. The second ray type is the distal, dark rays beyond the first type - truly a wonderful contrast with the bright rays when seen through the eyepiece. Made up of dark mare material similar to that in Mare Tranquillitatis (with minor amounts of highland material in the mix), these are believed to have come from an iron-rich layer in mare make-up - exposed on the surface before impact. The crater is Copernican in age (1.1 billions years old to the present), so the rim has a relatively sharp look to it. What has happened at the crater's eastern wall? A major slump of material might be presumed, but cross-section shows the sector to have a less levelled out slope to a more slightly convexed 'bump' between the floor level and up to the now-extended rim. If it isn't slump-related, then are we looking at some kind of oblique impact signature of an impactor that came in from the west (certainly, distribution of the ray materials don't point to such a type event). *'Remote Sensing Studies of the Dionysius Region of the Moon' by Giguere, T A. et al. *Journal of Geophysical Research,* Vol 111, E06009, 2006.

On 20 July 1969, Neil Armstrong, Michael Collins and Edwin Aldrin Jr. succeeded in the landing of Apollo 11 some 200 km away to the east.

Craters

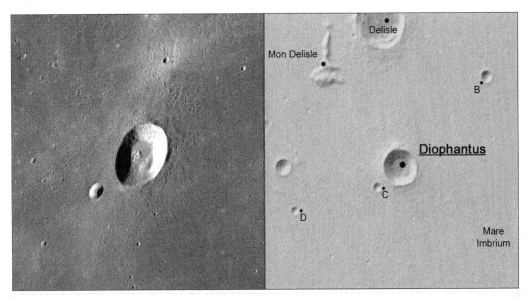

Diophantus	Lat 27.62N	Long 34.3W	17.57Km

Sub-craters:

Crater	Lat	Long	Size (km)	Crater	Lat	Long	Size (km)
B	29.08N	32.54W	6.19	C	27.29N	34.74W	4.62
D	26.91N	36.38W	4.24				

Notes:

Add Info: Like its brother, Delisle crater to the north, Diophantus lies in the Mare Imbrium. Both put on ejecta displays during low sun angle times and cast some nice shadows, too (the flat-ish terrain of Imbrium is simply wonderful for observing such apparitions, so make a note to try and catch the lowest angle possible). There is a slight difference in the make-up of material just north and south of the crater; where it is a bit brighter than its brother. Is the small, freshly-impacted crater to its north - some 25 km away and about a kilometre in diameter - responsible for the lighter material? This little crater looks like it was caused by an oblique-directed impactor from the southeast (there seems to be a 'zone of of avoidance' there), so if that is so, then the majority of its ejecta would have been thrown northwards - away from Diophantus. But that said, as Diophantus is of the Eratosthenian Period (3.5 to 1.1 billions of years ago) like Delisle, too, perhaps we are just seeing a younger-formed crater where its ejecta colour hasn't quite degraded (noting, of course, such light materials are known to reduce down to a lower albedo over time ~ after a billion years or so). The crater's interior is bright also, there is a small clump, or hill, in the centre, while towards the south, slumping seems to have occurred. Depth of crater Diophantus is around the ~ 3 kilometre mark approx..

Craters

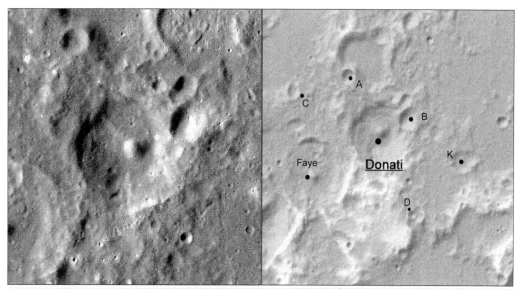

Donati	Lat 20.69S	Long 5.1E				35.84Km

Sub-craters:	Crater	Lat	Long	Size (km)	Crater	Lat	Long	Size (km)
	A	19.7S	4.49E	7.91	B	20.44S	5.68E	11.55
	C	19.97S	3.41E	7.77	D	22.12S	5.75E	4.48
	K	21.16S	6.77E	14.34				

Notes:

Add Info: Donati fits well here into this region of the Moon where old, dilapidated craters lie - e.g. Faye to its southwest, Airy to its north, and many others (note them!). Ejecta from numerous impacts - both big and small, from near and far - dominates the texture-type in the general region; making it hard at times to discern which crater is which, and what is the history in events of each. Donati has a crater depth of around 2 kilometres, its true rim appears best at its southern sector, and its central peak has just about survived the onslaught of the above-mentioned ejecta wars. It really doesn't have a floor to speak of, however, those areas that peek through have received numerous small impactors that gave rise to small craterlets formed - none bigger than a kilometre in size approx.. Look carefully also in the picture above and you'll see there is a bright strip that runs northwestwards from the peak to the inner rim. This isn't a change in the colour of the material there, but rather is due to light from the west striking a high piece of terrain. This 'strip', during a low sun from the east (around a 10-day-old moon), produces a nice little shadow that adds to the one being cast by the peak. And while you're at it, try and catch the light on Donati's peak, on Faye's, Airy's and Argenlader's peaks all at one time ~ after first quarter (a challenge!).

Craters

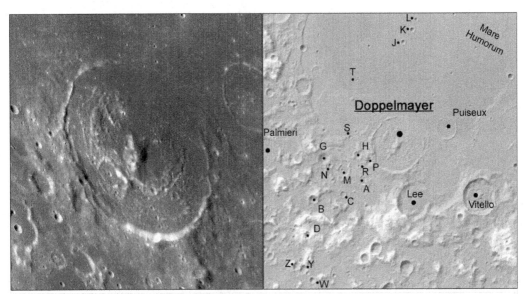

Doppelmayer Lat 28.48S Long 41.51W 65.08Km

Sub-craters:

Crater	Lat	Long	Size (km)	Crater	Lat	Long	Size (km)
A	29.84S	43.17W	10.14	B	30.5S	45.55W	11.62
C	30.33S	44.12W	7.18	D	31.84S	45.91W	8.64
G	28.93S	45.02W	13.36	H	28.85S	43.36W	10.53
J	24.52S	41.2W	5.68	K	24.04S	40.79W	5.68
L	23.65S	40.63W	4.59	M	29.39S	44.07W	13.78
N	29.21S	44.7W	5.37	P	29.09S	42.87W	9.9
R	29.26S	43.33W	3.74	S	28.14S	43.72W	4.63
T	25.97S	43.37W	3.02	V	29.82S	45.71W	6.66
W	33.59S	45.76W	7.48	Y	33.14S	46.17W	8.83
Z	33.02S	46.52W	10.48				

Notes:

Add Info: Lying some 50 km inside Mare Humorum's main ring (~ 400 km in diameter), Doppelmayer, like several large craters at mares (e.g. Fracastorius at Nectaris, Sinus Iridum at Imbrium...etc.,), has succumbed to the 'weighty' dynamics of subsidance within their centres. Its tilted region has been flooded by numerous layers of lava over millions of years, and its innards - peak, hilly terrain, terraces, small craters (note the buried one in the NE) and any other feature that existed - has been compromised under the strain. Two distinctive featues: firstly, Doppelmayer's hilly terrain surrrounding the peak has a concentric nature to it (possibly due to uplift of the floor); secondly, the peak may be higher than some portions of its remaining rim (a waning crescent time produces a peaky shadow that extends beyond the rim's diameter). Depth ~ 1 km approx..

Craters

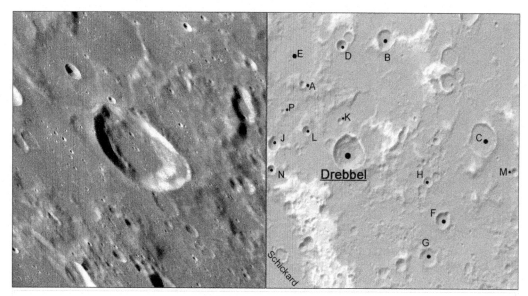

Drebbel   Lat 40.93S   Long 49.12W   30.23Km

Sub-craters:

Crater	Lat	Long	Size (km)	Crater	Lat	Long	Size (km)
A	38.99S	51.07W	7.13	B	37.78S	47.52W	17.49
C	40.46S	43.06W	28.32	D	37.95S	49.36W	10.77
E	38.21S	51.48W	59.8	F	42.79S	44.76W	15.02
G	43.84S	45.39W	16.58	H	41.74S	45.47W	9.57
J	40.62S	52.42W	12.99	K	40.06S	49.62W	32.51
L	40.3S	50.94W	9.26	M	41.29S	41.55W	7.08
N	41.34S	52.56W	8.7	P	39.72S	51.93W	4.88

Notes:

Add Info:

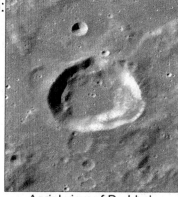

Aerial view of Drebbel

Is Drebbel a secondary crater produced by formation of the Orientale Basin to its northwest? Its oblong-ish shape certainly points towards the Basin's direction, and as it is in a region where many other secondaries from that main event lie, it may just be so. Or is that being too presumtious? Its easterly rim sector has undergone some serious slumping of material there (a consequence of the impact's direction?), while its westerly sector shows similar, but less dramatic events, too. But it then begs the question; what is the smooth-like material in the floor (also on the terraces)? Ejecta from Orientale? Can't be!

Craters

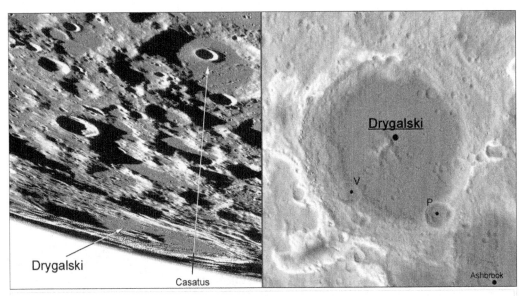

| Drygalski | Lat 79.57S | Long 87.18W | 162.49Km |

Sub-craters:

Crater	Lat	Long	Size (km)	Crater	Lat	Long	Size (km)
P	80.97S	99.69W	27.98	V	78.36S	93.47W	21.23

Notes:

Add Info:

Aerial view of Drygalski

It's a pity that both Drygalski and Ashbrook (a Farside crater) aren't better located for viewing needs, as the two together would make for some interesting observations. The most obvious feature would be the huge 'dollop' of ejecta (see aerial view, left) from Drygalski on to Ashbrook's floor. It literally looks like it flowed on to Ashbrook's floor afterwards; giving an insight of the behaviour of ejecta when it lands in large volume. Unfortunately, we can't see the dollop from our earth perspective, but we can see Drygalski with its 2-km-high peaks, the smooth half of its floor and its terraces (pre-Nectarian to Nectarian in age ~ 4.6 to 3.85 byo) D: 6.5km.

Craters

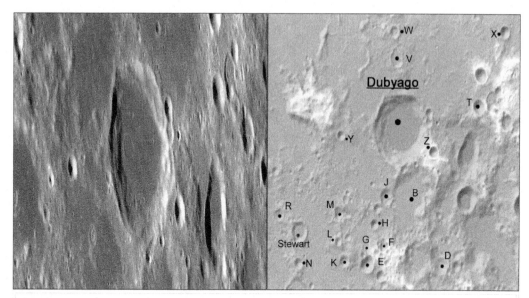

Dubyago	Lat 4.38N	Long 69.95E	48.12Km

Sub-craters:

Crater	Lat	Long	Size (km)	Crater	Lat	Long	Size (km)
B	3.08N	70.3E	33.96	D	1.46N	71.12E	14.09
E	1.47N	68.97E	12.6	F	1.89N	69.43E	9.89
G	1.84N	68.95E	7.56	H	2.33N	69.19E	6.88
J	2.92N	69.52E	11.67	K	1.51N	68.31E	8.63
L	1.94N	68.12E	6.11	M	2.56N	68.12E	8.65
N	1.46N	66.96E	7.15	R	2.52N	66.39E	9.07
T	4.87N	72.27E	10.71	V	5.89N	69.81E	10.97
W	6.51N	69.74E	8.87	X	6.5N	73.03E	8.41
Y	4.24N	68.19E	6.91	Z	3.89N	70.93E	8.81

Notes:

Add Info:

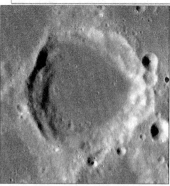

Aerial view of Dubyago

Before you take a look a Dubyago crater itself, note the general region in which it lies. To the northwest you have Mares' Crisium and Undarum, to the southwest, Mares' Fecunditatis and Spumans, and craters similar in size all around have dark-filled floors. Reason: the whole region is due to numerous lava-floods (over a billion years) that filled low spots created, perhaps, by the impact that was Crisium. Dubyago itself seems to have formed on slightly higher ground (ejecta of Crisium?), its flooded floor holds a ghost crater, its terraces and rim are well worn. Depth: 2.7 km.

Craters

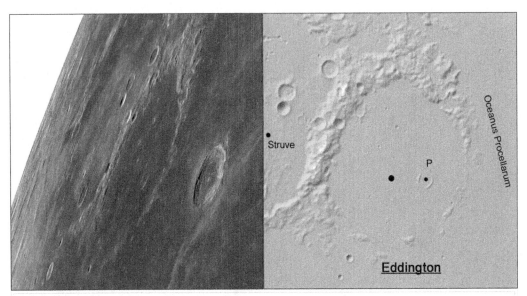

Eddington

| Eddington | Lat 21.5N | Long 72.02W | 120.13Km |

Sub-craters:	Crater	Lat	Long	Size (km)	Crater	Lat	Long	Size (km)
	P	20.03N	71.11W	11.43				
	Notes:							

Add Info:

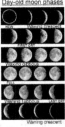

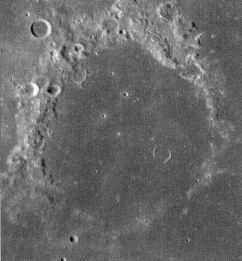

Aerial view of Eddington

A limb-hugger for sure, Eddington, like most large craters lying on the edges of mares (Oceanus Procellarum to its east), is one that suffered to the defeat of tilting and subsequent flooding. The orientation of the 'tilt' points southwards (as opposed to eastwards towards the general mare region); which says something about the area south of Eddington having, perhaps, thicker lava deposits than those in the east. No real rim is discernible, all that 'shines' through is its ejecta, that has encountered numerous impacts, floodings and debris by impacts near and far. Best observe at low sun times. Depth 1.3 km.

Craters

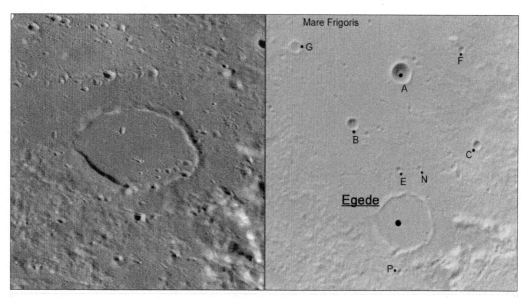

Egede   Lat 48.72N   Long 10.64E   34.18Km

Sub-craters:

Crater	Lat	Long	Size (km)	Crater	Lat	Long	Size (km)
A	51.57N	10.5E	12.27	B	50.59N	8.92E	7.27
C	50.19N	13.0E	5.79	E	49.74N	10.43E	3.62
F	52.02N	12.53E	4.31	G	52.0N	6.94E	7.34
M	49.78N	12.21E	4.06	N	49.76N	11.1E	3.48
P	47.79N	10.48E	3.58				

Add Info:

Notes:

From the outset, it's obvious that Egede has undergone some form of flooding - both inside and outside its rim. On the outside, lavas associated to Mare Frigoris have flowed around the rim entirely at first, and then filled the inside through possible breaching at the rim's western sector where it is lower. There seems to be a 'gap' here (note it at times when low eastern or western suns cast long shadows) in which the breaching may have occurred, but see also, to the right of the gap on the western part of the floor, an odd-looking piece of material (looks like a very small, straight mountain about 3 km long) that has connected to it a slump-looking deposit (very strange). Pull back a little in the scene, and you'll see that ejecta and secondary craters radiate in nearly every direction away from crater Aristoteles to the east (predominantly northwards as ejecta from crater Eudoxes south of Aristoteles seems to have covered the former). Is the small group of craters (all about ~ 1 km in size) in Egede's southern floor secondaries of Aristoteles? Also note the two oblong-looking craters midway on the floor, too, whose orientation suggest a different origin. Depth ~ 0.5 km.

Craters

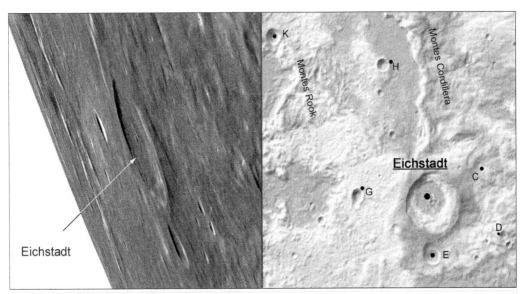

Eichstadt	Lat 22.63S	Long 78.42W	49.57Km

Sub-craters:

Crater	Lat	Long	Size (km)	Crater	Lat	Long	Size (km)
C	21.72S	76.86W	13.1	D	23.43S	76.12W	7.1
E	23.92S	78.46W	16.91	G	22.41S	80.85W	11.16
H	19.08S	79.99W	10.82	K	18.27S	83.4W	13.79

Notes:

Add Info:

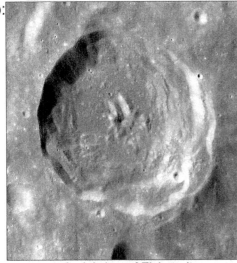

Aerial view of Eichstadt

Eichstadt presents a real challenge to anyone trying to glean any good detail from it - through our limited, earth-view perspective. Full lighting conditions are practically useless, while at low sun views the crater's interior is almost always filled in shadow (certain intermediate periods are thus best, but plan well for observation during favourable libration times). The crater formed on Montes Cordillera (aka: Orientale Basin), and has a wonderful series of peaks, and slumped terraces - particularly at its SE sector (the crater is therefore younger - possibly of Eratosthenian 3.15 to 1.1 byo). Depth ~ 4.5 km approx.

Craters

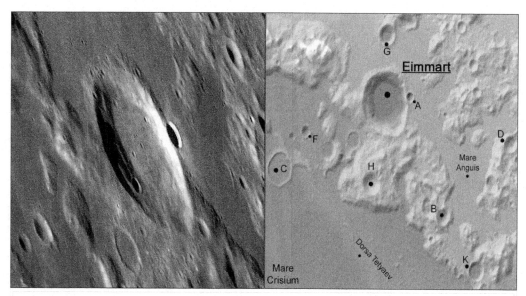

Eimmart		Lat 23.97N	Long 64.8E			44.99Km

Sub-craters:

Crater	Lat	Long	Size (km)	Crater	Lat	Long	Size (km)
A	24.11N	65.66E	7.34	B	21.37N	66.77E	12.19
C	22.41N	61.26E	22.68	D	22.98N	69.09E	11.78
F	23.29N	62.11E	8.3	G	25.52N	64.84E	14.23
H	22.13N	64.33E	16.89	K	20.12N	67.89E	13.81

Notes:

Add Info:

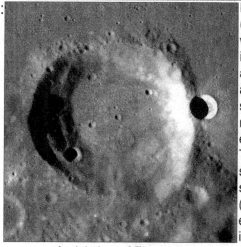

Aerial view of Eimmart

Eimmart formed on the inner ramparts (~ 500 km in diameter) created by the wonderful Basin that is Crisium. The impact mustn't have been very long afterwards as no real signs of ejecta are seen (did the impactor land on a semi-molten clump?), and look at its northern outer rim area where lava edges right up to it. A sign of flooding? The floor also may have experienced similar flooding, too, which blends in seamlessly to the terraced material (particularly in the SE) that looks too extended to suggest a possible oblique impact. Rays from sub-crater A cross the floor. Depth is about 3.5 km approx..

Craters

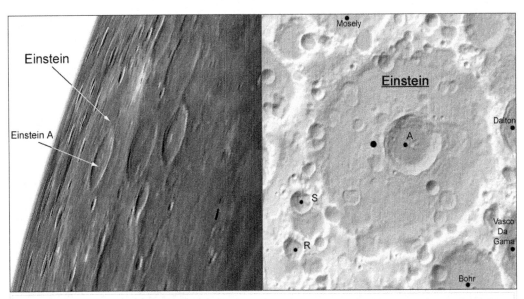

Einstein	Lat 16.6N	Long 88.65W	181.47Km

Sub-craters:

Crater	Lat	Long	Size (km)	Crater	Lat	Long	Size (km)
A	16.69N	88.25W	50.48	R	13.83N	91.88W	19.84
S	15.1N	91.67W	19.63				

Notes:

Add Info:

LADEE
Standing for Lunar Atmosphere & Dust Environment Explorer, the spacecraft impacted the lunar surface some 200 km away southwestwards from Einstein on 17 April 2014.

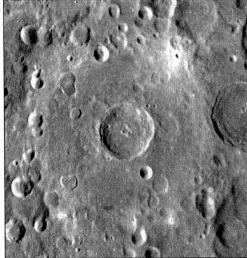

Aerial view of Einstein

A pity that Einstein is a hard crater to 'catch' any descent view of as, together with Einstein A, the two would make for a wonderful site. Obviously, favourable librations are the best times to observe, but even during such occasions all that can be gleaned is a small hint of its floor, and its western to eastern rim that usually becomes visible through shadows (especially just before full Moon occurs). Einstein A smacked dead-centre, and in doing so rebounded upward the floor, as well as adding ejecta deposits upon it (Einstein A is a good marker for finding Einstein itself). Depth is around ~ 1.85 km.

Craters

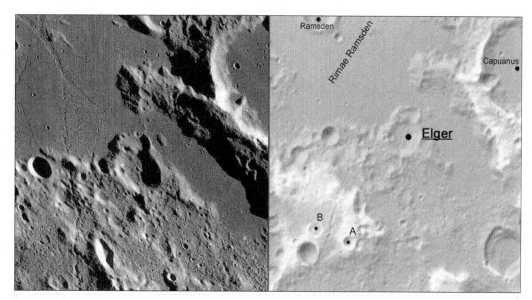

Elger	Lat 35.4S	Long 29.81W			21.51Km		
Sub-craters:	Crater	Lat	Long	Size (km)	Crater	Lat	Long	Size (km)
	A	37.37S	31.25W	8.11	B	37.12S	32.1W	8.8
	Notes:							

Add Info:

Aerial view of Elger

Take in a general perspective view of where Elger lies, and you'll see that two basins (Nubium to its northeast and Humorum to its northwest) fills the eyepiece. Each basin obviously produced impact ring/rim systems when they formed, which might then explain the isolated patch of lava at the crater's northern and eastern surrounds (connect also to Palus Epidemiarum). The crater hasn't much to offer in detail, and the only thing it has going for it is the odd, egg-shaped feature attached to its northern rim (what is it - does the small dome-like clump to its east also have something to do with it?). Depth ~ 1.3 km.

Craters

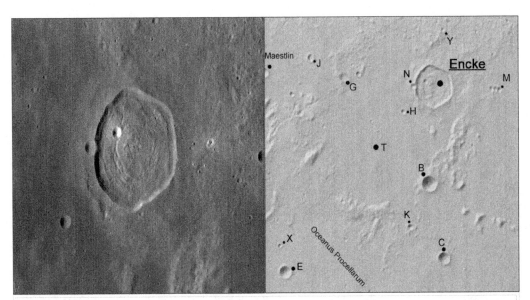

Encke		Lat 4.57N	Long 36.68W			28.27Km	

Sub-craters:	Crater	Lat	Long	Size (km)	Crater	Lat	Long	Size (km)
	B	2.36N	36.78W	10.89	C	0.66N	36.45W	8.72
	E	0.35N	40.2W	8.73	G	4.79N	38.79W	7.22
	H	4.0N	37.4W	3.51	J	5.24N	39.59W	5.13
	K	1.37N	37.24W	4.49	M	4.49N	35.12W	3.29
	N	4.59N	37.15W	3.88	T	3.23N	37.88W	94.68
	X	0.95N	40.33W	2.97	Y	5.86N	36.45W	2.88

Notes:

Add Info: When observing Encke, also take a look at its 'show-off' neighbour to the north, Kepler, as they sort of look the same. Each of their rims are sharp, have some crenulation to them, and while slumping is obvious in the latter, one has to look a bit harder to see such a feature in Encke (within the east side of the crater). But the floors of each is where the dividing line occurs - Kepler's looking more 'normal', while Encke's being completely different. This type of crater - with its concentric rilles, its upraised floor, are associated to some kind of volcanic activity underneath the crater after it formed. Perhaps, it is related to a weak zone created by the previous impact that was sub-crater T; as Encke has impacted on its (unseen) rim area. During low sun angle times, shadows produced from the rilles and hummocky hills in the floor show them up wonderfully, while high sun times highlights the bright surrounds of the fresh (~ 3km) crater in the NW sector. Note also sub-crater N at high magnification - its impact on Encke's rim must have blasted portions of it into space and onto the floor. Depth ~ 0.8 km.

Craters

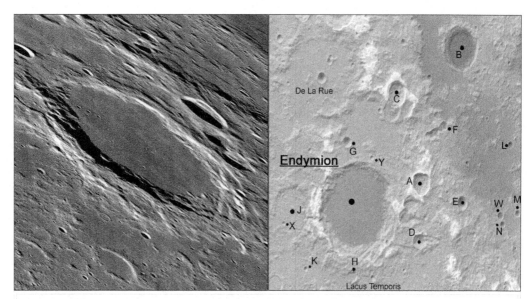

Endymion	Lat 53.61N	Long 56.48E	122.1Km

Sub-craters:

Crater	Lat	Long	Size (km)	Crater	Lat	Long	Size (km)
A	54.61N	62.56E	29.75	B	59.86N	67.75E	59.0
C	58.37N	60.55E	32.74	D	52.29N	62.15E	22.61
E	53.59N	66.24E	18.0	F	56.52N	63.6E	10.97
G	56.43N	55.59E	14.11	H	50.99N	56.15E	15.87
J	53.51N	50.71E	66.05	K	51.18N	52.2E	6.65
L	55.48N	71.39E	9.56	M	52.69N	71.08E	9.35
N	52.4N	69.63E	8.63	W	52.76N	69.54E	10.61
X	52.94N	50.03E	5.98	Y	55.78N	57.97E	6.57

Notes:

Add Info:

Aerial view of Endymion

Endymion looks, to all intents and purposes, like a 'skewed-view' version of crater Plato to its west. They've both got smooth floors filled with lavas, each shows off bright terracing all around, and either one is a challenge to the observer in counting how many small craterlets on their floors can be resolved in the eyepiece. They aren't, of course, the same, but their resultant look and formation must have had something to do with their locations near mares - Endymion/Mare Humboldtianum, and Plato/Mare Imbrium. Endymion has numerous more features (rays from crater Thales, ridges, crater chains...etc.,) than Plato. Observe them well! Depth ~ 4 km.

Craters

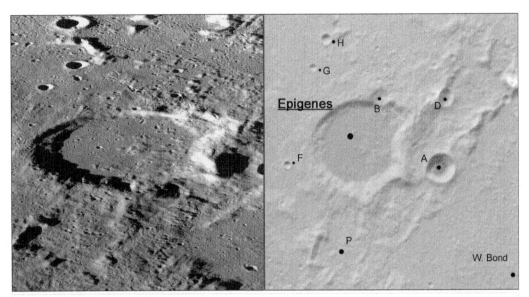

Epigenes	Lat 67.5N	Long 4.62W	54.51Km

Sub-craters:

Crater	Lat	Long	Size (km)	Crater	Lat	Long	Size (km)
A	67.04N	0.39W	17.57	B	68.45N	3.36W	11.27
D	68.35N	0.15E	9.79	F	67.11N	8.12W	4.6
G	68.97N	7.01W	4.67	H	69.53N	6.32W	6.79
P	65.53N	5.44W	38.39				

Notes:

Add Info:

Aerial view of Epigenes

Epigenes has lost nearly all its eastern rim portion due to an odd, double-clump of-a-mountain - concentric to crater W. Bond (not Bond's real rim). Ejecta from that feature certainly lies on Epigenes's floor as sparse hills, which subsequently have been covered by deposits from the Mare Imbrium impact to the SW. Peaks in its floor show up well at low sun angle times (are they original peaks or those of above-mentioned ejecta?), and note also the two rilles in the northern sector (the north-westerly one only shows up during an easterly sun aspect). Rays from crater Anaxagoras to north (~ 180 km away) cross its floor. Depth ~ 3 km.

Craters

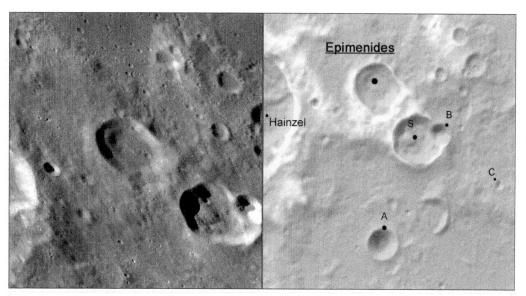

Epimenides   Lat 40.92S   Long 30.33W   22.56Km

Sub-craters:

Crater	Lat	Long	Size (km)	Crater	Lat	Long	Size (km)
A	43.27S	30.18W	14.9	B	41.62S	28.93W	10.22
C	42.34S	27.58W	4.27	S	41.71S	29.44W	24.09

Notes:

Add Info: What a 'messy' area does Epimenides (pronounced - 'Eppee-Men-Knee-Dees') lie in! Sometimes referred to as 'The Vitello Formation' - where ejecta from Mare Humorum (to the northwest) was the base on which Epinenides formed. The area has also seen secondaries from Mares' Imbrium (over 2000 km away to the north), Nubium (to the northeast), and Orientale (to the west). It's quite easy then to get lost here, however, good old Hainzel to the west with its odd shape stands as a good 'marker' and guide in to finding Epimenides. Having a depth of around 2.4 km, Epimenides's rim is well worn, its southeast sector seems to have undergone major slumping (or, are we looking at evidence of an oblique impact), and its smooth floor may consist of ejecta deposits from one of the above-mentioned major events. The terrace in its northwest is about the only feature that stands out during low sun angle times (easterly or westerly apparitions), but do try and catch hint of a central hill/mountain running NW/SE on the floor, too. Sub-crater A has been listed as one of the banded group - a crater exhibiting dark or light streaks of material (ejecta or dry debris flows??) on its inner walls, however, very high magnification is required to see them.

Craters

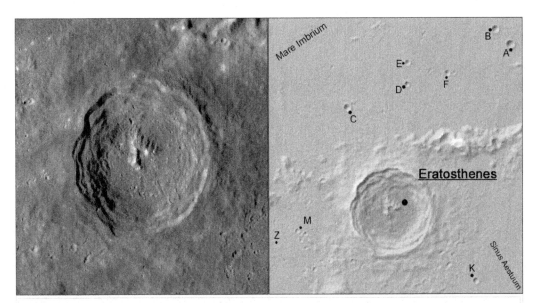

| Eratosthenes | Lat 14.47N | Long 11.32W | 58.77Km |

Sub-craters:

Crater	Lat	Long	Size (km)	Crater	Lat	Long	Size (km)
A	18.34N	8.33W	5.67	B	18.7N	8.7W	5.33
C	16.89N	12.39W	5.19	D	17.44N	10.9W	3.83
E	17.93N	10.89W	3.83	F	17.69N	9.91W	4.02
H	13.31N	12.25W	3.47	K	12.85N	9.26W	4.33
M	14.02N	13.59W	3.53	Z	13.75N	14.1W	0.57

Notes:

Add Info: Before you look at Eratosthenes, pull back your view in the eyepiece and try include crater Copernicus (over to the southwest). Eratosthenes looks like it is a 'pup' of Copernicus - both having similar features like wonderful terraces, peaks and crenulated, sharp-ish rims. Moreover, both are stratigraphic, geological unit markers in the lunar timescale - Eratosthenes (3.15 to 1.1 byo) and Copernicus (1.1 byo to the Present). Obviously, Copernicus is the younger (rays cover Eratosthenes), and while it most certainly steals away the limelight from Eratosthenes, the crater is well worth observing more. The crater formed just outside the main ring/rim system of Mare Imbrium, and its impact must have blown portions of Imbrium's ring/rim onto the Mare's southern floor, while at the same time mixing in its own ejecta deposits. Time has darkened the ejecta; which in its prime must have looked bright like Copernicus does today (at full Moon), but its effects, particularly during low sun views, makes for a wonderful display (better, it might be said, than Copernicus's does). The peaks cast long shadows too during these times, and try catch a view when the main peak (~1.2 km high) pops out brightly in a floor full of shadow. Depth of Eratosthenes is around 3.4 km approx..

Craters

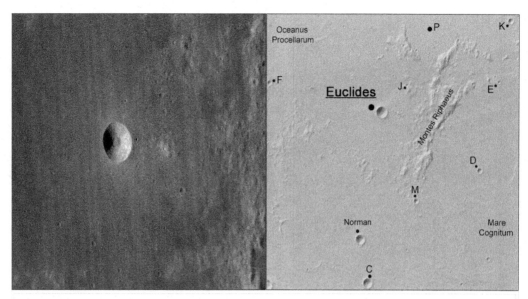

Euclides	Lat 7.4S	Long 29.56W	11.8Km

Sub-craters:

Crater	Lat	Long	Size (km)	Crater	Lat	Long	Size (km)
C	13.29S	30.08W	10.14	D	9.4S	25.81W	5.78
E	6.34S	25.13W	3.53	F	6.37S	33.73W	5.39
J	6.47S	28.59W	3.4	K	4.23S	24.72W	6.0
M	10.45S	28.23W	5.79	P	4.52S	27.69W	63.82

Notes:

Add Info:

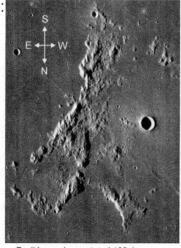

Euclides region - rotated 180 degrees

From the outset, it's clear to see that Euclides is a fresh, and young crater (Copernican - 1.1 byo to Present). Its bright ray material splashes out in every direction; reaching as far to Lansberg D to the northwest (~ 140 km away), and draping over, and beyond, Montes Riphaeus to the east. These rays 'shine' even under low sun apparitions, and it's presumed that the culprit responsible for their creation is due to Euclides being a secondary of crater Copernicus (some ~ 600 km away to the northeast). The crater itself hasn't very much to offer in terms of features, but its lofty rim around does cast some nice shadows at low sun times. Those of us using reflecting telescopes will see Euclides as the 'flashlight' in the running man's (Montes Riphaeus - see left) hand. Depth ~ 2.6 km.

Craters

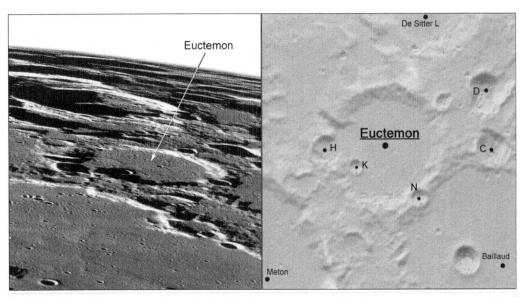

| Euctemon | Lat 76.26N | Long 30.57E | 62.7Km |

Sub-craters:

Crater	Lat	Long	Size (km)	Crater	Lat	Long	Size (km)
C	76.1N	38.68E	19.58	D	77.04N	38.7E	19.86
H	76.16N	26.21E	15.93	K	75.93N	28.43E	7.33
N	75.41N	32.95E	8.96				

Notes:

Add Info:

Aerial view of Euctemon

Though Euctemon looks as old as its neighbour, Baillaud (pre-Nectarian in age 4.6 to 3.92 byo) to the southeast, it's a hard call to say which formed first. All evidence of ejecta from each crater has been covered over by ejecta from the event that was formation of Imbrium Basin to the south. Though not very obvious from our earth-viewing perspective, the aerial view (left) shows how Imbrium's effect has also 'grooved' a portion of Euctemon's rim (see the features on several others nearby, too). The aerial view also shows a buried crater (~ 16 km in diameter north of sub-craters' Euctemon H and K). Depth of Euctemon 1. 6km.

Craters

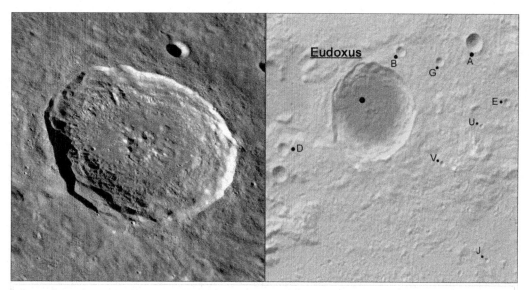

Eudoxus   Lat 44.27N   Long 16.23E   70.16Km

Sub-craters:

Crater	Lat	Long	Size (km)	Crater	Lat	Long	Size (km)
A	45.83N	20.09E	13.47	B	45.66N	17.35E	7.7
D	43.35N	13.22E	9.36	E	44.4N	21.17E	5.73
G	45.41N	18.78E	6.61	J	40.8N	20.29E	3.97
U	43.89N	20.27E	3.48	V	43.07N	18.81E	4.39

Notes:

Add Info: The first thing to notice, generally, from Eudoxus is that it looks very 'crusty-like'. This is a Copernican-aged (1.1 billions of years old to the Present) crater, so we would expect features like its rim, terraces, peaks etc., to be relatively sharp-ish - as they are. But why so 'crusty'? Is it due to how we observe the series of central, pity peaks in the floor that blend in seamlessly with other hilly lumps nearby? Is it due to how the lumpy material on the edges of the walls, which fills nearly a third of the crater overall, displays both concentric and radial features mixed - fractured and gooved, no doubt, due to the dynamics involved with initial impact? Take a look at crater Aristoteles (some 150 km away to the north), and you'll see these same latter features also exist. So, has it something to do with the material make-up - Imbrium ejecta - in which both craters landed? Whatever the answer, it's hard not to notice these features - especially during low sun times - as shadows that should appear continuous, look more mottled-like. Another odd feature to look out for is Eudoxus's western rim, where slumping has produced a very straight edge (almost 35 km in length) in the inner section, while the outer section appears sagged-like in its own ejecta - weird. Depth ~ 4.5 km approx..

Craters

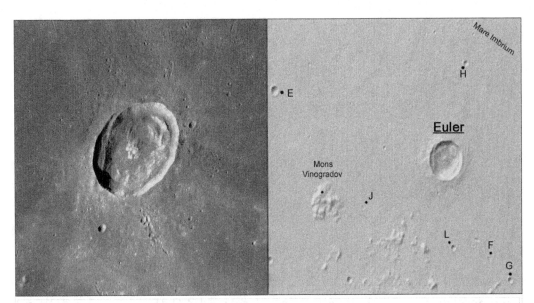

Euler	Lat 23.26N	Long 29.18W		26.03Km	

Sub-craters:

Crater	Lat	Long	Size (km)	Crater	Lat	Long	Size (km)
E	24.7N	34.07W	6.1	F	21.14N	27.96W	5.27
G	20.68N	27.39W	4.1	H	25.33N	28.57W	3.93
J	22.25N	31.51W	3.71	L	21.4N	28.97W	3.91

Notes:

Add Info: Euler is a crater of the Erastosthenian Period (3.15 to 1.1 billions of years old). Rays from crater Copernicus overlay it, but as to why there is a predominance of rays in the outer southeast portion of the crater is a puzzle. Did the ray material from Copernicus in landing disturb or expose lighter material underneath? Certainly, while it looks like slumped deposits cover about a third of Euler's floor in its eastern and southern sectors, the events have exposed a bright, ribbon-like wall of material all around. So, maybe, that might explain why the southeast is as we see it - Euler's own impact landed in material made up of a lighter constituent. Lava-flooding in the outer northwest sectors of the crater may have buried the brighter ejecta there (evidence of flooding is also seen within the floor of Euler), so would that be an answer to the 'puzzle' posed? Hmmm! Euler's rim is relatively sharp-ish all around, so formation of the crater may be on, if not close to, the border between the Eratosthenian and Copernican; making it a recent impact in geological terms. The crater has a depth of around 2.3 km, its peak rises to about 1.5 km in height (shows up nicely at low sun times). Images taken by cameras onboard the Lunar Reconnaissance Orbiter showed its crater walls exposed several layers of Imbrium lava flows. Pronounce it as 'Oil-er'.

Craters

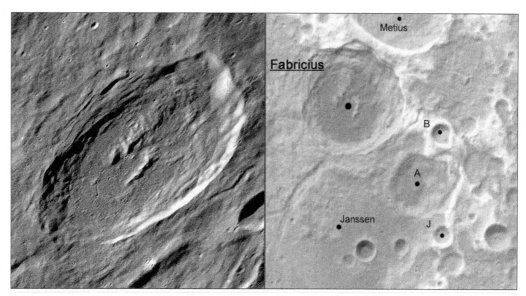

Fabricius Lat 42.75S Long 41.84E   78.9Km

Sub-craters:

Crater	Lat	Long	Size (km)	Crater	Lat	Long	Size (km)
A	44.64S	44.4E	55.02	B	43.56S	44.85E	16.55
J	45.83S	45.04E	15.59				

Notes:

Add Info: Fabricius lies in a region of the Moon known as the Janssen Formation; which is ejecta from the Nectaris Basin to its north (~ 850 km away). Eratosthenian in age (3.15 to 1.1 byo), Fabricius as it formed has just about clipped off its neighbour's southwestern rim (Metius - of Nectarian), and firmly landed within the crater Janssen (pre-Nectarian in age) - in which the formation is named after. The crater has a relatively sharp-ish rim all around, has a wonderful series of terraces, and has two main peak regions; whose northern-most one extends right in to slumped deposits in the northeast wall. Look carefully also at both the northwest and south sectors of the crater here, as its wall areas have extended into the outer terrain. Are these features a consequence in the dynamics of Fabricius's impact; which struck higher terrain in the northeast (in effect, Metius's southwestern rim) - sending shockwaves in a preferred (southwestwards) direction? The floor of Fabricius shouldn't be over-looked either as it contains several small concentric rilles interspersed amongst hummocky hills, where deposits and secondaries from the Imbrium Basin event lie, too. A good marker for finding the crater is noting Janssen firstly, and from there, it's easy-peasy. Depth of crater is around 2.5 km while main peaks are ~1.8 km approx..

Day-old moon phases

Note: As the Moon goes through various librations throughout the year, suggested times given in text for observations are only approximates.

Craters

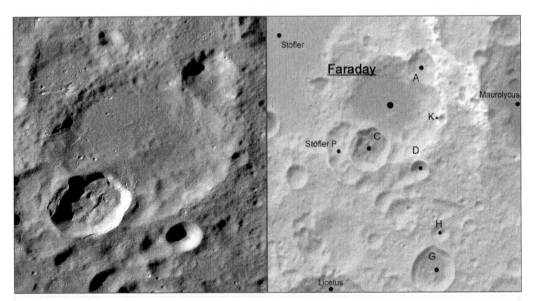

Faraday	Lat 42.45S	Long 8.75E				69.03Km	

Sub-craters:

Crater	Lat	Long	Size (km)	Crater	Lat	Long	Size (km)
A	41.54S	9.65E	19.99	C	43.32S	8.04E	28.49
D	43.78S	9.61E	13.47	G	45.92S	10.06E	29.62
H	45.12S	10.24E	11.91	K	42.68S	10.3E	6.71

Notes:

Add Info: We're in confusion territory again when looking for Faraday; as so many craters of similar size (across a range of ages) in where it lies can lead us to get easily lost. The only good marker around full moon times is bright-rayed Tycho to its west, where you then hop your way, carefully, eastwards from it to about five times its diameter away. But try the same marker method when Faraday and its jungle of craters lie in low sun views, then it's a different ball-game altogether (would Stöfler's large smooth floor to its northwest be a good marker?). Having found the crater, Faraday seems to confuse us more where, firstly, impact of sub-crater's A and C (and also Stöfler P on which C lies on to its west) have wiped away nearly half its rim, and, secondly, its own northwestern rim is comparatively smoother than its opposite southeastern one. Look closely at the former rim towards the northwest and you'll see a big clump (mountain) of material on Stöfler's floor. Is this Stöfler's own missing southeastern section thrown there by impact of Faraday (it isn't a Stöfler peak, nor the rim of another crater)? There are other similar features - ejecta clumps - like this to look out for in Faraday's floor, too (e.g. from A and C), and is that a hint of a peak at its centre? The area is covered in ejecta from Imbrium, and rays also from Tycho. Depth ~ 4km.

Craters

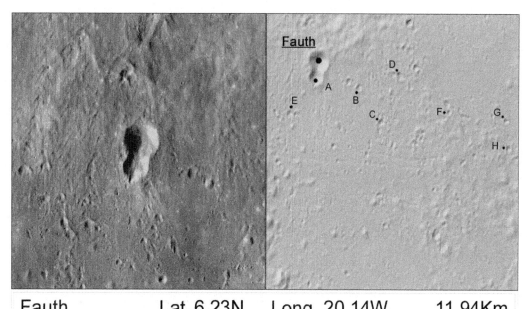

Fauth	Lat 6.23N	Long 20.14W	11.94Km

Sub-craters:

Crater	Lat	Long	Size (km)	Crater	Lat	Long	Size (km)
A	5.98N	20.19W	8.72	B	5.8N	19.32W	3.46
C	5.23N	18.85W	3.89	D	6.0N	18.46W	3.79
E	5.39N	20.72W	2.53	F	5.4N	17.45W	2.45
G	5.25N	16.27W	3.01	H	4.78N	16.22W	3.76

Notes:

Add Info:

Apollo 12, Luna 5, and Surveyor 3 all landed some 250 km away to the south of this crater

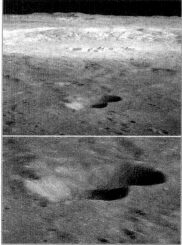

Apollo 12 orbital image - AS12-52-7739

Lying eastwards of Mare Insularum (pre-Nec.), and in a terrain covered with deposits and numerous secondary small craters from Copernicus to the north, Fauth, together with sub-crater A, is probably the real 'keyhole' crater on the Moon. It really doesn't have much to offer in terms of features - even at high magnification, but both Fauth and A (see left) are of interest when trying to determine which formed first. A's rim looks a bit sharper than Fauth's, so maybe it is younger. However, as no deposits from each are seen in either floors, perhaps, we are looking at a double-type impact event. During very low sun times the raised rims of each cast some nice shadows that defines the keyhole effect, and Fauth's NW inner rim suggest slumping. Depth ~ 2 km.

Craters

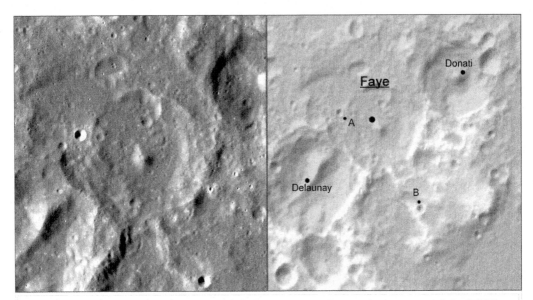

Faye	Lat 21.39S	Long 3.81E			38.02Km	

Sub-craters:	Crater	Lat	Long	Size (km)	Crater	Lat	Long	Size (km)
	A	21.2S	3.12E	3.55	B	22.6S	4.5E	3.87

Notes:

Add Info: Together, with its similar-looking neighbour, crater Donati, to the northeast, the two might, vulgarly, be referred to as the 'pair of nipples on the Moon'. Faye's nipple is slightly smaller but sharper to Donati's one, as with the rims, too (in those that we can discern), so Faye is probably younger. Portions of Faye's northwest rim is lost amongst a smooth plain of material (this whole area has been inundated with ejecta from the numerous craters around - not to mention from basins), while its southwest rim is just about observable - slightly lost upon the rim of Delaunay. And as we're at Delaunay, note the NE-SW-running ridge in the crater where its formation may be due to the dynamics of Faye's impact (is it a dump-load, or mixture, of both Delaunay's rim and Faye's ejecta?). The floor of Faye has a few small, but old craterlets each no larger than 2 kilometres in size, while sub-crater A is a more recent addition where it impacted just on the outer rim of Faye, and imparted some of its brighter material inwards. The peak (yes, the 'nipple') in Faye is about 1.5 km high, and at higher magnification a small impact crater (less than a kilometre in diameter) can be seen to atop it. This crater's formation is probably just the chance impact of an object hitting the peak's top, and not the vent of a volcano of which it might suggest. The nipples of both craters act as good markers for finding them. Depth of Faye ~ 2.7 km approx..

Craters

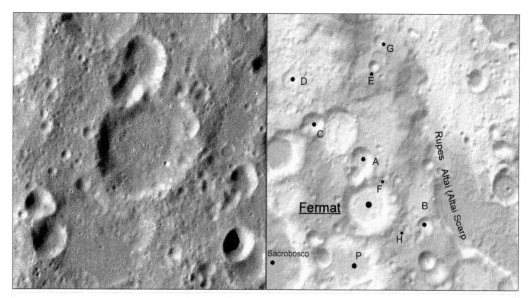

Fermat	Lat 22.71S	Long 19.79E	37.77Km

Sub-craters:

Crater	Lat	Long	Size (km)	Crater	Lat	Long	Size (km)
A	21.83S	19.6E	17.49	B	23.08S	21.09E	10.62
C	21.1S	18.47E	14.68	D	20.22S	17.98E	12.4
E	19.95S	19.87E	6.88	F	22.14S	20.14E	4.61
G	19.44S	19.99E	7.01	H	23.15S	20.66E	5.04
P	23.68S	19.44E	34.14				

Notes:

Add Info: Lying in the clumpy ring-region of Nectaris (its main ring/rim ~ 860 km in diameter) created by that event, Fermat, with its well worn appearance is probably of the pre-Nectarian age (4.6 to 3.92 billions of years old). Like most craters in this region (e.g. Zagut, Rabbi Levi to the south), Fermat is covered in ejecta from Nectaris, and later covered by more from Imbrium - 1900 kilometres away to the northwest. The floor really hasn't much to offer in terms of featues except for a few fresh craterlet impacts, and a small, barely perceptible, crater chain (catena) lined up to Fermat's northwest wall. Several impact craters dot its rim all around, and while the most northern one (that crater connected to sub-crater A) has just clipped the sector of Fermat's outer rim, its effect hasn't added any significant amount of material onto the floor area there (rather, it looks as if it just pushed Fermat's rim in a little). Look also at the 'thickness' of the rim, in ge- neral, for Fermat - the southeastern sector is almost twice as thick as that of the northwestern sector. Is this a consequence of the southwest sector being over a kilometre higher than the northwest, or, one of impactor direction? Depth ~ 2.0 km.

Craters

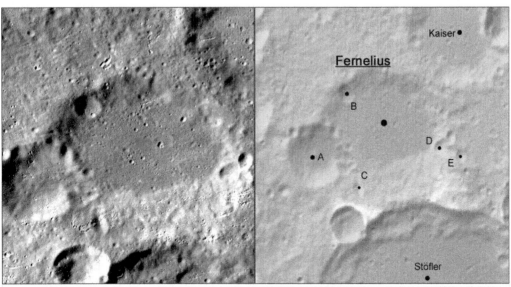

Fernelius   Lat 38.18S   Long 4.86E   68.42Km

Sub-craters:

Crater	Lat	Long	Size (km)	Crater	Lat	Long	Size (km)
A	38.4S	3.48E	29.63	B	37.45S	4.12E	9.12
C	38.94S	4.37E	6.65	D	38.28S	6.16E	7.72
E	38.4S	6.57E	5.57				

Notes:

Add Info: Fernelius, together with Fernelius A, looks, to all intents and purposes, like a 'ladybird'. The head (sub-crater A) is eating its daily diet of highland's material, while the body (Fernelius itself) - with its single dot (sub-crater B) on its back - crawls lazily behind. The body, essentialy the floor, has no legs, but looks more like a big 'D' on its side that has been filled with debris ejecta from the Imbrium Basin (to the northwest ~ 1200 km away). This D shape would have a more circular shape to it if it wasn't for the rubbly material to its south that fills the floor there, which from the looks of it might be ejecta-spill from Stöfler nearby. Stöfler is of the pre-Nectarian age (4.6 to 3.92 billions of years old), and so puts a constraint on the age of Fernelius as being older. However, taking into account the state of each of their rims respectively, the former would suggest it's formation is closer to the 3.92 byo mark, while the latter isn't far off from 4.6 byo. All that theorising aside, Fernelius is a battered crater, as nearly every section of its rim has been struck (particularly, sub-crater A, which is younger - note the intersection between the two). Modern impact events also show up, where debris from formation of Tycho to the east has 'sprayed' Fernelius's surrounds - creating a wonderful series of tiny crater chains and clusters. Depth of Fernelius ~ 3.3 km.

Craters

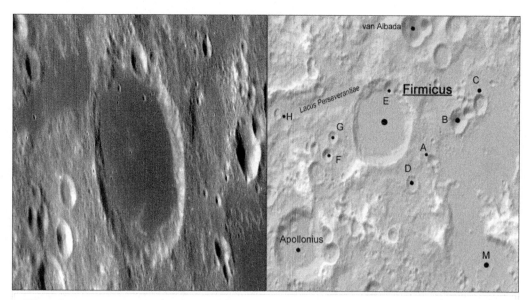

Firmicus	Lat 7.25N	Long 63.43E	56.81Km

Sub-craters:

Crater	Lat	Long	Size (km)	Crater	Lat	Long	Size (km)
A	6.44N	64.96E	8.37	B	7.31N	65.77E	17.23
C	7.71N	66.46E	12.95	D	5.9N	64.34E	11.48
E	8.08N	63.61E	9.08	F	6.53N	61.76E	9.24
G	6.93N	61.9E	9.11	H	7.44N	60.2E	7.83
M	4.08N	66.65E	44.25				

Notes:

Add Info:

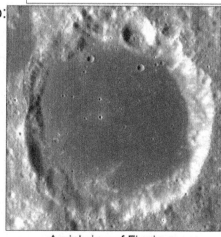

Aerial view of Firmicus

Firmicus lies within the main 740 km-diam., ring/rim of the Crisium Basin to the north. The effects of that giant impact fractured the rock in which Firmicus impacted upon; allowing lava to fill the crater from below, and thus burying any previous features - like peaks, craters, rilles etc., that might have existed on the original, true floor. Numerous small craters 'dot' the floor, and while the two northern ones and mid-centre ones are easily observed - a challenge for the observer is to photographically capture many of the less smaller ones. Take a 'bigger picture' also of this area to view the extent of Crisium's impact effects. D: 2.0 km.

Craters

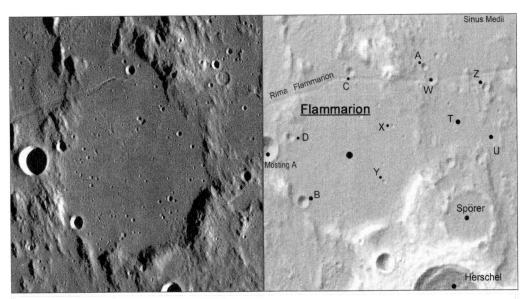

Flammarion	Lat 3.33S	Long 3.73W	76.18Km

Sub-craters:

Crater	Lat	Long	Size (km)	Crater	Lat	Long	Size (km)
A	1.95S	2.5W	3.58	B	4.04S	4.57W	5.88
C	2.02S	3.76W	4.29	D	3.03S	4.78W	4.48
T	2.8S	2.06W	33.11	U	3.0S	1.41W	11.38
W	2.13S	2.39W	6.13	X	2.87S	3.04W	2.38
Y	3.73S	3.19W	2.3	Z	2.26S	1.47W	4.01

Notes:

Add Info: Flammarion is a crater of the Nectarian Period (3.92 to 3.85 billions of years old). It lies just on the edge of a series of extensive lava flows stretching from Sinus Medii in the northeast, right down to Mare Nubium in the south - some 700 km away. These series of lavas flowed as thin sheets that meandered their way in and around hilly to low spots, and it looks like Flammarion didn't escape them - particularly northwards where breaches occurred - giving the crater the smooth floor we see today. Effects in formation of the Imbrium Basin to the northwest have left their mark on the crater's rim all around (note the striated appearance on the rim that point towards the Basin area), and fluidized ejecta from that event covers the floor, too (known as the Cayley Formation). Clusters of small craters pepper its interior, while Mösting A (outside its western rim) together with crater Lalande to its west (~ 150 km away) have both imparted ray material onto the floor (note these particularly at high sun times). Rima Flammarion seems to cut across a portion of the northwest rim, so it probably formed later, but note how its length extends to the east where it joins Rima Oppolzer.

Surveyors' 4 (failed before touchdown) and 6 landed some ~ 125 km away to the northeast of this crater.

Craters

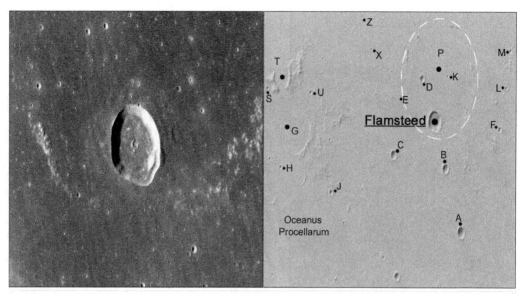

Flamsteed		Lat 4.49S		Long 44.34W		19.34Km	
Sub-craters:	Crater	Lat	Long	Size (km)	Crater	Lat	Long	Size (km)
	A	7.88S	43.0W	11.39	B	5.93S	43.81W	9.29
	C	5.52S	46.33W	9.1	D	3.19S	44.95W	6.65
	E	3.71S	46.1W	2.08	F	4.73S	41.15W	5.53
	G	4.74S	51.05W	42.9	H	5.88S	51.77W	4.29
	J	6.67S	49.33W	5.21	K	3.11S	43.73W	3.4
	L	3.46S	40.87W	3.76	M	2.33S	40.68W	4.52
	P	3.25S	44.25W	112.24	S	3.47S	52.39W	3.61
	T	3.14S	51.72W	23.1	U	3.61S	50.31W	4.62
	X	2.3S	47.4W	2.82	Z	1.3S	47.85W	2.73

Notes:

Add Info:

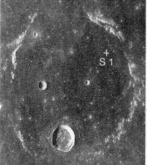

S1 marks the approximate location of Surveyor 1, which successfully landed on the Moon on 2 June 1966.

Flamsteed and sub-crater P

It's quite easy, and fast, to find and view Flamsteed in the eyepiece - simply use the much larger sub-crater, Flamsteed P, that cradles it. Eratosthenain in age (3.15 to 1.1 byo), it lies in Hi-Ti (High Titanium) basaltic lavas that flooded the region; which lapped at Flamsteed's outer rim and ejecta, but didn't quite spill over into the crater's interior. The floor area looks too flat, so perhaps some lavas seeped underneath - partially covering the wonderfully-looking ringed peak at centre. Features: slumped terraces in SE sector; layering in N sector; cuts a concentric ridge to P (low sun view); and note light and dark lavas in P. Depth 2.1 km.

Craters

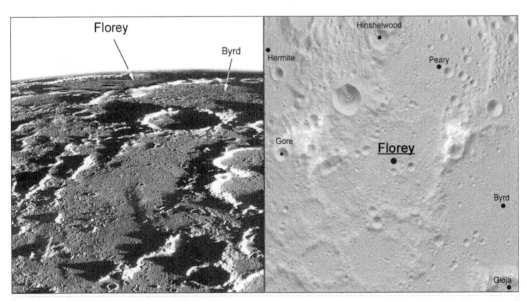

Florey	Lat 87.04N Long 19.75W	69.06Km

Sub-craters: No sub-craters

Notes:

Add Info:

Aerial view of Florey

Florey's floor level is some 200 metres above that of Byrd's to its southeast, but as it also has a tilt of similar amount towards Peary to its northeast, the crater really doesn't get that much of a 'look-in'. In effect, Florey's perspective becomes even more pronounced for observing any descent detail of its interior, and best we see is the odd lump just south of the two small craters in its centre (see left), and its rim in the northern sectors of the crater (Byrd's northwestern rim at over 2 km in parts doesn't help either). Florey, as with Byrd and Peary, are of the pre-Nectarian Period (4.6 to 3.92 byo). Depth ~ 1.8 km.

Craters

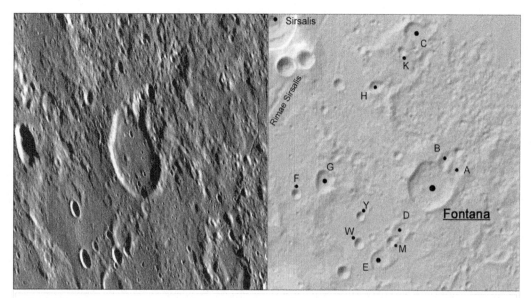

Fontana	Lat 16.04S	Long 56.79W	31.47Km

Sub-craters:

Crater	Lat	Long	Size (km)	Crater	Lat	Long	Size (km)
A	15.71S	56.17W	11.14	B	15.55S	56.46W	8.46
C	12.86S	57.19W	13.45	D	17.0S	57.49W	11.24
E	17.61S	57.98W	13.11	F	16.21S	59.95W	6.64
G	15.96S	59.29W	16.11	H	13.99S	58.07W	8.84
K	13.26S	57.45W	7.12	M	17.21S	57.71W	6.36
W	17.25S	58.47W	7.17	Y	16.68S	58.43W	6.06

Notes:

Add Info:

Aerial view of Fontana

The 'nick' just off the northeast sector of crater Fontana serves as a good marker for finding it easily. This nick looks like it's made up of several small impact craters, which at low sun times produces a slight 'valley-like' feature onto the floor. Several small craters, a rille, and a mound 'dot' the crater's floor (with evidence of slumping in the northwest), but it looks like all these have been covered by ejecta - from the various basins (Humorum, Imbrium and Orientale etc.,) around. Note also the 'crumbly-like' texture northeast of Fontana to the smoother look in the southwest (a consequence of Oceanus Procellarum's formation?). Depth 1.1 km.

Craters

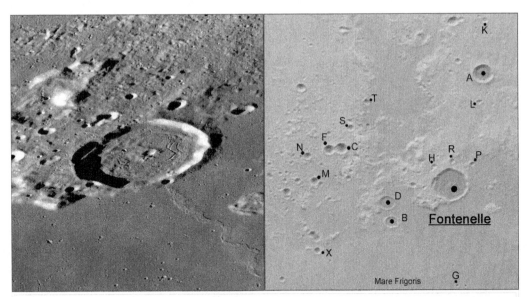

Fontenelle Lat 63.42N Long 18.96W 37.68Km

Sub-craters:

Crater	Lat	Long	Size (km)	Crater	Lat	Long	Size (km)
A	67.61N	16.15W	21.52	B	61.91N	22.99W	13.16
C	64.52N	27.32W	12.96	D	62.64N	23.44W	16.15
F	64.41N	28.21W	11.07	G	59.57N	18.37W	4.13
H	64.1N	20.15W	6.11	K	69.66N	15.77W	6.72
L	66.54N	16.62W	5.15	M	63.15N	28.86W	9.15
N	64.06N	29.75W	8.3	P	64.2N	17.26W	6.37
R	64.32N	18.77W	4.91	S	65.32N	26.85W	7.53
T	66.38N	25.88W	6.59	X	60.57N	27.81W	7.47

Notes:

Add Info:

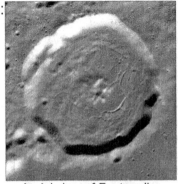

Aerial view of Fontenelle

Fontenelle is wonderfully situated when considered in terrain textures. Northwards of the crater the terrain is rubbly-looking, which has been additionally covered by ejecta from Imbrium to the south, while southwards the smooth lavas of Frigoris lap at its rim (the crater is aged in the Lower Imbrian - 3.85 to 3.75 byo, while the lavas may be around 3.6 to 3.4 byo). The crater cuts a ridge (possibly due to settlement of the mare) to the southeast, its interior has a series of stress (tectonic?) fractures close to its central peak, and slumping of the northwest inner wall has occurred. Depth ~ 1.5 km.

Craters

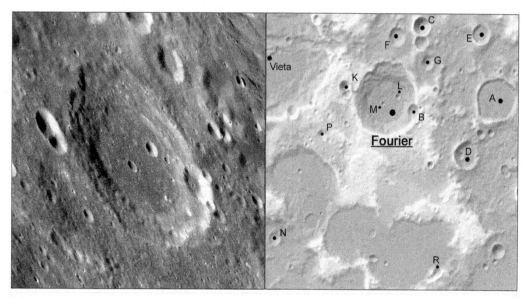

Fourier	Lat 30.31S	Long 53.1W	51.57Km

Sub-craters:

Crater	Lat	Long	Size (km)	Crater	Lat	Long	Size (km)
A	30.2S	49.67W	32.47	B	30.54S	52.21W	10.96
C	28.56S	52.04W	13.81	D	31.53S	50.56W	20.18
E	28.7S	50.23W	14.13	F	28.82S	52.79W	14.43
G	29.4S	51.84W	10.49	K	30.04S	54.35W	13.14
L	30.2S	52.74W	4.52	M	30.37S	53.2W	3.62
N	33.47S	56.58W	9.74	P	31.12S	55.07W	9.62
R	34.25S	51.38W	7.81				

Notes:

Add Info:

ESA's spacecraft - SMART-1, which was intentionally crashed on to the lunar surface, lies some 250 km away to the southwest of this crater.

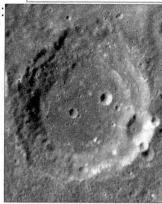

Aerial view of Fourier

Fourier lies on slightly higher terrain just outside one of the rings/rims (~ 800 km in diameter) produced by formation of Mare Humorum to its northeast (centred ~ 400 km away). Like its larger neighbour, Vieta to the northwest, both look almost the same: with levelled floors filled with ejecta deposits from Orientale (to the east ~ 1100 km away); 'slumpy' and 'bumpy' terraces; and numerous small secondary craters. Orientale's ejecta has also produced a 'gash' across Fourier's northern terraces (observe it around a 12-day-old moon period), but it isn't as obvious as the other 'gash' seen on crater Vieta's floor and rim, to its west. Sub-crater, Fourier C, during full moon times, acts as a good marker for finding Fourier.

Craters

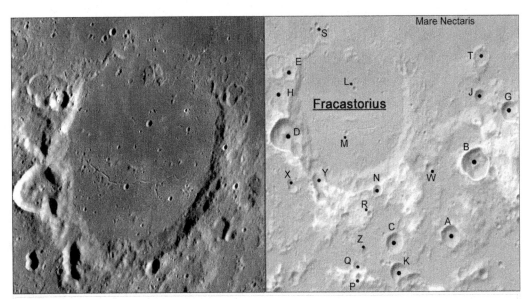

Fracastorius   Lat 21.36S   Long 33.07E   120.58Km

Sub-craters:

Crater	Lat	Long	Size (km)	Crater	Lat	Long	Size (km)
A	24.43S	36.44E	17.39	B	22.59S	37.25E	25.04
C	24.64S	34.53E	16.52	D	21.82S	30.87E	26.52
E	20.24S	31.02E	12.56	G	21.2S	38.35E	15.79
H	20.78S	30.57E	20.39	J	20.85S	37.38E	11.6
K	25.4S	34.68E	16.77	L	20.67S	33.21E	4.23
M	21.81S	32.92E	3.61	N	23.28S	33.96E	9.03
P	25.51S	33.34E	7.29	Q	25.21S	33.23E	7.81
R	23.88S	33.68E	5.23	S	19.03S	31.97E	4.47
T	19.85S	37.37E	13.73	W	22.7S	35.69E	6.65
X	23.06S	31.02E	6.87	Y	23.01S	31.93E	12.63
Z	24.89S	33.55E	10.08				

Add Info: It's believed that the reason why Fracastorius looks the way it does is because as the central part of the Nectaris Basin sank down under weight of lavas filling the interior (through fissures initially produced by that impact), the portion of the crater closer to the centre became tilted slightly in that direction. Looking at other craters close to, or on the extreme edges of other major impact regions, e.g. Letronne near Oceanus Procellarum, Sinus Iridum near Imbrium, and Le Monnier near Tranquillitatis...etc., the above reason may be the case indeed. As Fracastorius lies at Nectaris's ~ 400 km diameter ring/rim, additional lavas may have seeped up through other fractures there (note the main rille that follows the ring/rim's contour in a slightly raised area). Other features: note the change in colour of lavas north of the main rille (they are younger), as well as the crater count, and hint of a peak, too. Rays from Tycho cross the NW portion of the crater, but what is that large clump just within its northern rim section? Depth 1.7 km approx..

Craters

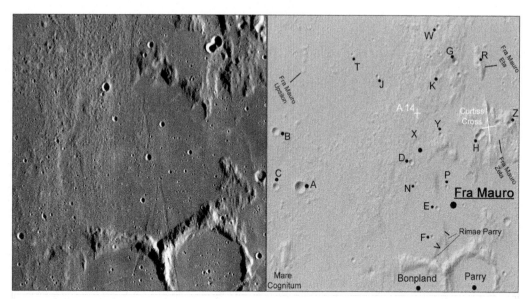

Fra Mauro   Lat 6.06S   Long 16.97W   96.76Km

Sub-craters:

Crater	Lat	Long	Size (km)	Crater	Lat	Long	Size (km)
A	5.47S	20.96W	9.29	B	4.03S	21.7W	6.96
C	5.45S	21.69W	5.92	D	4.82S	17.62W	4.46
E	6.0S	16.83W	3.42	F	6.73S	16.97W	3.07
G	2.21S	16.33W	5.67	H	4.15S	15.58W	5.61
J	2.62S	18.65W	3.28	K	2.54S	16.8W	6.08
N	5.35S	17.46W	3.18	P	5.44S	16.54W	2.95
R	2.27S	15.6W	3.8	T	2.06S	19.41W	2.5
W	1.33S	16.83W	3.7	X	4.47S	17.33W	19.65
Y	4.11S	16.74W	3.45	Z	3.81S	14.66W	4.75

Add Info:

Rimae Parry I crosses the floor of Fra Mauro, and is obviously younger (note how it 'cuts' Parry's rim, too). An optical phenomenon known as the 'Curtiss Cross' (see location in topography map above) occurs at the terrain north of this crater (around sunset terminator times).

Notes:

Mention Fra Mauro and one immediately thinks of the Apollo 14 mission. Sent to this area of the Moon particularly to retrieve samples of ejecta of the Imbrium Basin (Fra Mauro Formation) that covers the region in general, the 45 kg returned showed the impact occurred around 3.9 billions of years ago. Most of the effects of this material can be seen to cover nearly all of Fra Mauro's floor (with exception to the northeast), where additional signatures are seen to the north and south of the crater - extending like a long finger right down to the crater Guericke (~ 170 km away). Only a small section of Fra Mauro's rim exist today in the northeast. Lavas through a 'gap' in the east have wiped out nearly all signs of the rim there, while southwards both Bonpland and Parry spared no pity either (note how the lava has covered partially these crater's ejecta). Depth ~ 0.8 km.

Craters

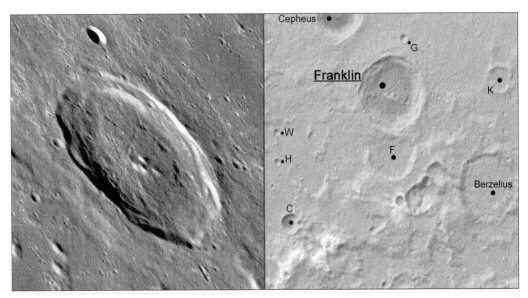

Franklin	Lat 38.73N	Long 47.64E	55.92Km

Sub-craters:

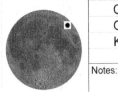

Crater	Lat	Long	Size (km)	Crater	Lat	Long	Size (km)
C	35.68N	44.26E	15.09	F	37.39N	47.69E	39.61
G	40.09N	48.12E	6.66	H	37.12N	43.74E	5.15
K	39.05N	51.48E	19.84	W	37.81N	43.69E	4.91

Notes:

Add Info:

Aerial view of Franklin

Franklin looks like a smaller version of crater Atlas to its northwest (~ 250 km away). Both have an over-extended series of terraces, each has peaks, and tectonic rilles run across their floors. Atlas is of the Upper Imbrian Period - (3.75 to 3.2 byo), but as Franklin's similar features look more worn-like than its neighbour, it must be older, but not by that much. Features: note the curious group of what looks like craters (or are they signatures of surface collapse) that 'mottle' the east and northwest portions of the crater's inner area, and the even spread of its own ejecta all around its exterior (best observed during low sun angle times). Depth 3.8 km.

Craters

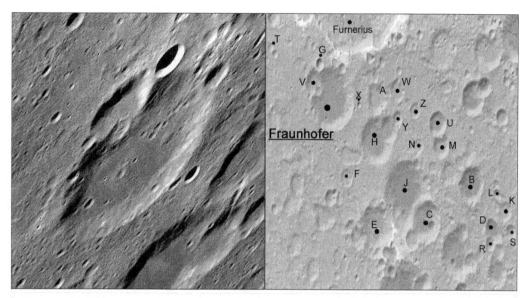

Fraunhofer   Lat 39.52S   Long 59.06E   57.75Km

Sub-craters:

Crater	Lat	Long	Size (km)	Crater	Lat	Long	Size (km)
A	39.58S	61.61E	29.17	B	41.93S	67.31E	33.08
C	43.02S	64.63E	36.66	D	43.16S	68.78E	24.99
E	43.2S	61.57E	44.73	F	41.76S	59.77E	15.92
G	38.51S	58.38E	10.61	H	40.76S	61.57E	42.77
J	42.31S	63.41E	62.94	K	42.56S	69.42E	17.2
L	42.1S	68.88E	9.0	M	40.89S	65.41E	22.41
N	40.91S	64.07E	12.59	R	43.51S	68.63E	12.52
S	43.11S	69.83E	13.28	T	37.89S	55.64E	8.12
U	40.27S	65.17E	22.84	V	39.0S	58.02E	23.79
W	39.36S	62.77E	19.45	X	39.72S	60.57E	5.75
Y	40.13S	62.81E	13.23	Z	39.92S	63.84E	14.1

Notes:

Add Info:

Aerial view of Fraunhofer

Fraunhofer is a crater of the pre-Nectarian Period - 4.6 to 3.92 byo). There's not a portion of this crater's rim and inner walls that hasn't escaped been struck by some impactor or other - just look at sub-crater V and the mess of smaller impacts it encountered. The floor of Fraunhofer is mostly smooth and dark-like to suggest some lava-filling from underneath, however, ejecta from the Nectaris Basin (~ 1000 km away to the northwest) has covered this region, in general. Features on floor: concentric-like crater (~ 3 km in diameter) on a hill (a peak?), and mound at the northeast. Depth ~ 2.2 km.

Craters

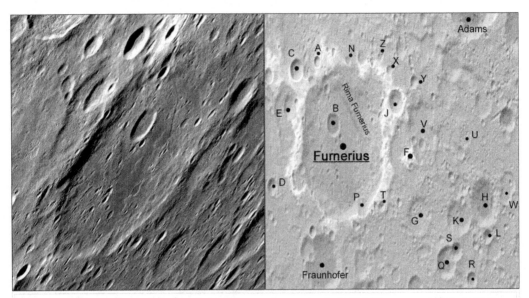

Furnerius	Lat 36.0S	Long 60.54E	135.03Km

Sub-craters:

Crater	Lat	Long	Size (km)	Crater	Lat	Long	Size (km)
A	33.54S	59.03E	11.21	B	35.44S	59.94E	21.75
C	33.74S	57.72E	21.17	D	37.06S	55.87E	16.39
E	34.78S	57.24E	22.27	F	36.1S	64.12E	52.06
G	38.14S	65.36E	33.72	H	37.59S	69.62E	44.01
J	34.84S	63.88E	26.8	K	38.12S	67.95E	36.19
L	38.6S	69.9E	13.98	N	33.57S	60.94E	9.97
P	37.92S	61.7E	17.56	Q	39.49S	67.24E	30.64
R	39.94S	68.97E	15.75	S	39.13S	67.91E	16.96
T	37.88S	62.93E	10.19	U	35.75S	68.29E	17.5
V	35.72S	65.5E	59.18	W	37.3S	70.99E	19.96
X	33.88S	63.46E	8.84	Y	34.2S	65.01E	11.78
Z	33.49S	62.94E	8.7				

Notes:

Add Info:

The Japanese spacecraft, Hiten (Muses-A) crashed some 140 km west of this crater, on the 10 April 1993.

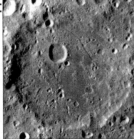

Aerial view of Furnerius

Like Fraunhofer crater to its southwest, Furnerius lies in a region where ejecta from the Nectaris Basin has covered it in some fashion or other. The floor area looks a right mess where patches of clumpy material 'melds' with what looks like lava deposits that may have seeped to the surface through fissures created from the initial impact event (main rille a possible source?). Sub-crater, Furnerius A, has imparted some bright ray material onto Furnerius's NW rim, but most went northwards due to it being an oblique impact. Depth ~ 3.9 km.

Craters

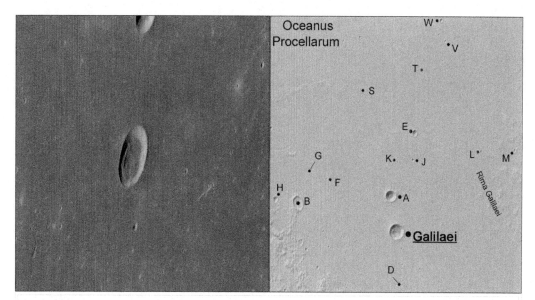

Galilaei		Lat 10.48N	Long 62.83W			15.99Km	

Sub-craters:	Crater	Lat	Long	Size (km)	Crater	Lat	Long	Size (km)
	A	11.69N	63.05W	11.17	B	11.43N	67.72W	15.91
	D	8.75N	62.75W	0.77	E	13.91N	61.94W	7.12
	F	12.29N	66.31W	2.87	G	12.65N	67.26W	0.81
	H	11.55N	68.81W	6.47	J	12.97N	62.04W	3.6
	K	12.93N	62.79W	2.65	L	13.25N	58.61W	3.37
	M	13.28N	56.93W	3.24	S	15.46N	64.77W	2.72
	T	16.16N	61.47W	1.58	V	17.11N	60.4W	2.72
	W	17.83N	60.61W	4.37				

Notes:

Add Info:

The word 'Galilaei', or more often written as 'Galilei', means 'of Galilei'. So, Galileo Galilei, reads as "Galileo, son of Galilei".

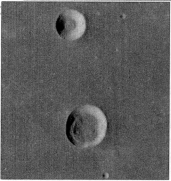

Aerial view of Galilaei (bottom)

You would think that as Galileo - sometimes referred to as the 'father of modern astronomy' (let alone also of physics and science) - that a much larger, more prominent crater would have been assigned to the man in honour of his contributions. But, alas, that wasn't to be the case, as seventeenth century politics, if not religiosity, too, got in the way. The crater itself really hasn't much to offer in terms of features, except at low sun times when its surrounding ejecta clashes wonderfully with several N-S ridges. The western, inner wall has slumped down, where it meets an ejecta-filled (Imbrium?) floor.

Craters

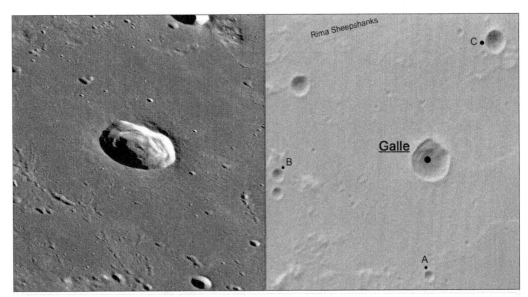

Galle	Lat 55.87N	Long 22.33E	20.96Km

Sub-craters:

Crater	Lat	Long	Size (km)	Crater	Lat	Long	Size (km)
A	53.99N	22.27E	5.77	B	55.56N	17.54E	7.01
C	57.8N	24.51E	11.34				

Notes:

Add Info:
Day-old moon phases

Note: As the Moon goes through various librations throughout the year, suggested times given in text for observations are only approximates.

N
W←→E
S

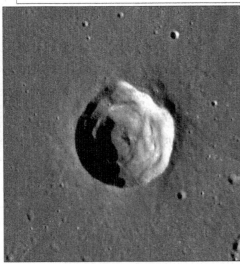

Aerial view of Galle

Having a relatively fresh-look (Eratosthenian in age - 3.15 to 1,1 byo) to it, Galle formed on Frigoris lavas of an older age (~ 3.8 to 3.5 byo). Signatures like low, contrasting ejecta exterior to the crater, as well as a 'puddle-like' floor in the centre, suggest the initial impactor struck lava deposits that may have been still in their semi-molten state. The northeast sector of the rim looks odd, too. Did Galle's impactor hit a higher clump of pre-existing material (of highland?); which then affected its final shape and interior? The inner walls all seem slumped, and a small peak is seen. Depth ~ 2.5 km.

Craters

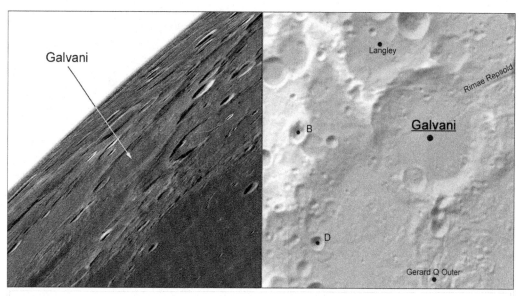

Galvani Lat 49.51N Long 84.56W 76.83Km

Sub-craters:

Crater	Lat	Long	Size (km)	Crater	Lat	Long	Size (km)
B	49.45N	88.82W	12.71	D	47.33N	88.07W	9.07

Notes:

Add Info:

A basin? A wider perspective view of the northern Oceanus Procellarum lavas east of Galvani almost suggest an old basin lies beneath (~ 1000 km in diameter, centred west of Mons Rumker).

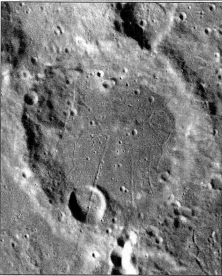

Aerial view of Galvani

Galvani impacted on the pre-existing, but huge crater that is Gerard Q Outer (~ 193 km) to its southeast. Galvani's rim is well-levelled, well worn, its general floor level is some ~ 500 m below that of Gerard Q Outer's, but it is also convexed upwards by that same ammount, too. As a consequence, several fractures developed across the floor; the most obvious (see aerial, left) is the unnamed, north-south-trending fracture that cuts through the small crater (~ 15 km in diameter) within Galvani's southern rim, and a less obvious, east-west-trending one that 'connects' to Rimae Repsold in the northeast. The eastern sector of the floor is slightly darker - hint, perhaps, of some partial flooding. Depth of Galvani ~ 2.5 km.

Craters

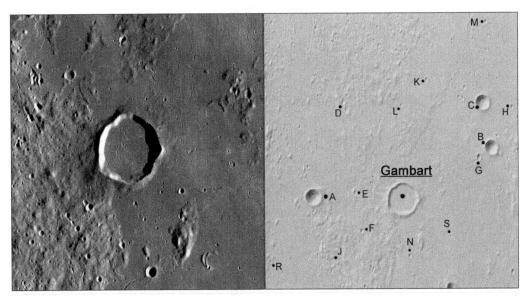

Gambart	Lat 0.92N	Long 15.24W	24.68Km

Sub-craters:

Crater	Lat	Long	Size (km)	Crater	Lat	Long	Size (km)
A	0.96N	18.76W	11.56	B	2.16N	11.59W	10.83
C	3.32N	11.82W	11.67	D	3.35N	17.72W	4.78
E	1.04N	17.24W	4.07	F	0.09N	16.96W	4.52
G	1.94N	12.04W	5.49	H	3.21N	10.62W	4.02
J	0.7S	18.2W	7.16	K	3.91N	14.22W	3.94
L	3.27N	15.28W	3.8	M	5.38N	11.72W	3.93
N	0.54S	14.94W	4.22	R	0.7S	20.8W	3.45
S	0.05S	13.24W	2.4				

Add Info:

Notes:

Gambart looks like it impacted just on the edge of the 'rubbly-looking' material to its west - believed to be ejecta from the Imbrium Basin (less than a 1000 km away to the north), and sometimes referred to as the Fra Mauro Formation. The crater has a relatively sharp-ish rim all around, so it doesn't look like the extensive lavas that flooded this region, generally eastwards, didn't quite rise up and over into Gambart's floor, but rather its inner lava source may be one from internal fractures. Impact ray material from Copernicus to the north (some 300 km away) crosses the crater's floor, so, together with its post-Imbrian impact signature (mentioned above), a constraint on its age of formation may be between 3.2 to 1.1 byo ago - therefore, of the Eratosthenian Period. As an exercise, have a look at other craters familiar to Gambart in this region (e.g. Kunowsky and Lansberg C - both 520 km and 430 km respectively away westwards). Depth 1.1 km.

Craters

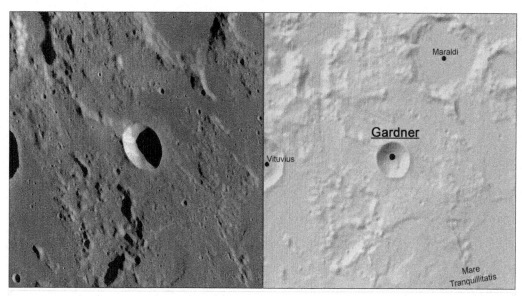

| Gardner | Lat 17.75N | Long 33.81E | 17.62Km |

Sub-craters: No sub-craters

Notes:

Add Info:

Not to be confused with crater Gärtner, found at 59.24N, 34.76E.

Aerial view of Gardner

One thing you'll notice while observing Gardner is how deep it appears at times. Shadows fill its central sectors too much at low sun times, while close to full moon times, or thereabouts, its bright walls look overly thick in width. The crater formed on uneven ground related possibly to the ejecta deposits from both the Tranquillitatis Basin, to its southeast, and the Serenitatis Basin, to its northwest, where it also kissed upon the lavas of Sinus Amoris to its east (ejecta from Gardner can just about be seen to overlie the lavas). As the crater has a depth of around ~ 3 km and its floor level is some 2 km below the level of the Amoris lavas, did Gardner's impactor 'tap' everso slightly into said?

Craters

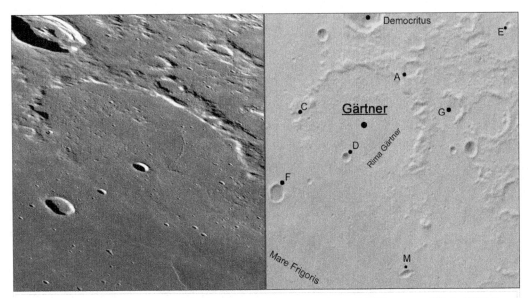

Gärtner	Lat 59.24N	Long 34.76E	101.79Km

Sub-craters:

Crater	Lat	Long	Size (km)	Crater	Lat	Long	Size (km)
A	60.71N	37.72E	12.73	C	59.5N	31.09E	8.47
D	58.51N	33.96E	7.66	E	61.58N	43.82E	4.74
F	57.53N	30.26E	13.85	G	59.65N	39.84E	23.42
M	55.53N	37.03E	10.85				

Notes:

Add Info:

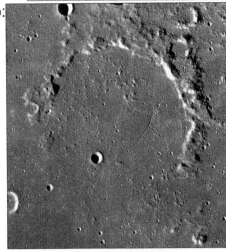

Aerial view of Gärtner

When observing Gärtner, note, mainly, the distinguishing character to the texture of the terrain both inside and southwards of this crater (it has a 'rubbly' look to it, is brighter, and is older, if not more eroded, than the slightly darker, mare of Frigoris near where it lies). The missing southern rim of Gärtner obviously lies underneath the mare, but note, secondly, its axis of tilt - suggesting the area here southwards is more shallower (note the wonderful ridges here, too, at low sun times). Gärtner's main rille is easily observable, but note also another branching off it to the northwest (a challenge). The crater is pre-Nectarian in age (4.6 to 3.92 byo). Depth ~ < 1 km.

Craters

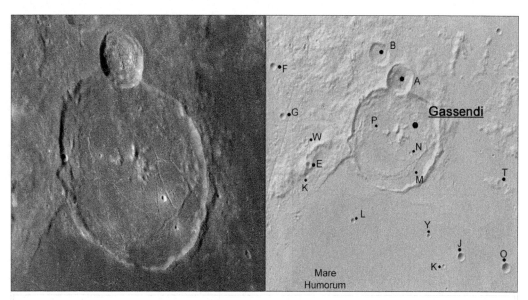

Gassendi   Lat 17.55S   Long 39.96W   111.39Km

Sub-craters:

Crater	Lat	Long	Size (km)	Crater	Lat	Long	Size (km)
A	15.55S	39.8W	32.22	B	14.66S	40.64W	24.66
E	18.45S	43.63W	7.14	F	15.03S	45.02W	7.79
G	16.75S	44.67W	7.35	J	21.62S	37.1W	9.11
K	18.78S	43.74W	5.62	L	20.39S	41.79W	5.32
M	18.61S	39.15W	3.07	N	18.07S	39.32W	3.74
O	21.96S	35.13W	10.34	P	17.29S	40.75W	2.34
R	21.94S	37.85W	4.28	T	19.05S	35.44W	9.53
W	17.66S	43.73W	6.36	Y	20.91S	38.51W	5.0

Notes:

Add Info:

This crater was once considered as an alternative landing site in place for the Taurus-Littrow site where Apollo 17 eventually set down.

Some of the rilles in Gassendi (in old Roman numeral reference)

Crater Gassendi is simply a spectacular view in the eyepiece at any time - during both high or low sun times. Its peaks, its rilles, and crenulated broken rim reveal different details missed by other observation times. Tectonic, volcanic effects underneath the crater are responsible for the rilles, its low depth floor (~ 1.4 km), and, no doubt, slumping (note the major slump in the west) of the inner rim. Obvious when viewing the crater is the brightness of its peaks (highest is about 1.6 km) and the two fresh impacts (sub-crater's Gassendi M and N) on the southeastern part of the floor. Take photos of Gassendi, drawings, too. Age: Nectarian (3.92 to 3.85 byo). Depth ~ 1.4 km.

Craters

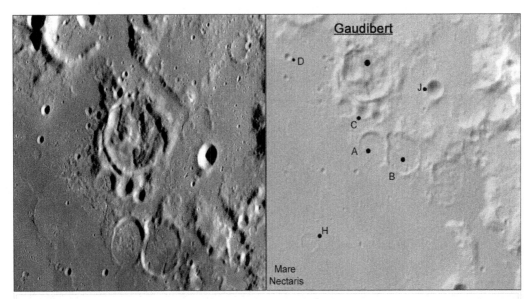

Gaudibert Lat 10.93S Long 37.82E 33.14Km

Sub-craters:	Crater	Lat	Long	Size (km)	Crater	Lat	Long	Size (km)
	A	12.2S	37.94E	17.71	B	12.32S	38.56E	20.96
	C	11.57S	37.8E	8.04	D	10.61S	36.34E	4.03
	H	13.83S	36.72E	10.33	J	11.18S	39.16E	9.98

Notes:

Add Info:

Concentric Craters (CCs): Thought to be the result of a normal impact event forming in a layered, lava region (each layer affecting, differently, the dynamics of the crater's material and its rebound properties).

Gaudibert crater looks, to all intents and purposes, like one of those 'knuckle-dusters' - a prohibited hand weapon used for causing extreme harm. Situated on one of Mare Nectaris's system of rings (its second one ~ 400 km in diameter), it must be one of the weirdest craters on the nearside of the Moon. Its interior looks completely 'out-of-whack' from what we'd expect to see from a crater, however, take a 'bigger-picture' view of several other craters in this region also (e.g. small crater west of sub-crater C, sub-crater A, sub-crater H, or Bohnenberger to its south some 170 km away), and you'll see that the crater isn't alone in its 'weirdness'. Whatever the cause to the look of craters at this side of the mare - volcanism, flooding, target rock in which the mare formed - those effects have gouged out Gaudibert's innards into a 'splurge' of mountains and hills. Closer examination shows two to three small rille-like features within the eastern sector of the rim, while a slightly thicker one runs across the main peak area in a NW-SE direction. These suggest possible uplift and volcanic acitivity underneath may be responsible for the 'bulbuous' look to Gaudibert, but then look at the southern rim sector and it, too, has gone 'weird' (is Gaudibert a crater within a crater, or, a unique large version of a concentric type crater?). Depth ~ 0.6 km.

Craters

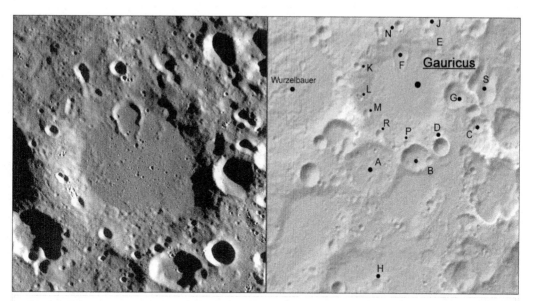

Gauricus Lat 33.91S Long 12.74W 79.64Km

Sub-craters:

Crater	Lat	Long	Size (km)	Crater	Lat	Long	Size (km)
A	35.64S	13.57W	37.22	B	35.39S	12.23W	21.76
C	35.26S	10.73W	9.58	D	35.14S	11.46W	12.3
E	32.58S	11.88W	6.69	F	33.11S	12.71W	12.27
G	33.99S	11.05W	18.41	H	38.24S	13.39W	8.48
J	32.34S	12.0W	8.97	K	33.35S	13.97W	5.33
L	34.06S	13.88W	4.2	M	34.42S	13.7W	5.25
N	32.44S	12.74W	6.99	P	35.09S	12.5W	5.15
R	34.87S	13.32W	5.87	S	33.91S	10.17W	14.59

Notes:

Add Info: Take away the numerous impact signatures of every size on Gauricus's rim (at its eastern and southern sectors), and we would be looking at an overly-curved feature to the crater walls and rim (like at it western sector). Why would that be? Looking at Gauricus's smooth-ish floor (and other craters nearby), and you'll immediatly note that ejecta has played a factor in this region. Secondary crater chains from the Imbrium Basin lie here, so that source, as well as earlier impacts closeby (e.g. Nectarian-aged Pitatus to north) must be responsible for the filled floor and 'curvy' rim. Gauricus is older then - possibly of pre-Nectarian (4.6 to 3.92 byo). Sub-crater F (oblique impact?) must be older than the Imbrium event (as it, too, is filled with ejecta, and has impacted upon Gauricus's inner rim), but as it's hard to see if ejecta from Pitatus covers it, F's formation is probably between these two events. Sub-crater G has imparted ejecta on to the floor, and rays from Tycho cross it from the SE. Depth 2.4 km approx..

Craters

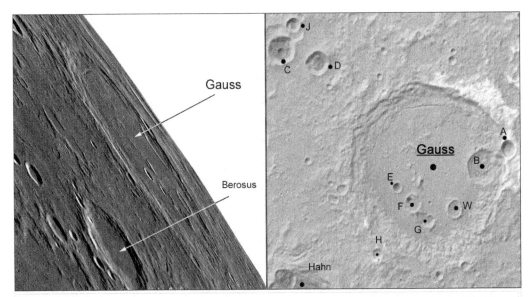

| Gauss | Lat 36.01N | Long 79.08E | 170.72Km |

Sub-craters:

Crater	Lat	Long	Size (km)	Crater	Lat	Long	Size (km)
A	36.54N	82.77E	20.7	B	35.99N	81.6E	35.46
C	39.72N	72.16E	30.0	D	39.37N	73.89E	25.11
E	35.35N	77.71E	10.45	F	34.83N	78.39E	18.0
G	34.26N	79.01E	18.01	H	33.15N	76.87E	12.13
J	40.57N	72.65E	15.82	W	34.63N	80.28E	18.5

Notes:

Add Info:

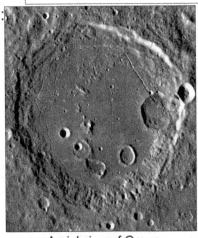

Aerial view of Gauss

It really is a pity that Gauss lies close to the limb, as from an aerial view shot (left), it certainly is a crater of note. Dated of the Nectarian (3.92 to 3.85 byo), as its ejecta overlies that of Humboldtianum's to the north (~ 600 km away), it is possibly younger (note how the rim and terraces look relatively sharp-ish). The interior contains several large rilles concentric to a slightly raised central floor, an off-centre peak (is it one?), and two dark regions suggest pyroclastic deposits - see: sub-crater G area and northwest of sub-crater E (note also the dark region between W and G). Favourable libration and terminator periods (~ 3-day-old moon and just after full) are thus the best times to view its features. Depth ~ 2.6 km approx..

Craters

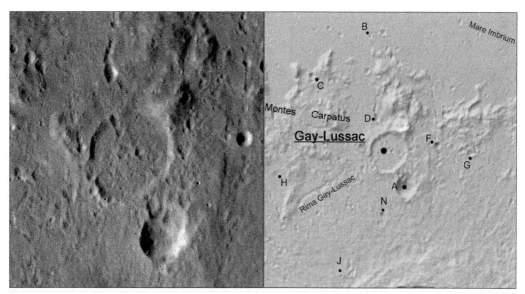

Gay-Lussac   Lat 13.88N   Long 20.79W   25.4Km

Sub-craters:

Crater	Lat	Long	Size (km)	Crater	Lat	Long	Size (km)
A	13.18N	20.38W	15.25	B	16.21N	21.18W	3.42
C	15.42N	22.55W	4.71	D	14.58N	21.06W	5.33
F	14.01N	19.65W	4.61	G	13.86N	18.89W	4.73
H	13.41N	23.33W	5.01	J	11.36N	21.85W	2.28
N	12.62N	20.92W	2.36				

Notes:

Add Info: Gay-Lussac doesn't have much to offer in terms of specific features, however, by looking at its location relative to both the Imbrium Basin to its north, and the crater Copernicus to its south, the following is an attempt in exercise for determining its approximate age. Firstly, we can see that the crater formed on top of a section of the Carpathian mountains - believed to be the result in formation of the Imbrium Basin (or, were pre-existing highlands affected by that main event) some 3.85 billions of years ago. Secondly, secondary signatures e.g. small craters and 'herringbone' patterns cover all of this area (as well as inside the crater itself) from the event that produced Copernicus crater less than 1.1 billions of years ago. We deduce, then, that Gay-Lussac formed some time between 3.85 and 1.1 billions of years ago. However, if we look further at the floor area that has been filled, the northern inner area is smooth-like while the southern looks as if it is ejecta from Copernicus. The smoother stuff, then, is older and is most likely volcanically-related e.g. mare lavas associated to Imbrium or the ancient, but controversial, Oceanus Procellarum Basin (of pre-Nectarian 4.6 to 3.92 byo). Given the crater has a relatively sharp-ish rim, its age then may be in between - perhaps, of the Eratosthenian (3.15 to 1.1 byo). Depth ~ 0.8 km.

Herringbone signatures are v-shaped patterns on the surface produced by low-angle, ejected material that lands as near-simultaneous impacts; whose secondary ejecta then interferes together to create the 'V'.

Craters

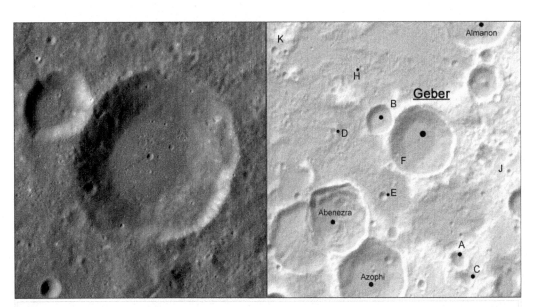

Geber	Lat 19.46S	Long 13.85E	44.68Km

Sub-craters:	Crater	Lat	Long	Size (km)	Crater	Lat	Long	Size (km)
	A	21.84S	14.7E	12.6	B	19.03S	12.91E	17.96
	C	22.18S	14.88E	10.02	D	19.25S	11.84E	4.53
	E	20.55S	12.96E	5.63	F	19.96S	13.15E	4.76
	H	18.0S	12.49E	2.91	J	20.06S	15.85E	3.62
	K	17.64S	10.55E	5.12				

Notes:

Add Info: Before you look at Geber itself, take a broader, more general view of the area in which it lies, and note how one or two other craters (e.g. Azophi and Playfair to the south and southwest respectively, or Abulfeda and Albategnius to the north and northwest respectively), show similar-like smooth floors with numererous small craters in them. As the make-up of this smooth material is believed to be fluidized ejecta from the Imbrium Basin some 1500 km to the northwest, many of these craters, then, must be older than that event, which occurred less than 3.9 billions of years ago (Geber, so, is probably Nectarian in age). There aren't many features to speak of in Geber, however, one of note would be the slightly thicker 'band' of slumped material seen at its east/southeast sectors. As there's also what looks like a fault running in a NE-SW direction across the floor, was the additional weight of the slump responsible (the terrain of this so-called 'fault' is lower in the SE sector and higher in the NW)? Sub-crater B may have formed around the same time Geber did as it doesn't seem to have imparted any of its eject on to the floor (a duo-like impact with Geber?) Depth 3.5 km.

Craters

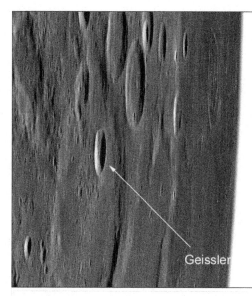

| Geissler | Lat 2.6S | Long 76.51E | 17.39Km |

Sub-craters: No sub-craters

Notes:

Add Info:

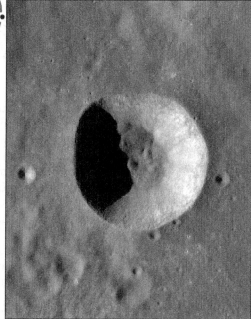

Aerial view of Geissler

Geissler isn't that hard to locate at most times when in light, as its bright walls 'shine'. Depth of the crater (~ 3 km approx.) also helps as at times the shadows produced contrast nicely against the shiny walls. The crater obviously has impacted on the floor of Gilbert (pre-Nectarian in age 4.6 to 3.92 byo). But what isn't so obvious is that Geissler also impacted upon a cluster of small craterlets on Gilbert's floor (possibly, secondaries from the Nectarian-aged Crisium Basin off to the northwest); confining down its own age beyond Nectarian. The aerial view (left) shows that slumping (perhaps, two separate slump events) occurred on its northwestern walls, which may be responsible for the hilly lumps on its floor, too.

Craters

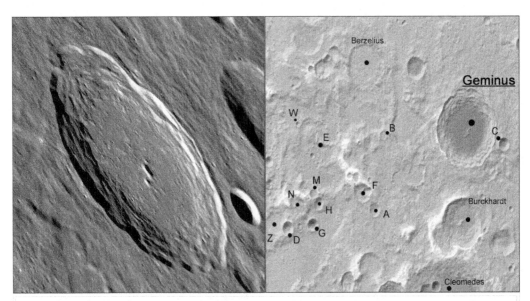

| Geminus | Lat 34.42N | Long 56.66E | 81.98Km |

Sub-craters:

Crater	Lat	Long	Size (km)	Crater	Lat	Long	Size (km)
A	31.51N	51.8E	14.72	B	34.14N	52.11E	10.24
C	33.88N	58.69E	14.57	D	30.58N	47.29E	15.99
E	33.84N	48.58E	55.48	F	32.08N	51.1E	21.78
G	30.8N	48.57E	14.84	H	31.57N	48.82E	13.5
M	31.88N	48.5E	10.68	N	31.41N	47.73E	25.13
W	34.27N	47.43E	5.62	Z	30.72N	46.64E	26.95

Notes:

Add Info:

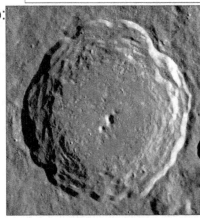

Aerial view of Geminus

Of Eratosthenian in age (3.15 to 1.1 billions of years ago), Geminus lies on one of Mare Crisium's main ring/rim system (the 1080 km in diameter one), and just on the edge of a trough-like zone (note the smooth patches either side of the crater some 250 km away in a SE and SW directions) created during that main event. The crater has a wonderful series of very sharp-looking terraces (note how much they extend into the floor) all within, and three main peaks whose highest is 1.1 km approx.. Of note, see the odd linear feature that crosses the outer, southern rim (it has a NW-SE orientation) sectors, and slightly 'cuts' into the inner rim walls. D: 5 km.

Craters

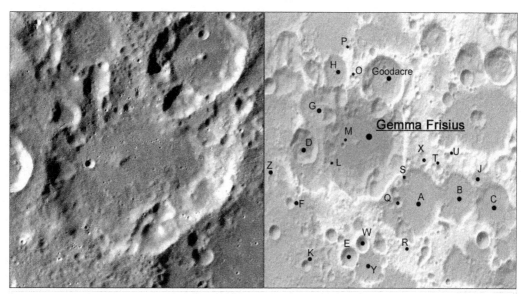

Gemma Frisius Lat 34.33S Long 13.37E   88.54Km

Sub-craters:

Crater	Lat	Long	Size (km)	Crater	Lat	Long	Size (km)
A	35.8S	15.15E	65.1	B	35.58S	17.22E	40.43
C	35.71S	18.74E	32.03	D	34.36S	10.9E	27.67
E	37.3S	12.72E	18.48	F	35.85S	10.29E	8.94
G	33.36S	11.44E	36.5	H	32.38S	12.17E	25.5
J	35.11S	18.06E	12.91	K	37.5S	10.92E	9.89
L	34.82S	11.8E	6.15	M	34.34S	12.46E	4.63
O	32.56S	12.81E	4.9	P	31.81S	12.78E	4.18
Q	35.83S	14.75E	9.35	R	37.15S	15.29E	4.47
S	35.28S	15.03E	4.59	T	34.92S	16.41E	7.97
U	34.61S	16.76E	8.04	W	36.95S	13.28E	14.04
X	34.69S	15.81E	14.79	Y	37.56S	13.53E	27.84
Z	35.2S	9.61E	9.7				

Notes:

Add Info: Gemma Frisius looks a right 'jumble' of all-sorts, doesn't it! Only the bottom half of the original crater has survived the onslaught of big and small impactors that have totally wiped out its top half. Those impactors cover a wide range of ages judging by the state of some of the craters (e.g. fresh-looking sub-crater D, and old-looking sub-crater A) produced by them, but one thing is for sure, Gemma Frisius is an old crater (~ Nectarian - 3.95 to 3.85 byo) having had plenty of time open to impact abuse. Oddly enough, what with all these going-ons, the crater is quite easily found - perhaps due to its ~ 5 km depth where deep shadows define its location, or, perhaps, using its 'Micky Mouse' ears of sub-crater D and Goodacre. It has a small peak. Smooth material on floor is possibly Imbrium ejecta.

Craters

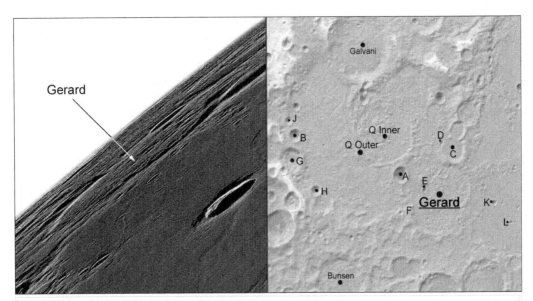

Gerard	Lat 44.54N	Long 80.51W				98.78Km

Sub-craters:

Crater	Lat	Long	Size (km)	Crater	Lat	Long	Size (km)
A	45.08N	82.27W	17.31	B	46.34N	88.07W	14.04
C	45.84N	79.33W	29.36	D	46.11N	79.97W	5.93
E	44.52N	80.99W	5.28	F	43.78N	82.28W	5.39
G	45.52N	88.22W	26.9	H	44.5N	86.81W	12.25
J	46.82N	88.47W	9.35	K	43.95N	77.28W	5.85
L	43.22N	76.48W	4.5	Q Inner	46.54N	83.13W	67.32
Q Outer	46.51N	84.55W	192.48				
Notes:							

Add Info:

Aerial view of Gerard

Gerard could easily take the title as the '*Venn set*' crater on the Moon, where it looks like the triplet set of Venn diagrams generally used in Mathemathics. The triplet lie on the southeastern rim of crater Gerard Q Outer (~ 193 km in diameter), but as to which of the triplet formed first isn't quite obvious as all three look similar in age and degradation. Gerard's floor looks a total mess and mix of possible ejecta from several impacts nearby and afar, with also some two to three big rilles radial to Gerard Q Outer (or are they related to Oceanus Procellarum in the east?).

Craters

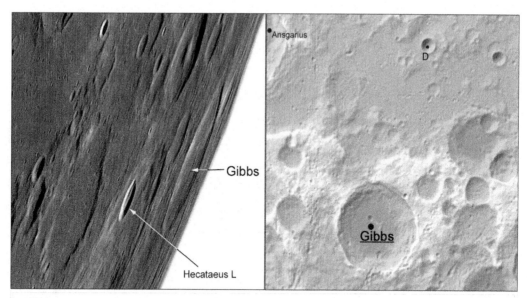

Gibbs	Lat 18.37S	Long 84.27E			78.76Km

Sub-craters:	Crater	Lat	Long	Size (km)	Crater	Lat	Long	Size (km)
	D	13.11S	85.89E	13.61				
	Notes:							

Add Info:

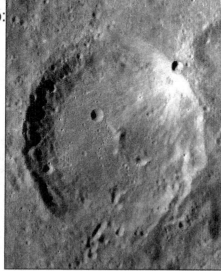

Aerial view of Gibbs

Just after full moon or a new moon are about the best times to observe Gibbs. When in fuller light, all signature of the crater becomes hard to see, as all that hints at its existance in view is the bright patch of ray material created by the small, unnamed crater on Gibb's northeastern rim (fresh-looking crater, Hecataeus L, may help, too). Gibbs lies on the ejecta, or the outer ring/rim (~ 1300 km in diameter) associated with formation of the Smythii Basiin, which lies ~ 500 km centred to the north. The crater has a small peak (~ 800 m high), smooth floor, and well-worn rim and terraces, where its southern sectors has been gouged by small secondaries from crater Humboldt (?) ~ 200 km to its south. Depth of Gibbs ~ 4.0 km

Craters

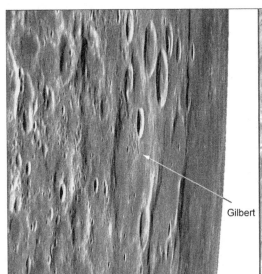
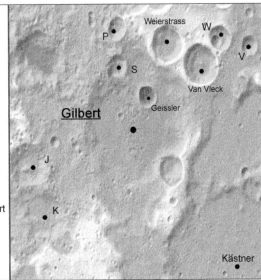

Gilbert   Lat 3.2S   Long 76.16E   100.25Km

Sub-craters:

Crater	Lat	Long	Size (km)	Crater	Lat	Long	Size (km)
J	4.32S	72.66E	35.44	K	5.68S	73.17E	39.36
P	0.85S	75.53E	19.53	S	1.79S	75.53E	17.27
V	1.29S	79.76E	15.82	W	1.11S	78.82E	21.78

Notes:

Add Info:

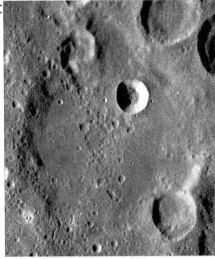

Aerial view of Gilbert

Gilbert can be a difficult target to find - even in the most favourable of librational periods. If it wasn't for the darkish patch of Mare Smythii to its east on the extreme limb, and the fresh, bright-walled crater Geissler (one time known as Gilbert D) on the floor, it's quite easy to get lost here. At full moon times it's near-impossible to see any definition to its well-worn rim or floor (use Geissler as a 'marker'), while during terminator times (just after full, or around a 3-day-old Moon), shadows obscure its 3.7 km depth. Smooth areas are most likely lava deposits extruded from within, while the material, and series of craterlets, that cross the floor in a N-S orientation are, perhaps, ejecta signatures from either Crisium or Humboldtianum.

Craters

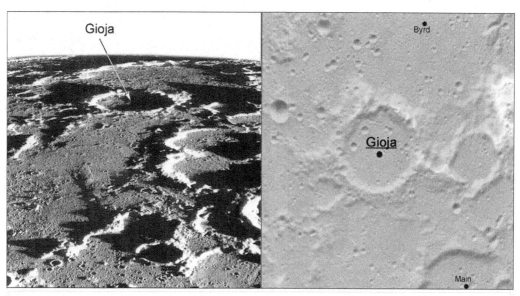

| Gioja | Lat 83.35N Long 1.76E | 42.47Km |

Sub-craters: No sub-craters

Notes:

Add Info:

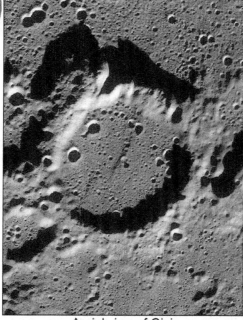

Aerial view of Gioja

Gioja has impacted upon the rim of Byrd to its north, and both have been covered by ejecta from the Imbrium Basin to their southwest. This same ejecta may also have 'chipped' off pieces of Gioja's rim (and in several other craters nearby), which has left a slight striated look to them all (one of the 'chips' on Gioja's northern rim serves as a good indicator at times for locating the crater). Gioja's floor looks odd; showing a small north/south trending ridge, or is it a scarp (?) that drops by about 200 m on the western side (it can be seen during most low sun times - easterly or westerly, while full moon times hints of the feature can be seen, too). Good librations are recommended for viewing Gioja. D:1.6 km.

Craters

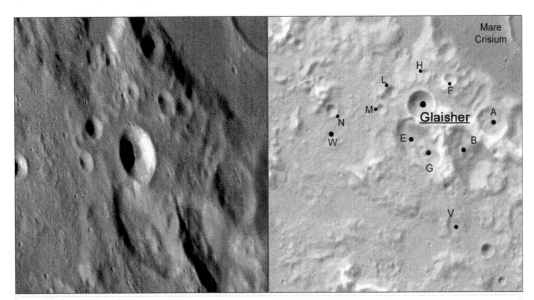

Glaisher	Lat 13.19N	Long 49.34E			15.92Km		

Sub-craters:

Crater	Lat	Long	Size (km)	Crater	Lat	Long	Size (km)
A	12.84N	50.73E	19.16	B	12.48N	50.15E	23.05
E	12.63N	49.19E	20.81	F	13.64N	49.92E	7.33
G	12.37N	49.49E	18.89	H	13.74N	49.45E	4.82
L	13.41N	48.77E	6.99	M	13.12N	48.56E	5.84
N	13.09N	47.58E	6.87	V	11.0N	50.0E	12.99
W	12.55N	47.64E	48.01				

Notes:

Add Info:

Aerial view of Glaisher

There's not much to Glaisher in terms of any significant features to observe. It looks relatively fresh, but unlike its wonderful neighbour to the north, crater Proclus (the 'show-off'), which is Copernican (1.1 byo) in age, the rim of Glaisher looks just that little bit more worn, and so it is probably of the Eratosthenian period (3.15 to 1.1 byo). The crater lies on a high piece of ground created as a result from formation of the Crisium Basin to its east; whose ejecta has covered some of Glaisher's larger-sized sub-craters - A, B, E, G (note the striated grooves that cross these) and W. Depth ~ 3.3 km approx.

Craters

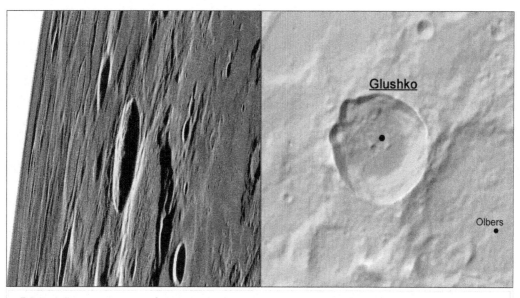

Glushko	Lat 8.11N	Long 77.67W	40.1Km

Sub-craters: No sub-craters

Notes:

Add Info:

Aerial view showing the extent to Glushko's ray system

It's a pity that Glushko is situated near the limb as from aerial close-ups of the crater it has some interesting features (slumped walls, peaks, oblique signatures, and a small impact-melt pond in its western sector). It's quite hard to get any descent views then of it - favourable librations or otherwise, so for now all we can do is admire its wonderful ray-system that extends for many kilometres to the east. Depth ~ 3.6km approx..

Craters

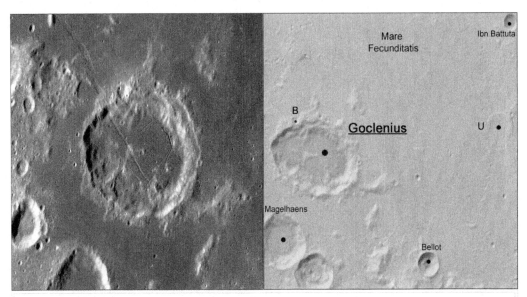

| Goclenius | Lat 10.05S | Long 45.03E | 73.04Km |

Sub-craters:	Crater	Lat	Long	Size (km)	Crater	Lat	Long	Size (km)
	B	9.22S	44.47E	6.22	U	9.33S	50.15E	22.14

Notes:

Add Info:

This crater also shows some nice pyroclastic vents (volcanic dark spot areas); found near the inner rim area of the floor at both the eastern and western sectors.

Aerial view of Goclenius showing its odd shape

From an initial view of Goclenius, it's quite obvious to see that physical and geological things have had their effects on the crater as a whole. Its floor is lower than the mare floor of Fecunditatis to its northeast (lavas within were sourced through 'cracks' - note the wonderful rilles). Its odd, rugby-ball shape is a puzzle, too (the result of an oblique impact?), and just what is going on at its northeastern rim (it looks too convex in opposite to the southwest rim that is more concave as expected). Note how parts of the rilles meet the slumped material, and also where they cut through hilly-mountain terrain. Depth ~ 2.2 km.

Craters

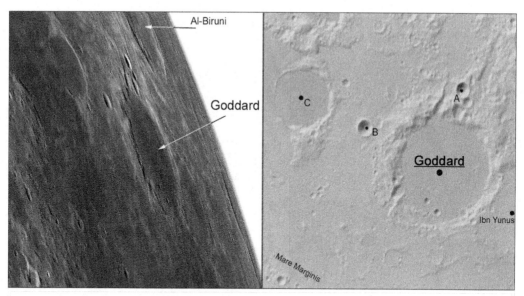

Goddard	Lat 15.15N	Long 89.13E	93.18Km

Sub-craters:	Crater	Lat	Long	Size (km)	Crater	Lat	Long	Size (km)
	A	17.07N	89.71E	11.06	B	16.13N	86.94E	12.22
	C	16.81N	85.21E	48.19				
	Notes:							

Add Info:

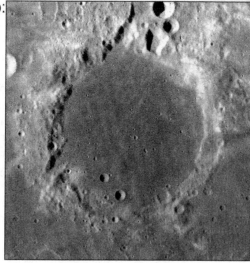

Aerial view of Goddard

Goddard lies just off-centre of the Mare that is Marginis (centred at some ~ 100 km to its southwest). Goddard's lava-filled floor is some 400 m below that of the Mare, and its albedo takes on a more slight darkness, too. From the aerial view (left), a large 'gash' on the crater's northern rim seems also to continue at the southwestern rim. Ejecta from Mare Smythii or crater Neper both to the southwest may be responsible (the gash is more radial to Neper). Signature of pyroclastic features are also seen in the aerial view - the most obvious just above the two small craters in the south. D: ~ 1.8 km approx..

Craters

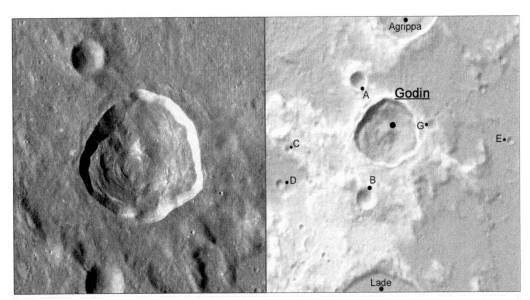

| Godin | Lat 1.82N | Long 10.16E | 34.25Km |

Sub-craters:

Crater	Lat	Long	Size (km)	Crater	Lat	Long	Size (km)
A	2.67N	9.66E	9.44	B	0.73N	9.79E	11.35
C	1.54N	8.38E	4.05	D	0.97N	8.24E	5.12
E	1.66N	12.37E	3.62	G	1.93N	10.97E	5.59

Notes:

Add Info:

Try to observe the 'tip-point' of Godin's peak as it stands out like a lighted 'dot' especially under low power and during a Full Moon (the dot also puts on shows during times when the floor is in shadow, say, around waning gibbous to last quarter).

If you ever wanted a perfect example in how slumping of material in a crater can change, or totally wipe out, its interior detail, then Godin is it. Aged of the Copernican (1.1 billions of years old to the Present), one would expect to see some of the floor for such a young crater, but this isn't the case as it has been entirely covered by the slumping effects. What would cause such effects - target rock?, proximity to other nearby impact events?, lunar quakes?. Slightly older Agrippa crater to its north shows some slumping of sorts in its west, but it isn't as extensive as Godin's. Both craters lie on high ground and on ejecta from the Imbrium Basin - some 1200 km away to its centre, so it may not be a target rock issue. Proximity can be ruled out also, as there isn't any other younger impact craters to each. So, quakes might be the possible answer. Another oddity of Godin is its central ~ 1.7 km-high peak - for such a young crater you would expect it to be that little bit sharper, but, it is more gentle (older, Agrippa's peak looks a tad sharper than it). The crater today has been measured as having a depth of around 3.2 km approximately, but this depth is only consigned to a small ring-like moat around the peak, which produces some nice but odd shadow effects when the crater is in a terminator position - easterly or westerly apparitions.

Craters

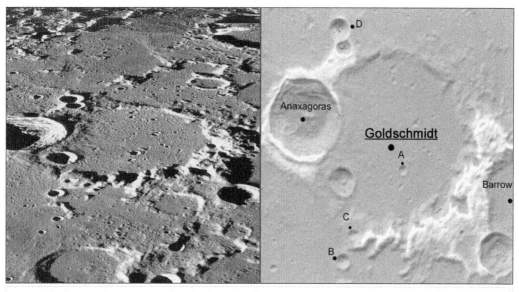

Goldschmidt	Lat 73.04N	Long 3.37W				115.26Km	

Sub-craters:

Crater	Lat	Long	Size (km)	Crater	Lat	Long	Size (km)
A	72.5N	2.45W	6.74	B	70.61N	6.78W	9.9
C	71.17N	6.1W	7.61	D	75.4N	7.74W	14.36

Notes:

Add Info:

Aerial view of Goldschmidt

When viewing Goldschmidt crater for the first time, it can be confusing as one or two other craters in the area look similar e.g. W. Bond south of it and Barrow to its east. All three are pre-Nectarian in age (4.6 to 3.92 byo), and are covered in ejecta from the Imbrium Basin to the southwest. As can be seen in the right-most, topograpic view above Goldschmidt's SE rim area is fairly high, so observe the wonderful shadows it sometimes casts on to the floor during easterly sun apparitions. Rays (and ejecta) from nearby Anaxagoras crosses the floor, too. Depth ~ 3.4 km approx..

Craters

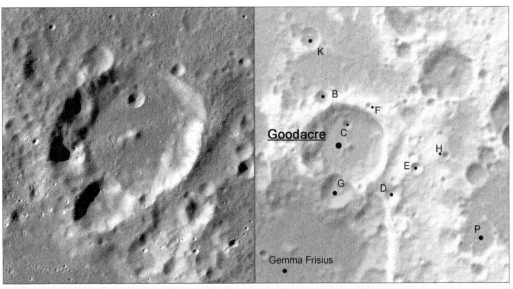

| Goodacre | Lat 32.67S | Long 14.08E | 44.09Km |

Sub-craters:

Crater	Lat	Long	Size (km)	Crater	Lat	Long	Size (km)
B	31.92S	13.64E	8.97	C	32.3S	14.13E	4.79
D	33.42S	15.01E	8.37	E	32.99S	15.49E	6.05
F	32.02S	14.55E	3.83	G	33.35S	13.91E	16.06
H	32.77S	16.06E	4.36	K	30.94S	13.41E	11.07
P	34.11S	16.68E	19.61				

Notes:

Add Info: With so much 'impact going-ons' in this, somewhat, confusing region in where Goodacre lies, a good marker for finding it is to use its much larger neighbour, crater Gemma Frisius, to its southwest. Obviously, Goodacre is younger than Gemma Frisius as it has impacted upon its northeast rim, but have a look at both of their rims in a single view, and you'll see that there's not much difference in the state of degradation between them - signature, perhaps, formation of each was relatively close. The border between the two shares the impact crater that is sub-crater Goodacre G, but note here how the ejecta from that event is distributed on to each of their floors - prominent in Gemma Frisius's deeper depth of ~ 5 km, and less obvious in Goodacre's 3.2 km depth. And what of the floors, which from the look to both of them aren't the real floors at all, but rather the ejecta-fill from numerous impacts nearby, and from basins like Imbrium to the northwest and Orientale to the west (such 'fills' have nearly covered what remains of Goodacre's peaks - the main one being around 0.6 km high). While it isn't obvious, ray material from crater Tycho to the southwest overlays the floor.

Craters

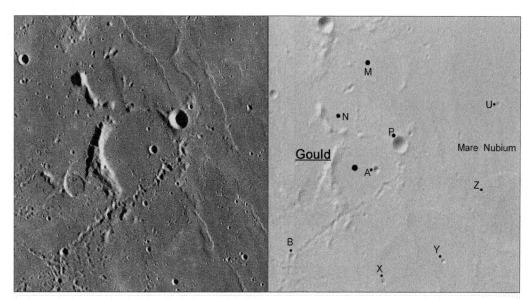

Gould   Lat 19.26S   Long 17.25W   32.99Km

Sub-craters:

Crater	Lat	Long	Size (km)	Crater	Lat	Long	Size (km)
A	19.23S	17.05W	3.34	B	20.53S	18.5W	3.55
M	17.67S	17.24W	41.54	N	18.35S	17.69W	17.33
P	18.85S	16.65W	7.55	U	18.22S	14.97W	2.69
X	20.91S	16.92W	2.31	Y	20.59S	15.88W	2.46
Z	19.56S	15.16W	1.97				

Notes:

Add Info: The series of lavas that extend right across Mare Nubium in which Gould lies have tried their best in attempting to wipe out the crater entirely. Only the western sector of its real rim defines that it was a complete crater at one time, however, between the structural changes in ridges that 'crimple' its eastern sector, not to mention impact of sub-crater, Gould P, at its northeast, Gould still succeeds to 'peek' through. From which direction did the lavas approach Gould from? Initial observations would suggest they predominantly came from an eastwards direction; as the main volume of the Mare lies approximately there, and the crater looks like it is 'tilted' towards that way. But, there's also a lot of lavas that extend right up to crater Copernicus in the north, also to Mare Cognitum in the northwest, and then over to the expanse of lavas in Oceanus Procellarum. The relatively long chain of craters running across the crater's southern sector are an obvious, wonderful feature of Gould (what was there initial source?) to look out for, however, try and view the long-peaked shadows cast by the high, western rim of the crater (and sub-crater, Gould P's, too) during terminator times - e.g. just after last quarter, or around a 11-day-old moon.

Craters

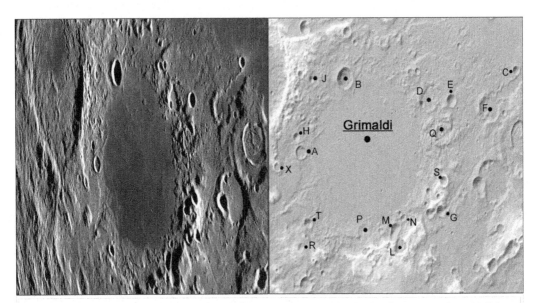

| Grimaldi | Lat 5.38S | Long 68.36W | 173.49Km |

Sub-craters:

Crater	Lat	Long	Size (km)	Crater	Lat	Long	Size (km)
A	5.39S	71.33W	14.52	B	2.94S	69.32W	22.16
C	2.65S	61.59W	9.31	D	3.66S	65.65W	20.65
E	3.67S	64.54W	13.23	F	3.91S	62.84W	28.98
G	7.33S	65.01W	11.8	H	4.88S	71.52W	9.11
J	2.93S	70.69W	15.38	L	8.53S	66.83W	18.27
M	8.02S	67.1W	18.68	N	7.56S	66.71W	7.97
P	7.96S	68.37W	8.26	Q	4.75S	64.87W	25.35
R	8.5S	71.36W	9.04	S	6.35S	64.88W	10.99
T	7.72S	71.01W	11.27	X	5.88S	72.41W	8.42

Notes:

Add Info:

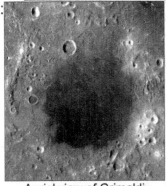

Aerial view of Grimaldi

You just can't avoid noting that Grimaldi's inner lavas look slightly darker than those to its east in the vast plains of Oceanus Procellarum (they really aren't, so it is probably just an optical illusion created by Grimaldi being surrounded by brighter terrain). Grimaldi, as such, isn't classed as being a crater, but more a basin with several rings to it (its main inner ring ~ 230 km in diameter can be seen in the aerial view at left). It's a pity that Grimaldi is at a limb zone as there are several wonderful features to be seen e.g. domes, rilles, ridges, rays etc., (some showing up better than most a day or two before full moon). Age: pre-Nec (4.6 - 3.9 byo).

Craters

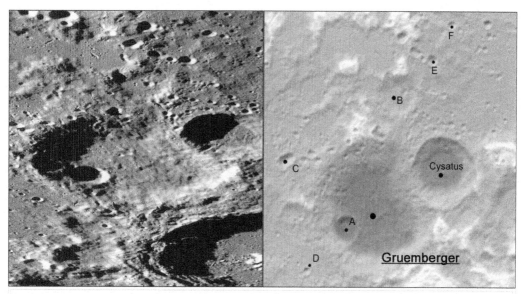

Gruemberger   Lat 67.04S   Long 10.3W   91.5Km

Sub-craters:

Crater	Lat	Long	Size (km)	Crater	Lat	Long	Size (km)
A	67.5S	12.21W	18.9	B	64.58S	9.15W	28.66
C	65.9S	15.4W	9.51	D	68.35S	14.68W	5.41
E	63.82S	7.32W	7.84	F	63.03S	6.43W	6.56

Notes:

Add Info:

Aerial view of Gruemberger

Observing this southern limb-area of the Moon is never easy. There's just so many craters 'clustered' together, an on-hand moon atlas is always advisable when observing. A good marker for remembering where Gruemberger is - when, say, in between changes of eyepieces, or referring to the atlas - is to use its bigger neighbour, Clavius, to its northwest. Because of its location, favourable latitudinal librations are always the best times to view the crater, but even then, some planning has to be done beforehand to avoid overly-filled shadows in the floor. Getting a good shot of Gruemberger, then, is a challenge for the serious astrophotographer. D: ~ 5 km.

Craters

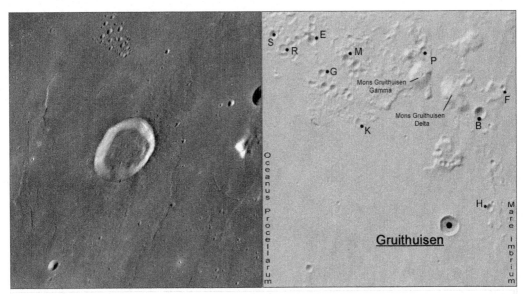

Gruithuisen   Lat 32.89N   Long 39.78W   14.98Km

Sub-craters:

Crater	Lat	Long	Size (km)	Crater	Lat	Long	Size (km)
B	35.65N	38.75W	9.4	E	37.39N	44.39W	7.5
F	36.27N	37.98W	4.19	G	36.59N	44.01W	6.01
H	33.32N	38.47W	5.57	K	35.39N	42.73W	6.02
M	36.94N	43.16W	8.97	P	37.17N	40.57W	9.13
R	37.15N	45.33W	6.79	S	37.47N	45.68W	6.65

Notes:

Add Info: Mention Gruithuisen amongst a group of lunar observers, and it's guaranteed that most will initially think of the domes (Gruithusien Gamma and Delta) firstly. The crater doesn't get that much-of-a-look-in, and why? It's because there isn't that much to be gleaned from it. Around full moon times, some features like its bright inner wall, its slumped deposits, and the one or two small clumps on its floor are a small reward (best during high power observations), but that's about it. Low sun angle times, say, around 12-day or 24-day-old moon times, do give some nice shadows - not really in its floor (as it's usually shadowed-filled during these times), but rather from the wonderful ejecta detail surrounding its outer rim, and from the height of the rim itself. Note also during such apparitions the ridges on either side of the crater running in a north-south direction, which are truly eye-candy when all - crater, ridges, shadows - are viewed together in a single view. Gruithuisen's rim is relatively sharp, so it probably is Eratosthenian in age (3.15 to 1.1 billions of years old), and a small, impact crater lies on the southeastern sector of the rim. Depth of crater is around the 1.9 km approx..

Distribution of the ejecta surrounding this crater suggests it may be an oblique, impact crater (the impactor coming in from the western direction).

Craters

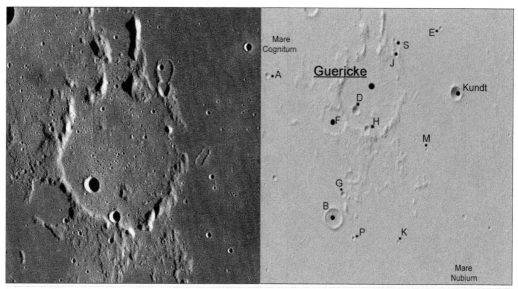

Guericke Lat 11.57S Long 14.19W 60.75Km

Sub-craters:

Crater	Lat	Long	Size (km)	Crater	Lat	Long	Size (km)
A	11.15S	17.29W	4.59	B	14.59S	15.31W	14.64
D	11.99S	14.62W	6.65	E	10.02S	12.06W	3.41
F	12.26S	15.33W	21.04	G	14.01S	15.02W	4.53
H	12.45S	14.29W	5.14	J	10.62S	13.43W	6.99
K	15.15S	13.31W	2.87	M	12.93S	12.47W	2.32
N	12.55S	9.93W	2.58	P	15.06S	14.7W	2.91
S	10.36S	13.4W	10.19				

Notes:

Add Info: Before you look at Guericke in general, take a broader view of the area in which it lies, and notice an odd-looking feature-effect to some, if not all, of the craters nearby. This feature-effect can be seen on craters Parry, Bonpland and in Fra Mauro to the north of Geuricke, while is also shows up in craters' Ptolemaeus and Alphonsus to the east. The effect, of course, is the groove-like appearance seen in nearly all of their rims and the surrounding terrain (it shows up wonderfully during low sun times). The grooves all point back to the central region of Mare Imbrium - some 1300 km away, and their formation was due to ejecta from that major event 'clipping', 'digging' and 'gouging' into high points like rims and terrain in its wake. Did this major, catastrophic event totally demolish portions of Guericke's missing rim, or simply dislodge big chunks of the rim into the central floor area (see the odd clump just north of sub-crater, Geuricke H)? Later, lavas may have breached through gaps in the crater, covering secondary craters formed by Imbrium (note the lack of these craters eastwards of Guericke in the lavas). Ray material from Copernicus covers the crater, too. Depth ~ 0.8 km approx..

Craters

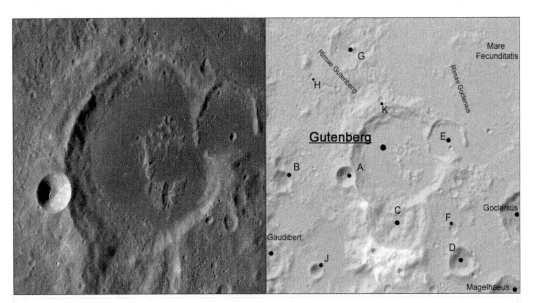

Gutenberg Lat 8.61S Long 41.25E 70.65Km

Sub-craters:

Crater	Lat	Long	Size (km)	Crater	Lat	Long	Size (km)
A	9.03S	39.91E	14.8	B	9.12S	38.32E	14.19
C	10.04S	41.13E	45.21	D	10.99S	42.84E	20.07
E	8.22S	42.42E	28.19	F	10.23S	42.61E	7.17
G	6.05S	40.04E	30.64	H	6.75S	39.06E	4.77
K	7.3S	40.82E	5.64				

Notes:

Add Info: Gutenberg lies in a clump of terrain related to Mare Fecunditatis in the northeast and to Mare Nectaris in the southwest. Like its neighbour, crater Goclenius, to its southeast, the two together are old ~ pre Nectarian (4.6 to 3.92 billions of years old); each a wonderful example in how lavas can engulf a crater's inner core. Such lavas within craters like these are usually sourced through underlying fractures, which act like channels to deeper deposits - allowing a greater understanding and history in to how the Moon formed. It's worth observing both craters in a single view as they each have similar features like rilles or peaks to look out for. Note how the rilles, in both craters, cut into portions of their respective rims and peaks (a signature that they must have occurred after the craters initially formed, but does not define if they are younger or older than the lavas). Look also at Goclenius's 'weird', convexed northeastern rim, and compare it to the remaining part of Gutenberg's rim (N and S of sub-crater E), which also seems to have this convexing-like feature. Is there a geological link, therefore, between the two craters that caused their NE sectors to be so? Depth ~1.8 km.

Rimae is the plural term of Rima (a single rille). Rimae named after a specific feature do not always imply they are physically related.

Craters

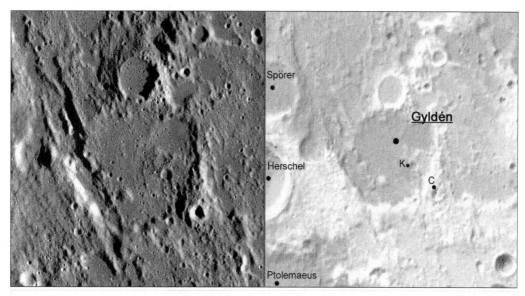

Gyldén	Lat 5.37S	Long 0.23E	48.15Km

Sub-craters:	Crater	Lat	Long	Size (km)	Crater	Lat	Long	Size (km)
	C	5.9S	0.99E	5.88	K	5.46S	0.6E	4.25
	Notes:							

Add Info: You might think that as Gyldén isn't that far off the central portion of the moon's nearside, it would be an easy target to observe. It isn't! It's old, worn-down appearance, along with its battered-to-death past, makes it quite difficult to glean any idea as to how the crater once looked in its prime. The whole area in which Gyldén lies has been affected by the impact that produced the Imbrium Basin - some 1200 km away to the northwest, and so we see lineated features all over the region; produced as a result of ejecta blocks obliterating nearly everything in their sights. This so-called 'Imbrium Sculpture' is perfectly seen on Gyldén's southwest sector of the crater, where a long gouged-like feature, over twice the diameter of the crater, has wiped out its rim there entirely. This 'gash' is actually deeper than the crater's floor, and during low sun times (just after first quarter, or before last quarter), a wonderful long line-of-a-shadow can be observed. If the floor's material is, presumably, ejecta fill from Imbrium, does this mean that the block that produced the 'gash' didn't follow that long after? Or, is previous material from some other source the main make-up of the floor (the crater did, after all, have plenty time to accumulate such deposits)? The small mound just east of the 'gash' doesn't look like a dome, or a peak, so it is probably part of the original block, or, is it part of Gyldén's rim? Depth ~ 1.1 km approx..

The impactor that produced the Imbrium Basin, and subsequently the Imbrium Sculpture, is estimated to have been 275 km in diameter. The impactor is also believed to have come in from a northwest direction; making the Basin an oblique-impact event.

Craters

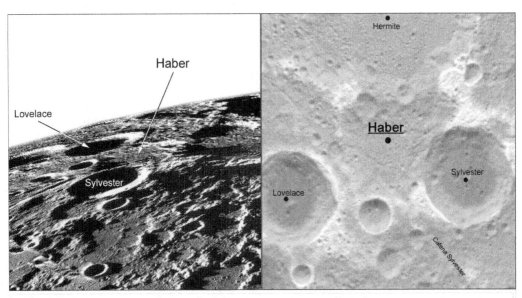

| Haber | Lat 83.4N | Long 94.63W | 56.79Km |

Sub-craters: No sub-craters

Notes:

Add Info:

Aerial view of Haber

You'd hardly believe Haber is a crater at all - given the lack of usual features (circularity, rim and walls etc.,) associated to them. A reason is because Haber lies approximately central between the three craters: Sylvester to its east, Lovelace to its west, and Hermite to its north. All three have impacted around Haber during different times (Hermite some 4.6 to 3.92 billions of years ago, Sylvester 3.92 to 3.85, and Lovelace younger again). In effect, all three have imparted their ejecta on to Haber's floor and rim; leaving its central level at ~ 1.5 km above both Lovelace's and Sylvester's floor, while ~ 1.0 km above Hermite's floor. Wonder, then, we 'see' Haber at all.

Craters

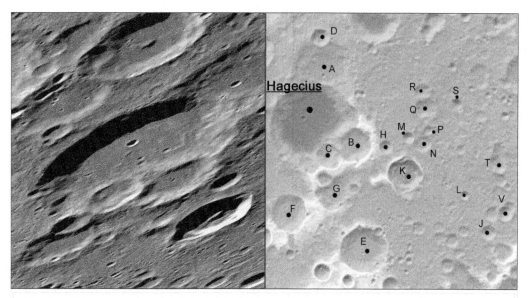

Hagecius Lat 59.92S Long 46.63E 79.55Km

Sub-craters:

Crater	Lat	Long	Size (km)	Crater	Lat	Long	Size (km)
A	58.26S	47.1E	54.33	B	60.48S	49.03E	32.42
C	60.78S	47.43E	23.15	D	57.32S	47.06E	16.32
E	63.5S	49.65E	42.98	F	62.38S	44.92E	32.32
G	61.86S	47.6E	30.56	H	60.48S	50.86E	12.8
J	62.69S	57.75E	13.6	K	61.26S	52.08E	31.36
L	61.7S	55.93E	8.11	M	60.18S	52.18E	10.5
N	60.33S	53.05E	15.82	P	59.91S	53.33E	7.84
Q	59.29S	53.09E	18.52	R	58.81S	52.84E	13.98
S	59.08S	54.87E	10.22	T	60.65S	57.73E	16.61
V	62.06S	58.64E	13.9				

Notes:

Add Info:

Aerial view of Hagecius

Hagecius is a 'tuffy' - not only to observe, but also to … find! It's guaranteed that if you come away from the eyepiece for a brief second and then try to locate it again, you'll have to go through the same task of firstly using the 'marker' you choose initially (Crater Janssen's definitive rille, to the northeast, points towards a cluster of similarly-sized craters in where Hagecius lies, so from there remember it is in the bottom of the group). Like its neighbours, Hagecius is an old crater with a worn rim, has well-levelled slumped deposits within, and has been the object of many-an-impact (see southeast). D: 3.97 km.

Craters

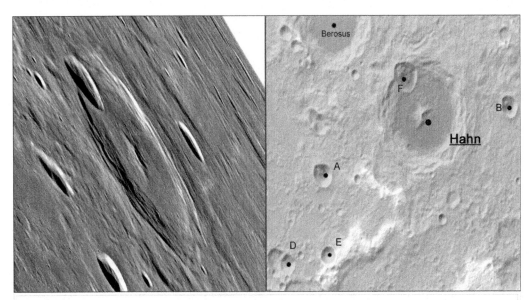

Hahn	Lat 31.22N	Long 73.55E		87.49Km			

Sub-craters:

Crater	Lat	Long	Size (km)	Crater	Lat	Long	Size (km)
A	29.67N	69.72E	18.31	B	31.38N	76.98E	17.87
D	27.41N	68.46E	14.86	E	27.68N	69.99E	15.24
F	32.16N	72.89E	25.63				

Notes:

Add Info:
Day-old moon phases

Note: As the Moon goes through various librations throughout the year, suggested times given in text for observations are only approximates.

Aerial view of Hahn

Like its similar-sounding fellow crater, Hayn, to the north lying some 1000 km away, Hahn is one for the 'limber' observers. The area, in general, is covered, with ejecta from two basins - firstly, by Humboldtianum to the north (near the other Hayn), and then by Crisium ejecta to the south that has superimposed upon it. But the crater is obviously younger - as its rim, its wonderful series of terraces and main peak all look relatively sharp. Sub-crater, Hahn F, is younger again (its ejecta lies on Hahn's floor), and the impactor that produced it has 'dug-down' into deep deposits in this crater with depth of 4.4 km approx..

Craters

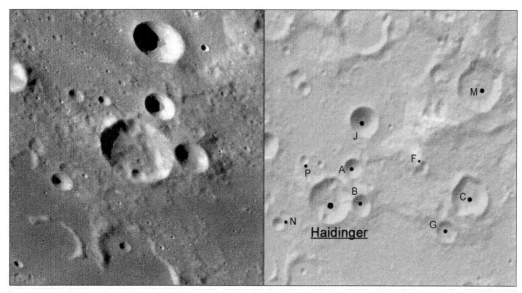

Haidinger   Lat 39.18S   Long 25.14W   21.33Km

Sub-craters:

Add Info:

Crater	Lat	Long	Size (km)	Crater	Lat	Long	Size (km)
A	38.69S	24.67W	8.55	B	39.25S	24.48W	10.08
C	39.1S	22.18W	18.25	F	38.63S	23.16W	4.6
G	39.6S	22.66W	10.12	J	37.99S	24.43W	14.22
M	37.46S	22.08W	20.96	N	39.52S	26.23W	6.26
P	38.57S	25.67W	4.49				

Notes:

Most of the lumpy, mountainy material here in which Haidinger lies is ejecta of Mare Humorum (Nectarian in age - 3.92 to 3.85 byo) to the northwest. Looking at the crater's features of worn rim, its levelled floor (containing slumped deposits, and possible ejecta clumps from the Imbrium Basin - some 2000 km away to the north), Haidinger, then, may have formed sometime between these two major events. Sub-craters' Haidinger A and B are much younger - the former shows some bright radial strokes, while the latter seems to have imparted a portion of Haidinger's eastern rim on to the floor. Not much, then, to be said about the crater, except that rays from Tycho crater to the southeast overlay it (best, and barely seen, during high sun times, and mainly covering the northeastern outer sector of the crater). There are quite a few craters here of similar size, so it's quite easy to get lost. However, a good location marker for finding the crater is to use that dark tongue of lava (Lacus Timoris) to its southwest, or its two small sub-craters, Haidinger A and B. Depth is approximately 2.3 km.

Craters

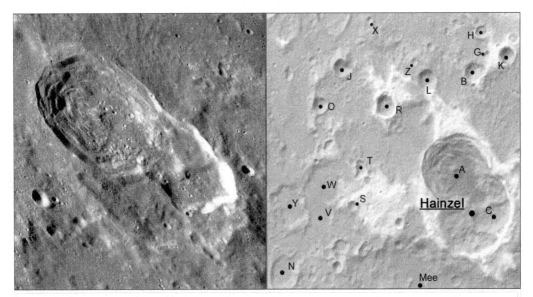

| Hainzel | Lat 41.23S | Long 33.52W | 70.56Km |

Sub-craters:

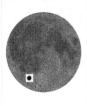

Crater	Lat	Long	Size (km)	Crater	Lat	Long	Size (km)
A	40.31S	34.02W	55.52	B	37.94S	33.45W	15.2
C	41.17S	32.79W	37.76	G	37.55S	33.0W	6.33
H	36.97S	33.17W	10.95	J	37.79S	37.88W	13.28
K	37.55S	32.31W	13.15	L	38.12S	35.01W	15.16
N	42.6S	40.19W	24.0	O	38.65S	38.65W	12.11
R	38.75S	36.42W	17.49	S	41.11S	37.68W	7.94
T	40.2S	37.32W	12.13	V	41.28S	38.8W	20.64
W	40.75S	38.67W	33.47	X	36.66S	36.89W	5.53
Y	40.97S	39.84W	29.65	Z	37.7S	35.43W	5.31

Notes:

Add Info:

Arial view of Hainzel

Lying in a region of extended basin deposits (formerly known as 'The Vitello Formation') from Mare Humorum to the northeast, the only portion of Hainzel's original crater that survived is at its southwest sector. Both sub-craters A and C have taken out the remaining portion of Hainzel now missing to the northeast. Sub-crater A may be the younger of the two, but as to what has happened at C's northeast sector with its overly-developed slumped deposits is an odd one - are they totally of A's? Hainzel, then, doesn't have much to say to the observer, so concentrate more on A and C when viewing it. Depth ~ 4 km approx..

Craters

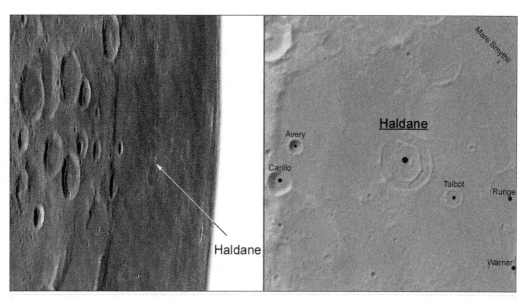

| Haldane | Lat 1.66S | Long 84.11E | 40.26Km |

Sub-craters: No sub-craters

Notes:

Add Info:

Aerial view of Haldane

Haldane looks more like two craters in one - both having very odd, hexagonal shaped rims off the normal circular form we're used to seeing in craters. Inner Haldane has a series of old-looking fractures in its west, while its centre is a mess of small peaky hills as dark pyroclastic deposits sidle up to them. Outer Haldane is of course the main rim of the crater, which lies in the Mare that is Smythii - centred ~ 100 km away to the east (note the series of concentric fractures outside the rim, too). Observing the crater can be quite difficult, so favourable librations are required, however, when seen, the hexagonality in the rim(s) show up more easily.

Craters

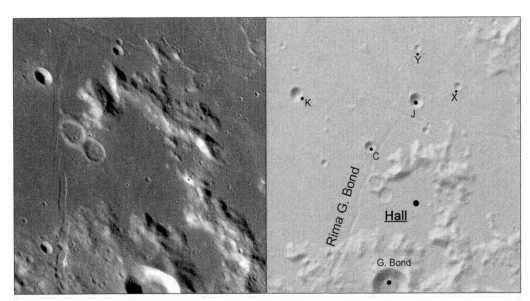

Hall	Lat 33.81N	Long 36.75E	31.77Km

Sub-craters:

Crater	Lat	Long	Size (km)	Crater	Lat	Long	Size (km)
C	34.71N	35.88E	5.7	J	35.47N	36.9E	7.55
K	35.57N	34.26E	7.21	X	35.7N	37.83E	4.0
Y	36.34N	36.94E	3.69				

Notes:

Add Info:

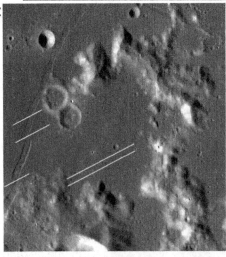

Aerial view of Hall (and series of faults)

Hall lies just outside one of the main rings/rims (the 920 km diameter one) of Mare Serenitatis. Looking more like a group of mountains in circular formation rather than an 'actual' crater (is it really one?), Hall has been breached by lavas to its southwest, and has had several faults cross its surrounds - see left. Note, how they are radial to Serenitatis (those on the Rima G. Bond rille are of horizontal 'strike-slip' faults, however, the one on the floor may be of a 'graben' type where a section has dropped down between two parallel faults). A nice bright fresh impact crater (~ 1 km) in the eastern sector of the rim is one that never fails to 'shine'. Depth ~ 1.2 km approx..

Craters

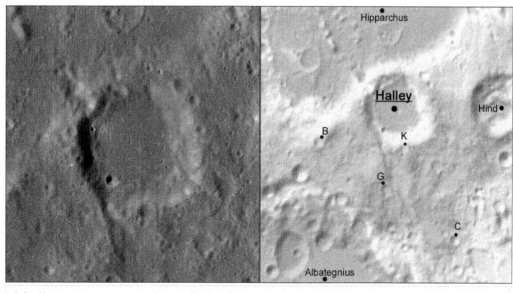

Halley	Lat 8.05S	Long 5.7E		34.59Km		
Sub-craters:	Crater	Lat	Long	Size (km)	Crater	Lat	Long	Size (km)
	B	8.48S	4.45E	5.45	C	9.88S	6.63E	4.7
	G	9.16S	5.51E	5.04	K	8.57S	5.82E	4.45
	Notes:						

Add Info: It must be a privilege as a regularly-sized crater, such as Halley, to lie in an area where some of the nicest 'biggies' (for example, craters Ptolemaeus, Alphonsus, Arzachel to the west, or, Hipparchus and Albategnius to the north and south respectively) formed on the Moon. Looking at its worn and battered appearance, Halley is obviously an 'oldie' like its neighbour, Hipparchus (pre-Nectarian in age 4.6 to 3.92 billions of years old), however, judging from the shape of its northwest rim sector, the two probably formed as a double impact event. Observe the area in general, and you'll notice that several craters have taken on a groove-like look to them. These grooves are the results of clumpy and rocky ejecta - created from the Imbrium Basin event to the northwest, and gouging into any high points, such as, rims and mountains that lay in its way (note the big gouge at Halley's western side, and how it, and other gouges, 'point' back to Imbrium). The crater's floor is smooth and hides, just about, a peak (hardly seen during low sun angle times), but as to what the material of the floor is made of (fluidized ejecta?, volcanic?), let alone its original source, is somewhat open to interpretation. Oddly enough, with so many craters nearby of similar size, Halley is quite easy to locate if using the three nicely-aligned craters (Hind, Hipparchus C and L) to its east. D: 2.5 km.

Craters

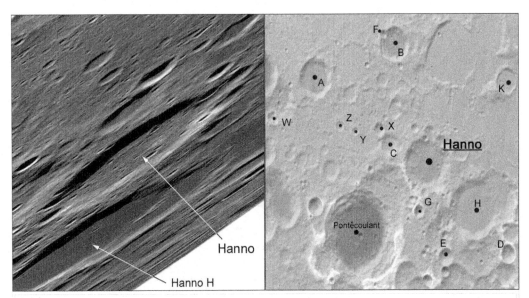

Hanno   Lat 56.46S   Long 71.38E   59.54Km

Sub-craters:

Crater	Lat	Long	Size (km)	Crater	Lat	Long	Size (km)
A	53.6S	63.47E	40.51	B	52.65S	68.95E	36.02
C	55.86S	68.54E	22.41	D	59.1S	78.03E	18.25
E	59.43S	73.04E	17.74	F	52.3S	68.22E	9.36
G	58.0S	70.77E	12.88	H	57.84S	74.96E	57.74
K	53.62S	76.74E	25.32	W	54.67S	60.13E	13.2
X	55.37S	67.81E	13.74	Y	55.39S	66.06E	8.41
Z	55.18S	64.97E	11.2				

Notes:

Add Info:

Aerial view of Hanno

Lying just on the outermost, slightly higher terrain not far from Mare Australe to the crater's southeast, Hanno's position make's it one for viewing during favourable librations. It's quite easy to get lost here in finding the crater; as so many of similar in appearance can lead to confusion (the darker deposits of Australe near the crater's eastern rim and sub-crater, Hanno-H, may help as a guide). The crater has a well worn look to it, has been pummelled by numerous impacts (old and recent), and a thin layer of ejecta from Nectaris to the north (some ~ 1500 km away) covers it. Depth ~ 3.3 km.

Craters

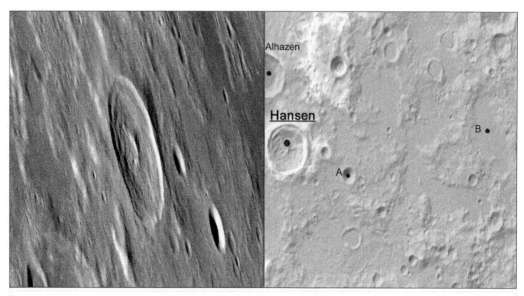

Hansen	Lat 14.04N	Long 72.54E					41.18Km

Sub-craters:

Crater	Lat	Long	Size (km)	Crater	Lat	Long	Size (km)
A	13.36N	74.67E	13.47	B	14.43N	79.52E	79.89

Notes:

Add Info:

Aerial view of Hansen

Hansen lies on ejecta deposits of the ring/rim system (the ~ 740 km in diameter one) associated to the formation of Mare Crisium to its west. The crater has a relatively fresh look to it; having a sharp rim whose series of terraces within produce some nicely-defined shadows during low sun angle times (a peak reflects the same effects, too). Together with its fellow neighbouring craters - Alhasen to its north, Condorcet to is southwest, and Crisium at its west, the 'quatre' make for a wonderful view in the eyepiece, or photo-shot. Peak height is ~ 1.8 km approx, depth is ~ 3.0 km.

Craters

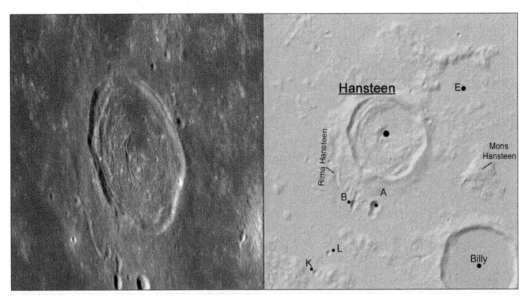

Hansteen		Lat 11.53S	Long 52.06W			44.99Km	

Sub-craters:	Crater	Lat	Long	Size (km)	Crater	Lat	Long	Size (km)
	A	12.76S	52.26W	6.54	B	12.73S	52.68W	5.38
	E	10.62S	50.6W	28.28	K	13.9S	53.41W	3.34
	L	13.55S	53.15W	3.11				

Notes:

Add Info: It's almost a given that when you observe crater Hansteen, you nearly always will include in your view its fellow brother, Billy, to its southeast. The two are of the same size (Billy is 45.57 km in diameter), of similar age - probably of Upper Imbrian - 3.75 to 3.2 billions of years old, have the same depth (Hansteen is ~ 1.4 km while Billy is 1.3 km), and each exhibits features related to volcanic activity (Billy's lava deposits probably welled up from beneath its interior, while the fractured appearance of Hansteen is due to same structural occurrences that didn't quite reach the surface). Hansteen may have went the same way as Billy, however, its impact was on slightly higher ground; leaving some 200 m difference in height finally between each of their floors. Besides the obvious concentric fractures in Hansteen, note how the several major slumpings of the rim's interior make-up have amalgamated to them - making it sometimes hard to distinguish where each starts and ends. Two other features to look for are the lava deposits in the floor's northeastern sector, and the odd-looking cleft at the rim's north. Are the two related? Are the northeastern lava deposits the results of a breach at the cleft (linking the crater to the much larger lava deposits of Oceanus Procellarum), which then closed, and blocked up the gap, afterwards?

Craters

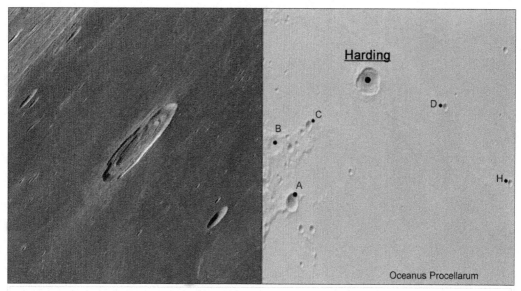

Harding   Lat 43.54N   Long 71.66W   22.57Km

Sub-craters:

Crater	Lat	Long	Size (km)	Crater	Lat	Long	Size (km)
A	40.39N	75.47W	14.53	B	41.81N	76.41W	16.73
C	42.39N	74.75W	8.49	D	42.85N	67.64W	6.29
H	40.75N	64.42W	5.53				

Notes:

Add Info:

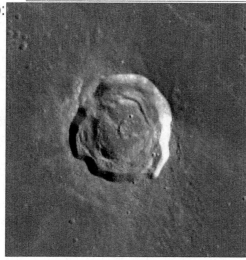

Aerial view of Harding

There are very few craters on the vast lava deposits that is Oceanus Procellarum. So when we see a relatively fresh-looking one like crater Harding, it's as well to note its uniqueness and rarity. The crater has a fine, but crenulated rim, has a wonderful series of terraces (due to slumping effects), and its outer ejecta deposits show up nicely during low sun angle times. Its location, of course, makes it a hard one to get any decent detail from; and that's even using the highest of magnification. Brighter ray material predominantly at its outer SE sector suggest's the impact may have been oblique (?).

Craters

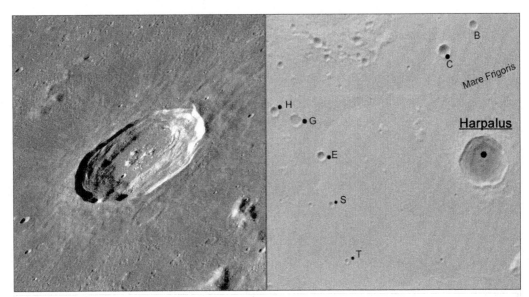

Harpalus   Lat 52.73N   Long 43.49W   39.77Km

Sub-craters:

Crater	Lat	Long	Size (km)	Crater	Lat	Long	Size (km)
B	56.24N	43.77W	7.71	C	55.59N	45.24W	10.0
E	52.77N	50.98W	7.52	G	53.63N	52.28W	10.37
H	53.8N	53.34W	7.85	S	51.55N	50.4W	4.42
T	50.12N	49.57W	4.35				

Notes:

Add Info:

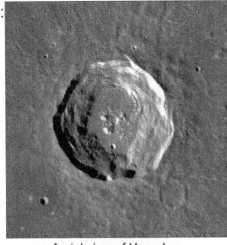

Aerial view of Harpalus

Harpalus observed under any lighting conditions never fails to disappoint. Its wonderful 'spray' of ejecta around its exterior produces some nice shadows that 'spoke' inwards to the crater's rim during low sun angle times, while high suns give a rewarding eyeful of detail to its inner surrounds. Eratosthenian in age (3.15 - 1.1 byo), Harpalus shows some nice terraces and slump effects (note, the two 'nicks' at the crater's northern and eastern rim, which suggest separate, major slump events), and three main peaks meet smoother material (is it of internal lava deposits or of impact melt?) on the floor. D ~ 3.5 km.

Craters

Hartwig	Lat 6.41S	Long 80.47W		78.46Km			
Sub-craters:	Crater	Lat	Long	Size (km)	Crater	Lat	Long	Size (km)
	A	5.76S	79.86W	9.95	B	8.42S	77.56W	10.22
	Notes:							

Add Info:

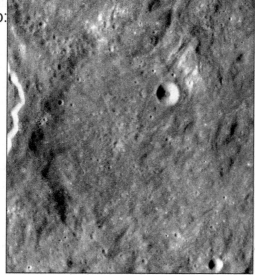

Aerial view of Hartwig

Locating Hartwig can be a little bit difficult at times, however, if Schlüter to its west happens to be in view, use it for finding the crater (when it isn't in view, the next best help is the small, young crater, Hartwig A, on its floor at the northeast). The location problem, and hence the observing problem, too, is due to the pasty-looking ejecta (from the Orientale Basin - centred ~ 500 km away to the southwest) that partially covers the crater. Schlüter also is younger than Hartwig, and its impact (and ejecta) has raised up Hartwig's western side by more than a kilometre than its eastern. Libration times help, but low sun views are better. Depth ~ 2.0 km.

Craters

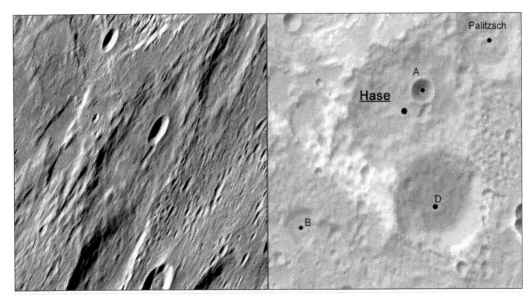

Hase	Lat 29.37S	Long 62.68E				82.08Km	

Sub-craters:	Crater	Lat	Long	Size (km)	Crater	Lat	Long	Size (km)
	A	29.06S	62.94E	14.55	B	31.51S	60.06E	19.98
	D	31.11S	63.3E	56.69				

Notes:

Add Info:

Aerial view of Hase

Lying close to crater Petavius in the northwest, Hase, from its appearance, is a crater of age (possibly of pre-Nectarian - 4.6 to 3.92 byo). Ejecta deposits from Petavius and Mare Nectaris overlie the crater's northern sectors (that also include sub-crater, Hase D), giving the two a hummocky, rubbly look to them. The rille-like feature crossing the floor's southwestern sector is part of a fracturing process - radially related to Nectaris, which forms part of the Rimae Hase network. Sub-crater, Hase A, is relatively fresh-looking, and together with the small, 'notch-like' crater that it has obviously impacted to its south, both serve as a useful reference. Depth ~ 3.5 km approx..

Craters

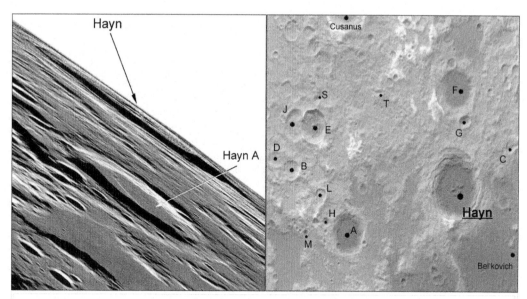

Hayn Lat 64.56N Long 83.87E 86.21Km

Sub-craters:

Crater	Lat	Long	Size (km)	Crater	Lat	Long	Size (km)
A	63.01N	71.11E	52.28	B	65.16N	63.86E	24.58
C	65.31N	91.45E	16.87	D	65.44N	61.64E	20.85
E	67.02N	66.1E	41.64	F	68.01N	85.75E	62.54
G	66.86N	86.31E	20.11	H	63.37N	68.61E	12.26
J	66.91N	63.71E	38.24	L	64.35N	67.7E	19.53
M	62.84N	66.28E	9.24	S	68.47N	66.14E	11.67
T	68.47N	74.58E	6.83				

Notes:

Add Info:

Hayn was formerly known as Strabo G. Note, be careful not to confuse with similar-sounding crater - that of, Hahn, at Lat 31.22N, Long 73.55E.

Aerial view of Hayn

Time your observation well when trying to catch a good view of Hayn. Like most features on the extreme limb portions of the Moon, favourable librations, and good lighting conditions are essential when viewing such huggers; where planning and timing can make all the difference in the detail finally seen. It is a pity that Hayn is where it is as it is quite a nice crater - having a wonderful series of peaks, a sharp rim all around (the crater is young and of the Copernican Period - 1.1 byo to Present), and terraces whose only ones we can see, from our earth perspective, are its eastern. Depth ~ 4.5 km approx..

Craters

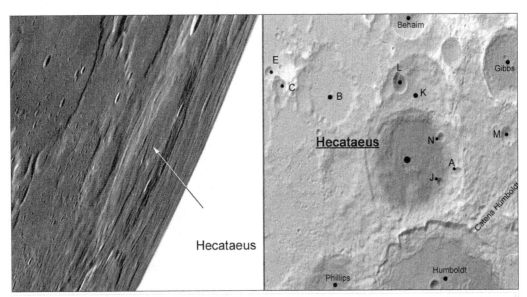

Hecataeus

| Hecataeus | Lat 22.06S | Long 79.68E | 133.67Km |

Sub-craters:

Crater	Lat	Long	Size (km)	Crater	Lat	Long	Size (km)
A	22.11S	81.78E	11.15	B	19.38S	75.62E	68.09
C	18.96S	73.09E	20.36	E	18.52S	72.64E	14.06
J	22.47S	80.9E	10.71	K	19.6S	79.69E	94.14
L	19.07S	78.95E	23.31	M	20.79S	84.19E	18.93
N	20.91S	80.94E	10.82				

Notes:

Add Info:

Aerial view of Hecataeus

A good finder's guide for Hecataeus is to firstly note its more obvious neighbour, Humboldt, to its south, and from there it's easily spotted. Hecataeus has got to be older than its neighbour (Upper Imbrian 3.75 to 3.2 byo) as that impact has deposited a huge clump of ejecta (and possibly rim material) onto Hecataeus's floor. The crater has a surprisingly deep depth of around ~ 5 km, which makes for some nice shadow-fills during low sun times, but as the crater's location is on the limb anyway, these times are likely to be avoided if any detail of the crater's interior is to be seen (together using a good libration, too).

Craters

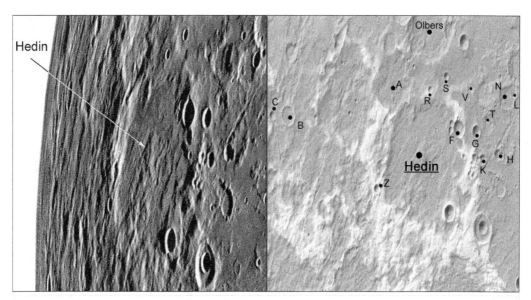

Hedin		Lat 2.87N		Long 76.57W			157.38Km

Sub-craters:

Crater	Lat	Long	Size (km)	Crater	Lat	Long	Size (km)
A	5.4N	78.06W	64.65	B	4.34N	83.85W	21.13
C	4.33N	84.69W	10.57	F	3.95N	74.5W	19.65
G	3.82N	73.51W	13.78	H	3.06N	72.33W	12.28
K	2.91N	73.16W	11.3	L	5.13N	71.47W	10.5
N	4.97N	71.88W	23.97	R	5.24N	76.09W	7.63
S	5.69N	75.22W	8.58	T	4.24N	72.96W	6.84
V	5.29N	73.85W	10.5	Z	1.91N	79.01W	10.22

Notes:

Add Info:

Aerial view - Lunar Orbiter Frame 4174 H1

There's a reason why Hedin looks like its been through the ejecta wars - it's because the huge event that created the Orientale Basin (whose centre is some 850 km away to the southwest) was the source. This ejecta - usually referred to as the 'Hevelius Formation' - sprayed out in every direction, which thinned in quantity the further it travelled from the main impact. At high magnification, another feature of Orientale's dynamic influence can be seen as a wonderful series of concentric rilles (tectonic cracks) that cross on Hedin's floor in a NW - SE direction. Observe these features during low sun times. D = 1.3 km.

Craters

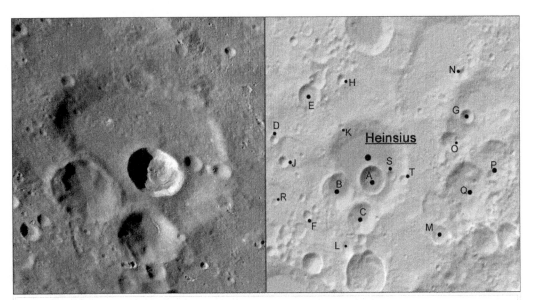

Heinsius   Lat 39.48S   Long 17.82W   64.87Km

Sub-craters:

Crater	Lat	Long	Size (km)	Crater	Lat	Long	Size (km)
A	39.77S	17.61W	19.53	B	40.01S	18.73W	23.71
C	40.64S	17.95W	22.42	D	38.82S	20.75W	6.54
E	37.87S	19.58W	16.76	F	40.55S	19.73W	7.2
G	38.32S	14.61W	10.09	H	37.51S	18.56W	7.29
J	39.32S	20.45W	7.9	K	38.58S	18.6W	4.89
L	41.25S	18.46W	7.23	M	40.98S	15.4W	12.45
N	37.35S	14.77W	7.06	O	38.84S	14.92W	4.08
P	39.4S	13.74W	35.52	Q	39.89S	14.54W	33.7
R	40.22S	20.79W	4.56	S	39.66S	17.0W	6.08
T	39.76S	16.6W	6.56				

Notes:

Add Info: Lying in a 'jumbled' region where craters of every sort, size, and age formed, it would be quite easy to get confused in finding Heinsius, if it wasn't for sub-crater, Heinsius A, on its floor. The sub-crater, at most times, always seems to stand out from others nearby; why? - because shadows fill its deep inner rim and floor whose bottom is some 2 km below that of the floor of Heinsius itself. In terms of appearance and impact dynamics, a nice exercise is to compare how sub-crater A looks to sub-craters' B and C (that have totally wiped out nearly all the southwestern rim of Heinsius), and work out the way rim and ejecta material from each must have been shifted and shoved around during impact. The crater lies only 180 km away from Tycho to the southeast, however, while ray material from that impact undoubtedly covered Heinsius, it isn't very obvious - even during high sun times.

Craters

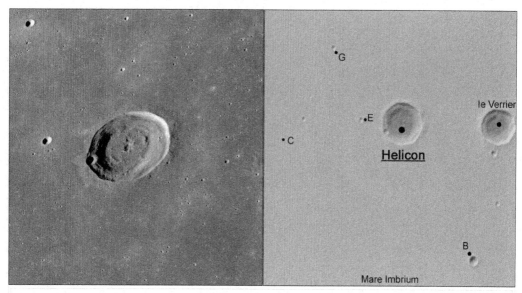

Helicon	Lat 40.43N	Long 23.11W			23.74Km	

Sub-craters:	Crater	Lat	Long	Size (km)	Crater	Lat	Long	Size (km)
	B	37.95N	21.3W	5.45	C	40.08N	26.26W	1.25
	E	40.49N	24.18W	2.69	G	41.77N	24.91W	2.52

Notes:

Add Info: When observing Helicon, be sure to also include its neighbour, le Verrier (to its east) in your view; as the two make for a nice comparison between similar features seen in each. One would say that Helicon is older than le Verrier as its ejecta has been covered more by lavas in Imbrium than le Verrier's (note the extent of its ejecta around the crater) has been. But this is not really a sufficient signature to infer a difference in age between the two - just by observing the ejecta; we also need to note how other features look, too. Observing, say, the rims of each; we can see a very slight sharpness to le Verrier's than Helicon's. The slumping of material within their interiors is also a help - Helicon's is more smoother and evenly distributed (suggesting it had time to develop so), while le Verrier's is more broken as a series of separate ridges and terraces. The floors of each aren't much help as they both look similar, however, if we know that le Verrier's is some 100 metres lower than Helicon's, then the impactor that caused it may have struck a more solid (cold lava) surface, which would then explain its ejecta volume and distribution. 'Eye-balling' features in neighbouring craters may not always work when trying to determine which is the youngest/oldest, but the method is something that should always be used during your observations. Depth ~ 2 km.

Craters

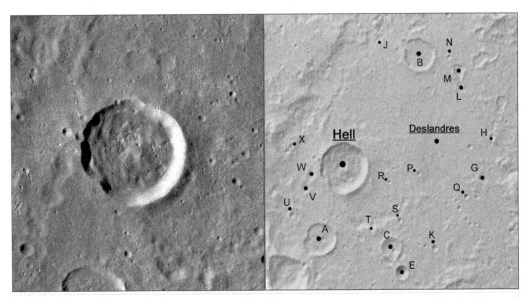

Hell   Lat 32.41S   Long 7.8W   33.31Km

Sub-craters:

Crater	Lat	Long	Size (km)	Crater	Lat	Long	Size (km)
A	33.91S	8.46W	20.97	B	30.01S	5.8W	21.55
C	34.11S	6.5W	13.96	E	34.63S	6.21W	9.32
H	31.73S	3.82W	4.87	J	29.7S	6.87W	5.19
K	34.09S	5.3W	4.82	L	30.65S	4.72W	5.33
M	30.35S	4.77W	9.68	N	30.06S	5.0W	3.45
P	32.54S	5.77W	3.42	Q	33.0S	4.47W	3.75
R	32.72S	6.56W	3.02	S	33.53S	6.27W	3.73
T	33.7S	7.04W	4.02	U	33.4S	9.2W	4.21
V	32.83S	8.83W	7.16	W	32.56S	8.67W	6.61
X	31.96S	9.2W	4.04				

Notes:

Add Info: Lying on the ejecta deposits of Mares' Nubium (pre-Nectarian - 4.6 to 3.92 byo) and Humorum (Nectarian - 3.92 to 3.85 byo) to the northwest and west respectively, Hell is an odd-looking crater. It's as if the top, northwestern half of the crater has undergone major changes due to slumping occurring there (extending the crater's size), while in the opposite, southeastern direction the crater avoided such structural dynamism. Why this dichotomy (did impact of the two small craters just off Hell's northwest rim produce the slumping/avalanching effect?)? The crater has a small peak of about 900 metres high (best observed during low sun times), however, not a hint of the original floor can be seen. Note how this crater relates to Deslandres, and the possible volcanic events both underwent.

Craters

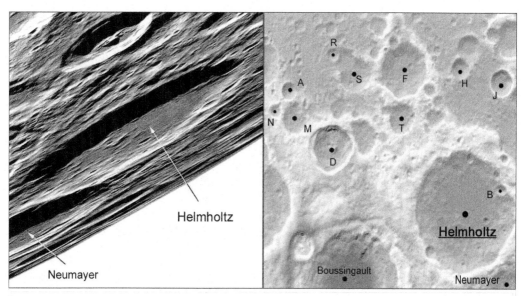

Helmholtz / Neumayer / Boussingault / Helmholtz

Helmholtz   Lat 68.64S   Long 65.34E   110.16Km

Sub-craters:

Crater	Lat	Long	Size (km)	Crater	Lat	Long	Size (km)
A	64.5S	51.52E	15.98	B	67.88S	68.66E	11.83
D	66.38S	54.09E	44.59	F	64.46S	60.55E	50.0
H	64.53S	64.97E	19.83	J	64.88S	68.24E	23.02
M	65.33S	51.37E	24.31	N	64.95S	50.2E	13.53
R	63.78S	55.27E	12.21	S	64.41S	56.7E	31.67
T	65.8S	60.08E	30.53				

Notes:

Add Info:

Depth of Helmholtz is around the 4.5 km mark, so deep shadows sometimes are the rule.

Aerial view of Helmholtz

Sometimes, it's okay to curse at the Moon's limb - especially when there are some nice big craters there that we just can't see because of our 'locked-in' view of them. Helmholtz, Boussingault, sub-crater Boussingault E, Neumayer and Boguslawsky form wonderful quintuplets that teases the viewer, but the problem is that they are only 'accessible' through favourable librations, good lighting conditions, and 'windows-of-opportunity' few have observed. These craters, therefore, don't get that much of a 'look-in', and the only ones doing any serious observations are the astrophotographers. Time to join them?

Craters

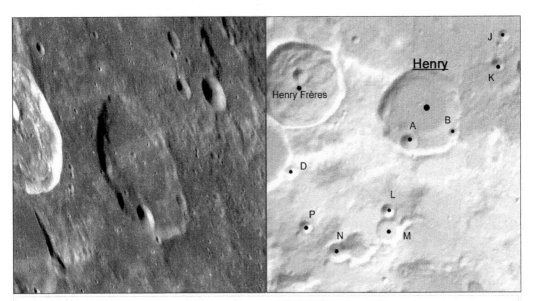

Henry   Lat 23.97S   Long 57.01W   39.06Km

Sub-craters:

Crater	Lat	Long	Size (km)	Crater	Lat	Long	Size (km)
A	24.42S	57.2W	8.31	B	24.28S	56.44W	5.14
D	24.87S	59.21W	6.78	J	22.79S	55.61W	5.89
K	23.22S	55.7W	6.4	L	25.46S	57.55W	6.48
M	25.77S	57.53W	12.84	N	26.11S	58.46W	8.77
P	25.7S	58.97W	6.38				

Notes:

Add Info:

Aerial view of Henry

When observing Henry be sure also to include its westerly neighbour, crater Henry Frères, in your view. From initial appearances, it's obvious to see that the two are of different ages - Henry is of the Late Imbrian, while Henry Frères is of the Eratosthenian. Henry's interior floor of depth 2.7 km is filled with ejecta from the Orientale Basin event to the southwest, its western sector also has been 'splashed' by ejecta from Frères, too (rays from sub-crater, Byrgius A, to the west - some 180 km away, crosses its floor). As Henry Frères has a depth of 3.8 km approx.,, low sun times produces some nice deep shadows in it, not seen in Henry's.

Craters

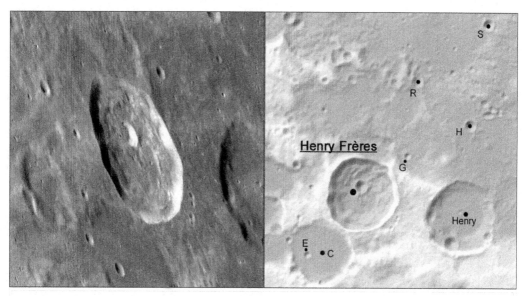

Henry Frères Lat 23.52S Long 59.02W 41.73Km

Sub-craters:

Crater	Lat	Long	Size (km)	Crater	Lat	Long	Size (km)
C	24.63S	59.83W	35.14	E	24.61S	60.13W	4.2
G	22.86S	58.1W	4.08	H	22.31S	56.83W	6.64
R	21.52S	57.9W	6.98	S	20.5S	56.48W	6.24

Notes:

Add Info:

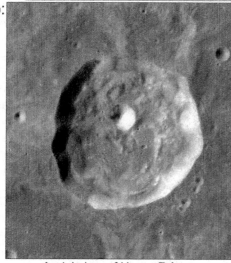

Aerial view of Henry Frères

Henry Frères is younger than its fellow neighbouring crater, Henry, to its east. Major slumping of material from nearly every sector of its interior rim has completely filled the floor, making it hard to define the original level. Was the small, fresh crater (about 8 km in diameter) just off centre on its floor a cause for some of the slumping, and did its impact also wipe out totally any semblance of a peak that once may have existed in the crater? Rays of Byrgius A (a Copernican-aged crater) to the west crosses H.Frères's floor at several points; some of which 'highlight' portions of the rim when viewed under favourable lighting conditions.

Craters

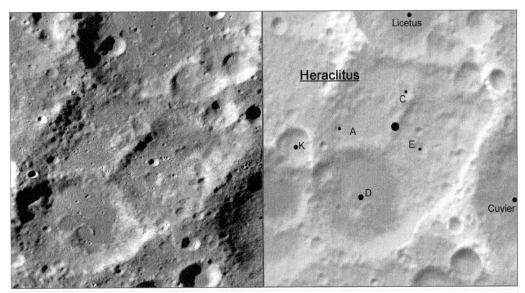

Heraclitus	Lat 49.31S	Long 6.42E	85.74Km

Sub-craters:

Crater	Lat	Long	Size (km)	Crater	Lat	Long	Size (km)
A	49.41S	4.54E	4.97	C	48.89S	6.22E	6.94
D	50.51S	5.14E	56.12	E	49.74S	6.55E	6.97
K	49.64S	3.45E	16.92				

Notes:

Add Info:

*Have a look also at crater Delaunay 22.26S, 2.62E, to see another divide-like feature.

Aerial view of Heraclitus

Heraclitus is quite easy to find in this area of numerous craters - simply because of the *divide-like mountain that crosses its floor. What is this feature? Is it the remains of a peak that once existed in Heraclitus before signature of it became confused, or lost, to nearby impact craters' Licetus and sub-crater Heraclitus D, to the north and southwest respectively? The feature casts some nice shadows during a terminator apparition of a ~ 19-day-old moon period; leaving it completely isolated in the crater's centre as a bright 'ridge'. The only remaining portion of the crater's original rim seems to exist at its west as nearly every other part has been obliterated by other impacts (the eastern rim is really of Cuvier's only). D ~ 3.3 km, peak ~ 1.3 km.

Craters

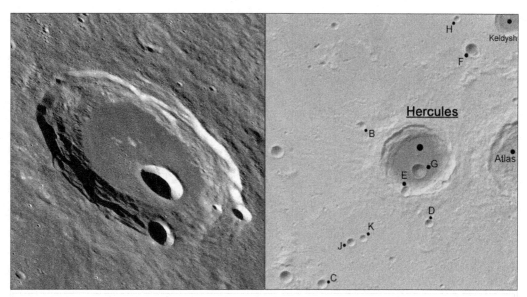

Hercules   Lat 46.82N   Long 39.21E   68.32Km

Sub-craters:

Crater	Lat	Long	Size (km)	Crater	Lat	Long	Size (km)
B	47.87N	36.7E	7.38	C	42.74N	35.36E	8.79
D	44.79N	39.73E	7.96	E	45.8N	38.7E	9.98
F	50.31N	41.75E	13.24	G	46.46N	39.28E	13.82
H	51.33N	41.04E	5.97	J	44.12N	36.44E	7.86
K	44.26N	36.96E	7.04				

Notes:

Add Info:

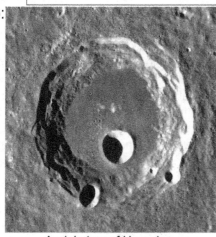

Aerial view of Hercules

Hercules is of the Eratosthenian Period - 3.15 to 1.1 billions of years ago. It has a wonderful series of terraces due to major slumping effects at the rim's interior, which meet a smooth floor whose makeup is of lava sourced from underneath. Several small bright 'dots' within the floor suggest peaks that possibly have been covered by the lava fill, but also note how ejecta from sub-crater, Hercules G, relates to this, too. Sub-crater, Hercules E, shows a nice impact effect where the event which struck on a sloped part of the inner rim produced a final crater having a more oval than round shape. Depth of Hercules ~ 3.2 km.

Craters

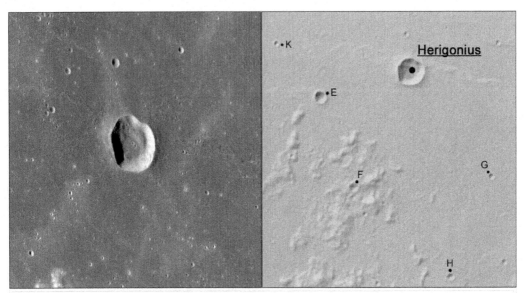

Herigonius	Lat 13.36S	Long 33.97W	14.86Km

Sub-craters:

Crater	Lat	Long	Size (km)	Crater	Lat	Long	Size (km)
E	13.82S	35.65W	6.63	F	15.48S	35.04W	5.23
G	15.27S	32.46W	3.17	H	17.12S	33.22W	4.15
K	12.83S	36.46W	3.07				

Notes:

Add Info: Before you take a look at Herigonius itself, take a 'bigger picture' view of its surrounds. Note: the vast lava deposits of Oceanus Procellarum to its north, the many signatures of partially-buried craters nearby, Herigonius's rille itself over to the west, and then there are the series of crinkle ridges where Herigonius actually interrupts a big one running in a northwest-southeast direction. Note also the group of small craterlets nearby to the crater - two nice sets at both the 11-o-clock and 8-o-clock positions, and another group a few kilometres away off to the east (note how all their individual impacts have lightened the ground - a sure signature of fresh material ejected from the regolith). Of course, you are not going to see all these in one view, or in one sitting at the telescope, but do make it a note to recognise them under several instances of different lighting conditions over time. Back to Herigonius crater. It obviously is a young crater - perhaps of the Eratosthenian Period (3.15 to 1.1 byo), its rim is relatively sharp. Something at its northern end has 'crinkled' the rim there; perhaps, due to the impactor hitting the above-mentioned ridge, but then, look at the southern end of the crater's rim whose edge is reversed to the north's direction (are these features of an oblique impact?). Depth of Herigonius is around the 2 kilometre mark.

Craters

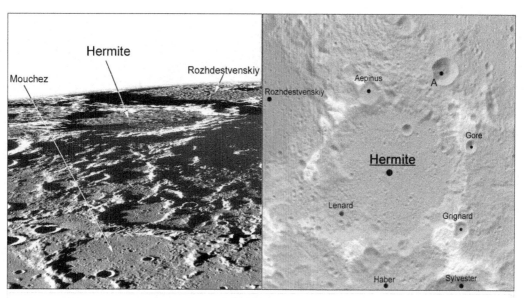

Hermite	Lat 86.17N	Long 93.32W	108.64Km

Sub-craters:	Crater	Lat	Long	Size (km)	Crater	Lat	Long	Size (km)
	A	87.94N	51.02W	19.86				

Notes:

Add Info:

Aerial views of Hermite

With such high latitude and longitude coordinates, it's a sure thing that good libration times, good lighting conditions, a good eye or camera, patience, are all that is required (for every observer's needs) for these elusive, and difficult limb-huggers. The crater's thick rim is just over four kilometres high in parts, its floor is slightly concaved (filled with deposits of ejecta and whatnot from several craters and basins, e.g. Imbrium, to the south nearby) - all of which leads to a narrow window of time to avoid catching shadows that will hide details. Lenard, off to its southwest, is slightly younger as it appears to have impacted on Hermite's rim there - both of which are probably pre-Nectarian in age (4.6 to 3.92 byo).

Craters

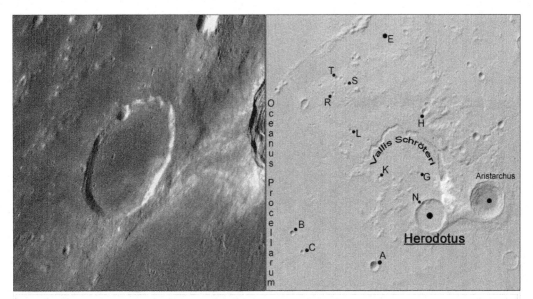

Herodotus Lat 23.25N Long 49.84W 35.87Km

Sub-craters:

Crater	Lat	Long	Size (km)	Crater	Lat	Long	Size (km)
A	21.52N	52.14W	9.98	B	22.57N	55.47W	5.64
C	21.94N	55.05W	4.68	E	29.35N	51.52W	36.55
G	24.68N	50.33W	3.51	H	26.77N	50.07W	6.22
K	24.51N	51.97W	4.72	L	26.11N	53.19W	3.93
N	23.64N	50.12W	4.32	R	27.37N	53.95W	3.94
S	27.68N	53.43W	4.1	T	27.88N	53.81W	5.03

Notes:

Add Info:
Vallis Schröteri is the largest sinuous valley on the Moon. Roughly 185 km long, 11 km wide and 1 km deep, this "rille" type feature is thought to be the remains of a collapsed, hollow lava tube that fell in on itself millions of years after the flow of lava had stopped.

Herodotus and the Aristarchus Plateau

Trying hard to ignore Herodotus's 'flashy' neighbour, Aristarchus to its east, the crater will always play second fiddle to it. Initially, it looks boring with its flat floor, worn rim, and non-featured attributes. But take a closer look - see how the crater's high, clumpy eastern sector has affected distribution of Aristarchus's bright ejecta across the floor and also at its western exterior. Look, too, at Herodotus's outer southeastern sector where a sharp, steep arc-like cut-off meets the lavas of Procellarum - what has happened there? Depth ~ 1.5 km approx..

Craters

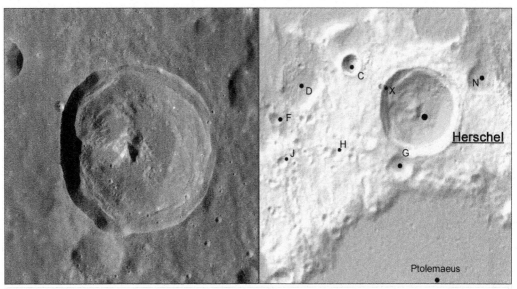

Herschel	Lat 5.69S	Long 2.09W				39.09Km

Sub-craters:	Crater	Lat	Long	Size (km)	Crater	Lat	Long	Size (km)
	C	5.01S	3.18W	9.89	D	5.32S	3.99W	19.24
	F	5.79S	4.39W	5.84	G	6.5S	2.41W	12.25
	H	6.31S	3.45W	4.7	J	6.42S	4.28W	4.85
	N	5.22S	1.09W	14.41	X	5.36S	2.72W	2.72
	Notes:							

Add Info: When informing to a fellow observer that you were 'looking at crater Herschel the other night', it's a given that he/she understands you mean it to be this one (it was named in honour of Frederick William Herschel (F. W. H.) - discoverer of the planet Uranus). There are two other Herschel craters on the Near Side: Caroline Herschel is at Lat 34.48N, Long 31.29W, while John Herschel is at Lat 62.31N, Long 41.86W. Frederick Herschel (this one) was born in 1738, Caroline Herschel, his sister, was born in 1750, while John Herschel, who is the son of F. W. H.) was born in 1792. Of all three craters Herschel is the most suitably-placed for observing, but each, in terms of their individual features, has a lot to offer the viewer. The crater has a slightly younger look to it than C. Herschel (both are possibly of the Eratosthenian Period - 3.15 to 1.1 byo) and it is obviously younger than J. Herschel, which is probably of the Nectarian Period (3.92 to 3.85 byo). The crater has wonderful terraces all within the rim - the eastern ones look like they formed in one event while the western ones may have occurred through several smaller events. Peak ~ 1 km high, depth of crater ~ 3.5 km approx..

Caroline Herschel is usually referred to as C. Herschel (see under craters, C), while John Herschel is referred to as J. Herschel (see under craters, J).

Craters

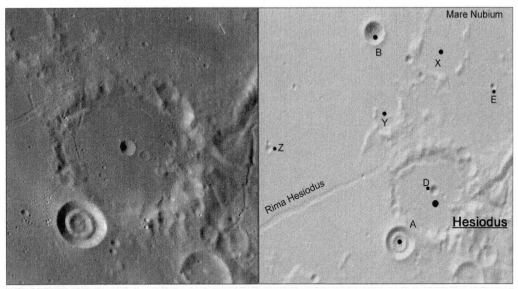

Hesiodus Lat 29.42S Long 16.42W 43.24Km

Sub-craters:

Crater	Lat	Long	Size (km)	Crater	Lat	Long	Size (km)
A	30.13S	17.07W	14.13	B	27.16S	17.54W	9.88
D	29.39S	16.43W	4.58	E	27.91S	15.35W	3.16
X	27.4S	16.28W	23.9	Y	28.32S	17.29W	17.8
Z	28.77S	19.51W	3.78				

Notes:

Add Info: Lying some 60 kilometres or so within the ~ 700 km-diameter ring/rim of Mare Nubium, Hesiodus is a crater of age. Nubium itself is of pre-Nectarian (4.6 to 3.92 billions of years old), and judging from the state of Hesiodus's battered, lava-strewn, ejecta-strewn (and any other strewn you care to think of) rim, it's a wonder that we see any semblence of a crater at all. The lavas that fill the crater's floor were more than likely sourced from underneath - the concentric fractures at its western and northwestern sectors hint at signatures of central uplift due to magma below trying to extrude upwards - like crater Pitatus over to its east (did additional lavas from Pitatus infill, too, through the small, 5-kilometre-wide gap that links both craters?). This 'gap' is an odd feature! Whilst it is known to produce some nice shadows and light effects particularly around 10/11-day-old moon periods, the puzzle is how did it form initially (the gap's channel doesn't 'point back' to the Imbrium Basin whose ejecta blocks in this area have changed the local landscape, so what caused it?). Of course, one can't leave a write-up on Hesiodus without mentioning Hesiodus A - a wonderful example of a concentric crater, and signature here, too, of possible layering of several lava fronts.

Concentric craters were once thought the result of of two impact craters - one lying within the other. While uncertainty stilll surrounds their formation, they may just be the product of a single impactor striking a layered surface. See crater Leakey for more.

Craters

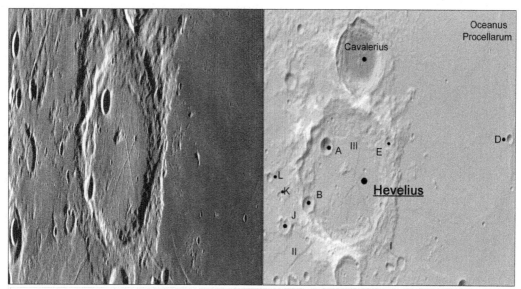

| Hevelius | Lat 2.2N | Long 67.46W | 113.87Km |

Sub-craters:

Crater	Lat	Long	Size (km)	Crater	Lat	Long	Size (km)
A	2.85N	68.23W	14.01	B	1.36N	68.98W	13.5
D	3.05N	60.91W	7.76	E	2.96N	65.78W	8.81
J	0.8N	69.9W	13.13	K	1.58N	70.08W	5.67
L	2.04N	70.48W	7.6				

Notes:

Add Info:

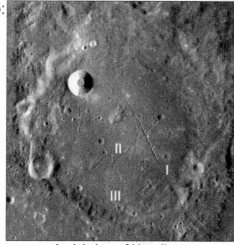

Aerial view of Hevelius

Criss-crossing rilles are the main features one sees immediately when viewing Hevelius. Their formation is probably due to the central floor doming upwards due to pressure from a nearby magma source underneath (but Rima I, which extends way outside the crater's southeastern sector, may be related to fracturing in the mare nearby). The eastern part of the floor is slightly darker; suggesting some magma may have extruded there, but this is only noticeable during high sun times. The whole area has been covered by ejecta from Mare Orientale to the southwest, whose event was named the 'Hevelius Formation'.

Craters

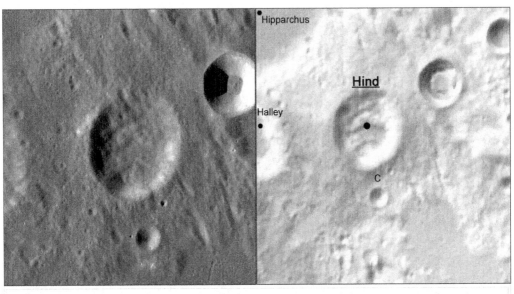

Hind	Lat 7.92S	Long 7.31E	28.5Km

Sub-craters:	Crater	Lat	Long	Size (km)	Crater	Lat	Long	Size (km)
	C	8.71S	7.44E	6.75				
	Notes:							

Add Info: From a 'bigger picture' perspective view of this area in which Hind lies, it's easy to see how the formation of the Imbrium Basin event to the northwest has left its mark. Gouged-out lineations that point back to the Basin are the product of huge blocks and chunks of ejecta from Imbrium 'striking' and 'digging' in to the terrain; obliterating everything in their wake. Such events have the ability to turn fresh-looking craters (predominantly smaller ones) into old-looking ones, as they simply can't survive the onslaught of material pummelling their rims. This may not be the case with Hind itself as its rim all around does have an old look to it, and also its floor - filled with ejecta (solid and fluidized) - shows signatures of same. One of the above-mentioned 'gouges' can be seen to run across just outside the crater's northwestern and southeastern rim area, however, while it may have been a smaller chunk of ejecta that caused it - eventually mixing into the local terrain, where would, for example, larger chunks finally end up if they survived? If their makeup allowed them to survive long enough through their destructive, gouging processes, then shouldn't some chunk or block be visible downrange? Do they eventually blend into the landscape as small mountains, small hills etc., covered by numerous dustings of ejecta by other impact events nearby, or, are such features over-looked at expense of the main feature e.g. crater, involved? D: 2.9 km.

Craters

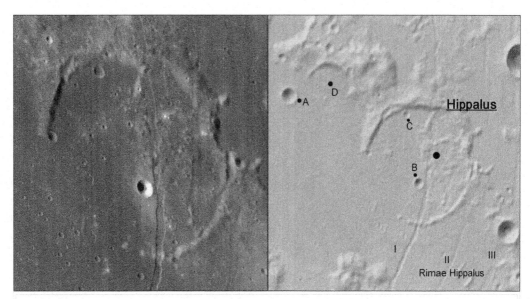

Hippalus	Lat 24.92S	Long 30.42W				57.36Km	

Sub-craters:

Crater	Lat	Long	Size (km)	Crater	Lat	Long	Size (km)
A	23.82S	32.85W	7.62	B	25.23S	30.39W	4.98
C	24.14S	30.64W	3.33	D	23.67S	32.1W	22.42

Notes:

Add Info: Hippalus lies on the 425-kilometre diameter ring/rim of Mare Humorum (a basin of the Nectarian Period - 3.92 to 3.85 billions of years old). True, the ring/rim is a little bit harder to see here if you were to compare it to the remaining portions of the basin, however, look at how nearby lava regions - in Nubium to the east or Cognitum to the northeast - have affected the terrain height in which Hippalus lies. It's obvious then why the crater looks the way it does. Humorum is the main culprit as to why Hippalus's western rim is missing; as its weighty volume of lava caused the central regions to sag over time, and thus 'tilting' in Hippalus. Note also how the wonderful series of rilles concentric to Humorum show up too in the area - signature, perhaps of a weaker zone between the two main basins. Hippalus, of course, is an old crater given its worn-down appearance, but is this simply due to time, or lavas that once rubbed against its remaining shores, and perhaps breaching them too? Rima Hippalus I produces a division in the look to the floor's regions - the eastern side is more rubbly-looking. But the question is what is this material? It really is a jumble and shows no signature related to fetures like, say, slumping of the rim, a peak...etc., so what are they - ejecta blocks from faraway basins? They are very wellcome in any case, as during low sun angle times they produce wonderful, long 'peaky' shadows. Depth ~ 1.0 km approx..

Craters

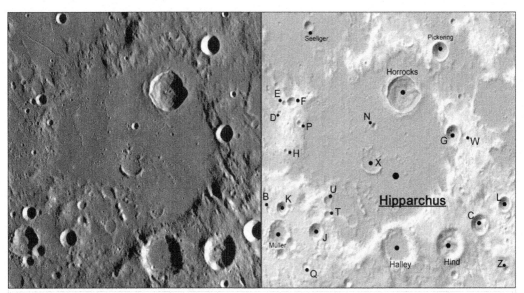

Hipparchus   Lat 5.36S   Long 4.91E   143.95Km

Sub-craters:

Crater	Lat	Long	Size (km)	Crater	Lat	Long	Size (km)
B	6.99S	1.73E	4.41	C	7.41S	8.21E	16.51
D	4.47S	2.1E	4.22	E	4.23S	2.23E	3.94
F	4.2S	2.47E	8.91	G	5.03S	7.4E	13.68
H	5.48S	2.28E	3.84	J	7.59S	3.2E	13.79
K	6.97S	2.16E	10.87	L	6.87S	9.0E	12.65
N	4.85S	4.98E	5.48	P	4.75S	2.7E	5.02
Q	8.5S	2.87E	6.89	T	7.15S	3.53E	6.01
U	6.77S	3.55E	7.73	W	5.08S	7.72E	5.21
X	5.85S	4.89E	17.3	Z	8.56S	9.06E	5.64

Notes:

Add Info: Hipparchus is one of those big craters that may have formed in a time when the Moon was still relatively young (it is aged of the pre-Nectarian 4.6 to 3.92 byo). Its appearance testifies to its age where numerous bombardment events over time have left only portions of its original rim peaking through. Like several other craters around Hipparchus (pull back for a wider view), all have a 'groove-like' look to them. The obvious culprit is the Imbrium Basin to the northwest where ejecta from that impact 'spat' huge chunks and blocks of material across the entire region and, literally, gouging out material at the highest points. Also note in the wider view the light-coloured hue of the material that fills Hipparchus's floor, and others. What is this material? It's too light to be basaltic, so it probably is a form of fluidized ejecta from Imbrium that 'mixed' with the ejecta in flight before it then landed on the surface (but note some darker patches, too). At full moon times Hipparchus and other major-sized craters nearby simply disappear.

Hipparchus doesn't show hint of a peak expected for a crater of this size, but it does have a series of very small 'dimples' (mounds) dotted in and around the floor (note them during low sun times).

Craters

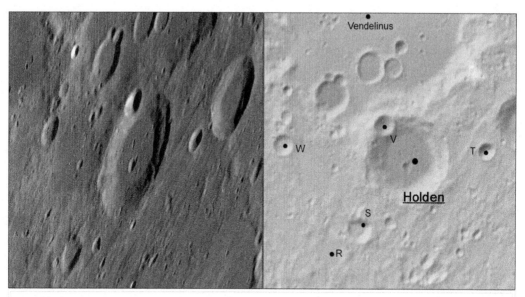

Holden		Lat 19.19S	Long 62.53E			47.6Km

Sub-craters:

Crater	Lat	Long	Size (km)	Crater	Lat	Long	Size (km)
R	20.82S	60.99E	18.3	S	20.41S	61.56E	13.83
T	19.02S	64.17E	9.46	V	18.57S	62.07E	11.57
W	18.91S	60.02E	11.53				

Notes:

Add Info:

Aerial view of Holden

Holden lies just outside the ~ 500-kilometre-diameter ring/rim of the old mare - a cryptomare - known as the Balmer-Kapteyn Basin (centred off the northwest corner of the Kapteyn B crater over to the east). Like its neighbouring craters, Vendilinus, Lohse, Lamé and others to the north, Holden has experienced flooding of lavas into its floor - due to an internal source. The crater is slightly younger than Vendilinus, but older than its neighbour, Petavius (Lwr. Imbrian - 3.85 to 3.75 byo) to its south. Secondary craters from Langrenus to its north lie on the floor, and ray material from oblique impact crater, Petavius B, to the west have 'splashed' it. Depth is around the 2.5 km mark.

Craters

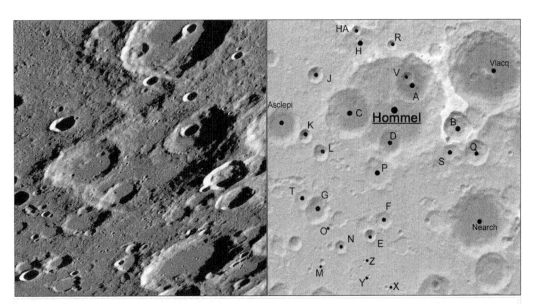

Hommel Lat 54.74S Long 32.93E 113.6Km

Sub-craters:

Crater	Lat	Long	Size (km)	Crater	Lat	Long	Size (km)
A	53.85S	33.97E	44.8	B	55.34S	36.94E	33.02
C	54.91S	29.82E	51.41	D	55.93S	32.55E	26.53
E	59.1S	31.04E	13.38	F	58.59S	32.03E	18.25
G	58.11S	27.4E	30.95	H	52.55S	30.67E	42.89
HA	52.08S	30.55E	10.43	J	53.53S	27.88E	17.63
K	55.57S	26.95E	14.86	L	56.18S	27.87E	17.28
M	59.89S	27.45E	7.29	N	59.43S	28.86E	13.51
O	58.7S	28.25E	6.01	P	56.98S	31.74E	33.52
Q	56.18S	38.31E	28.18	R	52.62S	32.67E	10.49
S	56.75S	36.13E	20.75	T	57.72S	26.49E	21.76
V	53.75S	33.65E	13.38	X	60.89S	32.17E	5.41
Y	60.48S	30.86E	3.79	Z	59.95S	30.51E	4.17

Add Info:

Aerial view of Hommel

It's always both frustrating but extremely rewarding when viewing the higher latitude craters of the moon's southeast. With so many to deal with all at once, a good tip is to break up the area into several sub-views, and try to group certain craters in to a cluster for easier recognition. The Hommel 'cluster' is one such group, and it's quite easy to locate the crater itself, once you use the above method (try it). Hommel is of the pre-Nectarian Period, its rim is battered to death by both old and new impacts. Rays from Tycho (to west) cross its floor.

Craters

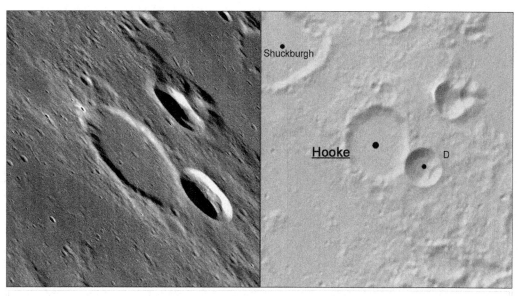

Hooke	Lat 41.14N	Long 54.86E	34.35Km

Sub-craters:

Crater	Lat	Long	Size (km)	Crater	Lat	Long	Size (km)
D	40.69N	55.86E	18.73				

Notes:

Add Info:

Aerial view of Hooke

Hooke lies in a slightly depressed zone (a basin?) where, if you look around, you'll see several patches of dark deposits which hint at lava flooding. The crater may have experienced such events itself as its floor is filled with similar-like material to that of Shuckburgh and Chevallier to its northwest (lying not far from the 3.7 byo lava deposits of Lacus Temporis). Sub-crater, Hooke D, is a relatively fresh crater that has just 'clipped' Hooke's southeastern rim, but no ejecta is seen on Hooke's floor, nor are the small craterlets in it related to it as secondaries. The small, 1-kilometre-sized crater on the northwest rim shows up nicely its bright ray material on Hooke. Depth 1.5km.

Craters

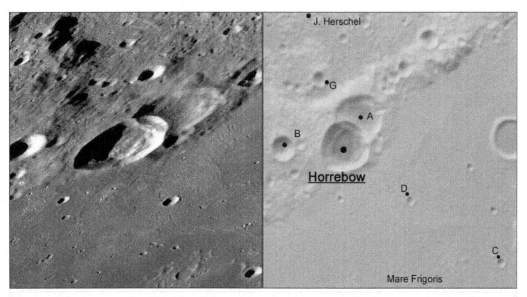

Horrebow   Lat 58.8N   Long 40.93W   24.98Km

Sub-craters:

Crater	Lat	Long	Size (km)	Crater	Lat	Long	Size (km)
A	59.22N	40.52W	24.66	B	58.73N	42.94W	11.9
C	56.96N	36.02W	4.69	D	57.96N	38.82W	4.34
G	59.82N	41.84W	7.07				

Notes:

Add Info: Several factors come into play that determine the final shape of a crater. Speed and angle of incidence of the impactor are two, but more importantly you also have impactor make-up and the target rock which dictate the final size. Given all these, could the initial topography (the surface shape) in which the impactor strikes, play a factor, too? It looks like it does when we look at Horrebow, which obviously has impacted upon sub-crater, Horrebow A. Where the two 'join', a depressed feature has formed that must have been the result of Horrebow's impactor striking an uneven surface (essentially, Horrebow A's southern rim). The feature isn't part of Horrebow A's missing rim as one might first believe, but more the slumping of material in Horrebow afterwards - possibly due to a weaker zone of material between the two giving way. Depth ~ 3 km approx..

Aerial view of Horrebow

Craters

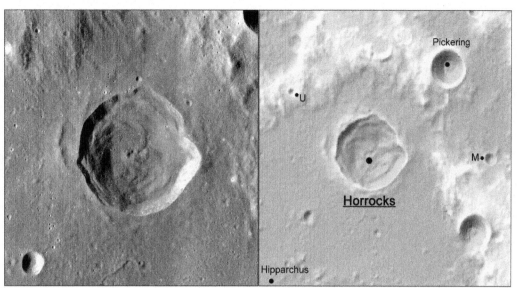

Horrocks	Lat 3.99S	Long 5.85E	29.65Km

Sub-craters:	Crater	Lat	Long	Size (km)	Crater	Lat	Long	Size (km)
	M	4.07S	7.63E	4.82	U	3.19S	4.74E	3.09
	Notes:							

Add Info: Horrocks lies on the floor of Hipparchus, which is a crater of the pre-Nectarian Period (4.60 to 3.93 billions of years old). Horrocks itself is believed to be Copernican in age (1.1 byo to Present), but judging from the worn look to some of its features - terracing, slumping and central mounds - it more than likely is on the extreme border between the next age, that of Eratosthenian (3.15 to 1.1 byo). One main feature of note is the odd 'kink' in the rim's eastern sector. This feature looks relatively small to how we view the crater overall, but when we imagine the extent to the volume of material that must have cascaded down in to the inner floor region, it really is huge (compare it to those landslides we sometimes see on Earth). A smaller version in the western sector of the rim has occurred, too, but it looks fresher and sharper than its counterpart, as mentioned above. There is hint of a peak, or a peak group, in the floor, but they aren't very prominant even under the lowest of lighting conditions. One other small feature that might catch the eye is the small mountain that lies to the west of the rim's outer sector. It looks like part of the rim of a buried crater, but if you look at how Horrocks's ejecta is distributed around it, one would have expected to see more of the buried rim continue further northwards. It then might just be an isolated 'chunk' of material related to Imbrium ejecta that covers this region in general.

Craters

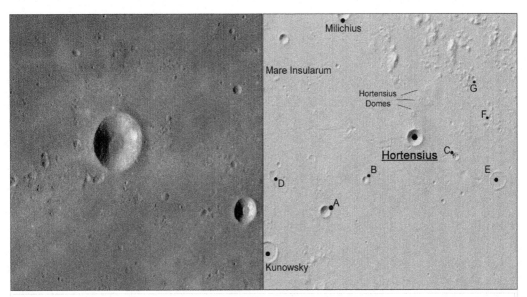

Hortensius   Lat 6.47N   Long 28.0W   14.16Km

Sub-craters:

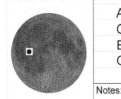

Crater	Lat	Long	Size (km)	Crater	Lat	Long	Size (km)
A	4.37N	30.75W	9.48	B	5.26N	29.5W	6.19
C	5.93N	26.73W	6.33	D	5.4N	32.38W	7.06
E	5.25N	25.43W	15.31	F	7.04N	25.66W	6.26
G	8.13N	26.17W	4.14	H	5.88N	31.18W	6.33

Notes:

Add Info:

There are more domes in this region - see west of crater Milichius and south-west of crater T. Mayer, which is some 170 km away northwards of Hortensius.

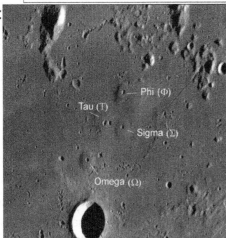

The Hortensius Domes

Hortensius itself doesn't have any main features to speak of, so instead spend your observing time on looking at the series of volcanic-like domes just north of the crater referenced to it. Five of the six stand out nicely under low sun angle times - each having a summit crater at their tops. These domes may have formed through underlying fractures possibly related to the Imbrium Basin event to the northeast, and 'grew' as highly viscous lavas extruded onto the surface. Southwards of Hortensius (~ two diameters' away), several more domes lie, but you need to time your lighting conditions to precisely 'catch' them. Depth ~ 2.9 km.

Craters

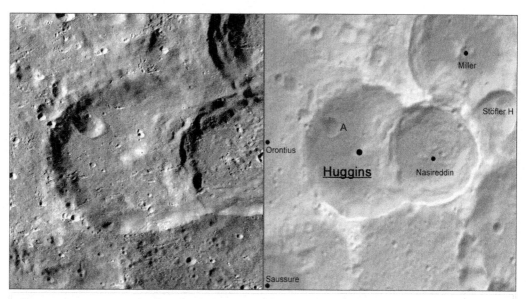

Huggins	Lat 41.07S	Long 1.52W	65.79Km

Sub-craters:

Crater	Lat	Long	Size (km)	Crater	Lat	Long	Size (km)
A	40.66S	2.27W	9.62				

Notes:

Add Info:

Aerial view of Huggins

Viewing Huggins and its neighbouring craters is like looking at impact history in motion. Huggins has impacted pre-Nectarian crater, Orontius to the west, Nasireddin has impacted Huggins and Miller to the north, and then sub-crater, Stöfler H (the smaller crater northeast of Nasireddin), has impacted Miller. All of Huggins's eastern rim is entirely gone, and probably lies mixed with ejecta from the impact of Nasireddin - now lying over half of Huggins's floor. The central mounds on the floor hint that Huggins once may have had a peak or peaks, but they haven't survived. At high magnification, note the secondary crater chains from Tycho - some 230 km away to the southwest. Depth ~ 1.8 km.

Craters

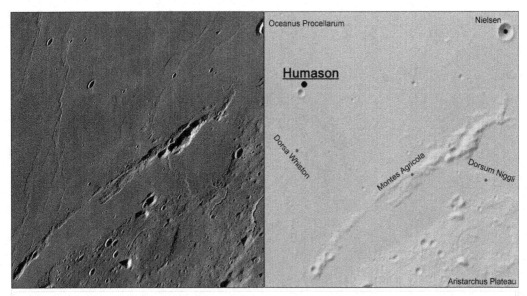

| Humason | Lat 30.73N | Long 56.66W | 4.34Km |

Sub-craters: No sub-craters

Notes:

Add Info:

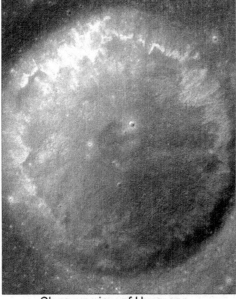

Close-up view of Humason

Like most small craters all over the lunar surface, Humason and its like don't get that much of a 'look-in' by the observer. Reason: because below a certain crater's diameter, amateur-sized telescopes just can't resolve any decent detail worth mentioning. It's a pity, really, as from the close-up view of Humason (left), a whole lot of wonderful features are being missed in these small craters. All we can do for now, however, is note how they fit into the region in where they lie, and observe their macro features - ejecta, albedo...etc. Humason is Eratosthenian in age. The lavas that lap up to its ejecta are younger. The crater kisses a small north-trending wrinkle ridge, and its depth is around 4 km approx..

Craters

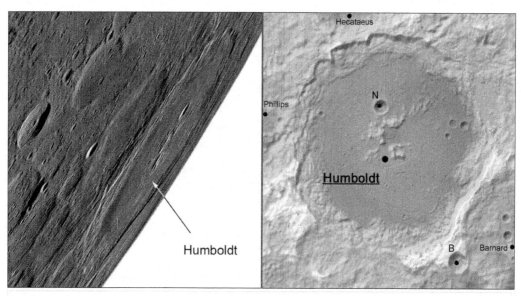

Humboldt	Lat 27.02S	Long 80.96E	199.46Km

Sub-craters:

Crater	Lat	Long	Size (km)	Crater	Lat	Long	Size (km)
B	30.83S	83.64E	20.88	N	26.05S	80.65E	14.49
Notes:							

Add Info:

Capturing a decent view of Humbold's concentric crater in the east - 26.55S, 83.35E - will be frustrating, however, it is possible to see the inner torus; given that all observing conditions are in your favour, and in sync.

Aerial view of Humboldt

What a 'beauty' - Humboldt has got it all in terms of a being a big 'show-off'. It's got peaks and terraces, rilles and fractures to die for, dark volcanic deposits surrounding nearly all of its outer floor, and it's size is bordering on it being thought of as more of a basin rather than a crater. But, alas, its a pity that Humboldt is on the extreme limb, as all of the above features prove a challenge to view. Plan well your times to observe - during favourable librations, good lighting, shadow conditions...etc. And make notes of what you see, as all will add in your admiration of this 'elusive' show-off. D: 5 km.

Craters

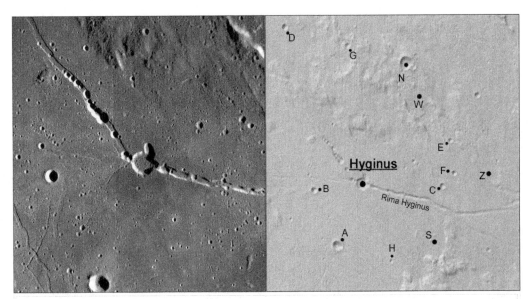

Hyginus   Lat 7.76N   Long 6.27E   8.7Km

Sub-craters:

Crater	Lat	Long	Size (km)	Crater	Lat	Long	Size (km)
A	6.32N	5.66E	7.69	B	7.61N	5.07E	4.85
C	7.69N	8.31E	4.15	D	11.4N	4.33E	4.71
E	8.75N	8.52E	3.78	F	7.99N	8.61E	3.81
G	10.97N	5.93E	3.84	H	5.94N	7.01E	3.42
N	10.55N	7.43E	10.57	S	6.41N	8.01E	23.61
W	9.62N	7.74E	21.02	Z	8.04N	9.44E	23.18

Notes:

Add Info: If it wasn't for the additional, odd-looking, rugby-shaped feature attached to the northern sector of Hyginus, one would immediately assume that this is a normal crater. But, Hyginus isn't considered a crater at all, and instead may be the result of a collapsed volcanic feature (a caldera) where overlying material on a possible magma chamber sunk down after it drained out its lava's load. If you look at the surrounding area in general you'll see other features - rilles, dark halo craters, dark deposits related to volcanic activity, so it may be that Hyginus is a caldera indeed (it doesn't have a raised rim either). Rima Hyginus is a wonderful sight also, where the northwest sector shows additional collapsed 'pits' that look, to all intents and purposes, like a crater chain. The eastern sector has one or two pits too, and the rille extends here for over a 130 km where it meets Rima Aridaeus. Many other rilles towards the south are seen also, but note the distinct character to those branching off from Hyginus itself and you'll see how the dark deposits 'gather' close to them. A hard-to-see rille branching off to the west-southwest (just under sub-crater B) clearly defines a distinct dark-light border between dark lavas and lighter material. Depth: 0.8 km.

The dark patch north of Hyginus is known as the Vaporum pyroclastic deposit in which it is found. The largest pyroclastic deposit on the Moon, however, lies west of crater Bode.

Craters

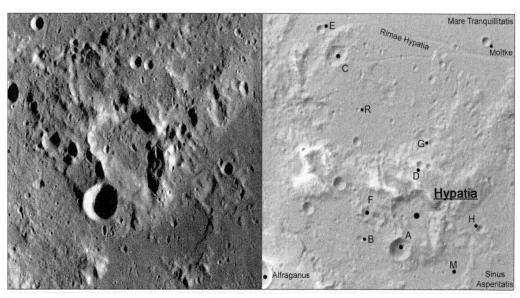

| Hypatia | Lat 4.25S | Long 22.58E | 38.82Km |

Sub-craters:

Crater	Lat	Long	Size (km)	Crater	Lat	Long	Size (km)
A	4.9S	22.22E	15.06	B	4.69S	21.32E	4.13
C	0.88S	20.76E	14.79	D	3.15S	22.68E	5.36
E	0.35S	20.42E	5.7	F	4.14S	21.47E	8.24
G	2.68S	22.96E	4.81	H	4.48S	24.06E	5.19
M	5.28S	23.41E	28.53	R	2.01S	21.25E	3.1

Notes:

Add Info: Hypatia is a weird-looking crater, isn't it! Its original shape and form has been altered beyond recognition through over-abuse of what looks like small impacts 'chipping' off a bit of the rim here and there (note the northern and eastern sectors of the rim in particular). Like most areas on the Moon, ejecta from the Imbrium Basin to the northwest has 'spat' over Hypatia and its surrounding terrain. However, its make-up may have been more fluidized rather than hard, rocky chunks as the general look to the area appears more mud-like rather than gouge-like, when we compare it to the striated look we see over to the west (e.g. near craters Hipparchus, Gyldén, Réaumur...etc.,). Note the odd shape to Hypatia as well. Is it the result of an oblique impactor that came in from the northwest? Also on this, if you look some 80 km to the northwest of the crater you'll see what looks like a smaller version of Hypatia with the same shape and same orientation. Both craters don't physically 'point' back to Imbrium as we might expect but more towards the Serenitatis Basin. Was Hypatia, then, a secondary crater event by Serenitatis, or was its ejecta more responsible for its odd shape?

Craters

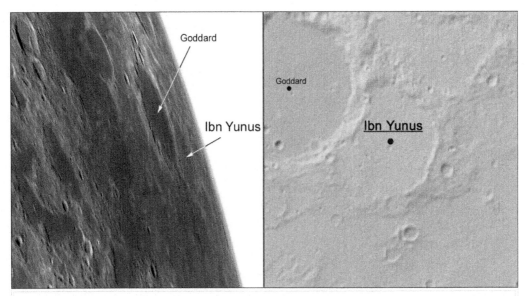

Ibn Yunus	Lat 14.14N	Long 91.14E	59.69Km

Sub-craters: No sub-craters

Notes:

Add Info:

Swirls: Magnetic anomalies are believed to be responsible for features like Reiner Gamma found at 7.39N, 58.96W, which may also be associated with areas of cometery impacts, as well as inferred electirc fields existing alongside the magnetic fields

Aerial view of Ibn Yunus

Being one for the extreme limb-hugging craters on the east side of the Moon, Ibn Yunus isn't, as you would expect, difficult to see. Favourable libations are required of course, however, together with its northwestwards neighbouring crater, Goddard, the dark floors of both show up quite easily. Mare Marginis is centred just off Goddard's southwest, and its effect is responsible for the filled floors of lava (Ibn Yunus's floor level is at some ~ 900 m above Goddard - perhaps, because Goddard's initial ejecta into Ibn Yunus raised it up?). The question to ask, however, is: 'did their floors fill due to internal flooding, or was breaching their cause?". Hint of a swirl-like marking on Ibn Yunus's northwest floor suggests a magnetic anomaly may apply to the crater. D: 0.3 km.

Craters

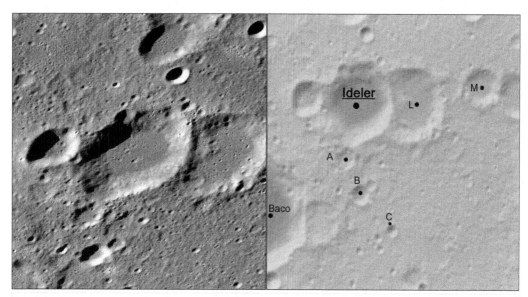

Ideler   Lat 49.32S   Long 22.24E   37.77Km

Sub-craters:

Crater	Lat	Long	Size (km)	Crater	Lat	Long	Size (km)
A	50.21S	21.93E	10.69	B	50.78S	22.35E	9.67
C	51.34S	23.23E	6.35	L	49.33S	23.65E	36.52
M	49.0S	25.57E	18.18				

Notes:

Add Info:
Day-old moon phases

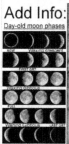

Note: As the Moon goes through various librations throughout the year, suggested times given in text for observations are only approximates.

Aerial view of Ideler

We're in the area of 'peppered with craters' again when we view Ideler. And it's not only of craters with similar size to Ideler we're talking about here, but the many thousands of small craterlets across the whole terrain - each around one kilometre in diameter. Just look at the effect that these craterlets have had on Ideler's rim, its inner walls, and floor (note it on other craters nearby, too). The area, in general, is known to hold secondaries from the Orientale Basin over to the west (of the 5 km in diameter range), but these smaller ones may not be related to it; nor, it looks like, are they of Nectaris Basin origin to the northeast. Whatever their original source, they probably were of a large cluster of particles that landed later after an impact event.

Craters

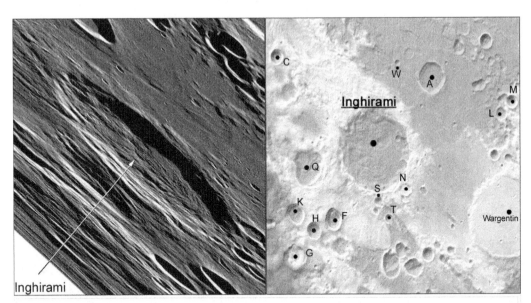

Inghirami — Lat 47.49S Long 68.95W 94.6Km

Sub-craters:

Add Info:

Crater	Lat	Long	Size (km)	Crater	Lat	Long	Size (km)
A	44.92S	65.45W	33.31	C	44.07S	74.59W	17.47
F	49.84S	71.59W	22.32	G	51.03S	74.29W	29.18
H	50.17S	72.89W	19.4	K	49.54S	74.05W	24.25
L	45.95S	61.1W	14.79	M	45.5S	60.52W	14.73
N	48.82S	66.93W	14.24	Q	47.99S	73.0W	44.03
S	49.26S	68.63W	11.71	T	49.83S	67.97W	10.1
W	44.45S	67.5W	6.99				

Notes:

Aerial view of Inghirami

If any other crater on the nearside of the Moon shows the dramatic effect that ejecta from the Orientale Basin has on them, then surely it has got to be Inghirami. The aerial shot (left) shows it best, where as fluidized ejecta from Orientale (coming from a direction off the top-left corner of the image) first struck the crater's northwest rim, it then 'plopped' on to the floor (with some force), creating a series of ripples that then solidified. This ejecta 'sludge' doesn't seem to have reached the southeastern rim, as a nice set of terraces there don't seem to have been affected. Age of crater is Nectarian, depth ~ 2.8 km.

Craters

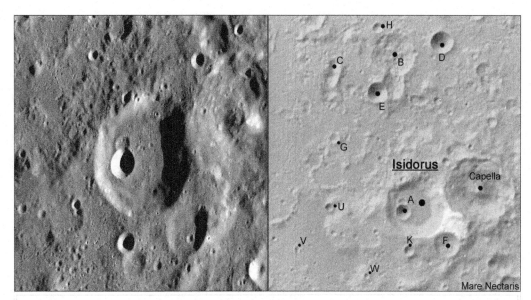

Isidorus		Lat 7.96S		Long 33.5E			41.39Km

Sub-craters:

Crater	Lat	Long	Size (km)	Crater	Lat	Long	Size (km)
A	8.06S	33.21E	9.92	B	4.54S	32.99E	29.51
C	4.8S	31.65E	7.93	D	4.28S	34.08E	14.95
E	5.39S	32.62E	13.53	F	8.79S	34.19E	17.18
G	6.42S	31.67E	6.53	H	3.95S	32.63E	6.74
K	8.92S	33.34E	7.15	U	7.95S	31.54E	5.56
V	8.88S	30.81E	3.85	W	9.49S	32.37E	3.98

Notes:

Add Info:

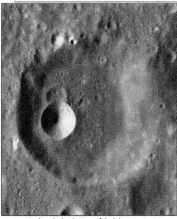

Aerial view of Isidorus

There are several 'pairs' of craters on the Moon that when you mention one of the pair, you immediately think of the other (other pairs include Theophilus & Cyrillus, Hevelius & Cavalerius). They aren't always connected to each other like we have here, as separated craters like Atlas & Hercules, or Hansteen & Billy also fit in to this 'pairing' too. Of course, it serves no useful purpose, but it's something that becomes unavoidable when observing craters close to each other. Isidorus lies on material from the Nectaris Basin. Ejecta from younger Capella covers the eastern sector of its floor (affected its rim, too), which was then covered by Imbrium ejecta. Depth ~ 2.5 km.

Craters

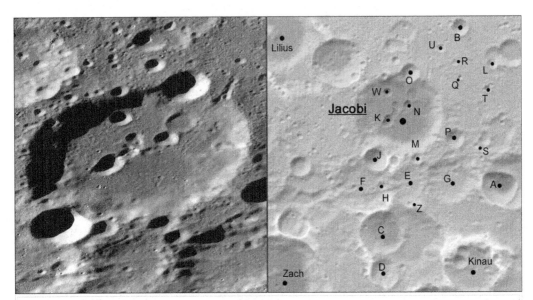

| Jacobi | Lat 56.82S | Long 11.3E | 66.28Km |

Sub-craters:

Crater	Lat	Long	Size (km)	Crater	Lat	Long	Size (km)
A	58.66S	16.03E	26.04	B	54.55S	13.94E	12.95
C	59.85S	10.49E	33.91	D	60.91S	10.55E	22.59
E	58.56S	11.72E	24.95	F	58.66S	9.56E	41.1
G	58.54S	13.67E	35.21	H	58.65S	10.55E	7.59
J	57.97S	10.28E	18.85	K	56.98S	10.84E	9.65
L	55.45S	15.38E	8.78	M	57.96S	12.16E	9.3
N	56.56S	11.75E	7.36	O	55.74S	11.85E	14.57
P	57.42S	13.86E	14.74	Q	55.85S	14.03E	3.92
R	55.47S	13.76E	5.35	S	57.6S	14.85E	4.78
T	56.11S	15.22E	5.52	U	55.1S	13.17E	7.12
W	56.24S	10.85E	6.89	Z	59.13S	11.82E	4.25

Notes:

Add Info:

Aerial view of Jacobi

While observing this over-cluttered area of craters on the Moon, the minute you take your eye away from the eyepiece, you're guranteed to be completly lost again as you return. So, next time you are observing craters in the region, be sure to make a note of some characteristic to the particular crater that is in your view. Jacobi's two craters at its northeast and southwest (sub-crater's O and J) are its key to finding it easily. O and J, and perhaps those craters on its floor, too, are likely secondaries from the Orientale Basin. Depth of Jacobi ~ 3. km.

Craters

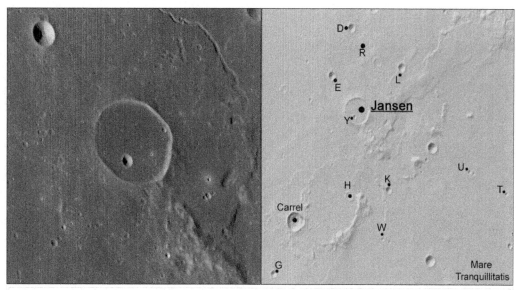

Jansen Lat 13.55N Long 28.64E 24.21Km

Sub-craters:

Crater	Lat	Long	Size (km)	Crater	Lat	Long	Size (km)
D	15.72N	28.46E	6.72	E	14.47N	27.83E	6.57
G	9.3N	26.02E	5.57	H	11.34N	28.39E	7.38
K	11.5N	29.67E	5.63	L	14.69N	30.09E	6.91
R	15.17N	28.79E	26.3	T	11.35N	33.48E	3.09
U	11.95N	32.29E	3.72	W	10.22N	29.47E	3.22
Y	13.4N	28.56E	3.04				

Add Info: Notes:

Day-old moon phases

Note: As the Moon goes through various librations throughout the year, suggested times given in text for observations are only approximates.

Sometimes while observing the full moon under low magnification, and close to high sun angle times, you'll notice the odd crater dotted here and there that looks a little bit darker to its surrounds. Craters Grimaldi, Billy and Plato are probably at top of the list, but you also have others like Crüger, Lassell, Firmicus, Endymion and many more to a lesser, darker extent that should be looked out for. Almost all of these craters stand out against the others simply because their floors are smooth, and filled with darker, possibly younger lavas sourced separately against older lavas in their general surrounds. But sometimes the 'dark-crater' effect can be due to brighter ejecta around their exteriors, or sunlight highlighting their inner rim walls - both giving a kind of optical-illusion effect. Jansen, however, is dark because it's probably a combination of all three effects. The lavas in its interior are younger than those in Tranquillitatis, its got a relatively bright rim, but its lighter ejecta close to its outer rim is a bit hard to see. Jansen's floor is higher than the mare's floor too, but the western sector of its rim is low enough to have acted as a breaching point where the younger lavas met the older. D ~ 0.4 km.

Craters

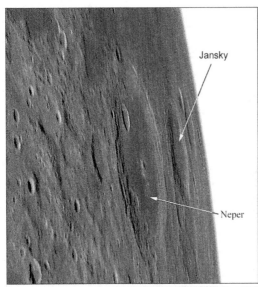
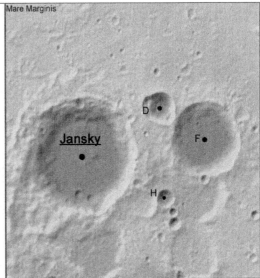

Jansky	Lat 8.63N	Long 89.5E		73.77Km	
Sub-craters:	Crater	Lat	Long	Size (km)	Crater	Lat	Long	Size (km)

Crater	Lat	Long	Size (km)	Crater	Lat	Long	Size (km)
D	9.56N	91.25E	23.02	F	8.94N	92.32E	47.49
H	7.71N	91.41E	11.84				

Notes:

Add Info:

Aerial view of Jansky

Jansky lies on slightly higher ground between two pre-Nectarian (4.6 to 3.92 byo) mares - Mare Marginis to its north and Mare Smythii to its south. Its floor level is some 2.3 kilometres below the level of Marginis's floor, and about a kilometre below that of Smythii's (the difference arises from Smythii's floor level being ~ 2 kilometres below Marginis's). Jansky's impactor thus struck deep, and as a result tapped into the similar lavas to the mares (crater Neper to its west also holds such deposits). Observing Jansky will prove difficult, however, favourable librations does bring it into view (use Neper, with its peak, as a good marker for finding/remembering its location). Depth of Jansky is around 4 kilometres.

Craters

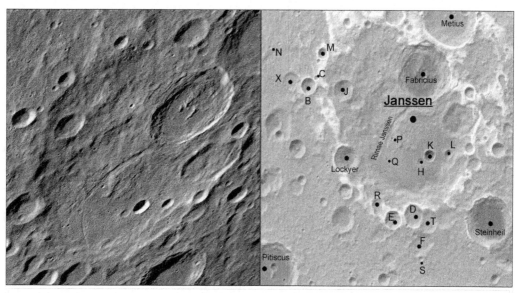

Janssen   Lat 44.96S   Long 40.82E   200.65Km

Sub-craters:

Crater	Lat	Long	Size (km)	Crater	Lat	Long	Size (km)
B	43.17S	34.37E	21.26	C	42.89S	34.91E	7.24
D	48.65S	41.22E	28.73	E	48.82S	39.77E	24.25
F	49.82S	41.92E	35.15	H	46.43S	41.71E	9.55
J	43.44S	36.58E	26.91	K	46.19S	42.31E	15.0
L	46.09S	43.58E	11.78	M	41.9S	35.44E	15.63
N	41.43S	32.12E	4.7	P	45.5S	39.76E	4.09
Q	46.35S	39.36E	4.94	R	48.28S	38.7E	22.9
S	50.42S	41.83E	7.04	T	48.88S	42.31E	29.01
X	42.93S	33.27E	24.29				

Notes:

Add Info:

Deslandres (227.02km)

Clavius (230.77km)

Schickard (212.18km)

```
      Des
Sch ---+--- Jan
      Cla
```

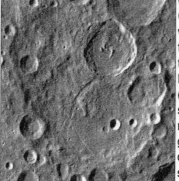

Aerial view of Janssen

Craters Janssen, Deslandres (to its northwest), Clavius (to its southwest) and Schickard (to its west) are some of the biggest craters found on the southern mid-latitudes. All sized just above the 200 km mark, the first three are of the pre-Nectarian Period, while Clavius is the youngest (Nectarian). They form a virtual, on-its-side cross when you connect all four - a useful tip for remembering them. Janssen is full of features: its got a wonderful main rille that cuts it in half, ejecta from the Nectaris Basin fills its top half, and several major impact craters have altered its face beyond original recognition. Explore it well!

Craters

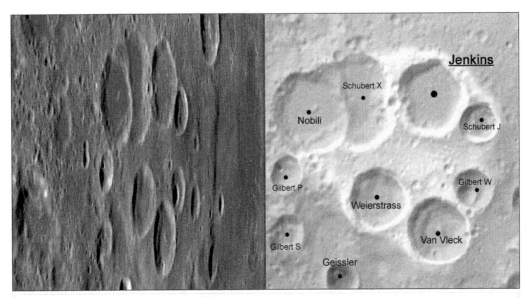

| Jenkins | Lat 0.37N | Long 78.04E | 37.77Km |

Sub-craters: No sub-craters

Notes:

Add Info:

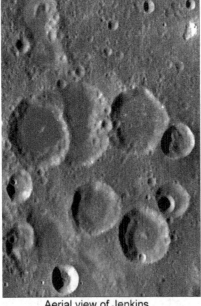

Aerial view of Jenkins

Lying on an isolated clump of ejecta related to Mare Smythii on the moon's eastern limb, Jenkins, like its neighbouring craters, fits in appropriately to this tight group all having depths of around three kilometres. It looks as if the crater formed on top of Schubert X - itself been impacted by Nobili to its west, however, as this latter event may have been later, Jenkins is probably the youngest of all three ("probably" as the series of small craterlets where the two craters join is adding confusion). All three craters are filled with lavas sourced through fractures created by their initial impactors (Nobili is the only one with a peak), but as Jenkins's floor is slightly higher, the extent of fill wasn't as so severe. High-rez views of its floor also shows tiny, fresh impact craterlets that 'point' back to the bright crater (top-right in left image).

Craters

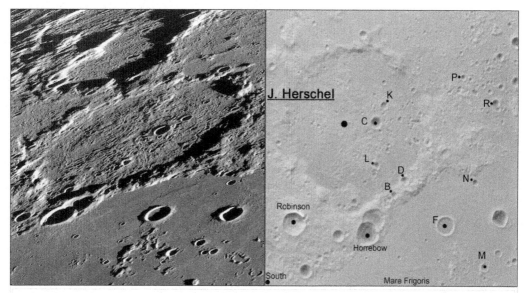

J. Herschel   Lat 62.31N   Long 41.86W   154.44Km

Sub-craters:

Crater	Lat	Long	Size (km)	Crater	Lat	Long	Size (km)
B	60.01N	39.0W	7.42	C	62.39N	40.15W	12.0
D	60.47N	38.17W	9.41	F	58.87N	35.5W	18.88
K	62.96N	39.47W	8.38	L	61.01N	40.22W	7.45
M	57.35N	33.11W	8.59	N	60.15N	33.04W	6.93
P	63.61N	33.01W	6.06	R	62.59N	30.7W	8.73

Notes:

Add Info:

Not to be confused with the other two Herschel craters on the Moon: C. Herschel (under craters, C) and Herschel (under craters, H).

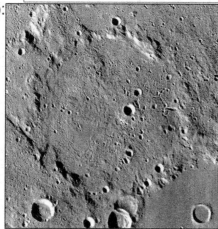

Aerial view of J. Herschel

Like several other craters (e.g. South or Babbage to its west) that lie in this general region, the distinctive rough-textured material dominating is said to be ejecta from formation of the Imbrium Basin to the southeast (centred ~1000 kilometres away). In fact, there is other evidence of its effect also seen as striations on [said] craters - most definitely seen on J. Herschel's northwestern rim. The central floor of J. Herschel is slightly convexed upwards - is this because of the additional dumpload of Imbrium's ejecta, or did volcanism underneath the crater have something to do with it (note the rille fractures at its southeast)? Depth ~ 1 km.

Craters

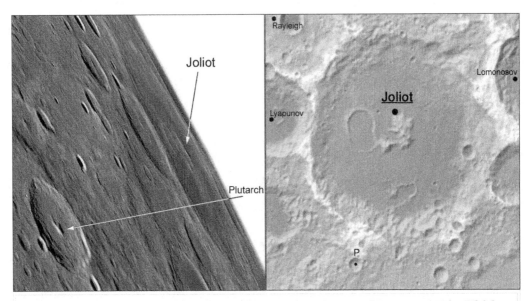

Joliot   Lat 25.79N   Long 93.39E   172.79Km

Sub-craters:

Crater	Lat	Long	Size (km)	Crater	Lat	Long	Size (km)
P	22.34N	91.99E	11.82				

Notes:

Add Info:

Aerial view of Joliot

Joliot is a big crater, and it's a pity that we can't get a good view of it from our earth-based perspective. The floor is partially lava-filled, and is some ~ 500 metres higher at its western side than at its east (did this slight 'tilt' dictate how the lavas filled the floor?). Crater Lyapunov has obviously impacted on Joliot's western rim, and, in effect, has 'squished it a bit there (compare it to the eastern side). Observing Joliot will obviously be a challenge, however, its high peaks (highest is over 2.0 kilometres), the two half-buried craters west of the peaks, and the bright patch southwards, act as good signatures to look out for. Age: Pre-Nectarian. Depth: 3.8km.

Craters

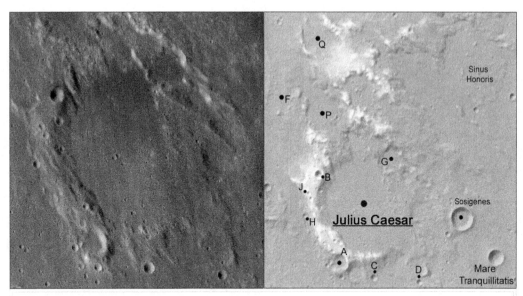

Julius Caesar   Lat 9.17N   Long 15.21E   84.72Km

Sub-craters:

Crater	Lat	Long	Size (km)	Crater	Lat	Long	Size (km)
A	7.66N	14.45E	13.75	B	9.76N	13.94E	6.72
C	7.3N	15.34E	4.97	D	7.19N	16.51E	4.54
F	11.64N	12.96E	25.06	G	10.16N	15.75E	20.15
H	8.77N	13.58E	3.48	J	9.31N	13.68E	3.12
P	11.17N	14.15E	36.1	Q	12.99N	13.95E	28.57

Notes:

Add Info: Before you look at crater Julius Caesar more closely, take in a wider view first so as to see how the area where it lies has been changed by the Imbrium Basin event (whose centre is some 1100 km to the northwest). Note the striated feel to the area that 'points' back to Imbrium's centre - due to huge ejecta blocks from that major event pummelling and gouging in to the terrain. The effect has thus taken its toll on Julius Caesar, and its only remaining western and northern rim looks like someone took a rake to it - dragging the tool from northwest to southeast. The bottom half of the floor area is some 800 metres higher than the northern half, and contains slightly rougher material - Imbrium ejecta again, but probably also a mixture of the northwest rim that was literally shaved off by the blast from Imbrium. Dark lava deposits in the northern sector of the floor were likely sourced separately through underlying fractures as there doesn't seem to be a breached path, or channel, linking them to Mare Tranquillitatis in the east. Observe Julius Caesar and its surrounds under low lighting conditions (before first quarter) and you'll be rewarded with the dramatic gouging effects. D: 1.3 km.

Craters

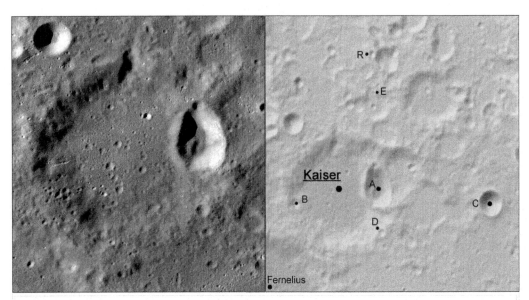

Kaiser	Lat 36.49S	Long 6.48E			53.15Km		
Sub-craters:	Crater	Lat	Long	Size (km)	Crater	Lat	Long	Size (km)

Crater	Lat	Long	Size (km)	Crater	Lat	Long	Size (km)
A	36.37S	7.28E	19.53	B	36.64S	5.53E	6.13
C	36.6S	9.63E	12.07	D	37.08S	7.37E	4.16
E	34.94S	7.14E	5.42	R	34.33S	7.16E	3.43

Notes:

Add Info: If it wasn't for sub-crater, Kaiser A, on this crater's eastern limb, Kaiser would be lost to the observer at the eyepiece in the blink of an eye. Its depth is over a kilometre, its rim all around has been levelled through the many numerous impacts nearby - big and small, and ejecta from basins afar, e.g. the Imbrium Basin some 2200 km away to the northwest, has covered nearly every aspect of its face and floor. Sub-crater A looks interesting - from its shape it might be an oblique impact event. However, as the southern sector of its floor seems to have an additional clump of material; is this clump the remnants from a smaller impact on its floor, which would then explain its oblong shape, or, simply it is of ejecta from elsewhere? The material within Kaiser has buried its true floor, but this is covered with a slight dusting of bright ray material (hard to see) from crater Tycho ~ 450 km away to the southwest, and also tiny, secondary crater clusters and chains related to that event, too (the small group of slightly larger craterlets in the central part of the floor may not be of Tycho's). There seems to be a small, odd-looking feature resembling a partially-buried crater in the floor's northern sector where the southern portion of its rim is missing, but is it really one? Low sun views shows the depth difference between Kaiser and A.

Day-old moon phases

Note: As the Moon goes through various librations throughout the year, suggested times given in text for observations are only approximates.

Craters

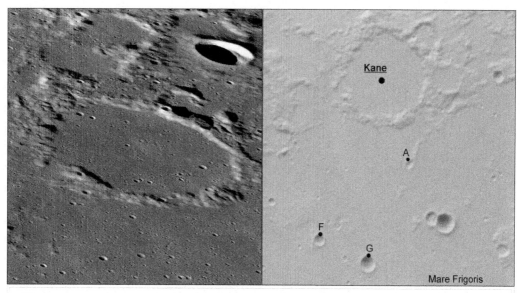

Kane	Lat 62.99N	Long 25.85E			54.96Km

Sub-craters:

Crater	Lat	Long	Size (km)	Crater	Lat	Long	Size (km)
A	61.24N	27.05E	6.07	F	59.66N	23.24E	7.2
G	59.25N	25.34E	9.71				

Notes:

Add Info:

The 'light-ray' effect (crater rays) usually occurs when low sunlight rays project through gaps in the rims of craters. There are hundreds of craters on the Moon that produce the effect - keep an eye out for them as some can be quite dramatic.

Aerial view of Kane

Kane is pre-Nectarian in age (4.6 to 3.92 byo). Like most other craters in this region, the influential effect of ejecta from the Imbrium Basin to the southwest has taken its toll on the crater. Its rim has a mottled, broken look to it, while the same material in its floor has similar characteristics to the material southwards as it meets the plainer mare (bright rays from Anaxagoras to the northwest also cross the floor). Both eastern and western terminator apparitions close to the crater produces some wonderful shadowed effects through 'gaps' in the rim - the former leaving a sliver of light in the bottom half of the floor, the latter producing a light-ray effect. D ~ 0.6 km.

Craters

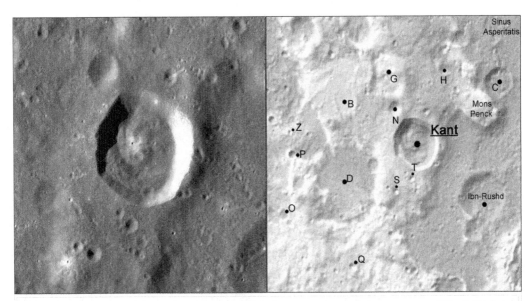

Kant Lat 10.62S Long 20.2E   30.85Km

Sub-craters:

Add Info:

Crater	Lat	Long	Size (km)	Crater	Lat	Long	Size (km)
B	9.76S	18.6E	14.58	C	9.39S	22.1E	17.97
D	11.36S	18.7E	52.4	G	9.21S	19.56E	33.54
H	9.16S	20.82E	6.69	N	9.91S	19.71E	9.05
O	12.03S	17.16E	6.6	P	10.82S	17.38E	5.71
Q	13.05S	18.72E	5.11	S	11.56S	19.73E	4.43
T	11.3S	20.1E	3.84	Z	10.38S	17.32E	3.0

Notes:

Before you take a look at crater Kant, pull back your view a little and you'll see immediately the extent of material on which it lies - Nectaris Basin ejecta. This ejecta forms part of the Basin's main diameter ring/rim (~ 860 km), whose most dramatic effect can be seen southeastwards of Kant where the infamous Altai Scarp lies. A section of the scarp northwards from the west of crater Catherina right up to Kant isn't as obvious or as high to its southeastern section, however, when we look at Kant we see the impactor that produced it must have struck a high part of the scarp (some 3 km high) again. Kant, then, is obviously younger than the Nectaris Basin, which formed 3.92 billion years ago. Kant has a sharp rim all around, its true floor seems only to occur at the crater's eastern inner rim, as the remaining area is filled with slumped deposits that meet a small mound (~ 1.2 km high) in the centre (possibly its peak?). This 'peak' has a fresh, 1 km-sized impact crater atop it whose bright ray material puts on a nice show during low and high sun views (the inner bright rim is also a 'shiner'). Depth ~ 3.5 km.

Craters

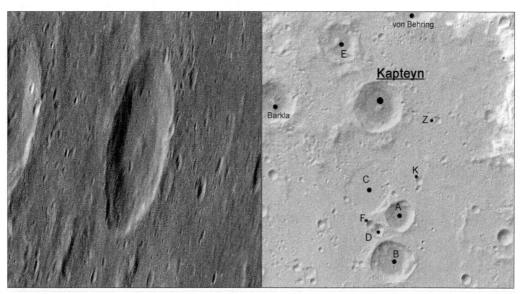

Kapteyn	Lat 10.79S	Long 70.59E				48.65Km

Sub-craters:

Crater	Lat	Long	Size (km)	Crater	Lat	Long	Size (km)
A	14.15S	71.28E	30.86	B	15.58S	71.0E	39.91
C	13.33S	70.25E	43.94	D	14.61S	70.57E	12.26
E	8.83S	69.26E	36.55	F	14.44S	70.32E	8.6
K	13.08S	71.95E	7.17	Z	11.21S	72.56E	6.65

Notes:

Add Info:

When viewing Kapteyn, make a note of the two other craters nearby - sub-crater Kapteyn E and Barkla - where all are similar in size, and each has a nice flattened peak in their centres. Kapteyn seems to be within ages of the other two; sub-crater E being oldest and Barkla the youngest. There's also a nice feature relationship to look out for between all three - E's peak is the highest (1.5 km) but it's the smallest and shallowest crater, while Kapteyn's and Barkla's peaks are both vying for being the lowest (Kapteyn's looks like the winner, while also being the deepest of these two at ~ 3.4 km).

Craters

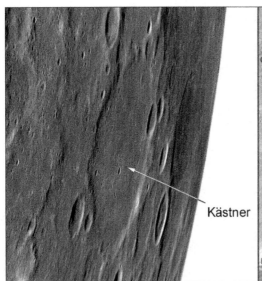
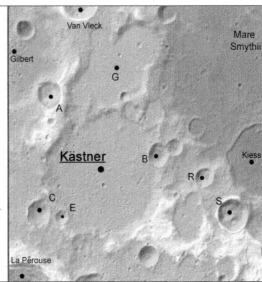

Kästner

Kästner	Lat 6.9S	Long 78.94E	116.7Km

Sub-craters:	Crater	Lat	Long	Size (km)	Crater	Lat	Long	Size (km)
	A	4.5S	77.27E	24.4	B	6.25S	80.71E	19.31
	C	7.94S	76.89E	22.12	E	8.06S	77.62E	11.88
	G	3.68S	79.35E	95.66	R	6.9S	82.17E	16.98
	S	7.99S	83.1E	29.18				

Notes:

Add Info:

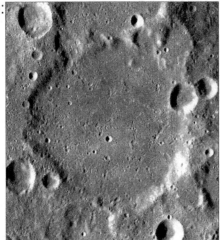

Aerial view of Kästner

A hard crater to observe because of its limb location, Kästner lies on ejecta deposits of the Mare Smythii to the east (it interrupts one of Smythii's main ring/rim systems - the 540 km diameter one). Its well worn rim is only really viewable at low terminator times, while at high sun views no real detail shows except for the slightly dark deposits on its floor. These deposits are more likely the result of lava flooding that occurred within the crater itself through underlying fractures, where the same may have happened in its sub-crater, Kästner G, too. Ejecta deposits - possibly from Mare Crisium to the northwest may also fill the floor. Depth ~ 3km.

Craters

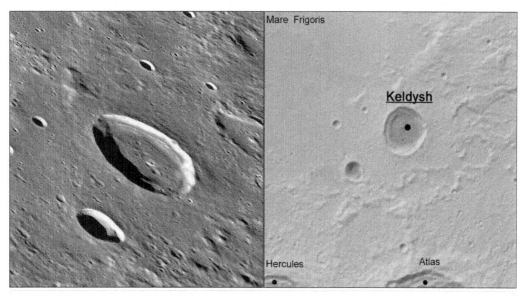

Keldysh	Lat 51.23N Long 43.65E	32.75Km

Sub-craters: No sub-craters

Notes:

Add Info:

Aerial view of Keldysh

Keldysh must be given the added title as being the easterly-most crater that kisses the older lavas of Mare Frigoris goodbye. The crater lies on a 'spit' of highland's material that interrupts a connection of Frigoris's lavas to those in LacusTemporis's in the east. The sharp rim of Keldysh suggest it's a young crater possibly of the Eratosthenian Period - 3.15 to 1.1 byo, and material on its floor looks like it is composed fully of slumped deposits - leaving a series of wide terraces all within the crater's inner rim. As the crater has a depth of ~ 2.7 km, does the floor also contain some deep-sourced lavas, too? Ejecta deposits around the crater's outer rim show up nicely during low sun views.

Craters

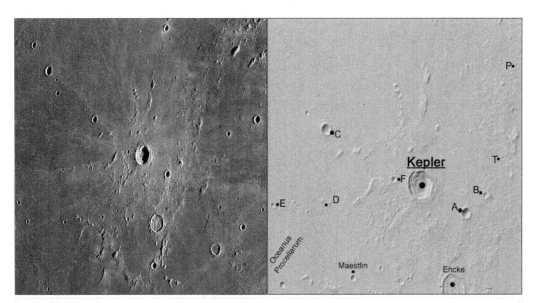

Kepler   Lat 8.12N   Long 38.01W   29.49Km

Sub-craters:

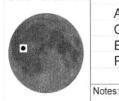

Crater	Lat	Long	Size (km)	Crater	Lat	Long	Size (km)
A	7.14N	36.12W	10.7	B	7.74N	35.33W	6.18
C	10.01N	41.86W	11.5	D	7.42N	41.89W	9.51
E	7.38N	44.02W	5.22	F	8.31N	39.09W	6.06
P	12.21N	33.99W	4.11	T	9.0N	34.61W	2.6

Notes:

Add Info:

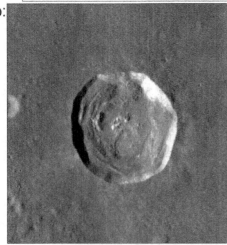

Aerial view of Kepler

For such a relatively small crater, Kepler makes its presence known mainly through its wonderful display of bright ejecta (ignore Copernicus's rays to the east for now). The rays spray out for some distance: ~ 500 km to the west (near to crater Reiner), and eastwards for about 120 km where the furthest trails are a bit hard to define as they mix with Copernicus's rays. The rays are also a signature of material depth excavated from Kepler - the furthest parts composing the topmost layer of the surface the initial impactor struck, while those near the outer rim are of the deepest layers. Depth 2.7 km approx..

Craters

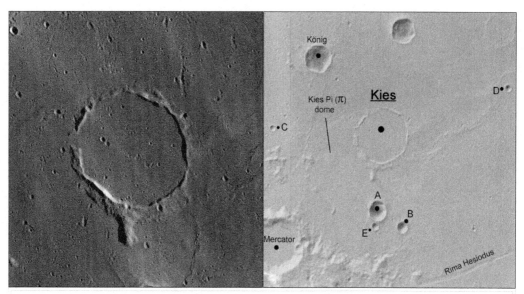

Kies	Lat 26.31S	Long 22.63W	45.54Km

Sub-craters:	Crater	Lat	Long	Size (km)	Crater	Lat	Long	Size (km)
	A	28.36S	22.76W	14.54	B	28.77S	21.93W	9.36
	C	26.07S	26.17W	4.7	D	24.93S	18.55W	5.61
	E	28.76S	22.83W	6.17				
	Notes:							

Add Info: There are many partially-buried, lava-filled craters on the Moon, but none so as distinctive as crater Kies - with its little 'handle' attached to the southern sector of its rim. The 'handle' doesn't look like it was initially part of Kies's ejecta before the lavas of Nubium lapped up to it, but more it was just an isolated clump of (or ejecta from some impact event) highland's material where Kies's own outer ejecta met it. The rim of Kies is well worn down, which would signify it is an old crater, however, if we look at how the lavas both around and within the crater have affected the crater's appearance, is the worn rim a consequence of lavas pouring into, up and over, it (unlikely?)? The small, five-kilometre-wide 'gap' at the rim's western sector was probably a breach-point where lavas infilled the crater, however, it could have went the other way as well - where interior lavas welled up in Kies's centre and breached outwards (as the gap has a circular character to it - was it originally an impact crater on Kies's rim?). The northern inner rim at Kies has an unusual look to it; where a very thin 'line' follows the rim's run (it is very hard to see even at high magnification). What this feature is is a puzzle as it doesn't look like a slumping effect, nor like a fault - which would be a very unusual place for one to have occurred. Rays from Tycho cross the crater. Depth 0.4 km.

Not to be confused with crater Kiess over in the east at 6.41S, 84.11E.

Craters

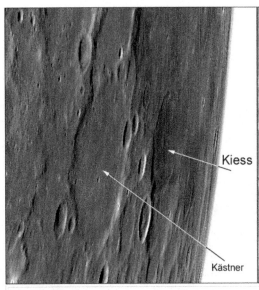

| Kiess | Lat 6.41S | Long 84.11E | 67.79Km |

Sub-craters: No sub-craters

Notes:

Add Info:

Aerial view of Kiess

Like so many other craters lying within Mare Smythii's interior, Kiess just about peeps above the plain of lavas. It is the biggest of all the other craters, and the deposits on its floor are darker, too. It looks like Kiess may have impacted on Widmannstätten to its east; causing Kiess's rim to slightly extend on to its neighbour's floor before it disappears (during favourable observations, the extensions of the rim are easily seen). Hint of a small peak (about ~ 200 metres high) appears off-centre on Kiess's floor, but it isn't obvious at times during westerly, low sun apparitions as shadows and the perspective hide it. Depth ~ 0.8 km.

Craters

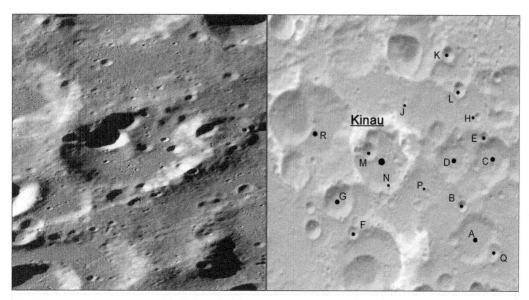

| Kinau | Lat 60.75S | Long 14.94E | 41.87Km |

Sub-craters:

Crater	Lat	Long	Size (km)	Crater	Lat	Long	Size (km)
A	62.3S	20.1E	32.69	B	61.66S	19.18E	7.75
C	60.68S	20.51E	30.07	D	60.71S	18.55E	26.33
E	60.22S	20.13E	6.71	F	62.26S	13.41E	8.3
G	61.61S	12.7E	22.02	H	59.82S	19.73E	5.96
J	59.59S	16.08E	4.85	K	58.6S	18.11E	9.26
L	59.31S	18.73E	9.59	M	60.62S	14.33E	11.44
N	61.41S	15.4E	4.62	P	61.42S	17.36E	4.95
Q	62.54S	21.02E	11.18	R	60.23S	11.55E	61.47

Notes:

Add Info:

Aerial view of Kinau

When looking for Kinau in this area where too many craters rule the same, use the two small craters within its northwestern rim (sub-crater Kinau M is one of them) as markers. Kinau is an old and pummelled-to-death crater that has seen a lifetime of small to large impacts on its rim, its floor, and surrounds. A small mound at its centre suggests hint of a peak, but as sub-crater M lies quite close to it, is the mound simply ejecta from it (note the slight linear features on the rim's south-eastern inner rim which point back to M). Depth ~ 2. 4 km approx..

Craters

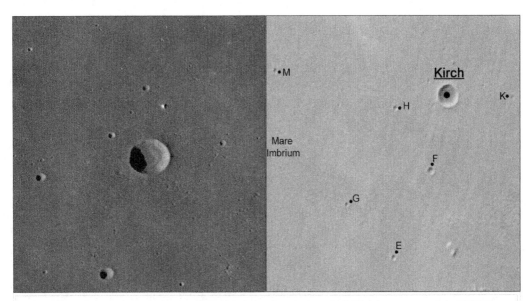

Kirch	Lat 39.27N	Long 5.62W	11.71Km

Sub-craters:

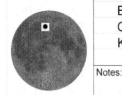

Crater	Lat	Long	Size (km)	Crater	Lat	Long	Size (km)
E	36.52N	6.93W	3.01	F	38.01N	6.05W	4.06
G	37.4N	8.09W	2.58	H	39.05N	6.94W	2.75
K	39.23N	4.0W	2.48	M	39.61N	9.96W	3.02

Notes:

Add Info:

Aerial view of Kirch

Kirch lying in its plain of Imbrium lavas looks, to all intents and purposes, like it 'bottomed-out' after impact. Having a depth of around ~ 1.8 km, its floor is more saucer-like in shape than bowl-shaped for a crater of its size. The central part of the saucer holds a series of low mounds, but as to what the remaining portion of the outer floor material is that meets the inner walls, it's anyone's guess as it doesn't look like slumped deposits (would it be of rebounded, or, disturbed material?). A small 'kink' at the crater's northeast inner wall (a failed small impact crater) is a challenge to see, but low sun times shows nicely another feature - the north-trending ridge on which Kirch has clipped.

Craters

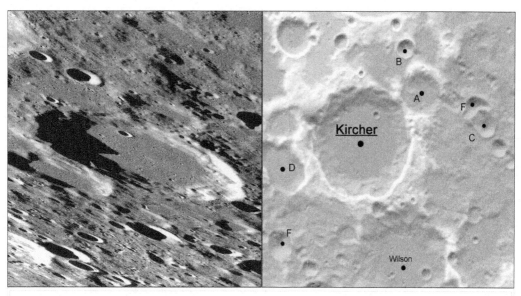

Kircher   Lat 67.01S Long 45.48W   71.2Km

Sub-craters:

Crater	Lat	Long	Size (km)	Crater	Lat	Long	Size (km)
A	65.95S	42.39W	30.83	B	65.08S	43.26W	12.13
C	66.87S	37.67W	11.17	D	67.45S	49.99W	39.03
E	69.06S	50.36W	19.12	F	66.11S	39.32W	12.63

Notes:

Add Info:

Aerial view of Kircher

Coming in at nearly 4.5 kilometres in depth, Kircher is a crater that when you go to observe it, nine times out of ten, shadows are almost always filling its floor. Its location, of course, is a contibuting factor, but its the high surrouding rim and terrain that is a curse to the observer (favourable librations are the best times to view - so plan well). The crater lies in a basin region: with the centre of the Schiller-Zucchius Basin some 350 km off to the northwest, Bailly Basin to the southwest ~ 280 km away, and the Orientale Basin over to its west ~ 1800 km away, whose influential ejecta possibly is responsible for its levelled floor. NW-SE rays (somewhat hard to see) cross the floor too, which most likely have their source from young crater Zucchius.

Craters

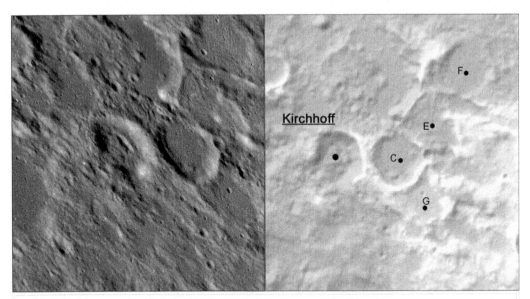

Kirchhoff	Lat 30.3N	Long 38.84E	24.38Km

Sub-craters:

Crater	Lat	Long	Size (km)	Crater	Lat	Long	Size (km)
C	30.31N	39.79E	22.75	E	30.61N	40.34E	26.18
F	31.47N	40.8E	27.75	G	29.79N	40.29E	17.68

Notes:

Add Info:

Aerial view of Kirchhoff

It's as well that Kirchhoff has a distinguishing, odd-looking feature within its inner floor area, as the crater hasn't anything to offer in terms of interest to the observer. What is that feature? It looks, to all intents and purposes, like it may be some kind of vent (note the crater isn't far away from the lava plains of Lacus Somniorum to the northwest). Certainly, there are signatures of volcanic activity in nearby sub-crater, Kirchhoff C (are the two little 'dimples' domes?) and lava-fill in craters like Römer E to the south. Looking closely at the northeastern inner rim of Kirchhoff, a major slumping of material seems to have occurred at the wall there. Moreover, it also looks like there is a small crater beneath the slumped material, so is it a vent at all? Depth ~ 2.6 km.

Craters

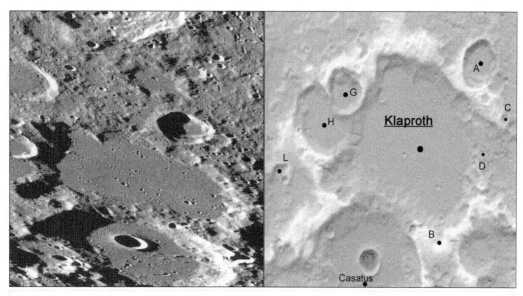

Klaproth   Lat 69.85S   Long 26.26W   121.37Km

Sub-craters:

Crater	Lat	Long	Size (km)	Crater	Lat	Long	Size (km)
A	68.2S	21.9W	34.87	B	71.89S	24.18W	15.51
C	69.22S	19.71W	5.91	D	70.35S	20.73W	7.73
G	68.69S	31.51W	30.67	H	69.23S	33.16W	49.33
L	70.12S	36.92W	10.23				

Notes:

Add Info:

Aerial view of Klaproth

Klaproth doesn't view 'crater-like' in the eyepiece - simply because several big impact craters (Casatus, sub-craters Klaproth H and G) that have replaced most of its rim, have given the crater a more rectangular, blocky look to it. Orientale Basin ejecta from the northwest some ~ 2000 km away overlies this area in general, however, the material in the crater's floor may not be of such deposits as that is quite a volume of fill for this distance (is the fill internally-sourced - covered by a dusting of Orientale?). The floor is about 1 km higher up than Casatus's, but 0.7 km lower of H's and G's levels.

Craters

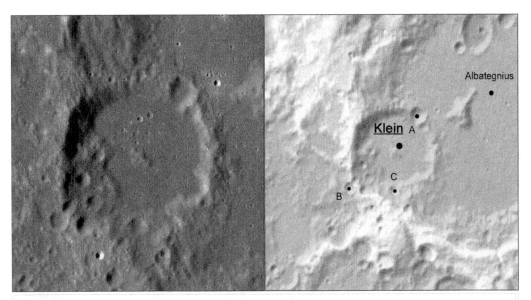

Klein	Lat 11.99S	Long 2.53E	43.47Km

Sub-craters:	Crater	Lat	Long	Size (km)	Crater	Lat	Long	Size (km)
	A	11.38S	2.98E	9.04	B	12.52S	1.78E	5.61
	C	12.58S	2.54E	5.21				
	Notes:							

Add Info:

This whole area holds many secondary craters from the Imbrium Basin event. Which are the most favourable?

Klein looks almost like a scaled-down version of Albategnius crater on which it has impacted upon. It's got a relatively flat floor like its bigger brother, its peak size has proportionally sized down to suit the floor area, and its western, inner rim wall material has overly extended into the crater's west - just like we see in Albtegnius. The floor level of Klein is some 0.6 kilometres lower than Albategnius's floor, however, as to whether both contain internal, volcanic lavas sourced from underneath, or the fluidized ejecta from the Imbrium Basin - whose centre lies some ~ 1500 km away, is uncertain. The whole area has been physically altered by that latter event (note the northwest-southeast striated grooves that cross several craters nearby, including Klein), so their floors certainly are filled with Imbrium's ejecta, however, we cannot rule out the volcanic option, too (perhaps, a mix of both?). Albategnius is of the Nectarian Period (3.92 to 3.85 byo) and it's obvious that Klein is younger. However, looking at their respective rims, which look quite similar in appearance, did the impactor that produced Klein follow not so long after the impactor that produced Albategnius? Sub-crater, Klein A, forms a 'link' to both craters, but also look at the small chain of craters crossing Albategnius's northwestern rim and the small cluster on Klein's southwestern inner rim. Was a single impactor responsible for both features?

Craters

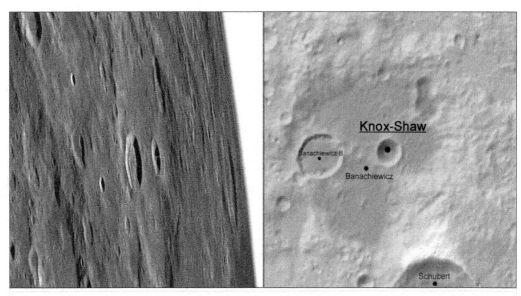

| Knox-Shaw | Lat 5.36N | Long 80.19E | 13.44Km |

Sub-craters: No sub-craters

Notes:

Add Info:

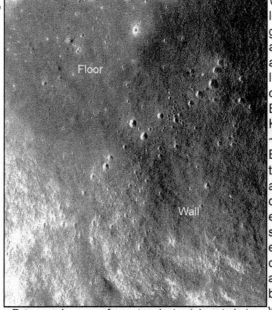

Extreme close-up of a crater cluster (almost chain-like) on Knox-Shaw's southeastern wall

When two relatively sharp-looking craters lie close together, like with Knox-Shaw and Banachiewicz B, they automatically act as good location markers for each other, and, in this case, for Banachiewicz, too. Level of Knox-Shaw's floor is some ~ 600 metres below that of Banachiewicz B's level, but their respective sizes, and ages (Knox-Shaw is oldest), dictated different, structural events (B has undergone some form of slumping) altered their look forever. The crater doesn't really have any distinguishing features, but its bright walls (as, too, with B's) always put on a good show at most times.

Craters

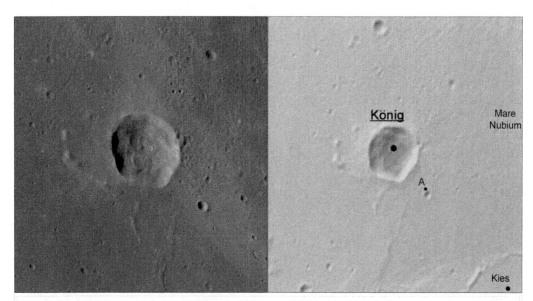

| König | Lat 24.23S | Long 24.68W | 22.86Km |

Sub-craters:	Crater	Lat	Long	Size (km)	Crater	Lat	Long	Size (km)
	A	24.78S	24.09W	3.32				
	Notes:							

Add Info:

Note the rays from Tycho to the southeast that cross the crater. See also the two partially-buried craters nearby. Is the small, dark 'spot' at König's centre volcanically-related - a vent, a last gasp to dark pyroclastic deposits?

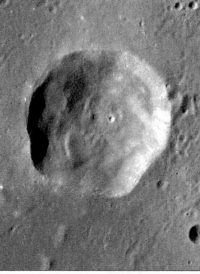

Aerial view of König

Like quite a lot of craters on the Moon that lie in a region where the terrain is predominantly of lava plains, the final floor of these craters have almost always a bottomed-out (saucer to flat) look to them. The flat ones are usually filled with lavas - sourced either from within, or through the breaching at some point that then levelled out. However, the saucer ones, like König, almost all have 'messed-up' slumped deposits on their central floor area. Why is this so? Is it due to the effect that the target rock's makeup has on their impactors? Is it due to hot, lapping lavas affecting the crater from outside - weakening its inner parts; its rim and walls? Such questions have yet to be answered, but they make looking at craters, like König, more interesting. Depth 2.4 km.

Craters

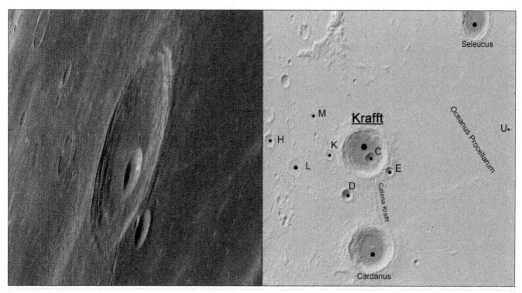

Krafft		Lat 16.56N	Long 72.72W			51.15Km

Sub-craters:

Crater	Lat	Long	Size (km)	Crater	Lat	Long	Size (km)
C	16.37N	72.49W	12.45	D	15.11N	73.38W	12.23
E	15.91N	71.8W	9.64	H	16.92N	77.91W	14.78
K	16.5N	74.68W	12.08	L	16.02N	76.51W	19.8
M	17.79N	75.61W	10.75	U	17.23N	64.76W	3.07

Notes:

Add Info:

Catena Krafft was at one time known as Rima Krafft, before it was later changed in 1976 to its current title.

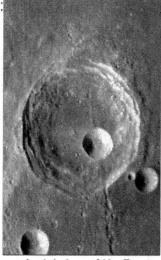

Aerial view of Krafft

Krafft is almost always associated with its twin, crater Cardanus, to its south. The two are similar in size (Cardanus is 49.57 km in diameter), their depths are each around the 3.5 km mark, and both are of the Upper Imbrium Period (3.75 to 3.2 byo) - with Krafft being that little bit younger. The wonderful feature of Catena Krafft connects both, and if you look closely during high magnification under suitable, low sun views, another smaller catena can be seen west of the main one. 'Catena' suggest we are looking at a 'chain of craters', but is this correct? What was their original source? It can't be Grimaldi crater (Basin) to the south, nor Eddington to the north as these are older formations. Are we then looking at a rima feature where a lava tube collapsed to produce a series of pits? The 'chain' or 'rille' continues into Krafft's floor (sub-crater, Krafft C, has impacted upon it), and emerges less obvious beyond its northern rim.

Craters

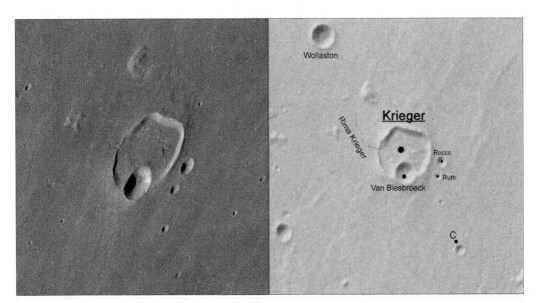

Krieger	Lat 29.02N	Long 45.61W	22.87Km

Sub-craters:	Crater	Lat	Long	Size (km)	Crater	Lat	Long	Size (km)
	C	27.73N	44.7W	4.15				

Notes:

Add Info:

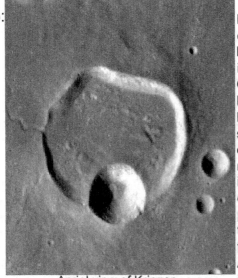

Aerial view of Krieger

The first thing that will strike you if looking at Krieger for the first time is its odd shape. The aerial view shows it best, where its northern rim looks like it has been pushed in by its own ejecta. Of course, this isn't so, so what might have caused it? Did the initial impactor hit higher, more stubborn terrain at that sector, or, is Krieger an oblique impact event (note its ejecta distribution both north and south). Another feature that is striking is the 'knick' groove on the crater's western rim - was it a breach point from which lavas within drained through. Van Biesbroeck ejecta lies on the floor, and ray material from crater Aristarchus to the southwest crosses it, too. Depth ~ 0.9 km approx..

Craters

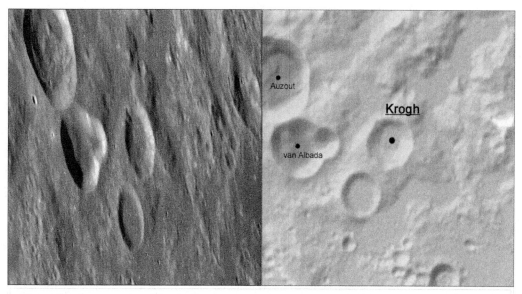

| Krogh | Lat 9.41N | Long 65.69E | 19.17Km |

Sub-craters:

No sub-craters

Notes:

Add Info:

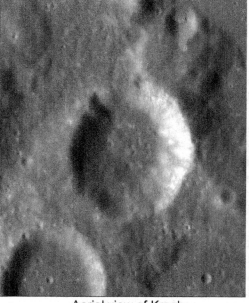

Aerial view of Krogh

While Krogh lies on a clump of ejecta possibly of Crisium Basin origin to its northwest, the crater also is situated just outside one of the Basin's ring/rim system - the 540 km in diameter one (the next ring outwards is 740 km). The crater really hasn't much to offer in detail except for its bright inner walls that show up nicely in full moon periods (Auzout's and van Albada's walls also put on a 'light' show, and are that little bit brighter than Krogh). The material in the floor's centre may be of lava deposits as the area here is also a 'moat' between the two above-mentioned rings (note the small patches of lavas northwest and south of the crater). D: 2. km.

Craters

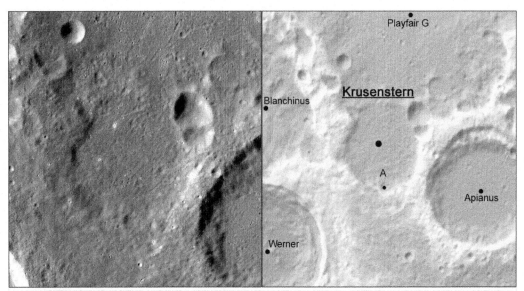

| Krusenstern | Lat 26.3S | Long 5.76E | 46.44Km |

Sub-craters:

Crater	Lat	Long	Size (km)	Crater	Lat	Long	Size (km)
A	26.98S	5.8E	4.89				

Notes:

Add Info: Low to mid-high sun angle times are about the best periods for spotting Krusenstern. At full moon times it simply is gone - just like several of the other craters surrounding it. The crater is of the pre-Nectarian Period (4.6 to 3.93 billions of years old), so it's had plenty of time to accept all manner of abuse - anything from ejecta to bombardment. Lying quite close to younger crater, Werner (Eratosthenian in age - 3.15 to 1.1 byo), to the southwest, this impact not only has covered Krusenstern entirely with impact ejecta, but also has splashed crater Blanchinus to [its] north, and less so in to Playfair G. As Krusenstern was nearer to Werner's impact, it took most of the ejecta too, so much so, that their respective floors differ in depth by about 800 metres (Playfair G is lower). Tycho lies some ~ 650 km away to its southwest also, however, while no ray material overlies Krusenstern directly, its eastern rim and inner walls there show signs that the crater did 'catch' some of it (best seen during high sun angle times). For such a 'practically-buried' crater, its depth is surprising coming in at nearly 2.2 km approximately, as one would have expected it to be less. However, looking at the high terrain in which the crater initially formed, this is possibly why the depth is thus so after all this time. Several small, fresh craterlets lie on the floor, with a very recent impact showing its presence in the northwest. See it?

Craters

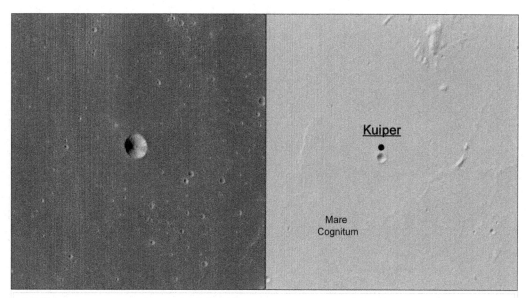

| Kuiper | Lat 9.78S | Long 22.68W | 6.28Km |

Sub-craters:

No sub-craters

Notes:

Add Info:

Apollo's 12 and 14, Luna 5 and Surveyor 3 all landed less than 250 km northwards away from this crater, while the Ranger 7 spacecraft lies some 55 km to its east. Don't bother attempting to try and see them as they are way too small.

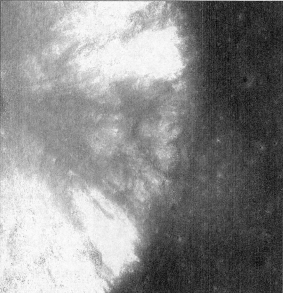

Close-up view of Kuiper's eastern sector

Kuiper is a crater that you will pass in the blink of an eye. It's got a relatively high rim that casts nice shadows during low sun apparitions, but that's about all it has to offer. At extremely high magnification, the small wedge of dark material (see aerial view) within its eastern wall can just about be made out, however, spotting the bright material that connects to it on the outer part of the rim is a challenge. The brighter material is probably a dusting of Kuiper's own ejecta, while the darker stuff may be impact melt that spashed onto the wall. Depth 1.5 km.

Craters

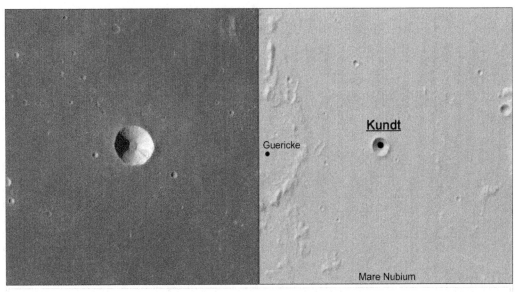

| Kundt | Lat 11.57S | Long 11.57W | 10.31Km |

Sub-craters: No sub-craters

Notes:

Add Info:

The bright material in Kundt's inner walls shows off the crater's presence easily - particularly during Full Moon times (note its bright ejecta, too).

Boulder tracks on Kundt's north-western wall and on its floor.

Kundt is a relatively-fresh impact crater that lies on lavas related to Mare Nubium in the south. At high magnification it's possible to make out its dot-like, dark floor, and many of the brighter, possibly, dry-debris flows (of small stones and rocks etc.,) that cascade down Kundt's inner rim and wall. Such features are due to seismic activity - both through moonquakes and nearby impacts. Another feature (not observable in the eyepiece) is boulder-rolls, which occur as large, individual rocks and boulders roll down the side of the crater - leaving a wonderful boulder track behind them.

Craters

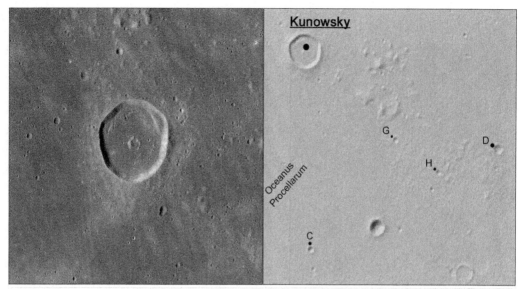

Kunowsky	Lat 3.22N	Long 32.53W	18.27Km

Sub-craters:

Crater	Lat	Long	Size (km)	Crater	Lat	Long	Size (km)
C	0.22S	32.39W	3.4	D	1.53N	28.84W	5.22
G	1.69N	30.8W	3.1	H	1.12N	29.99W	2.82

Notes:

Add Info: Kunowsky lies on a small, rubbly mound of ejecta related to the Imbrium Basin (whose centre lies some 1000 km away to the northeast), and in the lava plains of Oceanus Procellarum. The rim is slightly higher at its eastern side (~ 700 m above the mare) than at its western by about 200 m, and, except for the western side of the crater's outer sector where lavas have lapped in parts at Kunowsky's own ejecta, the remaining other outer sectors are evenly distributed with the ejecta, too. A large volume of the crater's northeastern wall has slumped down onto the floor; giving the rim a kind of 'kink' to it. However, while there looks like slumping may also have occurred west of the kink, its separate departure away from the rim there suggests it might not be slumping at all, but possible rebound material associated to the dynamics of how the crater formed (hard to say). The floor is filled with lavas sourced from underneath, which not only have they weaved their ways in and around the back of the possible rebound material, but also into the central area where a series of peaks (just over 100 metres high) lie. Ray material from both craters' Kepler to the northwest and Copernicus to the northeast cover this area, however, it looks like Kunowsky may only have got a slight dusting from each (see Copernicus's rays south of Kunowsky, and Kepler's southeast).

Craters

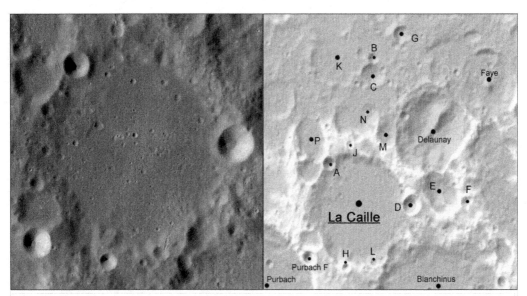

La Caille		Lat 23.68S	Long 1.08E			67.22Km	

Sub-craters:

Crater	Lat	Long	Size (km)	Crater	Lat	Long	Size (km)
A	22.87S	0.4E	7.82	B	20.96S	1.37E	5.95
C	21.25S	1.34E	13.5	D	23.64S	2.19E	11.14
E	23.45S	2.75E	27.66	F	23.6S	3.39E	8.25
G	20.52S	1.97E	10.15	H	24.79S	0.75E	5.35
J	22.54S	0.86E	4.51	K	20.97S	0.52E	29.38
L	24.69S	1.37E	5.04	M	22.34S	1.63E	15.74
N	21.96S	1.27E	11.2	P	22.41S	0.02E	22.69

Add Info:

Day-old moon phases

Note: As the Moon goes through various librations throughout the year, suggested times given in text for observations are only approximates.

Notes:

Like all the other craters: Purbach to the southwest, Blanchinus to the southeast and Delaunay to the northeast - all which surround La Caille, this is one of the pre-Nectarian (4.6 to 3.92 billions of years old) age. Its rim has been practically battered to death through small to large impact events, and its floor is covered with small craterlets all less than 2.5 km in diameter. As the floor is some ~ 700 metres above both Purbach and Blanchinus, and some 1200m above Delaunay's, La Caille may be the 'youngest' of all the other craters, as its impactor may have struck on higher terrain - due to additional ejecta deposits from these crater events. Of course, it's hard to detect if this is indeed the case, as nearly every piece of ground in the area has also been covered by additional ejecta deposits from Mare Nectaris to the east and Mare Imbrium to the northwest. Use the triangular arrangement of sub-craters La Caille A and D, and Purbach F for finding La Caille easily. Do Tycho's rays cross the floor?

Craters

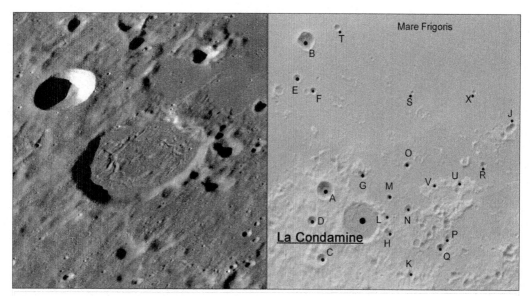

La Condamine Lat 53.54N Long 28.22W 37.83Km

Sub-craters:

Crater	Lat	Long	Size (km)	Crater	Lat	Long	Size (km)
A	54.43N	30.2W	17.1	B	58.87N	31.66W	16.44
C	52.4N	30.3W	10.18	D	53.51N	30.87W	9.97
E	57.73N	32.09W	7.38	F	57.39N	31.09W	6.73
G	54.91N	28.18W	7.17	H	53.17N	26.69W	7.02
J	56.06N	19.45W	7.19	K	51.92N	25.64W	7.77
L	53.64N	26.85W	6.68	M	54.24N	26.72W	5.51
N	53.88N	25.65W	8.34	O	55.18N	25.65W	7.0
P	52.93N	23.62W	8.76	Q	52.69N	24.0W	7.87
R	55.03N	21.33W	6.57	S	57.35N	25.21W	3.82
T	59.3N	29.76W	6.05	U	54.54N	22.72W	6.87
V	54.52N	24.14W	5.71	X	57.22N	21.51W	3.92

Add Info:

Notes:

Aerial view of La Condamine

La Condamine lies some ~ 80 km outside of the main ring/rim of Mare Imbrium to its south. Like Sinus Iridum to its southwest, the crater is of the Lower Imbrian Period (3.85 to 3.75 byo). The striated southwest-northeast features seen on the crater's outer sectors (not on the floor) all point back to Iridum, whose ejecta from the impact ripped and grooved the surface that we now see today. La Condamine's floor has underwent fracturing processes related to underlying volcanic activity, and as a consequence, its true floor is now covered with lava deposits, which then settled in and around higher terrain, hills and peaks (note the impact melt below mid-centre). D: 1.3 km.

Craters

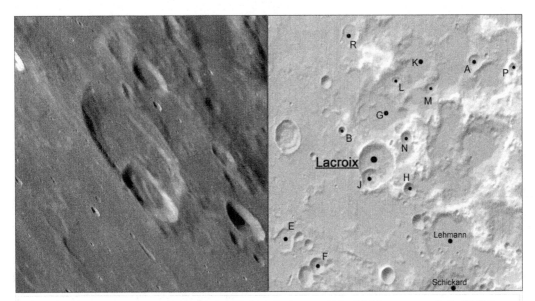

| Lacroix | Lat 37.93S | Long 59.2W | 36.07Km |

Sub-craters:

Crater	Lat	Long	Size (km)	Crater	Lat	Long	Size (km)
A	35.11S	55.34W	13.17	B	37.06S	60.56W	8.09
E	39.97S	63.03W	20.33	F	40.72S	61.73W	14.5
G	36.62S	59.25W	48.77	H	38.66S	57.84W	12.89
J	38.37S	59.43W	20.13	K	35.15S	57.76W	43.6
L	35.67S	58.41W	8.71	M	35.9S	57.06W	12.56
N	37.25S	57.98W	12.83	P	35.22S	53.85W	8.3
R	34.44S	60.22W	19.15				

Notes:

Add Info:

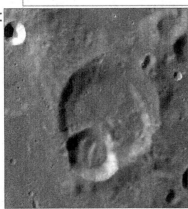

Aerial view of Lacroix

Lacroix lies in sub-terrain where the dominant influence is from the Orientale Basin (Early Imbrian) over to the northwest. Southwestwards of Lacroix, it's easy to see Orientale's effects (striations and aligned clumps point back to the basin), but near the crater itself the terrain looks plain as older, pre-Nectarian material from crater Schickard to the southeast meets the outer deposits of ejecta. Lacroix is easy to locate if using sub-crater Lacroix J on its southeastern flank as a marker, but Schickard is the better option. Bright ray material from Lacroix B ovelies Lacroix's floor, but only its northwest signature on the outer rim is really seen. D: 2.4 km.

Craters

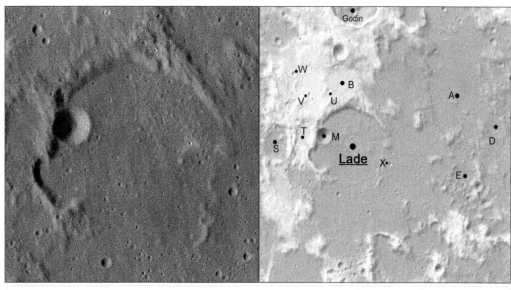

Lade	Lat 1.33S	Long 9.99E	58.11Km

Sub-craters:	Crater	Lat	Long	Size (km)	Crater	Lat	Long	Size (km)
	A	0.16S	12.73E	57.82	B	0.02N	9.8E	23.27
	D	0.89S	13.68E	14.61	E	1.89S	12.92E	19.77
	M	1.1S	9.38E	11.21	S	1.34S	8.21E	22.51
	T	1.06S	8.9E	20.43	U	0.13S	9.48E	3.43
	V	0.21S	9.03E	3.22	W	0.25N	8.63E	3.54
	X	1.72S	11.04E	2.94				

Notes:

Add Info: Initially, Lade must have formed on terrain that predominantly was higher in the northwest area of its outer rim than at its southeast. Clumps of rocks and of fluidized ejecta from the Imbrium Basin event to the northwest, at one time, rained down over the crater - leaving behind a 'grooved' rim (see how the grooves lineate back to the Basin's centre), and a full-to-brim floor. Like most of the craters nearby - from Ptolemaeus in the southwest right up to ~ 13 km Whewell crater to the northeast - the area is like a wide moat where the ejecta (mostly the fluidized type) infilled the lows that from a bigger picture (or in a low magnification) perspective is very easily seen. The floor of Lade is relatively plain - showing a rubbly-looking material (possibly of the rocky ejecta type from Imbrium) in the northeast sector of the crater, which was then covered by the fluidized stuff instantly (see also how this settled right up at the northwest, inner wall area - producing a wonderful sharp divide between the two). While the depth of Lade is just over 0.8 km, sub-crater, Lade M, is over two kilometres deep.

Craters

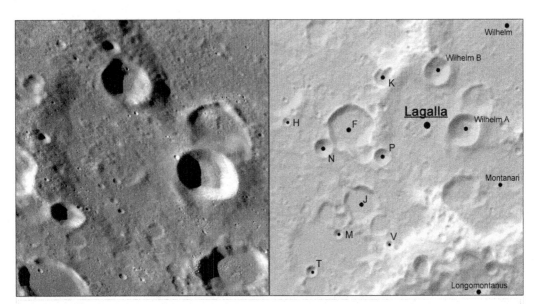

Lagalla   Lat 44.48S   Long 22.36W   88.78Km

Sub-craters:

Crater	Lat	Long	Size (km)	Crater	Lat	Long	Size (km)
F	44.63S	25.32W	29.14	H	44.43S	27.11W	5.52
J	46.1S	25.09W	21.82	K	43.64S	24.38W	10.3
M	46.68S	25.8W	6.16	N	44.98S	26.11W	11.56
P	45.2S	24.45W	10.09	T	47.39S	26.64W	6.98
V	46.92S	24.3W	5.18				

Notes:

Add Info:

Aerial view of Lagalla

Like its neighbour, crater Montanari to its southeast, Lagalla is probably of the late Nectarian Period (3.92 byo). Both are, firstly, covered with ejecta deposits from crater Wilhelm (probably of Late Nectarian age ~ 3.85 byo) to the northeast and, secondly, ejecta from crater Logomontanus (slightly younger than Wilhelm) to the south (compare the state of each of their rims and inner walls). This whole area has been subject to the Orientale Basin event over to the west, and so Lagalla and neighbours have all been covered with its ejecta (are sub-craters, Wilhelm A and B, secondaries of Orientale?). Depth ~ 1.4 km.

Craters

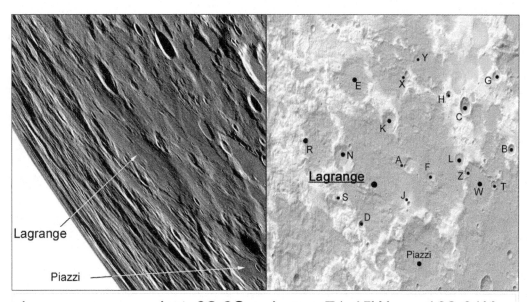

Lagrange   Lat 32.6S   Long 71.45W   162.21Km

Sub-craters:

Crater	Lat	Long	Size (km)	Crater	Lat	Long	Size (km)
A	32.51S	69.3W	6.68	B	31.42S	61.58W	15.18
C	29.88S	65.09W	22.82	D	34.94S	72.54W	11.69
E	29.01S	72.85W	58.12	F	32.91S	67.46W	15.1
G	28.53S	62.9W	17.0	H	29.48S	66.3W	10.61
J	34.04S	68.98W	7.56	K	30.73S	70.44W	29.31
L	32.12S	65.33W	17.09	N	32.05S	73.91W	32.51
R	31.33S	76.67W	126.59	S	33.83S	74.24W	11.35
T	32.97S	62.71W	11.99	W	32.97S	63.83W	57.39
X	28.72S	69.31W	8.22	Y	28.18S	68.56W	16.76
Z	32.55S	64.7W	12.96				

Notes:

Add Info:

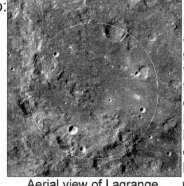

Aerial view of Lagrange

Near impossible to see even under the best of lighting conditions, the only appropriate time to actually observe any crater characteristic to Lagrange is when the sun is low and shadows give some outline to its central area (see image to left). The main reason why we don't see any detail to Lagrange at all is simply because if you live some 700 km away from one of the moon's largest impact on the west side - that of the Orientale Basin, then you're going to get covered by huge volumes of ejecta, and all sorts of destructive objects from blocks to rocks. Depth 2.4 km.

Craters

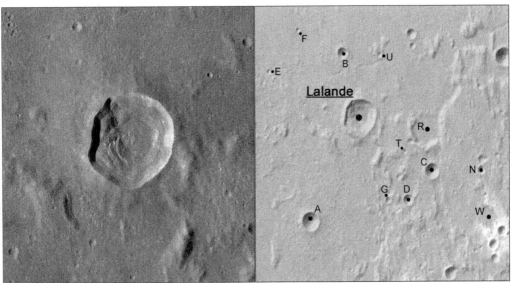

Lalande Lat 4.46S Long 8.65W 23.54Km

Sub-craters:

Crater	Lat	Long	Size (km)	Crater	Lat	Long	Size (km)
A	6.64S	9.81W	12.25	B	3.13S	9.02W	7.68
C	5.6S	6.91W	9.98	D	6.2S	7.48W	7.12
E	3.51S	10.77W	3.15	F	2.65S	10.08W	3.06
G	6.2S	7.96W	3.96	N	5.58S	5.74W	5.7
R	4.71S	7.02W	23.62	T	5.18S	7.56W	3.64
U	3.17S	8.16W	3.95	W	6.57S	5.61W	14.75

Notes:

Add Info: Before you observe Lalande have a look over at crater Mösting in the northeast and then come back to Lalande again. The two look almost completely the same, each having sharp rims, some wonderful landslide features of slumped material, formed terraces, and a hint of small peaks within their centres. Mösting is of the Copernican Period (1.1 billions of years old to the Present), so Lalande may be of this period, too. However, its above features look ever so slightly sharper than Mösting's. This area has been subjected to that of the Imbrium Basin event where its ejecta has covered and infilled into most of the nearby craters seen. However, as we're also close to highland material (the moon's original surface) that lies predominantly towards the southeast of Lalande, the crater formed on both. Obvious to Lalande is its rays, which look as if they have sprayed evenly all around the crater - some reaching as far as to crater Flammarion in the east (~ 200 km away). But look more closely, and you'll see the rays aren't as prominant southwestwards from the crater. Did the small mound nearby Lalande's outer SW rim prevent ray material from going that way, or, are we looking at Lalande as being an oblique-type, impact event? Observe at any time - it is a beauty. D: 2.6 km.

Craters

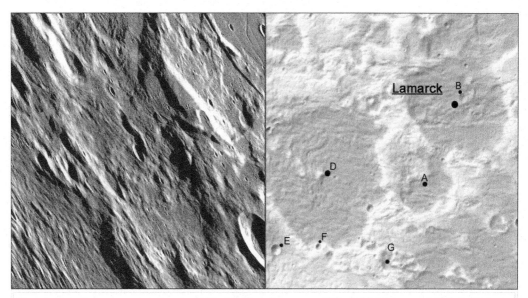

Lamarck	Lat 23.12S	Long 70.06W	114.65Km

Sub-craters:

Crater	Lat	Long	Size (km)	Crater	Lat	Long	Size (km)
A	25.15S	70.96W	53.4	B	22.85S	69.8W	7.6
D	24.9S	74.35W	131.28	E	26.8S	75.81W	8.74
F	26.76S	74.42W	6.21	G	27.27S	72.21W	14.71

Notes:

Add Info:

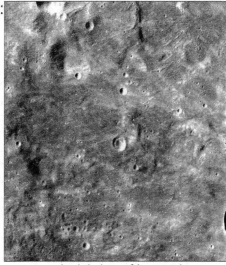

Aerial view of Lamarck

Lying not 700 km away from the Orientale Basin (and all that that had to offer in terms of ejecta coverage and effects) to the west, Lamarck is, surprisingly, not that hard to locate. Unlike crater Lagrange to its south that is a bit of a challenge to find, Lamarck's two main markers are: (1) sub-craters' Byrgius A (Lamarck is covered with the bright ray material from it) and D, to its east, 'line-up' towards its central regions; and (2), the tail-end of the Sirsalis rille to its northeast 'points' approximately at it. Lamarck B, is a concentric crater whose characteristic details are best observed during relatively high sun times (low sun times creates shadows that obsure). Depth ~ 1.1 km.

Craters

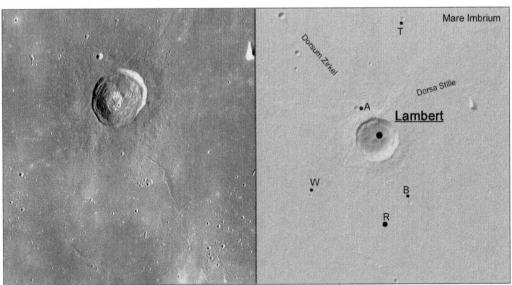

Lambert	Lat 25.77N	Long 20.99W		30.12Km		
Sub-craters:	Crater	Lat	Long	Size (km)	Crater	Lat	Long	Size (km)

Crater	Lat	Long	Size (km)	Crater	Lat	Long	Size (km)
A	26.46N	21.49W	3.73	B	24.34N	20.13W	3.85
R	23.88N	20.66W	55.71	T	28.47N	20.29W	2.89
W	24.49N	22.66W	2.33				

Notes:

Add Info: Lambert lies in the lava surrounds of the Imbrium Basin. With its wonderful display of ejecta observed best at low sun times, the crater 'interrupts' a series of ridges (dorsum), as its rim rises some ~ 600 metres above the level of the lava plains. The most obvious ridge, Dorsum Zirkel, to the north of the crater diminshes as it approaches Lambert's northern outer rim (it doesn't look like the ejecta is covering it up), but Dorsa Stille appears to have been a pre-existing feature (the crater 'cuts' it quite sharply), so ejecta may lie on it. Lambert, like similar-looking crater Timocharis over to its east is of the Erathosthenian Period (3.15 to 1.1 billions of years old). Both have sharp rims, both have terraces formed through slumping effects, and at the centre of each is a ring-like peak. It's a good idea to compare nearby craters that look relatively similar as they not only give you a bigger picture on each, but also say something about the geological make-up in which they formed. For example, Lambert's less-sharper rim and duller surrounds suggest it may be slightly older than Timocharis, while it also formed in a different lava flow. Lambert R looks like it is a buried crater, but note how a ridge close-by create's its northeastern rim, while the northwest portion is 'broken' (are these small, aligned ridges just giving us the appearance of a crater?). Rays from Copernicus to the south cross the crater. Depth of Lambert ~ 2.6 km.

Craters

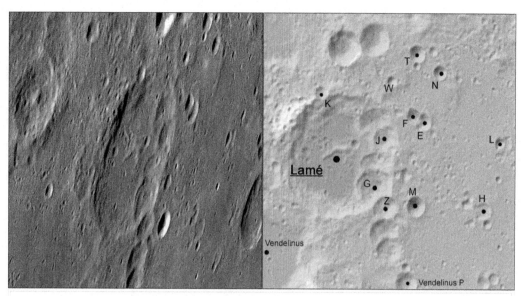

Lamé   Lat 14.76S   Long 64.56E   84.24Km

Sub-craters:

Crater	Lat	Long	Size (km)	Crater	Lat	Long	Size (km)
E	13.98S	66.72E	12.19	F	13.86S	66.47E	10.43
G	15.43S	65.48E	25.52	H	15.91S	68.32E	11.0
J	14.38S	65.77E	18.55	K	13.35S	64.14E	9.61
L	14.41S	68.68E	7.0	M	15.84S	66.56E	14.15
N	12.89S	67.13E	9.83	T	12.48S	66.55E	12.36
W	13.09S	65.9E	7.16	Z	15.89S	65.79E	19.82

Notes:

Add Info: Note how the series of seven craters on Lamé's eastern rim overlap each other - from the unnamed crater above sub-crater J right down to the second-most crater below sub-crater Z. Vendelinus P was probably in this group of impactors, too.

Aerial view of Lamé

Lamé overlies pre-Nectarian (4.6 to 3.92 byo) crater Vendelinus to its west, so it must be younger. However, as the wonderful series of craters that have impacted its entire eastern rim are possibly secondaries from the Nectarian-aged (3.92 to 3.85 byo) Crisium Basin to the north, Lamé is then of an age between the two. Lamé's impact has imparted a substantial amount of ejecta onto Vendelinus's floor, but it looks like it too was later impacted upon (seen at the northern portion of this ejecta). Lamé's floor is lava-filled like Vendelinus's is too, and a proportionally-sized peak lies at its centre. Depth of Lamé is around the 3.5 km mark.

Craters

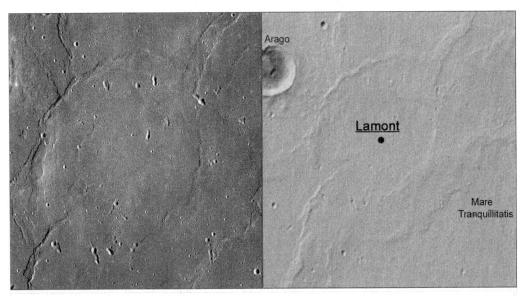

Lamont   Lat 5.14N   Long 23.32E   83.23Km

Sub-craters: No sub-craters

Notes:

Add Info:

Apollo 11 (20 July 1969), Ranger 8 (20 Feb 1965), and Surveyor 5 (11 Sept 1967) all landed in the Lamont area.

The titanium-rich mineral found at the Tranquility Base was given the name ARMALCOLITE: (ARMstrong, ALdrin, COLlins).

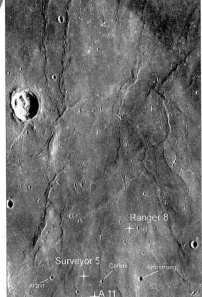

Aerial view of Lamont's surrounds

If it wasn't for the series of ridges that define Lamont's existance, it would be missed by the observer. These ridges show up Lamont better when the sun is low (during high sun views it is simply gone), and from its circular nature we may think that a crater lies underneath the lavas of Mare Tranquillitatis that cover it. But is Lamont really a crater at all? With so many ridges running here, there and everywhere (view their extent), it could simply be the alignment of these that are giving us the crater illusion (the area running south of Lamont right up to crater Plinius in the north has a valley-like look to it). Take a bigger picture view of Lamont's general area and notice the bright to dark colours of the lavas. The dark lavas are due to high concentrations of Iron Oxide + Titanium, while the light are lesser amounts of these in make-up.

Craters

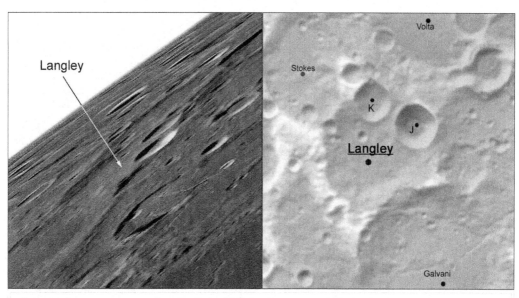

Langley	Lat 51.17N	Long 86.05W	59.17Km

Sub-craters:	Crater	Lat	Long	Size (km)	Crater	Lat	Long	Size (km)
	J	51.63N	85.05W	20.78	K	51.97N	86.23W	18.67

Notes:

Add Info:

Aerial view of Langley

Both crater Galvani to the southeast and crater Volta to the northeast have obviously impacted close to crater Langley (they are therefore younger). However, it isn't as clear when trying to figure out the formation interplay between crater Stokes to the northwest and Langley, as the connecting ridge (rims), and terrain/ejecta from Volta, too (which also impacted upon Stokes), is causing confusion. Sub-craters', Langley K and J, further add to the impact mess, however, their redeeming factor is that both act as good location markers for Langley. The crater is pre-Nectarian in age - 4.6 to 3.92 billions of years old, it has a depth of around 2.0 km, and views are best during favourable librations.

Craters

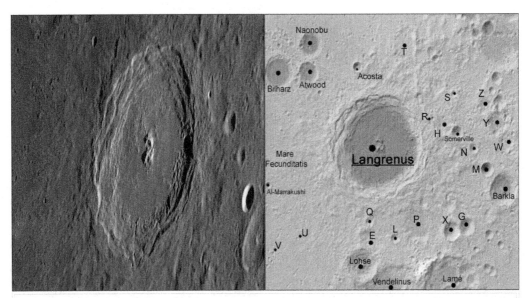

Langrenus   Lat 8.86S   Long 61.04E   131.98Km

Sub-craters:

Crater	Lat	Long	Size (km)	Crater	Lat	Long	Size (km)
E	12.73S	60.72E	31.03	G	12.17S	65.46E	21.9
H	8.03S	64.29E	25.24	L	12.67S	61.92E	12.64
M	9.82S	66.41E	18.17	N	8.98S	65.74E	12.54
P	12.04S	63.01E	42.27	Q	11.99S	60.69E	12.88
R	7.76S	63.75E	5.43	S	6.72S	64.79E	8.92
T	4.79S	62.3E	40.01	U	12.65S	57.15E	4.21
V	13.23S	55.94E	5.01	W	8.67S	67.32E	22.04
X	12.35S	64.71E	23.58	Y	7.88S	66.86E	28.99
Z	7.18S	66.26E	20.82				

Notes:

Add Info:
Langrenus's two main peaks can at times produce some wonderful long, spiky shadows across the floor - be sure to catch them.

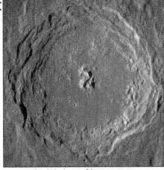
Aerial view of Langrenus

One could say 'it's a real pity that Langrenus lies where it lies', as we, the observers, are missing out on a wonderful treat. But that isn't all true, because if we look over to the more suitably-placed crater of Copernicus in the west, then we have, in a sense, an identical view of what Langrenus might look like from overhead. Langrenus is slightly older than Copernicus, but like it, it has a gorgeous set of terraces, two relatively-high peaks (northern 2.5 km, southern 2.0 km), impact melt on its floor, sharp rim, and rays that propagate westwards onto Mare Fecunditatis. What eye-candy! Depth ~ 4.5 km.

Craters

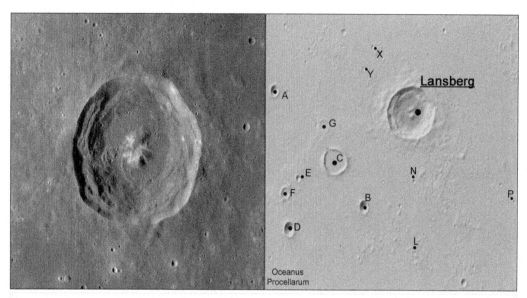

Lansberg Lat 0.31S Long 26.63W 38.75Km

Sub-craters:

Crater	Lat	Long	Size (km)	Crater	Lat	Long	Size (km)
A	0.18N	31.15W	7.97	B	2.49S	28.14W	9.0
C	1.47S	29.21W	17.91	D	3.01S	30.64W	10.46
E	1.84S	30.35W	4.95	F	2.19S	30.78W	8.32
G	0.64S	29.48W	9.54	L	3.56S	26.41W	4.1
N	1.91S	26.44W	3.48	P	2.36S	23.03W	2.43
X	1.2N	27.87W	2.99	Y	0.7N	28.19W	3.56

Notes:

Add Info:

Apollo 12 (19 Nov 1969), Luna 5 (12 May 1965) and Surveyor 3 (20 April 1967) all landed in the general region southeast of Lansberg. Apollo 14 (5 Feb 1971) also landed, too ~ Lat 3.67S, Long 17.46W.

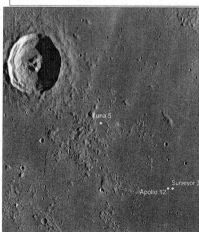

Aerial view of Lansberg region

Lansberg is of the Upper Imbrian Period (3.75 to 3.2 byo). A wonderful display of ejecta (seen best at low sun times) overlies lavas of Mare Insularum to the north and Oceanus Procellarum to the west; all of which have been spattered by ray material from crater Copernicus to the northeast. The crater has a relatively sharp rim all around, some well defined terraces, two main peaks - each about one kilometre high. Several small, wispy rilles lying northwestwards and eastwards to Lansberg require low sun views and high magnification to see them, but the rille (or is it fault feature?) south of sub-crater N is an easy target. Depth ~ 2.74 km.

Craters

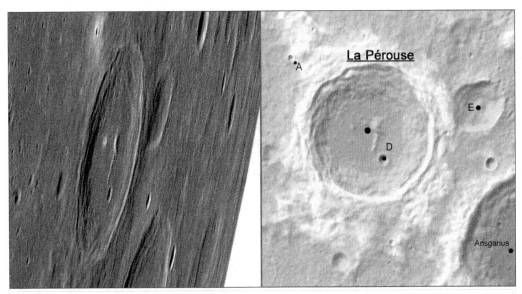

La Pérouse	Lat 10.67S	Long 76.28E	80.4Km

Sub-craters:

Crater	Lat	Long	Size (km)	Crater	Lat	Long	Size (km)
A	9.27S	74.7E	4.08	D	11.17S	76.57E	6.57
E	10.17S	78.51E	35.22				

Notes:

Add Info:

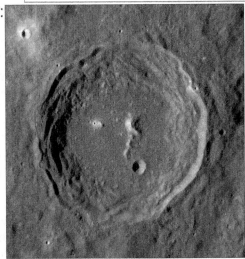

Aerial view of La Pérouse

La Pérouse lies on high terrain due to the confluence of where two ring systems meet (the 750 km diameter Mare Smythii ring to the northeast and the 500 km diameter Balmer-Kapteyn ring to the southwest). The crater is probably of the mid Eratosthenian age (~2 byo), it has a relatively sharp rim all around whose terraces look the same, and one main peak (~ 2 km high) with a tail of mountains attached southwards. Westerly sun apparitions (and suitable librations) are the best times for viewing the crater's details. La Pérouse A serves as a good marker for finding the crater. Depth ~ 4 km approx..

Craters

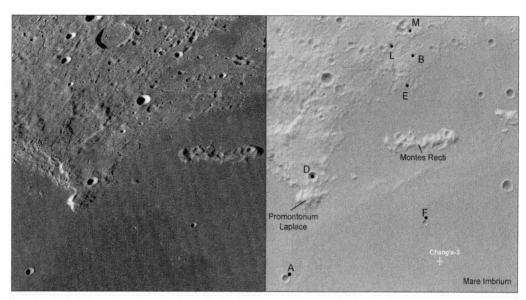

Laplace   No crater named

Sub-craters:

Crater	Lat	Long	Size (km)	Crater	Lat	Long	Size (km)
A	43.74N	26.93W	8.16	B	51.35N	19.86W	5.04
D	47.27N	25.58W	10.57	E	50.33N	20.3W	6.09
F	45.55N	19.86W	5.87	L	51.74N	21.06W	6.45
M	52.22N	19.93W	5.0				
Notes:							

Add Info:

Chang'e-3: On 14 Dec., 2013, China successfully soft-landed Chang'e-3 some 40 km southeast of crater, Laplace F. Several hours later, a small rover called Yutu, (translated as 'Jade Rabbit') deployed from the lander on to the lunar surface.

There isn't, as of writing, a particular crater on the Moon which has the official title, or name, of 'Laplace'. Naming features for planetary bodies in our Solar System is the responsibility of the 'Working Group for Planetary System Nomenclature' associated to the International Astronomical Union (IAU). Names are usually requested by investigators mapping or describing specific surfaces or geological formations on the body in question, which are then reviewed by the IAU workgroup for possible approval (note: public submissions are also considered). The first systematic listing of lunar nomenclature occurred in 1935 in a report "Named Lunar Formations" compiled by Mary Blagg and Karl Müller, which later led to further listings that included changes to areas such as features' diameters, sizes, heights, latitudes and longitudes. Initially, named lunar features were attributed to deceased scientists and philosophers only, however, today, artists, writers, mythical characters are included, as well as living people: craters' Armstrong, Collins and Aldrin have been named in honour of the Apollo 11 crew. For Pierre-Simon Laplace (French mathematician and astronomer), however, the promontory Laplace is named in *his* honour. During medium to high sun times its bright head 'shines', while in low easterly sun times its 2.8 km-high mountain produces a relatively long shadow across the floor of Sinus Iridum to the west.

Craters

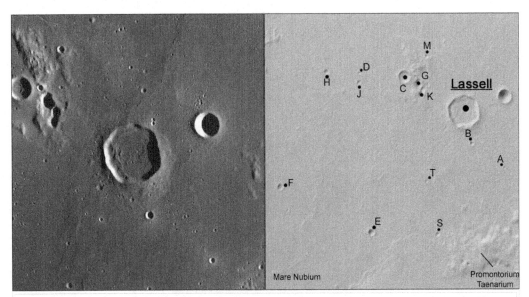

Mare Nubium   Promontorium Taenarium

Lassell   Lat 15.49S   Long 7.9W   21.82Km

Sub-craters:

Crater	Lat	Long	Size (km)	Crater	Lat	Long	Size (km)
A	16.77S	6.88W	2.03	B	16.22S	7.69W	2.97
C	14.68S	9.36W	8.74	D	14.59S	10.48W	1.96
E	18.24S	10.22W	4.82	F	17.15S	12.49W	4.67
G	14.82S	8.99W	6.19	H	14.54S	11.27W	3.76
J	14.78S	10.47W	3.22	K	15.08S	8.95W	3.94
M	14.21S	8.8W	2.82	S	18.29S	8.57W	3.34
T	17.07S	8.83W	2.16				

Notes:

Add Info: Is Lassell a crater? It's got all the usual features: rim, ejecta display around its exterior, a floor etc.,, that we would expect to see from such an event. From a bigger picture perspective, the feature cuts across a northeast-southwest trending ridge (best seen during low easterly sun periods), and lies on younger lavas than those to its northwest (around the area where the cluster of sub-craters K, C, G, M, D, H and J lie). The material in its floor, which looks slightly darker to the above-mentioned younger lavas, probably represents deeper lavas that sourced through fissures into its interior. So, why ask the question: "Is Lassell a crater?" The reason is that Lassell could also be a caldera: a collapsed magma chamber following the eruption of fluidic basaltic lavas, and drainage of the chamber leading to collapse. The ejecta signature around Lassell may just be 'pushed-out' material as the caldera grew, the floor material represents the fluidic lavas, and the terraces seen in Lassell's western sector are resultant products as the chamber collapsed. Of course, these are all unproven propositions, and Lassell may just be a crater after all, however, when looking at features on the Moon, be sure to be open to other explanations for their formations.

Within the western sector of Lassell's floor - near the collapsed terraces - is a feature that looks similar in appearance to the Cobra Head of the Aristarchus Plateau. As it is higher up than the floor, would it at one time been a source for the fluidic lavas?

Craters

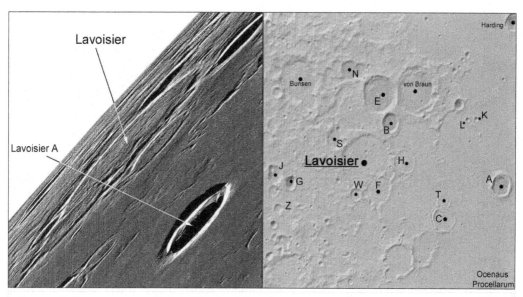

Lavoisier   Lat 38.17N   Long 81.25W   71.01Km

Sub-craters:

Crater	Lat	Long	Size (km)	Crater	Lat	Long	Size (km)
A	36.97N	73.26W	28.51	B	39.76N	79.75W	24.55
C	35.78N	76.73W	34.9	E	40.88N	80.39W	50.56
F	37.03N	80.47W	35.63	G	37.21N	85.61W	18.19
H	38.17N	78.94W	29.12	J	37.49N	86.58W	21.67
K	39.74N	74.43W	6.67	L	39.74N	74.98W	6.42
N	41.9N	82.34W	25.21	S	39.05N	83.08W	24.62
T	36.5N	76.62W	18.57	W	36.82N	81.85W	16.2
Z	36.12N	86.29W	12.45				

Notes:

Add Info:

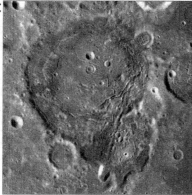

Aerial view of Lavoisier

Though difficult to see in this region close to the vast lava plains of Oceanus Procellarum, Lavoisier, along with several other craters - von Braun, Bunsen, Lavoisier B, E, F & H, is one where volcanic activities have left their mark. Each shows a series of fractures due to volumes of liquid magma (under pressure) that didn't quite reach the surface, but forceful enough to push up against a cold crusted floor. Lavoisier is a challenge to view always, and libration times are recommended, if not terminator periods, too (the dark pyroclastic patches show up during high sun times).

Craters

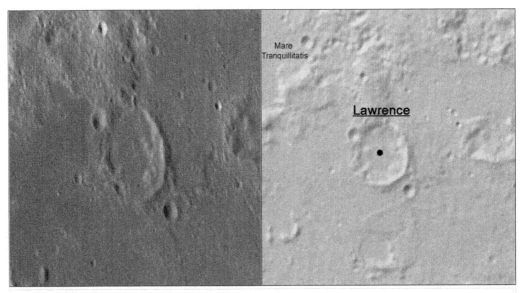

Lawrence   Lat 7.35N   Long 43.3E   24.02Km

Sub-craters: No sub-craters

Notes:

Add Info:

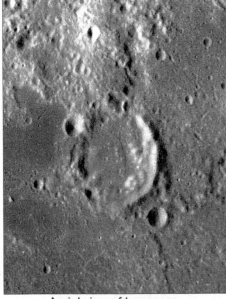

Aerial view of Lawrence

Lawrence lies in an outer trough region related to Mare Fecunditatis in the southeast. Possibly of the Nectarian Period (3.92 to 3.85 byo), lavas seem to have played a dominant role as to its final look. What with darker, younger lavas lapping at its western outer shores, its eastern shores may have experienced a breach of older lavas through a small 'gap' seen in the northeastern rim. The rim here is some 400 m higher than the west rim, but in contrast, the lava plain is nearly 600 m below that of the younger lava plain in the west. Given this. Was there also a breach in the west too as Lawrence's northwestern rim looks strange (a small rille lies within the inner wall there)? A rille, or is it a fault, also run across the central floor.

Craters

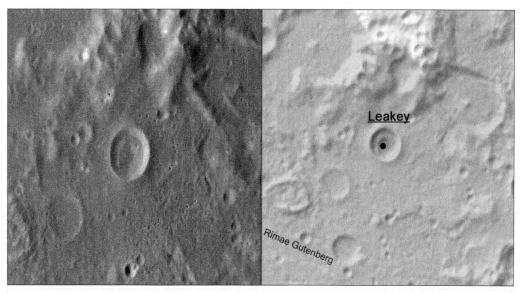

Leakey	Lat 3.19S	Long 37.46E	12.48Km

Sub-craters: No sub-craters

Notes:

Add Info:

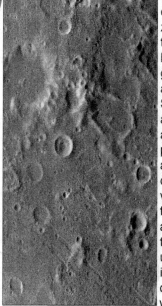

Aerial view of Leakey

Considered a concentric crater (CC), Leakey, along with others like Hesiodus A (30S,17W), Crozier H (14S, 49E) or Marth (31S, 29W), must be the top four in CCs that are easy to observe. Others, like in crater Humboldt (26S, 83E) or at Lagrange T (33S, 63W), are a challenge, however, all are wonderfull and unique to their surrounds. So far todate, upto 60 CCs have been officially stamped, but there are a lot more new ones found awaiting approval. Formation of CCs was initially put down to as an impactor striking layered surfaces where each deeper layer affected the impactor's dynamics and rebound properties. However, three different types of CCs have led to other theories where they might be products of simultaneous impacts, volcanic extrusion or igneous intrusion. Whatever their formation, most CCs are usually found near mare-highland boundaries, in isolated mare ponds, and within floor-fractured craters (like, Humboldt). CCs usually have shallower depths than other craters, shorter rim flanks, and smaller rim heights. Depth of Leakey ~ 1.8 km.

Craters

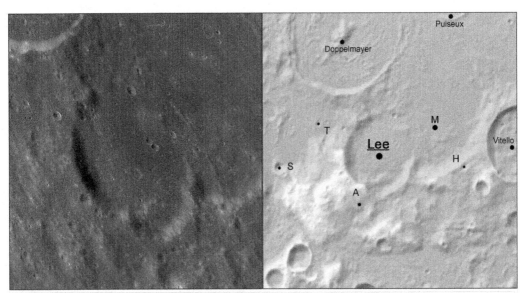

| Lee | Lat 30.66S | Long 40.76W | 41.17Km |

Sub-craters:

Crater	Lat	Long	Size (km)	Crater	Lat	Long	Size (km)
A	31.5S	41.24W	17.44	H	30.85S	38.85W	4.53
M	29.83S	39.7W	72.33	S	30.83S	42.96W	6.89
T	30.05S	42.1W	3.5				

Notes:

Add Info:

Aerial view of Lee and environs

Lee lies in the edge-zone of one of Mare Humorum's ring/rim system (the 420 km-diameter one). It's obvious that Lee has impacted upon the southwestern rim of sub-crater, Lee M, but what is going on at Lee's own southeastern sector - it appears that two rim signatures exist (which one relates to Lee, did another crater lie there originally?) The dark lava signatures in the northeast part of Lee's floor contrast to the lighter in the southwest. They're not as dark as the pyroclastic, volcanic materials of Doppelmayer's northeastern floor, but are they similar in their formation (several NE-SW trending rilles - one very obvious - crosses the floor of Lee). Height of lee's southwest rim is nearly 2.6 km approx..

Craters

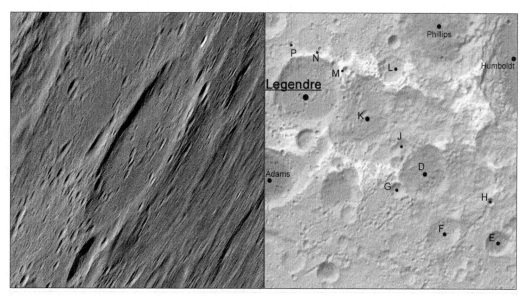

Legendre   Lat 28.92S   Long 70.02E   78.08Km

Sub-craters:

Crater	Lat	Long	Size (km)	Crater	Lat	Long	Size (km)
D	31.6S	75.15E	57.54	E	33.86S	78.57E	26.08
F	33.66S	76.12E	40.09	G	32.23S	73.92E	16.61
H	32.44S	78.17E	8.44	J	30.63S	74.1E	12.49
K	29.79S	72.21E	91.91	L	28.12S	73.5E	31.06
M	28.17S	71.61E	9.56	N	27.47S	70.5E	5.82
P	27.31S	69.28E	6.99				

Notes:

Add Info:

Aerial view of Legendre

Legendre isn't that hard to locate; once you use the much more obvious crater of Petavius to its west as a handy marker. Mottled with secondary craters (each about 3 km in diameter) - radially and concentrically directed from, possibly, both the Nectaris Basin and Petavius (Late Imbrian in age) to the northwest, and Humboldt (Lower Imbrian) over to its east, Legendre is a crater that is often overlooked. Having impacted on sub-crater Legendre K to the east, the two join another smaller set of less-obvious craters to their east - all best observed during low sun times. Depth of Legendre ~ 3 km.

Craters

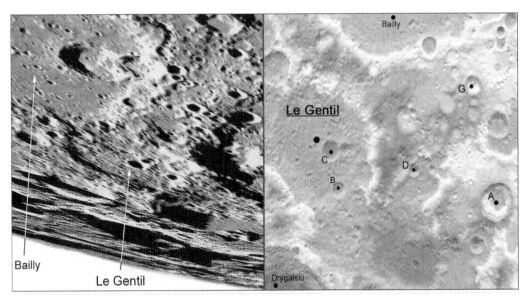

Le Gentil   Lat 74.27S   Long 76.02W   125.38Km

Sub-craters:

Crater	Lat	Long	Size (km)	Crater	Lat	Long	Size (km)
A	74.63S	52.63W	37.0	B	75.48S	74.05W	15.72
C	74.36S	75.28W	18.68	D	74.6S	64.1W	12.38
G	71.74S	59.12W	17.35				

Notes:

Add Info:

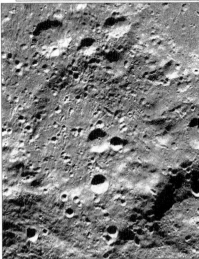

Aerial view of Le Gentil

Unless you've got plenty of time and endless patience, Le Gentil will prove a challenge to 'catch'. Its much larger neighbour, Bailly, to its northeast dominates any view you care to make, and even when you do get all the right conditions going - libration and lighting...etc., Le Gentil still is elusive. Bailly seems to have impacted on its northeastern rim - imparting substantial amounts of ejecta onto the floor, but so too has crater Drygalski to the southwest of a lesser extent. The crater may be of pre-Nectarian (4.6 to 3.92 byo), is peppered with ~ 2-km-sized craterlets (secondaries from Orientale to the northwest?), and lies just on the main 2500 km-diameter ring/rim of the South Pole Aiken (SPA) Basin to its southwest. D: 2 km.

Craters

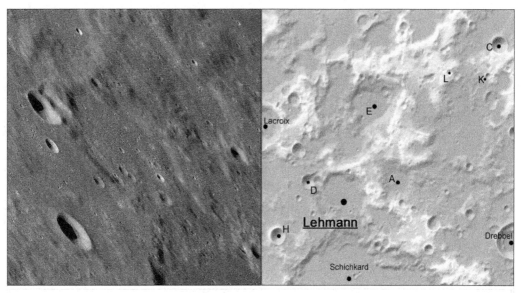

Lehmann   Lat 39.96S   Long 56.17W   53.85Km

Sub-craters:

Crater	Lat	Long	Size (km)	Crater	Lat	Long	Size (km)
A	39.51S	53.87W	29.64	C	35.57S	50.17W	15.08
D	39.54S	57.49W	12.08	E	37.44S	54.96W	44.88
H	41.01S	58.73W	13.81	K	36.44S	50.53W	5.58
L	36.42S	52.04W	5.85				

Notes:

Add Info:

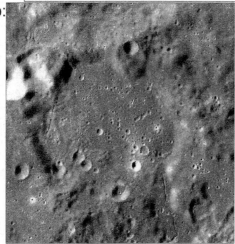

Aerial view of Lehmann

Ignoring, if you can, the wonder of crater Schickard to the south, Lehmann is often overlooked when the two are in view. Both craters are of the pre-Nectarian Period (4.6 to 3.92 byo), but it looks like Lehmann may have been the earlier of the two (are its features more worn-looking than Schickard's?). Lehmann's floor is some 800 m above the floor of Schickard, and a small rille connects them both (did lava flow from the former into the latter?). The dark material in Schickard's floor is of older mare basalts, while the lighter, also seen in Lehmann's floor, is a covering of Orientale (to NW) material. D:~1.3 km.

Craters

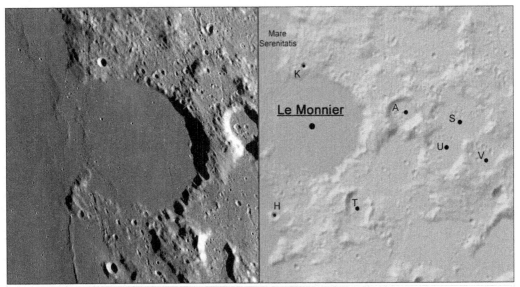

Le Monnier   Lat 26.66N   Long 30.5E   68.4Km

Sub-craters:

Crater	Lat	Long	Size (km)	Crater	Lat	Long	Size (km)
A	26.83N	32.49E	20.87	H	24.97N	29.62E	4.7
K	27.75N	30.26E	4.02	S	26.78N	33.78E	38.32
T	25.17N	31.46E	16.21	U	26.11N	33.41E	25.14
V	26.03N	34.25E	23.1				

Notes:

Add Info:

Luna 21 (or Lunik 21) landed in this region on the 15 Jan 1973. The lander then deployed a rover called Lunokhod 2, which took TV images, photos, as well as mechanical studies of the local soil (regolith). Apollo 17 lies some 200 km away south of this crater.

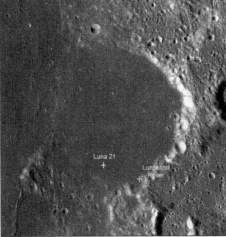

Aerial view of Le Monnier

Lying on the eastern edge of the Serenitatis Basin, Le Monnier has succumbed to the older, darker lavas of the Late Imbrian Period - 3.85 to 3.75 byo (lighter-coloured lavas are of the Late Imbrian Period 3.75 to 3.2 byo) associated with the Mare's initial formation. Two wrinkle ridges (extensions of Dorsa Aldrovandi to the south?) replace Le Monnier's missing western rim, and the floor within is slightly tilted downwards by about 100 metres towards the east - resulting in some nice shadows off the inner rim during low easterly sun times. The crater is famous for the Russian, Luna 21 mission. D: 2.5 km.

Craters

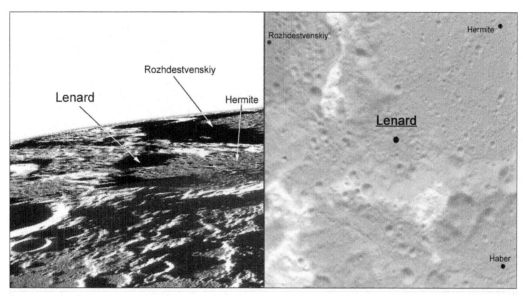

| Lenard | Lat 85.19N | Long 109.69W | 47.65Km |

Sub-craters: No sub-craters

Notes:

Add Info:

Aerial view of Lenard

Lenard 'connects' to Hermite by it having impacted on its rim. The real floors of both are hidden underneath a hummocky, pitted deposit that is said to be from several major impact events - near and afar e.g. the Imbrium Basin to the south (centred ~ 1700 km away), but also minor impacts from the many craters that dominate this northern region of the Moon. Rozhdestvenskiy to the northwest of Lenard takes on the same characteristic hummocky-pitted texture, and why not, as all three craters have been around for a long time (pre-Nectarian in age 4.6 to 3.92 byo). A small, scarp-like feature (see left) hugs Lenard's northwest floor which continues up on to its wall. Is it observable from our perspective?

Craters

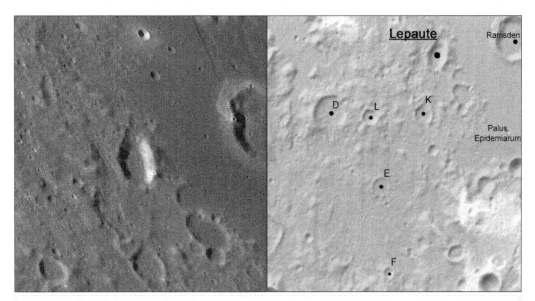

Lepaute		Lat 33.3S		Long 33.69W			16.36Km
Sub-craters:	Crater	Lat	Long	Size (km)	Crater	Lat	Long	Size (km)
	D	34.36S	36.3W	20.41	E	35.75S	35.06W	10.48
	F	37.34S	34.88W	6.23	K	34.39S	34.01W	10.54
	L	34.44S	35.32W	9.05				
	Notes:							

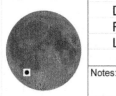

Add Info:

Aerial view of Lepaute

Lepaute is an odd-looking crater! It's east-west width is some 4 kilometres less than its north-south, suggesting that it might just be an oblique impact created crater (the longer axis 'points' to the Imbrium Basin - so is it a secondary from that event?). Lying near a depressed zone (Palus Epidemiarum) - possibly produced as a result in formation of the Humorum Basin to its northwest, the associated lavas nearby may have breached its rim at some point in time (note the 'gap' at the northern part of its rim - did it serve as a breach-point?). Material of the crater's inner rim has a high albedo signature; easily seen at high sun times. D: 2 km.

Craters

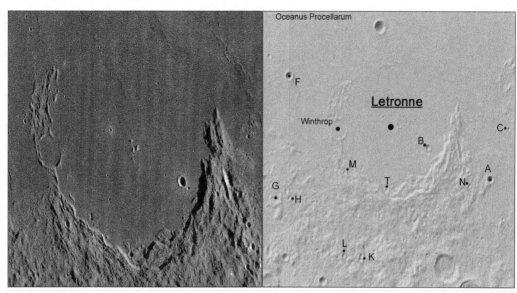

| Letronne | Lat 10.5S | Long 42.49W | 117.59Km |

Sub-craters:

Crater	Lat	Long	Size (km)	Crater	Lat	Long	Size (km)
A	12.15S	39.1W	7.11	B	11.27S	41.3W	5.5
C	10.71S	38.49W	3.44	F	9.23S	46.13W	7.45
G	12.69S	46.64W	10.08	H	12.67S	46.17W	4.19
K	14.37S	43.69W	7.39	L	14.32S	44.29W	4.17
M	11.96S	44.24W	3.91	N	12.33S	39.83W	3.46
T	12.48S	42.77W	3.77				

Notes:

Add Info:

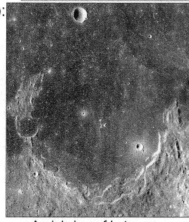

Aerial view of Letronne

If you were to look, firstly, at Fracastorius crater over to the east (21.36S, 33.07E), and then come back to look at Letronne, you would almost believe that you were seeing mirrored versions of each. Both are missing their respective northern rims due to subsidence, each has a small group of peaks at their centres, and their floors are dotted with 'splotchy' fill of lavas dark and light. Letronne, of course, doesn't have the major fissure that Fracastorius does, however, the two are still connected, in sorts, by lying on the edges of major mare deposits. Low sun views of Letronne are always a catch as ridges that cross its floor produce some wonderful effects. Depth 1.2 km.

Craters

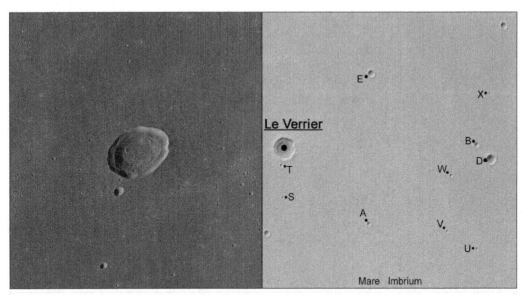

Le Verrier Lat 40.33N Long 20.61W 20.52Km

Sub-craters:

Crater	Lat	Long	Size (km)	Crater	Lat	Long	Size (km)
A	38.17N	17.32W	4.11	B	40.19N	12.88W	4.84
D	39.74N	12.35W	8.74	E	42.42N	17.01W	6.5
S	38.96N	20.67W	2.32	T	39.87N	20.74W	3.47
U	37.26N	13.18W	3.23	V	37.83N	14.29W	3.23
W	39.39N	13.98W	3.25	X	41.6N	12.16W	2.71

Notes:

Add Info:

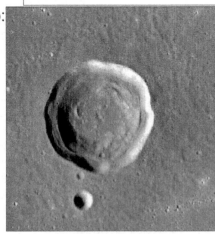

Aerial view of Le Verrier

Le Verrier may be younger than the Imbrian-aged lavas that surround it, as most of its mottled-like ejecta lays on top (compare crater Helicon's ejecta to the west). The mottled look to Le Verrier's ejecta is notable only on the outermost sectors - why? Reason: as ejecta is shot out in all directions from the initial impact, its distribution thins and breaks up into individual smaller clumps the further away as it lands. As a result, you get some areas of highs - the clumps, while the lower are of areas where the ejecta was thin to nonexistent. Hence, the mottled-look! Times of low sun views recommended. D: 2 km.

Craters

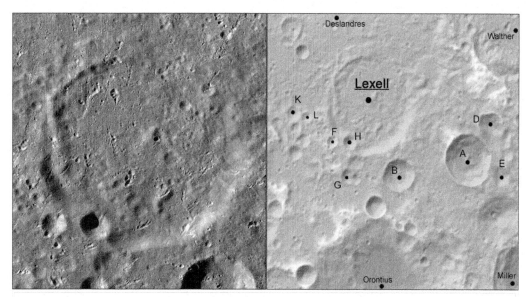

| Lexell | Lat 35.78S | Long 4.27W | 63.7Km |

Sub-craters:

Crater	Lat	Long	Size (km)	Crater	Lat	Long	Size (km)
A	36.92S	1.39W	33.59	B	37.27S	3.41W	21.63
D	36.18S	0.75W	18.47	E	37.23S	0.42W	13.0
F	36.56S	5.38W	7.46	G	37.3S	4.94W	9.26
H	36.58S	4.88W	8.98	K	35.98S	6.48W	10.39
L	36.04S	6.12W	7.18				

Notes:

Add Info:

Herringbone signatures are usually seen in crater chains radial to the primary impact crater: in Lexell's case, Tycho The apex of each will always point in a direction to the primary crater's location.

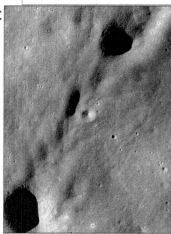

Herringbone signatures

Judging by looks alone at the decrepid state of Lexell, one would think it was one of the latest pre-Nectarian Period (it's actually of the Eratosthenian - 3.15 to 1.1 byo). The crater has clipped pre-Nectarian Deslandres's southern rim, but look at its own western rim which appears strange, where it seems like two separate rims have joined. Are we looking at some kind of double impact event in Lexell as a whole? Its northeastern rim is an oddity, too - just what happened to it (during first-quarter, a NW-SE trending trough about 8 km can be seen to cross into the floor there)? At high magnification, note the wonderful series of secondary craters (and herringbone signatures) on the floor - all from Tycho. D: 1.6 km.

Craters

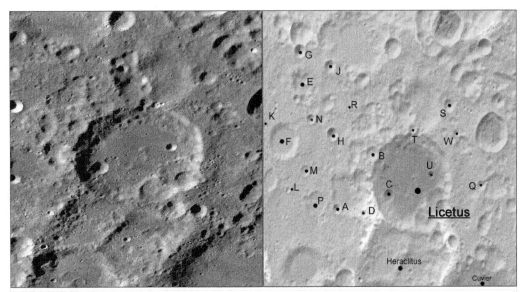

| Licetus | Lat 47.19S | Long 6.54E | 75.42Km |

Sub-craters:

Crater	Lat	Long	Size (km)	Crater	Lat	Long	Size (km)
A	47.85S	3.2E	8.1	B	46.5S	4.83E	11.75
C	47.56S	5.53E	10.31	D	47.99S	4.4E	5.21
E	44.62S	1.84E	18.76	F	46.07S	0.94E	28.89
G	43.85S	1.87E	10.42	H	45.98S	3.11E	10.34
J	44.27S	3.11E	10.36	K	45.59S	0.01W	5.55
L	47.3S	1.05E	4.15	M	46.84S	1.85E	8.69
N	45.57S	2.21E	7.92	P	47.7S	2.27E	20.18
Q	47.25S	9.68E	8.25	R	45.29S	3.84E	6.39
S	45.28S	8.17E	10.43	T	45.89S	6.61E	7.33
U	47.03S	7.44E	6.56	W	45.99S	8.5E	7.21

Notes:

Add Info:

Compare how shadows fill Licetus to crater Cuvier at its south-east (both look similar in view, but as Licetus's floor-level is some 0.5 km lower down, it makes for a slightly deeper contrast between the two.

Aerial view of Licetus

Lying in territory where numerous craters can lead to confusion, Licetus, is found more easily if one uses the odd 'ridge' in crater Heraclitus to its southwest as a marker. Licetus's impact has wiped out entirely Heraclitus's northeastern rim, leaving a relatively sharp 'edge' to Licetus's rim on the zone between the two (Licetus's ejecta here must have something to do with the weird 'ridge' in Heraclitus). The crter has a small, smoothed peak about 0.4 km high, while depth of the crater is 3.5 km approx.. Tycho's rays from the west cross its floor.

Craters

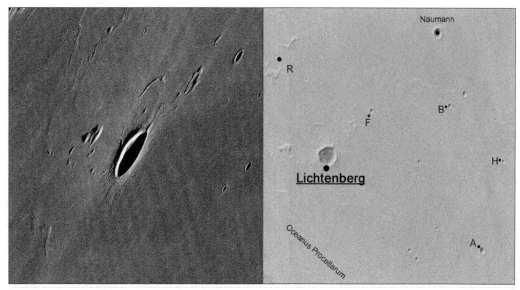

Lichtenberg	Lat 31.85N	Long 67.72W	19.53Km

Sub-craters:

Crater	Lat	Long	Size (km)	Crater	Lat	Long	Size (km)
A	28.97N	60.11W	6.87	B	33.25N	61.52W	4.86
F	33.19N	65.45W	5.19	H	31.52N	58.9W	4.18
R	34.58N	70.06W	30.26				

Notes:

Add Info:

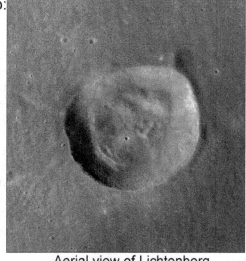

Aerial view of Lichtenberg

Lichtenberg is a Copernican-aged crater (1.1 billions of years old to the Present). Its sharp rim rises to some 0.8 km above the surrounding lavas of Oceanus Procellarum, and at most times its bright display of ejecta at the northwestern outer sector acts a distinctive marker for finding the crater. Did darker, younger lavas at its southeastern outer sector cover the brighter rays there, or was Lichtenberg an oblique-type crater? Lichtenberg impacted upon an unnamed crater nearly twice its size to its northwest, which now remains a buried crater with a peak showing through. Depth ~ 2.8 km.

Craters

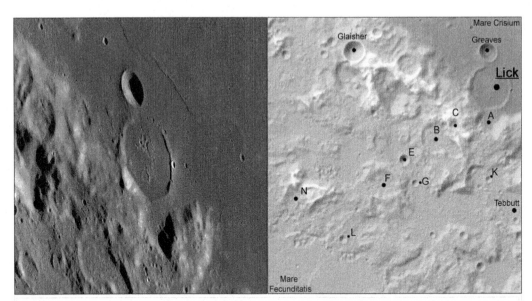

Lick	Lat 12.36N	Long 52.84E				31.63Km	

Sub-craters:

Crater	Lat	Long	Size (km)	Crater	Lat	Long	Size (km)
A	11.48N	52.84E	24.4	B	11.14N	51.41E	21.16
C	11.48N	51.92E	8.28	E	10.62N	50.65E	7.82
F	10.04N	50.06E	22.14	G	10.07N	50.92E	4.78
K	10.17N	52.79E	4.79	L	8.72N	49.1E	5.91
N	9.67N	47.9E	23.51				
Notes:							

Add Info:
Rays from the small, bright impact crater at Lick's eastern rim are easily observed during high sun times (the impactor struck on a sloped rim, which led to most of the material being directed eastwards).

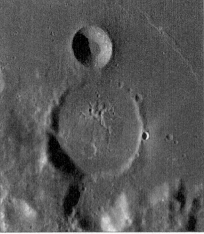

Aerial view of Lick

Lying just within one of Crisium's ring/rim system (the 555 km-diameter one), Lick is a crater of the Lower Imbrian Period (3.85 to 3.75 byo). Surrounded by younger lavas than those found over at Crisium's eastern sector, these lavas have flooded through a low point at the crater's northeast rim (the four small craterlets(?) here at the missing rim can just about be seen at high mag.,). The mounded, fractured floor (volcanic effects) is close in height as Lick's eastern rim (0.4 km), but at the southwestern rim area the mountain there is some 4.5 km high (when the terminator, 17-day-old moon, is on Crisium's floor, shadows cast by this mountain can hide Lick in parts).

Craters

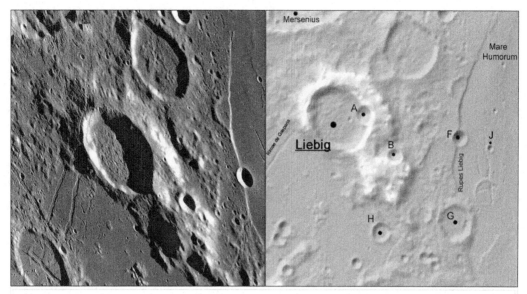

Liebig		Lat 24.35S	Long 48.3W				38.96Km

Sub-craters:	Crater	Lat	Long	Size (km)	Crater	Lat	Long	Size (km)
	A	24.28S	47.75W	11.09	B	25.01S	47.15W	8.7
	F	24.67S	45.78W	8.5	G	26.14S	45.83W	20.18
	H	26.32S	47.41W	10.32	J	24.83S	45.13W	3.62

Notes:

Add Info:

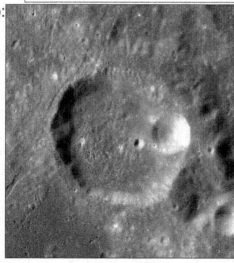

Aerial view of Liebig

Liebig has formed on the Humorum's main ring/rim, and partly on a trough region beyond. As a result, the eastern rim's outer sector is higher than the western, and perhaps this might also explain Liebig's slightly oval-like shape. The floor is level, and the material within is probably a mix of lavas tapped into after impact and target rock of the ring/rim (the texture has a slight mottled to mild undulating look). Liebig's ejecta at the southwest is covered with later lavas that filled in the trough region there, where the wonderful rilles of de Gasparis formed afterwards (one of the rilles actually crosses Liebig's outer northwestern rim).

Craters

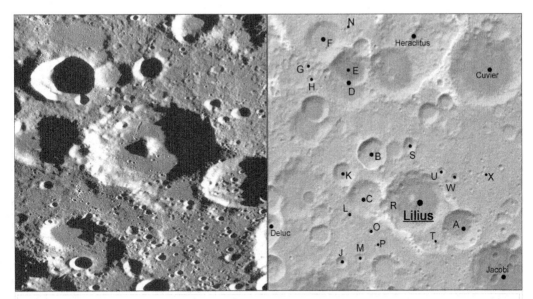

Lilius	Lat 54.6S	Long 6.09E			61.18Km

Sub-craters:

Crater	Lat	Long	Size (km)	Crater	Lat	Long	Size (km)
A	55.43S	8.78E	38.53	B	53.02S	3.76E	29.32
C	54.42S	3.25E	35.13	D	50.57S	2.95E	58.78
E	50.26S	2.86E	38.35	F	49.38S	1.55E	40.39
G	50.14S	0.66E	6.22	H	50.55S	0.75E	7.81
J	56.4S	1.71E	12.03	K	53.63S	2.15E	22.5
L	54.94S	2.4E	6.33	M	56.29S	2.81E	9.25
N	49.05S	2.7E	4.53	O	55.5S	3.54E	6.54
P	55.97S	3.77E	3.52	R	54.7S	4.31E	9.35
S	52.8S	5.82E	16.52	T	55.94S	7.4E	5.0
U	53.6S	7.58E	7.64	W	53.81S	8.33E	8.77
X	53.63S	9.84E	4.07				

Notes:

Add Info:

Aerial view of Lilius

In this area of clusters and clusters of craters that all look alike to lead one astray, Lilius stands out with its ~ 1.2 km-high peak (use it as a marker). It looks like the crater has just clipped, firstly, Lilius C's eastern rim to its west (or did it - Lilius R may be responsible for the ejecta in C and Lilius), and, secondly, on top of another, less obvious crater whose remaining rim can just about be seen to its southeast (Lilius T lies within it). Lilius A looks younger with its sharper rim, and adds to the confusion over the less obvious crater mentioned. D: 3.7 km.

Craters

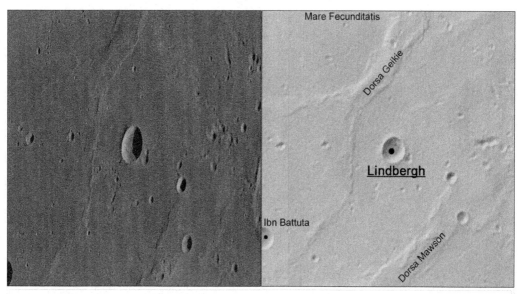

| Lindbergh | Lat 5.41S | Long 52.9E | 13.34Km |

Sub-craters: No sub-craters

Notes:

Add Info:

Chang'e 1: On 1 March 2009, the Chinese lunar orbiter, Chang'e 1, was intentionally crashed onto the surface - some 120 km away north of this crater.

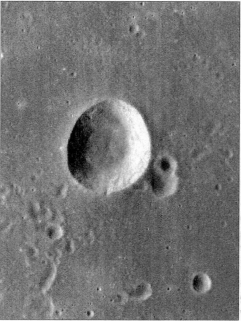

Aerial view of Lindberg

During low sun views when shadows rule, it's easy to locate Lindbergh as it lies right in the middle of two great ridges - Mawson and Geikie respectively. However, during high sun views to full moon times, the ridges disappear, and Lindbergh takes on a more bright rimmed feature (like Ibn-Battuta, too), which is still easily found. Its western rim is lower by about 300 metres than its eastern rim, and so too are the level of lavas (eastern are higher by about 200 metres). The crater, thus, has a kind of tilt to it towards the central part of Mare Fecunditatis. These aspects are about all that the crater has going for it in terms of features, as they cast some nice shadows across the Mare. Depth ~ 1.9 km.

Craters

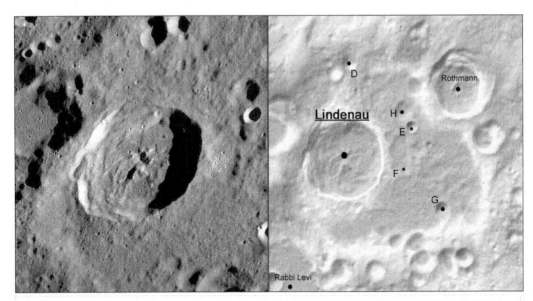

Lindenau	Lat 32.35S	Long 24.77E	53.08Km

Sub-craters:

Crater	Lat	Long	Size (km)	Crater	Lat	Long	Size (km)
D	30.44S	24.93E	9.78	E	31.68S	26.49E	7.36
F	32.5S	26.36E	7.34	G	33.28S	27.3E	9.44
H	31.37S	26.28E	11.87				

Notes:

Add Info:

Highest peak (southwards of the group) is ~ 0.8 km high, while depth of the crater is ~ 3 km.

Aerial view of Lindenau

Lindenau is aged of the Upper Imbrian Period (sometimes called the Late Imbrian Period ~ 3.75 to 3.2 byo). A bigger picture view of its location shows that it may have impacted upon an old, unnamed, pre-Nectarian crater having a diameter of around 130 km. Lindenau's inner rim walls all seem to have underwent some form of slump effects, particularly in the west, which reached so far into the floor where it met its series of peaks (highest is around ~ 0.8 km). Ejecta from Lindenau overlies the unnamed crater's southeastern floor, while full moon times show Tycho's rays cross its northern rim.

Craters

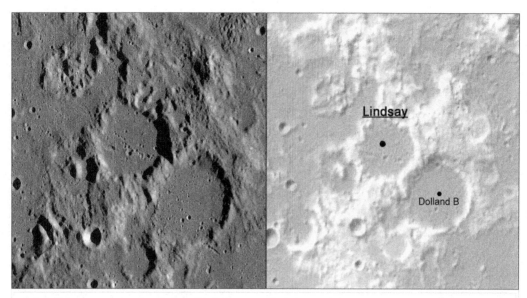

| Lindsay | Lat 7.0S | Long 13.01E | 32.23Km |

Sub-craters: No sub-craters

Notes:

Add Info:

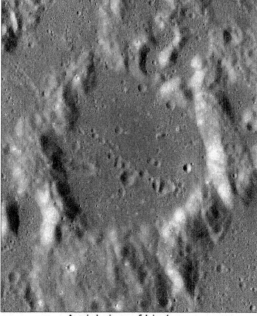

Aerial view of Lindsay

Lindsay lies in a clumpy, lumpy region of the Moon known as the Descartes Formation. The crater, like others in the region, is heavily grooved thanks to blocks of Imbrium ejecta that gouged and bull-dozed into anything in their way (note how the striations 'point' back to the Imbrium Basin). The floor materi- is probably of Imbrium ejecta, too - much less blocky, however, it may lie on earlier ejecta from the Nectaris Basin down to the southeast (together, these fills have lowered Lindsay's depth to just over a kilometre). What of the chain of craters that cross Lindsay's floor in the NW-SE direction? Are they secondaries created by Theophilus?

Craters

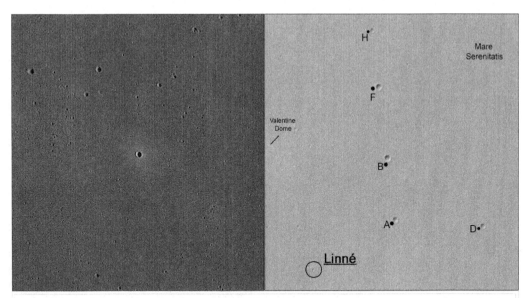

Linné	Lat 27.75N	Long 11.8E	2.23Km

Sub-craters:	Crater	Lat	Long	Size (km)	Crater	Lat	Long	Size (km)
	A	28.98N	14.37E	4.15	B	30.54N	14.17E	5.35
	D	28.73N	17.11E	4.28	F	32.34N	13.95E	4.94
	G	35.9N	13.3E	4.21	H	33.77N	13.77E	3.38

Notes:

Add Info: Don't sub-craters Linné A, B, F and H make for a wonderfully aligned series of craters? Are they circumferentially-distributed secondaries from the Imbrium Basin to the west, or just chance alignment?.

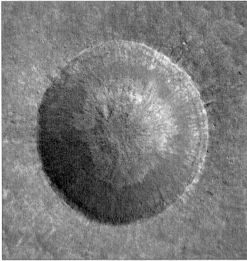

Close-up aerial view of Linné

Observing features on the Mares is always rewarding. At one moment you can be looking at macro features like wrinkle ridges to dark deposits to bright rays crossing their plains, while at the next moment seeing more the micro features like small craters to rilles to isolated clumps peaking through. These latter features can sometimes be passed over in favour of viewing the former, but noting how they relate to the whole will be time well spent. Linné is one of those features and, oddly enough, for its size, it does, in a way 'jump out' at you (due to its bright outer ejecta) in the eyepiece. D: 0.5 km.

Craters

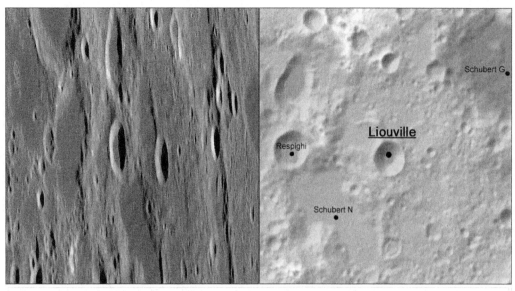

Liouville   Lat 2.73N   Long 73.57E   16.12Km

Sub-craters: No sub-craters

Notes:

Add Info:

Not to be confused with crater Louville at 44.12N, 46.04W (see crater Sharp page).

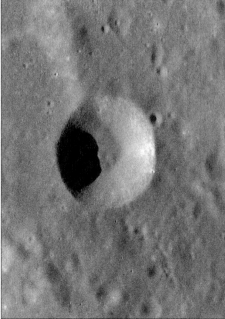

Aerial view of Liouville

Southeast of Mare Crisium lies a trough-like zone related to the basin's original formation. Lavas have filled the troughs (e.g. Mares' Undarum and Spumans), and it's there that Liouville (like its neighbour, crater Respighi, to it west) formed on the edge of a more isolated section. The crater hasn't much to offer in terms on any detail, however, it does act as a nice comparison to its neighbour, whose wall appears brighter - a signature, perhaps, that it is younger than Liouville (or, that it formed on material of a different makeup, but doubtful). For such a crater on the extreme limb, it's easy to find, and should really be thought of as the duo partner of Respighi - likewise, vice-versa. Depth ~ 2.9 km.

Craters

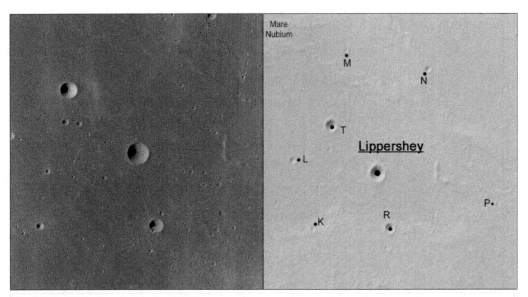

Lippershey   Lat 25.96S   Long 10.38W   6.74Km

Sub-craters:

Crater	Lat	Long	Size (km)	Crater	Lat	Long	Size (km)
K	26.72S	11.45W	2.22	L	25.77S	11.78W	2.58
M	24.28S	10.9W	1.84	N	24.53S	9.57W	2.65
P	26.37S	8.38W	1.91	R	26.71S	10.18W	4.18
T	25.3S	11.13W	4.93				

Notes:

Add Info:

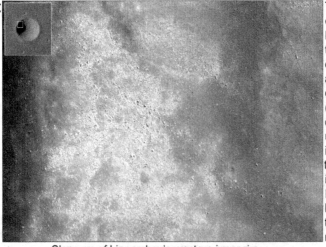

Close-up of Lippershey's western inner rim

There's not much to glean from Lippershey! Like other smaller craters nearby (e.g. Lippershey T or R), it probably formed as a result of a small ejecta block (created by large impactors nearby) that shot up at a high angle, landing back down again at the same angle; resulting in a near-vertical impact. Image (left) shows rubble and boulder-rolls in the crater, and the many materials excavated at various depths. D: 1.4km.

Craters

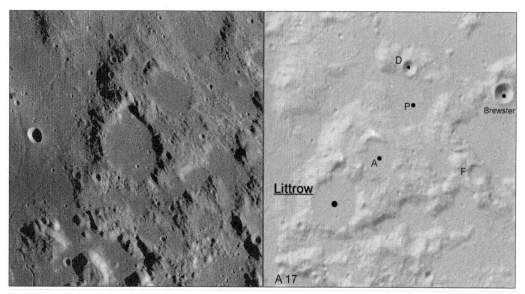

A 17

| Littrow | Lat 21.5N | Long 31.39E | 28.52Km |

Sub-craters:	Crater	Lat	Long	Size (km)	Crater	Lat	Long	Size (km)
	A	22.3N	32.19E	23.56	D	23.7N	32.81E	7.76
	F	21.97N	34.12E	10.03	P	23.2N	32.88E	36.1

Notes:

Add Info:

Aerial view of Littrow

It's an old crater possibly of the Early Nectarian (3.93 byo). Lying on the southeastern ring/rim of Mare Serenitatis, Littrow's own rim, and in fact the whole terrain around, takes on a more mottled to undulating look to it (extending for a 100 km either way north and south of the crater). Material on Littrow's floor probably contains a mix of darker buried lava deposits similar to those in Serenitatis, which now are overlain by lighter deposits of fliudized ejecta from the Imbrium Basin to the northwest. A small, one-kilometre-sized crater north of Littrow crater is responsible for the bright display there, which shows up best during westerly sun apparitions. D: 1.25 km.

Craters

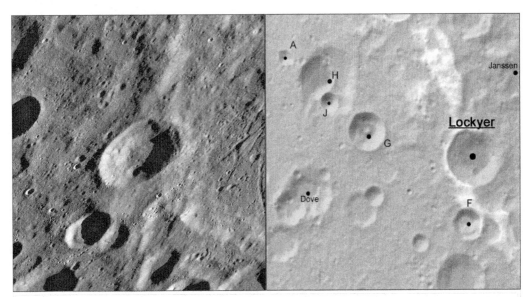

Lockyer	Lat 46.27S	Long 36.59E				35.08Km

Sub-craters:

Crater	Lat	Long	Size (km)	Crater	Lat	Long	Size (km)
A	44.12S	30.99E	9.64	F	47.59S	36.42E	19.85
G	45.68S	33.32E	24.02	H	44.48S	32.38E	31.75
J	45.06S	32.22E	11.89				

Notes:

Add Info:

Aerial view of Lockyer

Relate Lockyer to the huge crater of Janssen to its east on whose western rim it has impacted upon (you won't have any troouble in finding it again as you leave or return to the eyepiece). At high sun times to full moon, don't bother to try and find it as it simply disappears. The crater has to be younger than Janssen (pre-Nectarian), but from the state of its worn-down look, its age isn't far off the same (probably Early Nectarian ~ 3.93 byo). The crater's western rim is about a kilometre higher than at its eastern side, which may be due to the impactor that initially produced the crater striking a weaker zone of target rock (Janssen's impact causing such a zone). Don't forget to note Lockyer G's unusual form. D: 1.5 km.

Craters

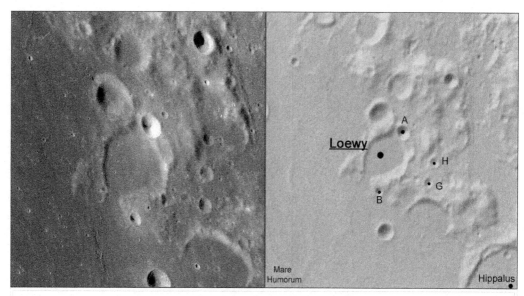

Loewy		Lat 22.69S	Long 32.85W			22.45Km	

Sub-craters:	Crater	Lat	Long	Size (km)	Crater	Lat	Long	Size (km)
	A	22.32S	32.54W	6.66	B	23.21S	32.98W	4.09
	G	23.07S	32.08W	4.48	H	22.78S	31.99W	4.7

Notes:

Add Info: Loewy from an aerial perspective looks more oval-like than round for what we would expect from a crater. Its NE-SW width is some 5 kilometres greater than its NW-SE width, while its rim at the southeast is nearly 0.5 kilometres higher than at its northwest. The crater has obviously responded tilt-wise to the structural, weighty effects of the Humorum Basin as its centre sagged, but when we look at the tilt in Loewy's floor, the slope isn't in the direction of the Basin's centre but is opposite-wise towards Loewy's eastern sector. Why is this? Perhaps the answer lies with the unnamed wrinkle ridges that lies just west of the crater (at low sun times, say, around a 11-day-old moon, these ridges show up wonderfully, and deserve proper names like the other ridges that have names e.g. in Mare Serenitatis and like)? One of these ridges just kisses the southern end of Loewy, and as such features usually produce a slight 'bulge' to the normal level of the lavas at where they form. Is this then why Loewy's floor is tilted in the opposite direction to the Mare's centre? One obvious feature to Loewy is the 5-km-wide gap at the crater's southwestern sector. The rim there has been totally demolished by lavas, which may have initially flowed through a small gap that later developed into a wider one. Note how Loewy A's impact has dusted the floor of Loewy, which shows up nicely during high sun to full moon views. D: 1.1 km.

Craters

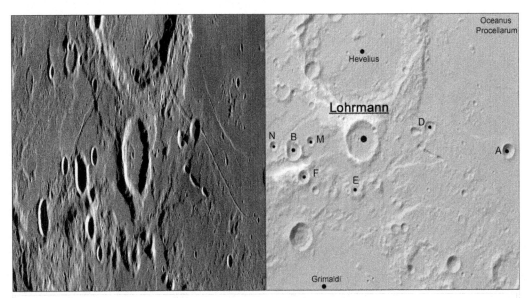

Lohrmann	Lat 0.44S	Long 67.38W	31.25Km

Sub-craters:

Crater	Lat	Long	Size (km)	Crater	Lat	Long	Size (km)
A	0.75S	62.78W	11.94	B	0.75S	69.54W	13.57
D	0.14S	65.27W	10.28	E	1.76S	67.59W	9.43
F	1.39S	69.23W	10.54	M	0.48S	69.0W	7.23
N	0.6S	70.19W	7.66				

Notes:

Add Info:

Aerial view of Lohrmann

Lohrmann must be slightly younger than its northern neighbour, Hevelius (Nectarian in age 3.92 to 3.85 byo), as it's obvious that it has impacted upon its ejecta. Less obvious, however, is that the crater also overlies a ring/rim of Grimaldi (pre-Necarian) that is harder to see, because it is 'broken up' (see the four or five blocks south of Lohrmann). Lohrmann's own ejecta in the north has heightened the area where it joins with Hevelius, but its ejecta in the south has been covered by late-flooding lavas possibly associated to formation of Grimaldi (or are they of Procellarum?). D: 1.3km.

Craters

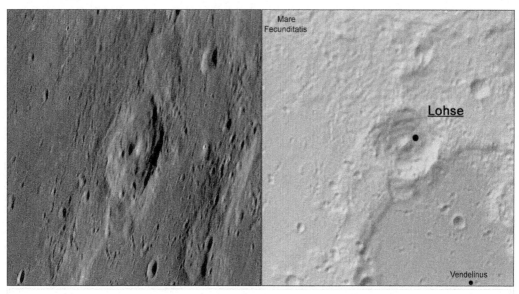

| Lohse | Lat 13.76S | Long 60.31E | 43.34Km |

Sub-craters: No sub-craters

Notes:

Add Info:

Aerial view of Lohse

Lohse has obviously impacted upon the northwestern rim of pre-Nectarian Vendelinus. The crater looks odd from a texture perspective, where its innards appear as if numerous, small landslides have occurred all at once. If you look to Lohse's northern exterior, this texture can also be seen to reach right up to Langrenus, which hints then that the material may be ejecta from that crater. Perhaps, a large volume of this ejecta (in a molten state?) fell onto Lohse, which then slid down towards the centre, covering its 1.3 km-high peak in the process, too. Underneath this material may lie intruded lavas, like we see in Vendelinus, that relates to the Balmer-Kapteyn Basin, whose centre lies some 300km eastwards.

Craters

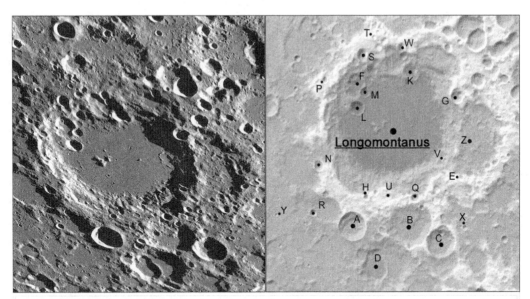

Longomontanus Lat 49.55S Long 21.88W   145.5Km

Sub-craters:

Crater	Lat	Long	Size (km)	Crater	Lat	Long	Size (km)
A	52.94S	24.15W	28.61	B	52.93S	20.92W	45.89
C	53.49S	19.11W	31.54	D	54.38S	22.92W	26.13
E	51.35S	18.11W	6.17	F	48.31S	23.71W	18.42
G	48.69S	18.51W	13.42	H	52.02S	23.32W	6.78
K	47.91S	20.95W	15.13	L	49.09S	23.7W	15.01
M	48.61S	23.29W	9.65	N	50.93S	25.86W	10.39
P	48.13S	25.31W	6.48	Q	52.06S	20.6W	9.87
R	52.45S	26.32W	7.17	S	47.34S	23.36W	12.1
T	46.8S	22.75W	5.43	U	51.98S	22.09W	5.82
V	50.8S	19.0W	5.23	W	47.13S	21.36W	9.91
X	52.97S	17.77W	5.04	Y	52.41S	28.38W	3.91
Z	49.91S	18.0W	96.13				

Notes:

Add Info:

Aerial view of Longomontanus

Longomontanus doesn't fail in competing for being one of the 'show-offs' on this part of the Moon. What with Schiller, Tycho and Clavius (all pompous snobs) lying not very far from it, the crater catches the observer's eye through its wonderful series of terraces, its off-central peaks, and the additional pre-extension of sub-crater Z to its east that makes it look bigger than it really is (Z's floor is some 300 km higher than Longomantanus's). Rays from Tycho obviously cross its floor, but they only are seen at high sun times. D:4.8 km.

Craters

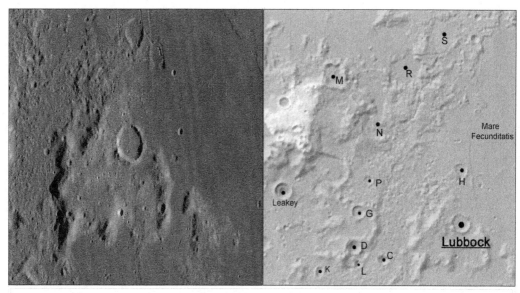

Lubbock	Lat 3.99S	Long 41.79E	14.09Km

Sub-craters:

Crater	Lat	Long	Size (km)	Crater	Lat	Long	Size (km)
C	4.89S	39.88E	7.46	D	4.58S	39.17E	10.55
G	3.71S	39.3E	10.33	H	2.67S	41.79E	9.24
K	5.15S	38.35E	6.05	L	4.99S	39.29E	6.28
M	0.41S	38.69E	22.16	N	1.57S	39.73E	25.8
P	2.93S	39.53E	7.22	R	0.17S	40.45E	23.79
S	0.63N	41.31E	24.96				

Notes:

Add Info:

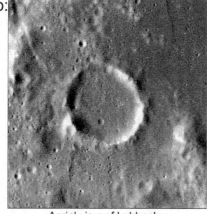

Aerial view of Lubbock

Lubbock lies upon an isolated plot of land related to one of Mare Fecunditatis's ring/rim system - the 690 km-diameter one. The crater is possibly Imbrian in age (3.85 to 3.2 byo), it has a relatively sharp rim all around, and is 'cut-through' by Rimae Gloclenius I. This latter feature is hard to see as it peters out towards the crater's outer southern rim, but at low sun times and high magnification, hint of it can be seen. The rille probably crosses Lubbock's floor, but internally-sourced lavas (similar to Fecunditatis) have flooded over it. Work out the sequence of events for Lubbock, rille and mare.

Craters

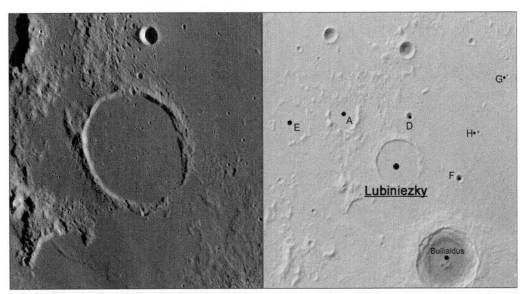

Lubiniezky   Lat 17.88S   Long 23.89W   42.97Km

Sub-craters:

Crater	Lat	Long	Size (km)	Crater	Lat	Long	Size (km)
A	16.51S	25.68W	28.16	D	16.52S	23.49W	7.05
E	16.61S	27.38W	37.24	F	18.34S	21.86W	7.02
G	15.36S	20.28W	3.65	H	17.01S	21.21W	4.29

Notes:

Add Info: Lubiniezky lies on an odd patch of land that isn't quite obvious as to which of the surrounding Mares - Nubium eastwards, Humorum westwards, or Cognitum northwards - is responsible (perhaps, it is original highland's crust?). The crater has obviously been flooded by lavas - either related to Nubium or Cognitum, which from recent research are about 3.2 billions of years old. These lavas can be seen to cover over ejecta signatures of crater Bullialdus to the southeast (note the lack of ejecta between Bullialdus and Lubiniezky), in Lubiniezky's own floor where they've breached in at its south-eastern rim, and in outer surrounding craters like sub-craters Lubiniezky A and E to the northwest. The rim is some ~ 400 metres higher in the west than in the east, but it has a relatively sharp look to it in parts easily observable. Like crater Plato, there are several small impact craters (around the 1 km mark) to test the magnification limits of the telescope you are using, but those smaller will prove a challenge. At high sun times the dark lava-filled floor contrasts well against the brighter material of the above-mentioned odd patch, but don't forget to note, around the same time, the bright ray just off the crater's southern rim that originates back to Tycho in the southeast - some 800 km away. Depth ~ 0.8 km approx..

Craters

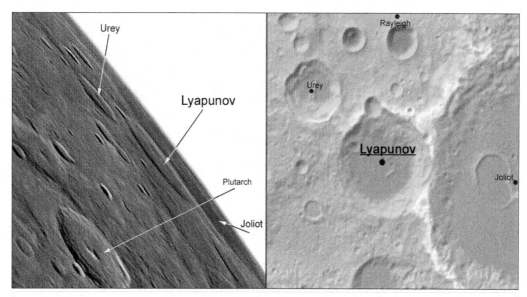

Lyapunov	Lat 26.43N	Long 89.36E	67.58Km

Sub-craters: No sub-craters

Notes:

Add Info:

Aerial view of Lyapunov

Lyapunov formed on two major-sized craters - Joliot to its east and Rayleigh to its north. Both are of the pre-Nectarian Period (4.6 to 3.92 byo), however, as Lyapunov is obviously younger, and the state of its appearance is similar-looking, its age may not be far off that of the above. Level of Lyapunov's floor is at some ~ 0.5 km below Joliot's floor level, and ~ 1.0 km below Rayleigh's level. Flooding has occurred within Lyapunov, but it isn't as extensive-looking as that seen in Joliot (is this due to the crust being that bit thicker under Lyapunov?). There is hint of a ring-type peak at the crater's centre, and its northern and eastern walls reflect on the above impact effects. D: 4.0 km.

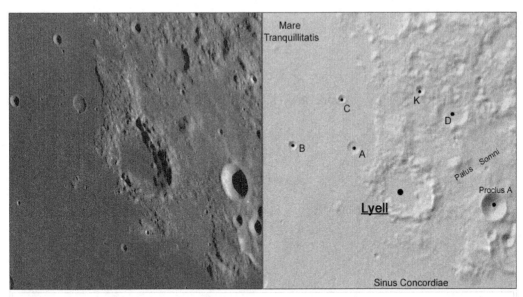

Lyell		Lat 13.63N	Long 40.56E		31.17Km

Sub-craters:

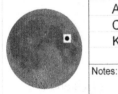

Crater	Lat	Long	Size (km)	Crater	Lat	Long	Size (km)
A	14.32N	39.6E	6.98	B	14.36N	38.46E	4.66
C	15.14N	39.39E	4.15	D	14.91N	41.46E	17.4
K	15.28N	40.86E	5.0				

Notes:

Add Info:

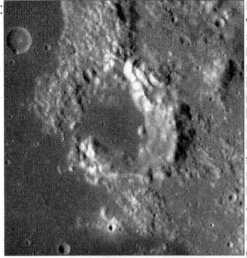

Aerial view of Lyell

Lyell lies in a mish-mash of sorts, where deposits from nearly all the surrounding Mares -Tranquillitatis, Fecunditatis, Serenitatis and Crisium 'dumped' their ejecta - producing the region known as Palus Somni. The lavas here are old ~ 3.5 byo! They have obviously flooded into Lyell at its southwest (those lavas towards Tranquillitatis's centre are older again), but subsequent deposits of ejecta from Mare Imbrium that also overlie the area has added to the complexity in geology. High sun to full moon views show up Lyell's location easily, where you get the brightness of the Palus contrasting against the dark lavas.

Craters

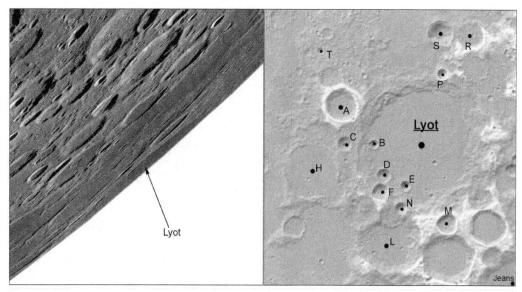

Lyot	Lat 50.47S	Long 84.8E	150.6Km

Sub-craters:

Crater	Lat	Long	Size (km)	Crater	Lat	Long	Size (km)
A	48.99S	79.94E	38.39	B	50.51S	81.9E	10.84
C	50.48S	80.11E	17.39	D	51.69S	82.54E	16.69
E	52.07S	83.91E	13.08	F	52.34S	82.34E	20.73
H	51.5S	78.0E	63.59	L	54.25S	83.15E	80.08
M	53.45S	86.68	26.32	N	52.93S	83.67E	14.8
P	47.7S	85.81E	13.47	R	46.17S	87.34E	32.47
S	46.12S	85.57E	29.84	T	46.88S	78.83E	8.65

Notes:

Add Info:

Aerial view of Lyot

At longitude 84.8 East, Lyot is obviously going to be a challenge to see. Favourable librations are, of course, the best time, but as they don't sometimes correlate with suitable lighting conditions, several attempts in observation will be the norm. At times you will get a glimpse of it on the extreme limb, however, as a broad expanse of dark lavas in Mare Australe, in which the crater lies, lay before you in your view, observations can sometimes be confusing - smaller crater Oken (79 km in diameter) northwards is mistakenly reported as being Lyot. Age: pre Nectarian. Depth: 2.9 km.

Craters

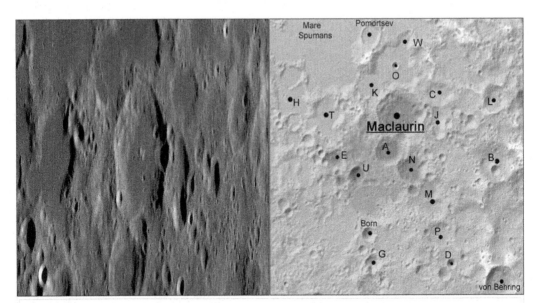

Maclaurin   Lat 1.92S   Long 67.99E   54.33Km

Sub-craters:

Crater	Lat	Long	Size (km)	Crater	Lat	Long	Size (km)
A	3.23S	67.56E	29.1	B	3.68S	71.49E	36.07
C	1.25S	69.4E	30.6	D	7.07S	69.88E	11.17
E	3.5S	65.61E	25.08	G	7.08S	66.98E	18.76
H	1.58S	64.1E	41.96	J	2.57S	69.38E	14.9
K	0.97S	66.81E	34.7	L	1.41S	71.48E	28.28
M	4.88S	69.18E	42.13	N	3.82S	68.39E	30.67
O	0.14S	67.56E	41.47	P	6.11S	69.45E	31.93
T	1.86S	65.34E	33.33	U	4.04S	66.42E	18.81
W	0.59N	68.22E	22.19	X	0.09N	68.71E	18.55

Add Info:

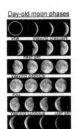

Day-old moon phases

Waning crescent

Note: As the Moon goes through various librations throughout the year, suggested times given in text for observations are only approximates.

Notes:

Aerial view of Maclaurin

Maclaurin lies just on the outer edge of Mare Spumans - a moat-like region related to formation of Mare Crisium to the north of the crater and Mare Fecunditatis to the west. Both Mares and outer depressed areas are filled with old lavas ~ 3.4 byo, and Maclaurin's floor, which is some 2 kilometres lower than Spuman's, may contain older lavas still. The general shape of the crater takes on a more pentagonal look. Slumping of the crater's walls is more prominant at the northwest sectors, while southwards Maclaurin A (with another crater on it) has impacted the rim. D: 3.4 km.

Craters

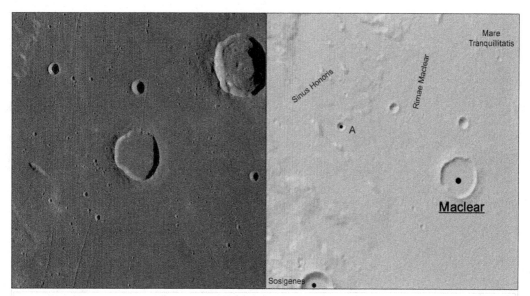

Maclear	Lat 10.52N	Long 20.1E	20.34Km

Sub-craters:	Crater	Lat	Long	Size (km)	Crater	Lat	Long	Size (km)
	A	11.3N	18.0E	4.11				

Notes:

Add Info:

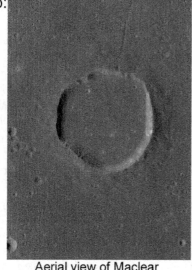

Aerial view of Maclear

In some older publications you'll come across Maclear spelled as 'Mac Clear' or 'MacLear'; all referenced back to Thomas Maclear, an Irish-born, South African astronomer of the 18th century. Though the Tranquillitatis Basin, in which Maclear lies, is an old one of the late Nectarian (~ 3.92 byo), the lavas it impacted upon are slightly younger (~ 3.7 byo). Both Maclear's own ejecta and the faint tectonic fault at its northern outer sectors are easily seen, but what is less seen is the continuance of the fault at the crater's outer southern sectors. This fault, like others that parallel it in the region (e.g. Rimae Maclear or Rimae Sosigenes southwards), is circumferential to Tranquillitatis, and formation of it was probably responsible for the lavas within Maclear's own floor (note, how the fault isn't seen on the floor). Depth: 0.6 km approx..

Craters

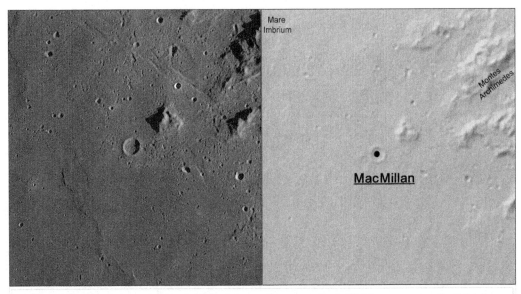

MacMillan   Lat 24.2N   Long 7.85W   6.88Km

Sub-craters: No sub-craters

Notes:

Add Info:

Apollo 15 lies some 340 km away east from this region.

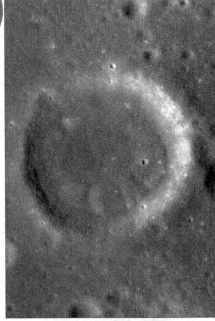

Aerial view of MacMillan

Lying in a relatively bright patch of material that is believed to be a late volcanic feature of the Imbrium Basin, MacMillan doesn't look like it has impacted upon this material, but rather succumbed to the volcanic effects. Material in the crater's floor has the same darkish colour, generally, as those younger lavas to its south, but looking to the surrounds at the crater's northern outer sectors, it explains why MacMillan hasn't ejecta signatures as these lavas probably covered them. On close-up (left), several bright patches with greater crater counts suggests the floor originally contained older materials, which were then later intruded-upon by younger lavas - sourced from within through fissures etc. The rim's walls are bright at high sun views, while low sun views show low level shadowing. Depth 0.4 km.

Craters

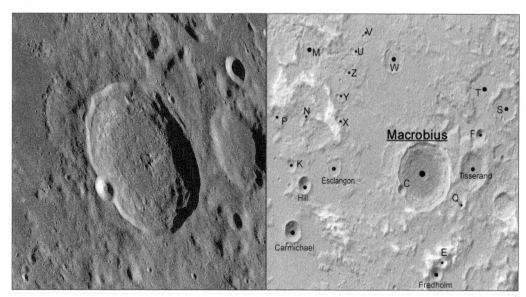

Macrobius   Lat 21.26N   Long 45.97E   62.79Km

Sub-craters:

Crater	Lat	Long	Size (km)	Crater	Lat	Long	Size (km)
C	20.84N	45.03E	10.66	E	18.71N	46.81E	9.86
F	22.47N	48.53E	11.43	K	21.46N	40.23E	12.01
M	24.97N	40.94E	41.08	N	22.8N	40.84E	4.98
P	22.95N	39.52E	16.31	Q	20.4N	47.63E	8.24
S	23.27N	49.59E	27.16	T	23.78N	48.55E	32.13
U	24.94N	42.78E	6.49	V	25.39N	43.27E	4.47
W	24.72N	44.62E	25.53	X	22.92N	42.17E	6.16
Y	23.58N	42.15E	5.04	Z	24.32N	42.54E	5.21

Notes:

Add Info:

Aerial view of Macrobius

Did impact of Macrobius C have anything to do with the odd-looking slumped deposits seen as Macrobius's north-western outer sectors?

Macrobius lies just on the outer edge of one of Mare Crisium's ring/rim system (~ 750 km in diameter) - believed to form when the impact is so great it causes the target rock to ripple out like waves, leaving several rings and rims in place. The crater is of the Lower Imbrian Period (3.85 to 3.75 byo). While several well-defined terraces occur at the crater's eastern sector, those in the west look different - taking on a more 'broken-up' look. Off-centre peaks on the floor may be part of a ring of peaks, and ejecta distribution outside the northwest sector looks like it slumped afterwards onto a 'moated', lava-filled area. Depth ~ 4 km.

Craters

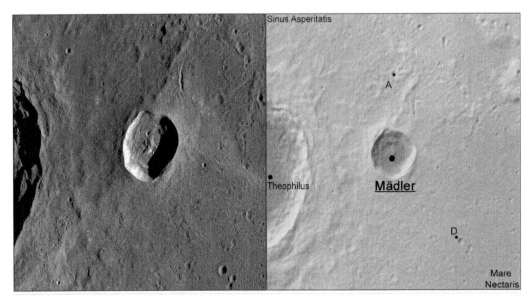

Mädler	Lat 11.04S	Long 29.76E	27.58Km

Sub-craters:

Crater	Lat	Long	Size (km)	Crater	Lat	Long	Size (km)
A	9.53S	29.79E	4.48	D	12.68S	31.17E	3.76

Notes:

Add Info:

Aerial view of Mädler

From the state of Mädler's sharp rim it's safe to say that it is a relatively young crater - younger than Eratosthenian-aged Theophilus to its west. Several major slumps in the crater have altered its shape - the most prominent ones seen at its northern and eastern sectors. These slumps have added complexity to material on the crater's floor, where they meet a series of peaks (highest is around 600 m) and impact melt deposits. A wider view of the crater's surrounds show an unusual distribution of odd rays (southwards, they are prominant - is Mädler an oblique impact?). But what is happening with those over to the east - has topography contained their direction? Is the north-east slump responsible for them only, and not the original impact that created Mädler? Depth 2.8 km.

Craters

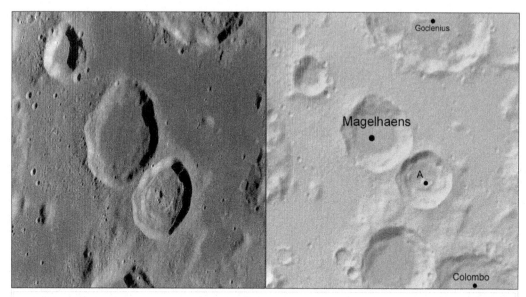

Magelhaens	Lat 11.98S	Long 44.07E				37.2Km

Sub-craters:

Crater	Lat	Long	Size (km)	Crater	Lat	Long	Size (km)
A	12.73S	44.99E	29.35				

Notes:

Add Info:

Aerial view of Magelhaens

Lying between a clumped region of material from both the Fecunditatis Basin (within the 690 km ring) to the northeast and the Nectaris Basin (within the 620 km ring) to the west, Magelhaens is a crater of the Late Nectarian (~ 3.85 byo). Lavas in its floor suggests possibly sourcing through fissures created by the initial impact, but as the northern rim is roughly at the same level as the level of lavas seen at its outer sector there, are the floor's lavas then really a consequence of 'breaching'? Magelhaens A is younger, and ejecta from it lies on Magelhaens's south-eastern floor sector. Magelhaens crater and its surrounds make for a good study in sequential, geological events. Depth:1.7 km.

Craters

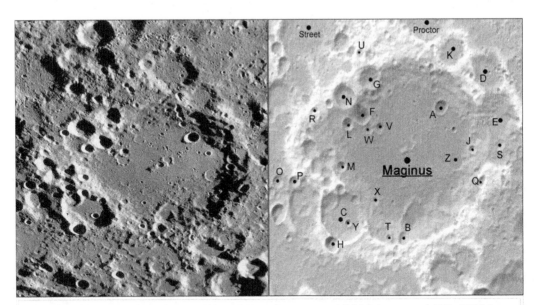

Maginus   Lat 50.03S   Long 5.89W   155.58Km

Sub-craters:

Crater	Lat	Long	Size (km)	Crater	Lat	Long	Size (km)
A	48.87S	4.42W	13.36	B	52.42S	6.29W	11.66
C	51.75S	9.5W	47.27	D	47.92S	2.29W	38.3
E	49.08S	1.72W	38.37	F	49.03S	8.3W	16.83
G	48.03S	7.85W	22.06	H	52.48S	10.08W	14.75
J	50.0S	2.87W	7.55	K	47.32S	3.92W	29.76
L	49.26S	8.98W	10.68	M	50.4S	9.38W	9.09
N	48.49S	9.08W	23.43	O	50.66S	12.65W	10.78
P	50.7S	11.81W	10.48	Q	50.85S	2.33W	9.24
R	48.83S	10.55W	8.26	S	49.82S	1.5W	10.28
T	52.39S	7.1W	5.77	U	47.38S	8.21W	9.37
V	49.41S	7.41W	9.52	W	49.42S	7.94W	8.32
X	51.39S	7.76W	7.66	Y	51.96S	9.18W	6.4
Z	50.33S	3.61W	15.82				

Notes:

Add Info:

Aerial view of Maginus

Like its 'bigger brother', crater Clavius, below it, Maginus is a feast full of rewards to the viewer's eye. The main thing that will hit you when viewing Maginus through the eyepiece is how 'battered-to-death' the crater is. Practically all its western rim has been obliterated by numerous impacts of all sorts and sizes; giving the crater an extension to its west (eastern-wise, impacts by sub-crater's D and E have had their share too). Explore: its terraces, its floor, its lava, and rays from Tycho, its peaks and buried craters, and subsequent ejecta(s) from events nearby. Depth ~ 5 km approx..

Craters

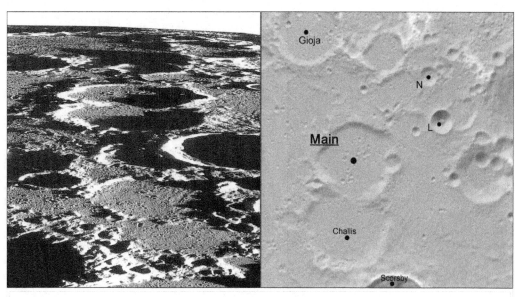

Main	Lat 80.87N	Long 10.41E	47.43Km

Sub-craters:

Crater	Lat	Long	Size (km)	Crater	Lat	Long	Size (km)
L	81.44N	22.73E	14.32	N	82.36N	22.13E	11.56

Notes:

Add Info:

Use crater Scorsby as a librational marker for judging when Main is in best view. That is, if you can see more of Scorsby's northern rim and its inner wall, then a favourable libration for observing Main is occurring.

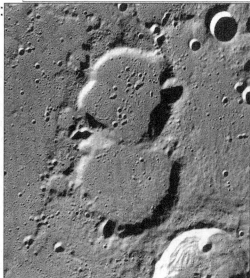

Aerial view of Main

As per usual with features and craters, like Main lying on the moon's extreme limb, patience in capturing any descent view of it is essential. It's easily located simply by remembering that it, together with crater Challis south of it, 'line-up' nicely to give a very squished number 8. Both craters are of the Nectarian age (3.92 to 3.85 byo), but as to which one impacted first, it's a hard call - Main looks like it has clipped upon Challis's northern rim. The floors of each are covered with Imbrium ejecta, but what of those craterlets on Main's own floor - are they secondaries from impact of Scorsby to the southeast? Depth of Main ~ 1.7 km.

Craters

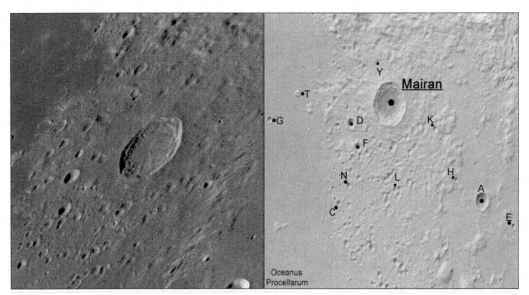

Mairan	Lat 41.6N	Long 43.5W	39.49Km

Sub-craters:

Crater	Lat	Long	Size (km)	Crater	Lat	Long	Size (km)
A	38.64N	38.79W	15.83	C	38.64N	46.1W	6.35
D	41.0N	45.54W	10.04	E	37.81N	37.24W	5.62
F	40.31N	45.17W	8.4	G	40.9N	50.83W	5.61
H	39.31N	40.06W	4.53	K	40.83N	41.1W	5.99
L	39.07N	43.22W	5.98	N	39.19N	45.59W	5.9
T	41.79N	48.39W	2.98	Y	42.8N	44.14W	5.71

Notes:

Add Info:

Note how the ejecta surrounding Mairan's outer rim area looks smoother to the surrounding terrain (at places in which it has partially-buried small craterlets in the region.)

Aerial view of Mairan

Mairan lies on both the ejecta remains of the Imbrium Basin to its southeast and Sinus Iridum to the east. The crater formed not quite long after those two main impact events occurred (it's of the Upper Imbrian - 3.75 to 3.2 byo), leaving behind a relatively sharp rim, terraces, small peak, and lavas predominantly in the northwestern part of the floor (tilted towards that direction), which possibly were sourced through early-formed fissures during impact. Best times to observe the above details in the crater are around a day or two before full moon and before last quarter. D: ~ 2.7 km.

Craters

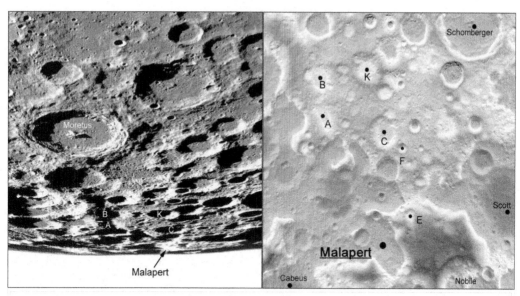

Malapert   Lat 85.0S   Long 11.4E   72.38Km

Sub-craters:

Crater	Lat	Long	Size (km)	Crater	Lat	Long	Size (km)
A	80.18S	3.79W	33.33	B	78.79S	2.93W	32.72
C	80.98S	10.08E	38.99	E	83.94S	19.83E	19.47
F	81.53S	14.73E	11.41	K	78.7S	6.37E	39.16

Notes:

Add Info:

Malapert

Aerial view of Malapert

When you do happen to see a good view of Malapert in your eyepiece, there are only two occassions in which it will occur: (1) either you planned well ahead to view it during a suitable libration, with suitable lighting conditions, and Luna was in your sky in the right place, (2) you were lucky to happen upon it by chance. Every now and then a hint of it will come through during neither of the above, however, if you don't have a good atlas (with South Pole aspects) at your side during such occassions, then it will prove extremely challenging as to which rim, which peak, which mountain feature relates to what. Observe well!

Craters

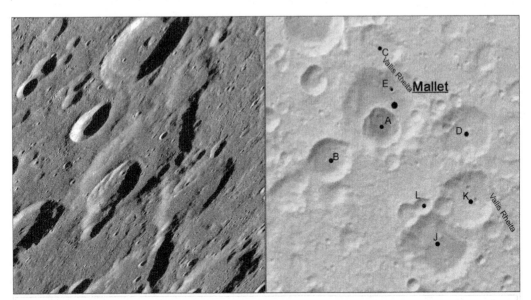

Mallet	Lat 45.41S	Long 54.05E			58.92Km	

Sub-craters:

Crater	Lat	Long	Size (km)	Crater	Lat	Long	Size (km)
A	45.99S	53.98E	28.19	B	46.71S	52.17E	32.04
C	44.02S	53.9E	34.78	D	46.09S	56.9E	41.25
E	45.07S	54.33E	4.49	J	48.79S	55.99E	55.87
K	47.62S	57.15E	42.67	L	47.83S	55.48E	13.03

Notes:

Add Info:

Aerial view of Mallet

Mallet lies in an area where the wonderful structure, Vallis Rheita, dominates the general view – it 'cuts' through several craters there. The valley is believed to be a secondary crater chain radial to the Nectaris Basin to the northwest, as huge ejecta blocks 'plopped' onto the surface in sequence (these decreased in size, in directional speed and force, the further they landed away from the initial impact). Mallet's northeastern sector bore the blunt of several of those blocks, giving the rim there a fine 'straight edge' look to it. The valley looks like it bends at Mallet. Why? D:~ 5 km.

Craters

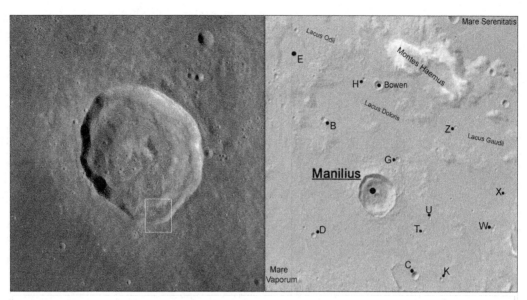

Manilius   Lat 14.45N   Long 9.07E   38.34Km

Sub-craters:

Crater	Lat	Long	Size (km)	Crater	Lat	Long	Size (km)
B	16.61N	7.28E	5.55	C	12.07N	10.34E	6.94
D	13.24N	6.99E	4.72	E	18.38N	6.35E	48.37
G	15.47N	9.76E	4.7	H	17.8N	8.63E	3.08
K	11.96N	11.18E	3.34	T	13.35N	10.62E	2.67
U	13.74N	10.81E	2.51	W	13.41N	12.9E	3.83
X	14.43N	13.35E	2.04	Z	16.4N	11.69E	2.97

Notes:

Add Info:

Like crater Menelaus over to its east (~ 200 km away), Manilius and it show up nicely as bright craters during near to full moon times.

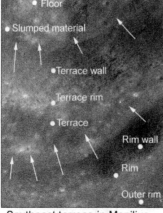

Southeast terrace in Manilius

Lying on a bright patch of, possibly, original highlands (the patch actually lies in a trough-like region related to formation of the Serenitatis Basin to the northeast), Manilius is a crater of the Imbrian Period - perhaps of the Late Epoch (3.75 to 3.2 byo). The crater has a relatively sharp rim to suggest it's younger still, however, partial coverage of its ejecta at the western sector by lavas of the same age may be 'pushing' it into the Late Epoch. A large volume of the inner rim's southern sector has slumped to create a significant terrace that is easily observed, where subsequent slumping occurred afterwards. Main peak about ~ 1 km high. Original floor is only partially seen. D: 3 km.

Craters

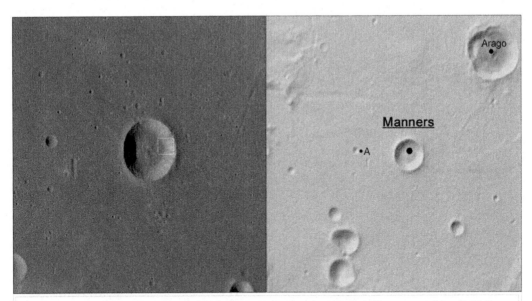

Manners	Lat 4.57N	Long 19.99E	15.05Km

Sub-craters:	Crater	Lat	Long	Size (km)	Crater	Lat	Long	Size (km)
	A	4.64N	19.11E	3.18				

Notes:

Add Info:

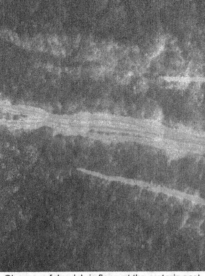

Close-up of dry-debris flows at the crater's east

Though Manners lies in a region where some of the oldest lavas once filled Mare Serenitatis, from the state of its exterior rim, later, younger lavas may have partially covered its ejecta (seems particularly missing in the west). Compared with crater Arago (Eratosthenian in age) to its northeast, its slightly more worn look suggests it has to be older than that crater. The crater seems to have formed on a E-W trending ridge that seems to parallel Rima Ariadaeus (~ 30 km away to its northwest), but catching a good view of it requires low sun views. The crater's rim is some 500 metres above the Mare, and its floor is slightly convexed by 200 m (high sun views contrasts the rim's bright inner wall nicely against the darker floor).

Craters

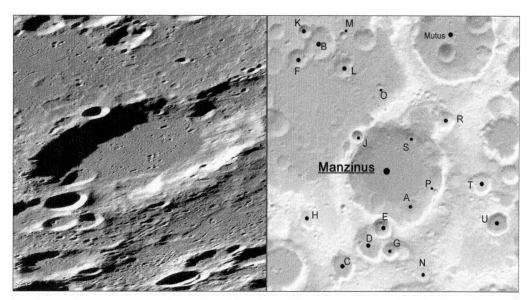

Manzinus	Lat 67.51S	Long 26.37E			95.97Km

Sub-craters:

Crater	Lat	Long	Size (km)	Crater	Lat	Long	Size (km)
A	68.51S	27.45E	19.75	B	63.7S	21.12E	27.71
C	69.99S	21.68E	24.2	D	69.38S	24.21E	32.86
E	68.98S	25.15E	18.14	F	64.08S	19.64E	17.4
G	69.64S	25.77E	16.41	H	68.64S	19.14E	17.94
J	66.4S	23.41E	11.38	K	63.35S	20.28E	12.29
L	64.44S	22.63E	17.15	M	63.54S	22.7E	6.57
N	69.98S	27.47E	9.45	O	65.0S	25.06E	6.08
P	67.93S	29.33E	5.8	R	65.92S	29.85E	15.34
S	66.46S	27.29E	10.88	T	67.6S	32.79E	19.53
U	68.65S	34.4E	20.72				

Notes:

Add Info:

Aerial view of Manzinus

In this over-populated area of craters of all sizes and shapes, Manzinus isn't that hard to locate if one uses the impact crater (unnamed and about 40 km in diameter) just off its northeastern rim as a marker for finding it. This impact crater overlies ejecta from crater Mutus to the northeast and Manzinus itself - both of which are of the pre-Nectarian Period ~ 4.6 to 3.93 byo. As this area is also well known for secondaries from the Imbrium Basin (some ~ 3000 km away to the northwest), would sub-craters E and J be such ones? D: 5 km.

Craters

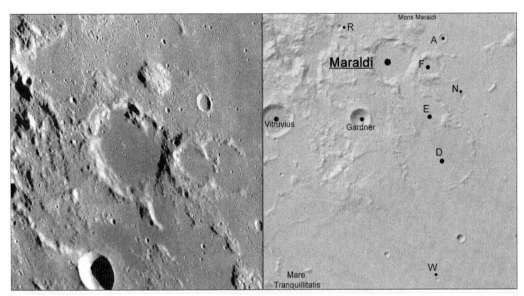

Maraldi	Lat 19.36N	Long 34.8E	39.62Km

Sub-craters:	Crater	Lat	Long	Size (km)	Crater	Lat	Long	Size (km)
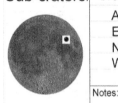	A	20.02N	36.3E	7.4	D	16.74N	36.08E	66.49
	E	17.87N	35.76E	31.82	F	19.19N	35.85E	18.36
	N	18.37N	36.88E	4.5	R	20.32N	33.19E	4.89
	W	13.17N	36.09E	4.22				

Notes:

Add Info:

Maraldi has a small peak that is easily seen against the dark lavas on its floor.

Aerial view of Maraldi

Situated in a trough-like zone related to formation of basins Serenitatis and Tranquillitatis to the west and south respectively, Maraldi shows off its Nectarian age (3.92 to 3.85 byo) through its dishevelled, broken appearance - just look at all its remaining rim. The crater has undergone some major slumping events at its western sectors, while at its northeast such slumped material has almost become isolated from the rim as lavas flowed in and around it. These lavas may be internally sourced, however, a drop in elevation (about 1 km) from sub-crater E to Maraldi, might make them externally sourced (note a breach point). D: ~1.5 km.

Craters

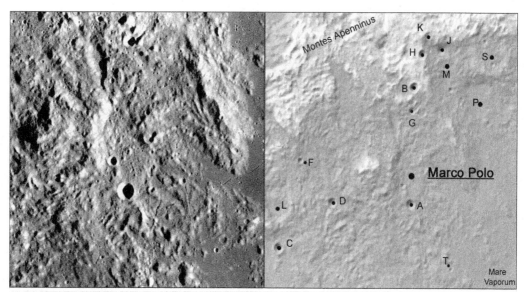

Marco Polo   Lat 15.52N   Long 2.05W   28.25Km

Sub-craters:

Crater	Lat	Long	Size (km)	Crater	Lat	Long	Size (km)
A	14.9N	1.96W	6.38	B	17.18N	1.88W	6.36
C	14.03N	5.01W	6.33	D	14.93N	3.79W	5.81
F	15.73N	4.53W	3.71	G	16.72N	1.93W	4.42
H	17.81N	1.69W	5.66	J	17.9N	1.27W	4.43
K	18.13N	1.5W	9.49	L	14.82N	5.02W	18.47
M	17.54N	1.12W	38.1	P	16.97N	0.32W	28.06
S	17.76N	0.01E	22.05	T	13.63N	1.02W	2.73

Notes:

Add Info: Marco Polo would easily be overlooked as simply a depressed piece of relief in this overly complicated region of lunar terrain. However, it is, and was at one time, a normal, fully-formed crater. That is, before ejecta from the Imbrium Basin event covered over most of it, and put it into the category of 'hard-to-find' craters on the Moon. The crater itself probably formed initially on original highland crust, along with ejecta deposits from the Serenitatis Basin to the east, however, besides the Imbrium ejecta, and subsequent after-effects from lateral to vertical movements of said material, the crater became superfluous to all these events. What then is left is a partially buried crater lying on the outer ejecta deposits that now is called the Apennine mountains with valleys and hills similar to like we see on Earth. Later lava flows in Mare Vaporum to the southeast tried their best to cover up Marco Polo, but they didn't reach in far enough through (note how close they did flow just to the east of sub-crater A). Depth of crater is around 0.4 km.

Craters

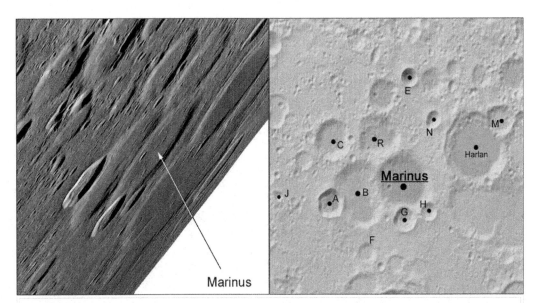

| Marinus | Lat 39.38S | Long 76.57E | 56.55Km |

Sub-craters:

Crater	Lat	Long	Size (km)	Crater	Lat	Long	Size (km)
A	39.92S	73.27E	25.24	B	39.67S	74.62E	58.24
C	38.05S	73.47E	37.4	E	36.28S	76.82E	17.52
F	41.4S	74.88E	18.96	G	40.44S	76.59E	23.38
H	40.2S	77.68E	17.08	J	39.62S	70.97E	9.11
M	37.44S	80.77E	28.39	N	37.51S	77.84E	17.88
R	38.02S	75.48E	44.12				

Notes:

Add Info:

Aerial view of Marinus

A good marker for finding this crater on the southeastern limb is to look firstly for the two craters, G and H, on its southern rim. The crater lies within two main rings/rims (with diameters ~ 660 km and 880 km respectively) of the Australe Basin whose centre is some ~ 400 km away to the southeast. Marinus's eastern rim is about a kilometre higher than its west, which would be expected as the inital impact upon B allowed more lateral movement of ejecta in that direction onto B's floor. One of the rilles of Rimae Hase crosses Marinus's floor, but catching a view of it will be hard. D: ~ 3 km.

Craters

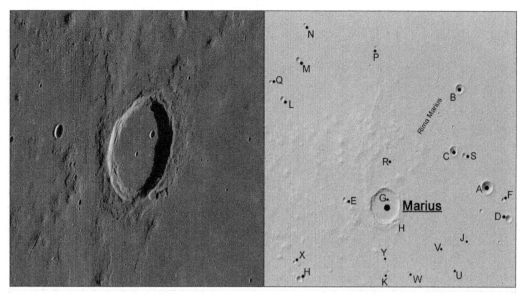

| Marius | Lat 11.9N | Long 50.84W | 40.09Km |

Sub-craters:

Crater	Lat	Long	Size (km)	Crater	Lat	Long	Size (km)
A	12.6N	46.04W	15.23	B	16.33N	47.35W	11.11
C	13.98N	47.64W	11.08	D	11.4N	45.07W	8.72
E	12.13N	52.73W	5.55	F	12.11N	45.3W	5.53
G	12.09N	50.6W	3.34	H	11.32N	50.39W	4.69
J	10.45N	46.91W	2.98	K	9.4N	50.7W	3.61
L	15.88N	55.7W	6.93	M	17.3N	55.01W	6.41
N	18.72N	54.74W	3.99	P	17.9N	51.34W	3.94
Q	16.5N	56.28W	4.89	R	13.64N	50.31W	4.81
S	13.81N	47.13W	6.34	U	9.56N	47.7W	2.3
V	9.87N	48.31W	1.63	W	9.4N	49.73W	2.82
X	9.73N	54.99W	4.82	Y	9.78N	50.77W	2.52

Notes:

Add Info: Technical problems with the Soviet Union's Luna 7 spacecraft forced early end to the mission. It impacted the surface some 130 km east of this crater.

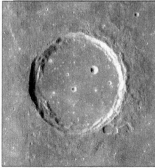
Aerial view of Marius

Marius lies on the edge of one of the largest group of volcanic domes and cones on the Moon - that of the Marius Hills. The crater has obviously been subject to lava flows lapping at its outer shores; where those easterly are some 800 metres lower than its rim there, while westerly they are at 400 m lower (lavas within Marius give a crater depth of around 1.6 km). Marius's ejecta display is very obvious at low sun times, as, too, is the odd-looking 'chink' (a rille?) at its outer southeast. The floor has been dusted by Kepler's rays to the east.

Craters

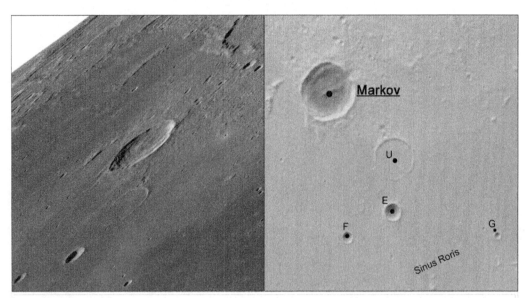

Markov	Lat 53.43N	Long 62.84W	39.92Km

Sub-craters:

Crater	Lat	Long	Size (km)	Crater	Lat	Long	Size (km)
E	50.66N	60.28W	12.25	F	50.1N	62.04W	7.86
G	49.99N	56.25W	5.12	U	51.9N	60.16W	29.02

Notes:

Add Info:

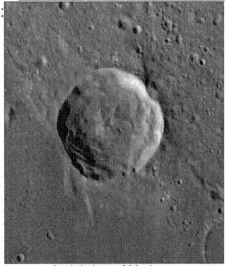

Aerial view of Markov

Markov formed on highlands material, with an additional mix of ejecta produced by the formation of the Imbrium Basin event to the southeast. Much of Markov's ejecta can be seen from our Earth perspective, however, beyond its western rim, later lava flows have covered it up there (see aerial view), and also south of the crater. An odd-looking ridge-like feature east of Markov's rim extends from sub-crater U up towards the north (shows up nicely around a 25-day-old moon period), and it seems its existance affected the rim producing a kind of 'dint' (it doesn't look impact-related). Markov has a depth of around 3.4 km, and its levelled peaks reach only ~ 800 metres approx..

Craters

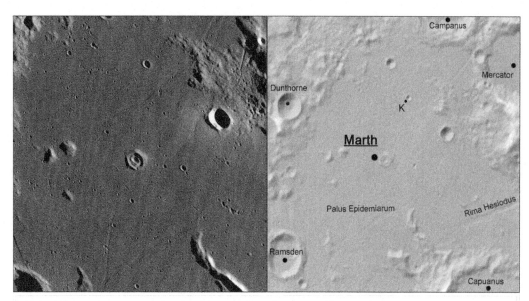

| Marth | Lat 31.16S | Long 29.35W | 6.54Km |

Sub-craters:

Crater	Lat	Long	Size(km)	Crater	Lat	Long	Size(km)
K	29.96S	28.77W	3.02				

Notes:

Add Info:

When viewing Marth in the eyepiece, also try to include the other concentric crater in the region nearby; that of Hesiodus A (30S, 17W) over to the east at some 300 km away.

Aerial view of Marth

Always an eye-catching feature under all lighting conditions, Marth is considered as a single impact event where separate layers of the target rock responded differently according to their individual make-up and hardness (such concentric features of a much smaller scale were once used to determine the regolith thickness from orbital photos taken during the Apollo 16 mission). Low sun views show a small rille which occurred later affected Marth's outer torus rim at its south, and that an apron of material (slightly lacking at the southwest) surrounds the crater. For more about concentric craters in general, refer to crater Leakey in this book. D: 1.1 km.

Craters

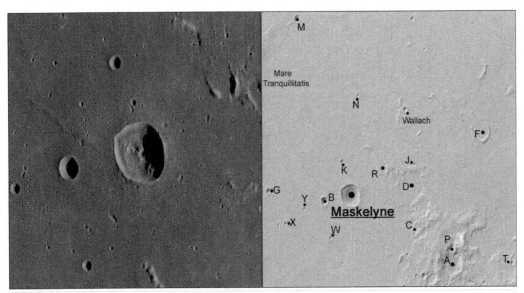

| Maskelyne | Lat 2.16N | Long 30.04E | 22.42Km |

Sub-craters:

Crater	Lat	Long	Size (km)	Crater	Lat	Long	Size (km)
A	0.03N	34.09E	29.42	B	1.97N	28.96E	8.34
C	1.07N	32.67E	9.1	D	2.41N	32.45E	31.66
F	4.18N	35.3E	20.93	G	2.3N	26.7E	5.85
J	3.14N	32.63E	3.69	K	3.23N	29.69E	5.36
M	7.8N	27.88E	7.3	N	5.35N	30.36E	4.4
P	0.47N	34.14E	8.46	R	3.0N	31.33E	12.43
T	0.04S	36.59E	5.18	W	0.8N	29.21E	4.17
X	1.27N	27.39E	3.71	Y	1.73N	28.16E	4.27

Notes:

Add Info: It's always nice to view in the eyepiece craters of similar size, having similar characteristic features, and where both are similar in age. So, before looking at Maskelyne itself, take a look also at crater Arago, which lies some ~ 280 kilometres off to its northwest. From the worn-down states of each of their crater's rims, interiors and outer ejecta, Maskelyne looks just that little bit younger than the two (note the ejecta albedos of each, too). Several series of slumped banks of material occur in each also; so much so that their floors are covered with them, making it really hard to see the original floor as the remaining portions are occupied by low-level peaks. At low sun views the wonderful, crinkled ridges within Tranquillitatis show that each crater has impacted nearby some: in Maskelyne's case, its overlying ejecta defines the ridge it clipped as being pre-formed. Query: Is Maskelyne an impact crater created by an impactor that came from the northwest (look at its shape). Depth ~ 2.5 km.

Missions: Apollo 11, Surveyor 5 and Ranger 8 all landed some ~ 200 kilometres west of this crater. Three small craters also lie west of Mason: Armstrong, Aldrin, and Collins (named after the Apollo 11 astronauts).

Craters

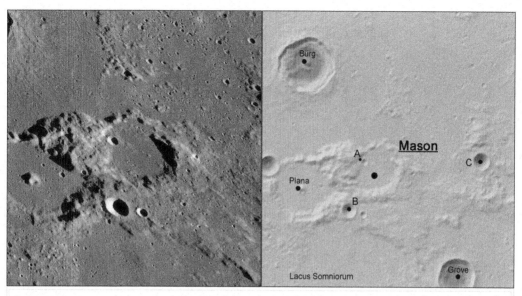

| Mason | Lat 42.7N | Long 30.51E | 33.33km |

Sub-craters:

Crater	Lat	Long	Size (km)	Crater	Lat	Long	Size (km)
A	42.92N	30.13E	4.66	B	41.84N	29.66E	9.81
C	42.9N	33.87E	11.59				

Notes:

Add Info:

Aerial view of Mason

Mason is situated on the edge of a lava-flow zone related to those as similarly seen at the eastern end of Mare Frigoris (~ about 3.7 byo). But distinct also to lavas associated with formation of Lacus Mortis, which make up Mason's floor. The rim of Mason is hard to define as much of it is covered over by lavas in the north, while the rest lies underneath clumps of ejecta deposits from other nearby crater impacts (the large, 2-km-high clump of material on its west rim suggests it might have come from bigger impact events, for example, like basins in the region). Low sun views show up Mason best. Depth ~ 1.4 km.

Craters

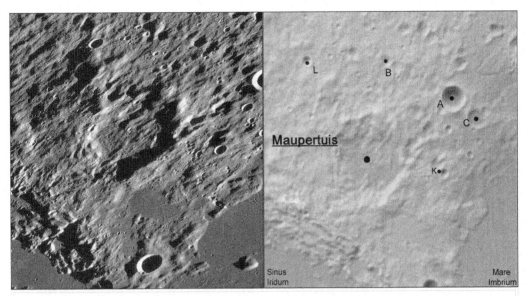

Maupertuis	Lat 49.7N	Long 27.28W	45.49Km

Sub-craters:

Crater	Lat	Long	Size (km)	Crater	Lat	Long	Size (km)
A	50.66N	24.75W	13.91	B	51.37N	26.8W	6.07
C	50.3N	24.09W	10.47	K	49.33N	25.1W	5.28
L	51.34N	29.24W	6.21				

Notes:

Add Info:

Aerial view of Maupertuis

At times of full moon you can just about make out Maupertuis with its small, relatively dark patch of material, which presumably came through formation of Sinus Iridum to the southwest. This huge impact dumped a whole lot of its ejecta across the region in general - note the NE-SW trending striations, and on the main 1123 km ring/rim of the Imbrium Basin on which now partially-buried Maupertuis lies (note the large clump at its northeast sector). Additional clumps cover the floor, and at very high magnification several small rilles and crater chains can be seen to criss-cross them.

Craters

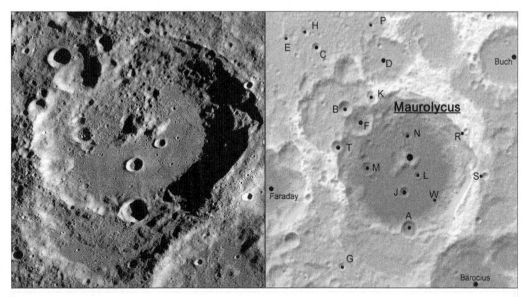

Maurolycus   Lat 41.77S   Long 13.92E   115.35Km

Sub-craters:

Crater	Lat	Long	Size (km)	Crater	Lat	Long	Size (km)
A	43.56S	14.14E	15.08	B	40.34S	11.74E	11.97
C	38.71S	10.76E	7.96	D	39.18S	13.19E	43.88
E	38.43S	9.71E	5.65	F	40.66S	12.22E	25.03
G	44.51S	11.51E	6.95	H	38.3S	10.39E	7.35
J	42.57S	13.94E	8.84	K	39.99S	12.66E	7.71
L	42.11S	14.44E	5.79	M	41.94S	12.53E	10.57
N	41.08S	14.02E	7.22	P	38.13S	12.69E	3.94
R	40.91S	16.22E	4.5	S	42.08S	17.01E	6.43
T	41.34S	11.42E	9.63	W	42.86S	15.15E	4.08

Notes:

Add Info:

The two mains peaks on the floor come in at around 1.8 km high for the westerly peak and 1.3 km high for the easterly peak.

Aerial view of Maurolycus

Maurolycus is quite easy to find in this populated region of craters when one uses the big 'lip' (a ~ 100 km-diameter-sized unnamed crater at the southwest on which it has impacted upon) as a marker. Another unnamed crater (~ 65 km in diameter) at Maurolycus's northwest was impacted upon too, where several less-sized impact craters 'wiped' out the remaining rim (note the wonderful ejecta displays on the floors of all three craters). Full moon times makes Maurolycus near-invisible (rays from Tycho to its west that overlie it don't help either). Depth ~ 4.8 km.

Craters

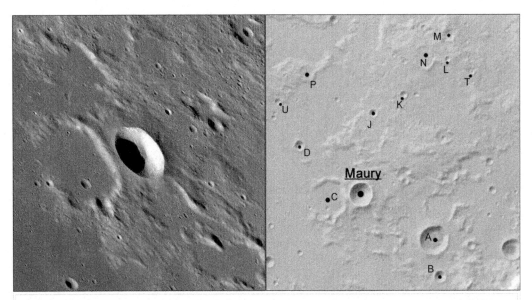

| Maury | Lat 37.11N | Long 39.69E | 16.58Km |

Sub-craters:

Crater	Lat	Long	Size (km)	Crater	Lat	Long	Size (km)
A	35.99N	41.9E	21.37	B	35.11N	42.05E	8.62
C	36.98N	38.68E	28.82	D	38.27N	37.78E	7.61
J	39.04N	40.09E	5.97	K	39.47N	41.07E	5.2
L	40.29N	42.5E	4.23	M	40.86N	42.55E	9.25
N	40.36N	41.77E	17.6	P	40.01N	37.98E	11.91
T	39.97N	43.29E	2.74	U	39.31N	37.07E	4.69

Notes:

Add Info:

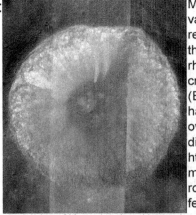

Aerial view of Maury (sectioned images)

Maury lies just on the eastern edge of the lava plain known as Lacus Somniorum - a depressed zone caused possibly by formation of the Serenitatis Basin to its southwest and, perhaps, the Crisium Basin to its southeast. The crater has a relatively sharp rim, so it is young (Eratosthenian in age?), but not so young to have had its resultant ejecta smoothed down over time, if not rays from the event remain on display, too. High sun times shows up its bright inner wall material (see left) that possibly is made up of Anorthosite - an intrusive igneous rock containing a minimum of 90% plagioclase feldspar, and which also makes up the true crustal highlands of the Moon. Depth ~ 3.4 km.

Craters

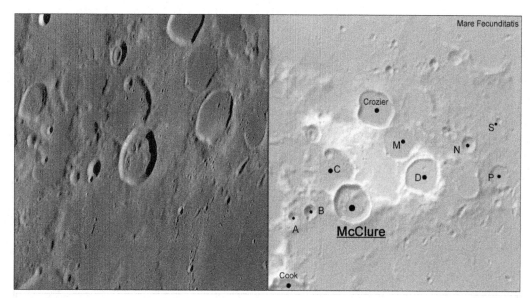

McClure   Lat 15.32S   Long 50.24E   23.96Km

Sub-craters:

Crater	Lat	Long	Size (km)	Crater	Lat	Long	Size (km)
A	15.6S	48.95E	6.12	B	15.49S	49.33E	8.09
C	14.73S	49.86E	25.94	D	14.8S	51.74E	24.24
M	14.19S	51.24E	20.53	N	14.19S	52.71E	9.09
P	14.77S	53.37E	12.26	S	13.73S	53.36E	3.52

Notes:

Add Info:

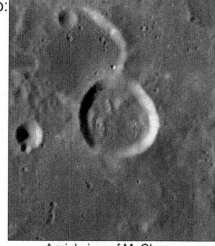

Aerial view of McClure

McClure lies on material possibly related to formation of the Fecunditatis Basin event to its northeast. The crater is slightly oval with its main axis pointing ominously back to the Imbrium Basin from the north west (is it a secondary crater from that event?). McClure has obviously impacted upon sub-crater, McClure C, however, as signatures of its ejecta look very sparse, did lavas within C cover it up (C's floor level is some 600 metres above that of McClure's). The material in McClure's floor is believed to be a mix of ejecta from the Imbrium Basin and lavas similar to those at its outer south. Full moon times shows up McClure's bright wall easily. Depth: 1.2 km.

Craters

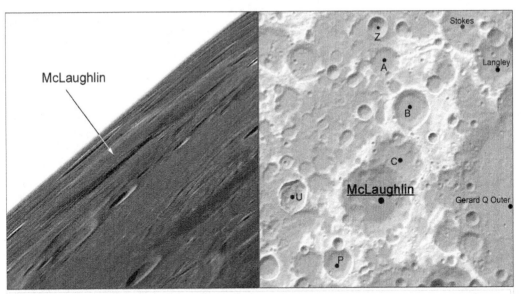

| McLaughlin | Lat 47.01N | Long 92.83W | 75.29Km |

Sub-craters:

Crater	Lat	Long	Size (km)	Crater	Lat	Long	Size (km)
A	51.32N	92.32W	31.6	B	49.82N	90.99W	40.84
C	48.17N	91.79W	57.55	P	45.02N	94.66W	32.72
U	47.06N	96.99W	31.63	Z	52.45N	92.79W	20.71

Notes:

Add Info:

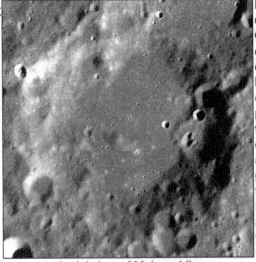

Aerial view of McLaughlin

McLaughlin has impacted both on the ejecta deposits of Gerard Q Outer to its east, and on sub-crater, McLaughlin C, to its north-east (C's floor level is some 1.5 kilometres above the floor level of McLaughlin). Hint of a small peak (?) is seen on the floor too, which should be looked out for during observations as it acts as a good marker for noting McLaughlin's location (can be difficult at times - even during favourable librations). The crater is Nectarian in age (3.92 to 3.85 billions of years old); putting an age constraint on Q Outer and C, and has a depth of around ~ 1.8 km.

Craters

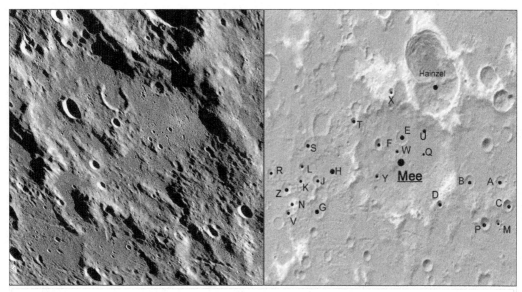

Mee   Lat 43.63S   Long 35.19W   134.11Km

Sub-craters:

Crater	Lat	Long	Size (km)	Crater	Lat	Long	Size (km)
A	44.49S	29.27W	13.3	B	44.58S	31.17W	14.99
C	45.28S	28.72W	12.85	D	45.4S	33.02W	8.62
E	43.15S	35.41W	15.71	F	43.37S	36.83W	11.85
G	45.48S	40.73W	22.09	H	44.21S	39.62W	48.05
J	44.49S	40.7W	10.04	K	44.45S	41.62W	8.99
L	44.0S	41.55W	8.01	M	45.91S	29.25W	7.81
N	45.26S	42.26W	6.21	P	45.97S	30.05W	12.88
Q	43.61S	33.81W	1.2	R	44.08S	43.44W	9.5
S	43.27S	41.08W	11.93	T	42.56S	38.31W	9.09
U	42.96S	34.02W	8.47	V	45.53S	42.56W	7.39
W	43.64S	35.69W	5.45	X	41.55S	36.04W	7.6
Y	44.46S	36.93W	6.4	Z	44.7S	42.6W	12.76

Notes:

Add Info:

Not to be confused with crater Mees, which lies on the Farside at 13.57N, 96.18W (never seen from our earth-based perspective).

Aerial view of Mee

From the outset it's clear that Mee is an old crater of the pre-Nectarian Period (4.6 to 3.92 billions of years old). Nearly all its rim has been affected by small to large impact events nearby: for example, formation of crater Hainzel took out quite a lot of Mee's northeastern sector, while to its southeast a huge volume of ejecta - possibly related to the Humorum Basin - covered Mee's rim giving it a frothy look to it. Orientale (over to the northwest), of course, played its part too (secondaries etc.,), and rays from Tycho overlie the crater. Depth: 2.7 km.

Craters

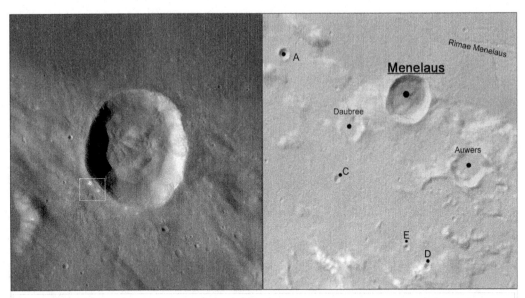

Menelaus Lat 16.26N Long 15.93E 27.13Km

Sub-craters:

Crater	Lat	Long	Size (km)	Crater	Lat	Long	Size (km)
A	17.07N	13.39E	6.18	C	17.07N	14.48E	4.02
D	13.23N	16.31E	4.11	E	13.23N	15.89E	3.48

Notes:

Add Info: Fresh impact (right) events near a crater can give a lot of information about the crater's surrounds. In this case, these two small impacts have excavated immature (younger) deposits of Menelaus's ejecta onto a mature (older) surface.

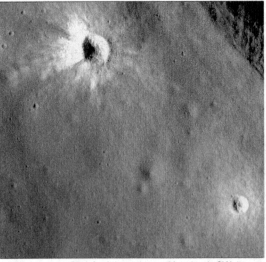

Close-up view of two fresh impacts near Menelaus's SW rim

Menelaus formed on the main 700 km ring/rim of the Serenitatis Basin. The crater has a relatively sharp-looking rim, and its floor contains material from what looks like the result of a single slump at its southern sector (indication it may be a recent event after the crater formed). Ray distribution suggest the crater formed obliquely, but these rays aren't a maturity signature of the crater, but rather a compositional one of its surrounds (less mature, brighter material overlies older, darker material). Depth ~ 2.6 km.

Craters

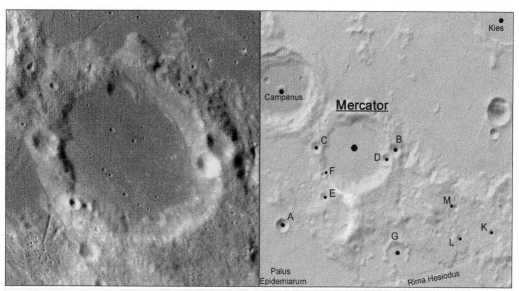

| Mercator | Lat 29.25S | Long 26.11W | 46.32Km |

Sub-craters:

Crater	Lat	Long	Size (km)	Crater	Lat	Long	Size (km)
A	30.64S	27.83W	8.18	B	29.16S	25.2W	7.72
C	29.15S	27.04W	7.72	D	29.33S	25.37W	6.55
E	30.1S	26.85W	5.33	F	29.67S	26.88W	3.06
G	31.13S	25.07W	14.0	K	30.63S	22.78W	3.88
L	30.77S	23.55W	3.73	M	30.21S	23.7W	3.77

Notes:

Add Info: Mercator should never be looked at on its own really as its brother to its northwest, crater Campanus, makes for wonderful comparisons in how and where they formed. Mercator is the older of the two - Nectarian in age (3.92 to 3.85 byo) as opposed to Campanus which is of the Early Imbrian (3.85 to 3.75 byo). The two impacted upon ring/rim basin's material associated to pre-Nectarian Nubium to the northeast, and to a lesser extent that of younger material from formation of Late Imbrian Humorum Basin to the northwest. Note the differences in each of their ejecta displays, their rims and inner walls, too - Mercator's looks that bit more worn than Campanus's. Both obviously have undergone some sort of internal flooding, however, these floods occurred as separate events and times - Mercator's floor level is some one kilometre above that of Campanus's, whose lavas look slightly darker (a young signature). This difference in level might also explain why Mercator doesn't have a peak like that seen in Campanus as it's probably been buried by the additional volume of flood (Campanus's floor is also some 500 metres lower at its southeast than at its northwest).

Craters

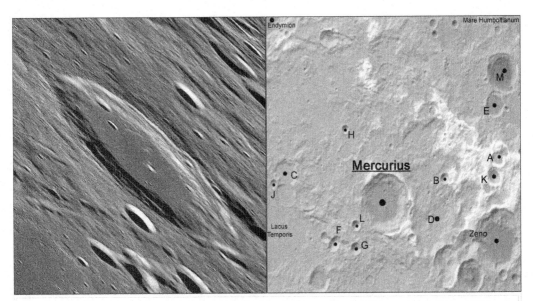

Mercurius	Lat 46.66N	Long 66.07E		64.3Km			

Sub-craters:

Crater	Lat	Long	Size (km)	Crater	Lat	Long	Size (km)
A	47.87N	73.27E	19.4	B	47.43N	69.82E	11.76
C	47.59N	59.66E	29.26	D	46.13N	69.0E	69.48
E	49.76N	73.43E	24.83	F	45.19N	62.85E	13.8
G	45.11N	64.22E	13.91	H	49.19N	63.48E	10.03
J	47.15N	58.99E	10.38	K	47.32N	73.01E	20.0
L	45.9N	64.24E	11.56	M	50.85N	74.12E	41.52

Notes:

Add Info:

A slight dusting of ejecta from the Crisium Basin to the south also overlies this area.

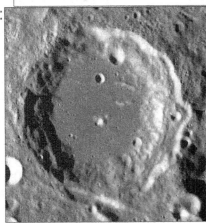

Aerial view of Mercurius

Mercurius formed on the outer ejecta deposits of the Humboltianum Basin to the northeast. Nectarian in age (3.92 to 3.85 byo), the crater features worn down terraces, a small 0.3 km-high peak, a buried crater (northeastern part of floor), and a rim that has been subject to major slumping effects at it east. The southern part of its outer rim shows an obvious 'gash' that probably was created by ejecta from Imbrium over to the west, which itself is seen to cross a similar-like feature on the crater's western sector seen to radiate back to the above-mentioned Basin. D: 3.6 km.

Craters

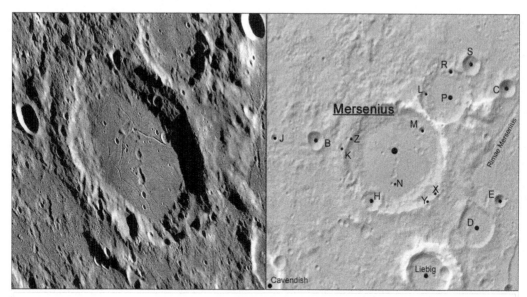

Mersenius   Lat 21.49S   Long 49.34W   84.46Km

Sub-craters:

Crater	Lat	Long	Size (km)	Crater	Lat	Long	Size (km)
B	21.07S	51.68W	13.33	C	19.76S	45.99W	13.6
D	23.16S	46.84W	31.0	E	22.51S	46.13W	9.53
H	22.53S	50.02W	15.32	J	20.99S	52.92W	6.03
K	21.25S	50.94W	4.52	L	19.94S	48.41W	3.72
M	21.34S	48.56W	5.28	N	22.13S	49.41W	3.09
P	20.02S	47.87W	39.44	R	19.4S	47.68W	5.23
S	19.22S	47.06W	15.62	U	23.01S	50.12W	4.74
V	22.93S	50.67W	4.23	W	23.05S	50.96W	4.65
X	22.45S	48.01W	4.45	Y	22.68S	48.31W	4.01
Z	21.04S	50.75W	3.72				

Notes:

Add Info:

Aerial view of Mersenius

At full moon times the only feature of Mersenius to be seen is the bright 'ray' (possibly caused by clusters of small craterlets - Byrgius A related? - disturbing the soil) which crosses its floor. Low sun views (around last quarter) are thus recommended, but even with these the crater still proves challenging in this bright region of ejecta related to the Humorum Basin to its east. One would expect to see the obvious rilles on its floor during such low sun views, too, but only hint of the north-south trending one in the western sector shows up. D: ~ 2.9 km.

Craters

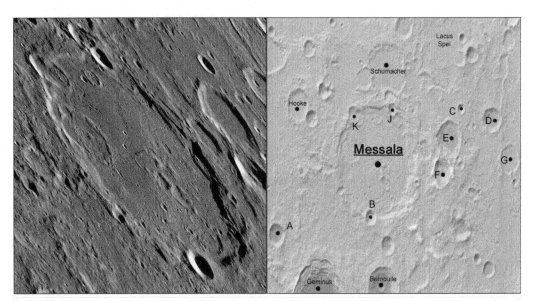

Messala   Lat 39.31N   Long 60.06E   122.4Km

Sub-craters:

Crater	Lat	Long	Size (km)	Crater	Lat	Long	Size (km)
A	36.57N	53.82E	24.88	B	37.37N	59.81E	16.44
C	41.0N	65.84E	11.48	D	40.52N	67.89E	28.69
E	39.99N	64.95E	38.83	F	38.85N	64.38E	32.25
G	39.07N	68.91E	30.7	J	41.12N	61.2E	13.3
K	40.98N	58.51E	13.43				

Notes:

Add Info:

During full moon times the crater is hard to find, however, two good location markers are the bright inner wall of Messala B, and the dark deposits on the floor's western sector.

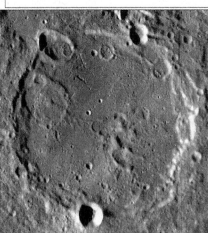

Aerial view of Messala

For such a big crater Messala's low depth of around 2 km suggests the crater has undergone some sort of major infill over its pre-Nectarian (4.6 to 3.92 byo) lifetime. Basins' Crisium to its south and Humboldtianum to its northeast are the main culprits that filled Messala with their ejecta products (Crisium more than Humboldtianum), but volcanic related fills (note the floor's northwest area) are responsible, too. Several obvious buried craters lie under these fills also, but what isn't so obvious is the group of thin, hard-to-see rilles everwhere on the floor, which probably existed before being covered.

Craters

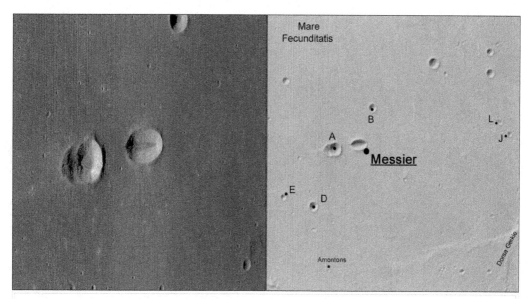

Messier	Lat 1.9S	Long 47.65E	13.8Km

Sub-craters:	Crater	Lat	Long	Size (km)	Crater	Lat	Long	Size (km)
	A	2.03S	46.94E	11.03	B	0.9S	48.06E	6.87
	D	3.59S	46.32E	7.79	E	3.35S	45.43E	5.02
	J	1.6S	52.16E	3.69	L	1.26S	51.87E	5.36

Notes:

Add Info:

Messier's rays extend on either side of its main axis for about 110 km northwards and 120 southwards, while those from Messier A and the unnamed crater are at 180 km.

Aerial view of the Messier ray system

From an initial view of Messier, it's unavoidable to note its companion, sub-crater Messier A, to its west. Messier obviously formed from an oblique impact event, but how did Messier A form, and the unnamed crater to its west? Did a piece of the impactor that formed Messier continue on downrange to produce A and/or the unnamed crater? Was Messier a separate event to Messier A and the unnamed crater? The distribution of rays may be a key - Messier's are at right angle to its long axis, while those at Messier A and the unnamed crater are parallel. Depths: Messier and Messier A ~ 2.2 km, unnamed crater ~ 1.2 km.

Craters

Metius   Lat 40.42S   Long 43.37E   83.81 Km

Sub-craters:

Crater	Lat	Long	Size (km)	Crater	Lat	Long	Size (km)
B	40.18S	44.37E	14.49	C	44.22S	49.07E	10.35
D	42.66S	48.48E	10.6	E	39.8S	42.84E	6.73
F	39.12S	42.74E	8.5	G	40.37S	45.25E	9.88

Notes:

Add Info:

Aerial view of Metius

Metius is Nectarian in age (3.92 to 3.85 byo) and lies in a region known as the Janssen Formation - ejecta from the Nectaris Basin to its north. Crater Fabricius to its south has just clipped its southwestern rim; imparting, in the process, ejecta onto Metius's floor whose level is some one kilometre higher up. Hint of a ring or peak (~ 500 m high) in Metius's floor is also seen, where several clusters of small craterlets (hard to see) relate to nearby impact events. A 'gash' at its northeast may be the result of a huge block of ejecta striking the rim, but what was its source–Fabricius, or Stevinus to its northeast? D: 4 km.

Craters

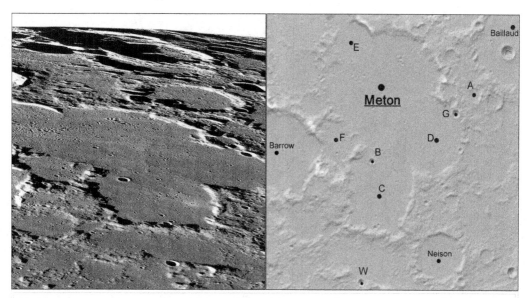

Meton   Lat 73.57N   Long 19.63E   124.7Km

Sub-craters:

Crater	Lat	Long	Size (km)	Crater	Lat	Long	Size (km)
A	73.23N	30.67E	14.65	B	71.39N	18.17E	6.69
C	70.45N	18.87E	86.12	D	72.16N	24.46E	78.77
E	75.19N	15.25E	43.28	F	72.01N	14.03E	49.61
G	72.76N	28.12E	8.95	W	67.46N	17.49E	7.74

Notes:

Add Info:

All craters in this region, including Barrow, are believed to be of the pre-Nectarian period (4.6 to 3.92 byo).

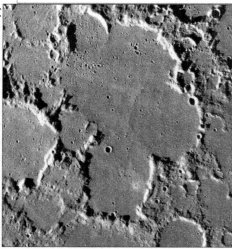

Aerial view of Meton

Remember when looking at crater Meton that it isn't actually the 'clover-like' group of craters, as seen in the image left, but rather the single-most, biggest one northwards within them all. The 'clover' is made up of five craters in all (C, D, E, F and Meton itself), and it looks like Meton is the oldest (note how the rims, and partially-covered parts of the others, 'cut' into Meton's rim). Mateial within the group's floor is believed to be ejecta fill from Imbrium to the southwest, and rays from Anaxagoras to its west cross these. Favourable librations dramatically changes the detail seen on the floor. Depth ~ 1.8 km.

Craters

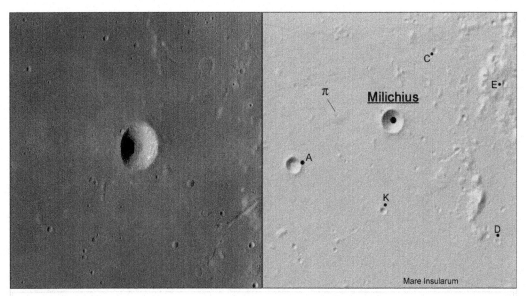

Milichius		Lat 10.01N	Long 30.23W				12.19Km
Sub-craters:	Crater	Lat	Long	Size (km)	Crater	Lat	Long	Size (km)
	A	9.25N	32.07W	8.23	C	11.2N	29.44W	2.93
	D	7.97N	28.27W	3.47	E	10.62N	28.15W	2.54
	K	8.49N	30.4W	3.76				
	Notes:							

Add Info: Milichius lies at the extreme, outer ring/rim of the Imbrium Basin to its northeast (at Montes Carpatus), and in the pre-Nectarian-aged mare that is Insularum. The crater hasn't much to offer in terms of features, and the only thing it has going for it is its ejecta - best observed at low sun times, which overlies the mare (see this at Milichius A, too). Obvious during these low sun times is the wonderful dome over to Milichius's west (some 30 km away). The dome is usually refferred to as Milichius Pi (π) - an 'effusive' dome formed by lavas that effused out of its main vent (the small pit-like crater on its top) at a slow rate, of low temperature, and of high viscosity (more stickier, like, say, as honey is). Mare domes are believed to have formed from materials and mineralogies of the upper mantle during the late stages of volcanism on the Moon, while those that formed during the early stages had lavas more fluidy, of higher temperature, and of massive volumes and mineralogies of the lower crust. Observe more domes to the north midway between Milichius and crater Tobias Mayer; they are a little harder to see, so look under lower suns and terminators). There are many other domes to observe all over the Moon, so make sure to check them out. Depth of Milichius ~ 2.5 km.

Other Domes
Effusive:
Kies Pi (π)
26.9S, 24.3W.

Intrusive
(those formed
from
Laccoliths):
Valentine
31.0N, 10.0E.

Highland:
Gruithuisen
36.5N, 40.0W.

Craters

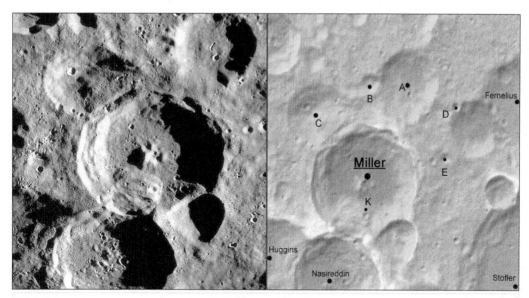

Miller	Lat 39.37S	Long 0.78E					61.37Km

Sub-craters:

Crater	Lat	Long	Size (km)	Crater	Lat	Long	Size (km)
A	37.69S	1.79E	36.65	B	37.67S	0.92E	11.32
C	38.2S	0.35W	35.06	D	38.03S	2.99E	4.79
E	38.93S	2.75E	5.87	K	39.92S	0.83E	3.91

Notes:

Add Info:

Aerial view of Miller

During certain low sun times it's impossible to ignore the large clump of material lying on the southwest sector of Miller's floor. The feature obviously is related to formation of Nasireddin to the southwest, but is it as a result of ejecta from this crater, or was it caused by slumping of Miller's wall afterwards (note the sharp dividing ridge between the two). Was Miller and Nasireddin a dual-impact event? The two craters are similar in appearance, in age (mid-Imbriun ~ 3.5 byo?). Ray material from Tycho (~ 300 km away over to the southwest) overlies Miller, but it isn't obvious. Peak:1.3 km. D: 3.5 km.

Craters

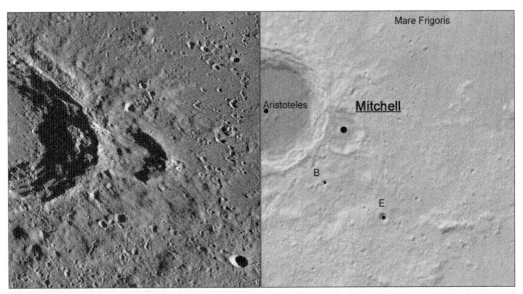

Mitchell	Lat 49.77N	Long 20.17E				32.15Km

Sub-craters:

Crater	Lat	Long	Size (km)	Crater	Lat	Long	Size (km)
B	48.53N	19.44E	5.4	E	47.7N	21.73E	7.72

Notes:

Add Info:

Aerial view of Mitchell

Mitchell is generally believed to be an old crater. Is this correct? Look at the size of nearby Aristoteles to its west, and think of that impact event which threw huge volumes of ejecta in Mitchell's direction. The ejecta not only filled its floor, but splashed well beyond its eastern rim. All signatures of younger features were thus instantly wiped out through coverage, and so Mitchell could still be a relatively young crater hidden beneath. What is the slight 'gash' at its northwestern rim? It looks like an ejecta feature of Aristoteles onto Mitchell's floor, but is it really (was Bürg to the southeast responsible?)? Depth ~ 1 km.

Craters

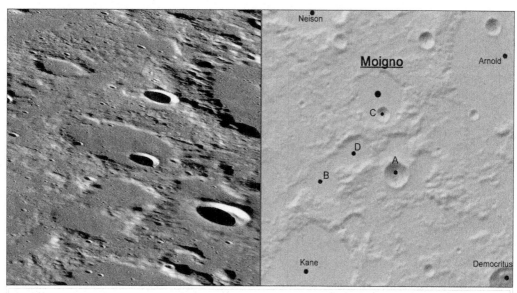

Moigno	Lat 66.27N	Long 28.8E					36.83Km

Sub-craters:

Crater	Lat	Long	Size (km)	Crater	Lat	Long	Size (km)
A	64.81N	29.72E	15.77	B	64.64N	26.11E	24.27
C	65.98N	29.06E	9.22	D	65.19N	27.59E	23.88

Notes:

Add Info:

Aerial view of Moigno

Like many of its neighbouring craters, Moigno may be of the pre-Nectarian Period (4.6 - 3.92 byo). This area is generally accepted as one where ejecta from the Imbrium Basin event played a domnant role in terms of the majority of material seen in all these craters. But seen under full moon conditions, Moigno's floor, and also in the floors of other craters nearby, darker material shows up - a signature, perhaps, of late volcanic activity (note the dark patch on the outer north-eastern rim of Moigno C in image left). Easy to get lost here between observing stints at the eyepiece, so either use Moigno C or Arnold to easily find it again. D: 0.8 km.

Craters

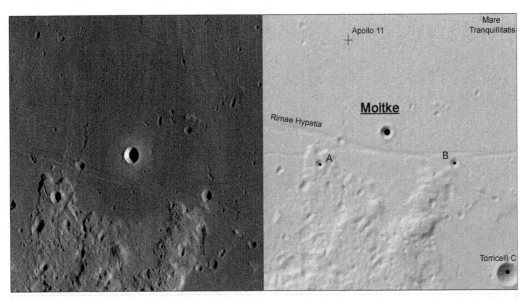

Moltke	Lat 0.59S	Long 24.16E		6.15Km			
Sub-craters:	Crater	Lat	Long	Size (km)	Crater	Lat	Long	Size (km)
	A	1.05S	23.17E	4.47	B	1.05S	25.19E	4.25
	Notes:						

Add Info:

Apollo 11 Landed on the lunar surface on 24 July 1969. Four days later, all three astronauts involved with the mission successfully, and safely, returned to Earth.

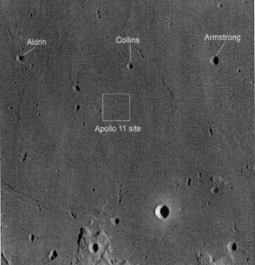

Aerial view of Apollo 11 site (and named craters)

Moltke's bright outer surrounds and inner wall material makes it an easy crater to see at almost any time of viewing. Some of its bright ejecta partially overlies Rimae Hypatia to the south, while its rim (some 250 metres above the mare plain) casts some nice shadows during low sun times. Apollo 11 landed some 50 km away to the north, and while the crater can be used as a location marker for this wonderful mission, it can also be used as a guide in noting the three craters (difficult to see during full moon times) - named after the three astronauts who flew. Depth ~ 1.3 km.

Craters

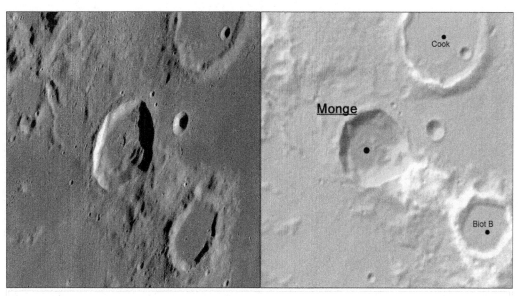

| Monge | Lat 19.24S | Long 47.54E | 36.6Km |

Sub-craters: No sub-craters

Notes:

Add Info:

Aerial view of Monge

From the aerial view of Monge (left) it's obvious to see why this crater has an odd shape - major slumping has occurred at the southeastern sector. This feature may be related to how the original impactor that produced Monge struck an uneven ground surface at its southeast (it is some 2.5 km higher today). Did this result in more material at its southeast being ejected in that direction? Did it weaken that sector also, which then led to the slumping? Low sun times show some radial signatures of the ejecta at the outer crater's southwest, but elsewhere it isn't seen. At full moon times a bright ray, having a NW-SE orientation, crosses the crater's floor and these outer sectors respectively. Its source may be that of Furnerius A (~ 500 km away to the southeast). But is it? D: 1.5 km.

Craters

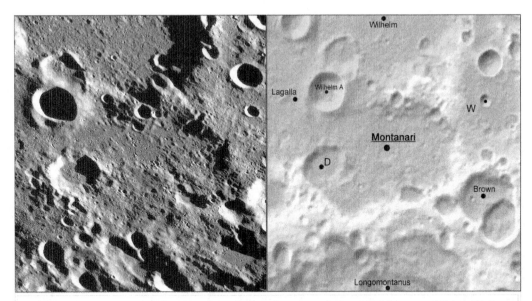

Montanari	Lat 45.83S	Long 20.76W		77.05Km		
Sub-craters:	Crater	Lat	Long	Size (km)	Crater	Lat	Long	Size (km)
	D	45.95S	22.21W	26.51	W	44.82S	18.1W	6.35

Notes:

Add Info:

Aerial view of Montanari

If you weren't told initially that an actual crater existed at one time where now Montanari lies, you would pass it over as simply an accumulation of debris from other craters nearby. It's hard then to see the sequence of impact events, however, it looks like Lagalla formed first, then Montanari, then Wilhelm, and finally Longomontanus. It gets a bit messy between Lagalla and Montanari where it could be the other way of sequence (Montanari first and then Lagalla), however, as Wilhelm A seems to have impacted on a higher piece of ground, this may be the remaining rim of Montanari that then lies on Lagalla. Mares' Nubium, Humorum, Imbrium etc., reflect the regional landscape. Tycho rays overlie. D: 2 km.

Craters

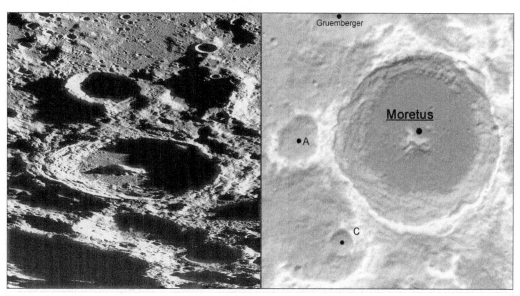

| Moretus | Lat 70.63S | Long 5.95W | 114.45Km |

Sub-craters:

Crater	Lat	Long	Size (km)	Crater	Lat	Long	Size (km)
A	70.42S	13.98W	34.57	C	72.62S	11.63W	16.38

Notes:

Add Info:

The small rille (a fracture feature?), on the eastern sector of the floor can just about be seen at higher magnification.

Aerial view of Moretus

Moretus is considered as a crater that formed some time in the Eratosthenian Period (3.15 to 1.1 byo). It almost looks similar to younger Tycho (Copernican) to its northwest - both showing wonderful sharp crenulated rims, terraces, and peak (highest part ~ 2.2 km), however, it's obvious lack of rays defines it as slightly older. The crater has a thick ejecta 'band' around most of its exterior rim - the southern sectors especially, whose clumpy height produces shadows that at terminator times fill the crater's innards (the peak's tip sometimes 'peeps' through during such occassions). Taking into account the southern ejecta thickness, and that the width of Moretus's terraces in the south are wider, did the crater form as an oblique impact event?

Craters

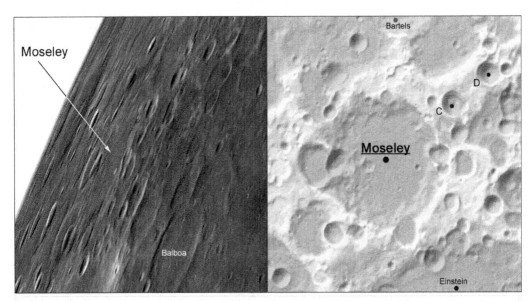

Moseley	Lat 20.95N	Long 90.2W	88.89Km

Sub-craters:

Crater	Lat	Long	Size (km)	Crater	Lat	Long	Size (km)
C	22.21N	88.52W	18.89	D	22.87N	87.65W	17.51

Notes:

Add Info:

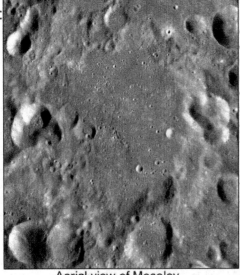

Aerial view of Moseley

From the aeriel view (left) of crater Moseley, it's clear from the numerous, small craters covering nearly every section of its rim that some major impact event was responsible (the Orientale Basin lies ~ 1200 km due south, so they are most likely secondaries from it). Crater Einstein also lies due south right next to Moseley, however, as it looks like it may have formed on [its] pre-Nectarian (4.6 to 3.92 byo) ejecta, Moseley, then, is younger (but not by much - given its decrepit-looking state). Moseley's floor level is some ~ 500 metres below Einstein's floor level, so shadows fill its interior when Einstein's interior is in light (libration times help, though). Depth ~ 3.2 km.

Craters

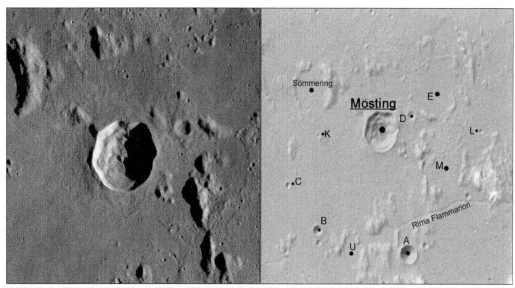

| Mösting | Lat 0.7S | Long 5.88W | 24.38Km |

Sub-craters:

Crater	Lat	Long	Size (km)	Crater	Lat	Long	Size (km)
A	3.21S	5.19W	12.19	B	2.73S	7.39W	6.79
C	1.81S	8.11W	3.81	D	0.37S	5.12W	6.7
E	0.18N	4.59W	36.32	K	0.77S	7.37W	3.12
L	0.69S	3.43W	3.02	M	1.37S	4.36W	31.59
U	3.15S	6.58W	18.22				

Notes:

Add Info:

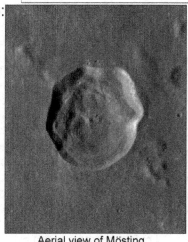

Aerial view of Mösting

Craters like Mösting (pronounced 'Muhs-ting'), for example, look at Lalande ~ 140 km away to its southwest, or, say, Konig, Bessel and Lambert etc.,, and you'll see they have one main thing in common - they all formed on mares. This is not to say that all craters which form on mares end up looking this way, but some, like those above, indicate that the depth of mare material may have shallower depths. It's believed that as the impactor stikes the surface rock below the shallow mare material, some of the impact force that is absorbed is affected differently to a non-mare surface; resulting in less rebounding, less excavation of material, and less chance of a peak forming (note the hint of one in Mösting). Depth of Mösting is around 2.5 km approx..

Craters

Mouchez	Lat 78.38N	Long 26.82W			82.78Km	

Sub-craters:

Crater	Lat	Long	Size (km)	Crater	Lat	Long	Size (km)
A	80.89N	30.37W	49.9	B	78.35N	23.05W	7.72
C	77.42N	26.12W	12.52	J	79.52N	38.59W	17.11
L	78.71N	40.78W	19.36	M	80.26N	50.13W	17.26

Notes:

Add Info:

Aerial view of Mouchez

In this region where the texture of the terrain looks monotonously the same (presumed to be unconsolidated ejecta from the Imbrium Basin, to the south), Mouchez is easily found by looking for the two small craters (B and C) on its floor. Most of Mouchez's eastern rim is missing where it connects to a ~ 40 km-wide valley that runs right down to Philolaus G (270 km away) in the south, while its southern rim has been demolished where an unnamed crater (~ 45 km in diameter) lies. At full moon times a wonderful ray from crater Anaxagorus to the southeast is easily seen to cross over C and the floor, which in extreme close-up is made up of a long chain of small secondary craters (all ~1.5 km in size) that extend just beyond L.

Craters

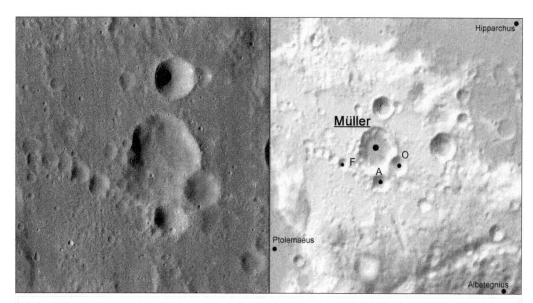

Müller   Lat 7.64S   Long 2.04E   23.4Km

Sub-craters:

Crater	Lat	Long	Size (km)	Crater	Lat	Long	Size (km)
A	8.14S	2.13E	8.68	F	7.86S	1.48E	5.78
O	7.89S	2.41E	9.66				

Notes:

Add Info: You'd miss Müller in the eyepiece, generally, as the wonderful crater of Ptolemaeus off to its southwest is the 'stealer'. The crater has a saucer-like shape to it, and there seems to be a slight 'ridge' divide running northwest-southeast across its floor. This 'ridge' hints that Müller may not be the result of a single impactor, impact event, but possibly a feature where two impactors crashed down onto the surface simultaneously (did an additional impact event extend the crater's southwest sector again, or was it part of the above possible double impact scenario?). Sub-craters, Müller O and A (which appear slightly younger), have impacted the south and southeast sectors of Müller's rim, while the southwest suffered a series of small cluster 'bombs' giving it a kind of mottled look. One obvious feature off to Müller's west is the chain of craters running in a northwest-southeast trend. What was the source of this chain? It doesn't look like it was crater Herschel, some ~140 km off to the northwest, as the chain's orientation doesn't exactly 'point' back to its centre. However, two craters do 'line' up - Reinhold and Kepler (some 800 km and 1300 km away respectively). Looking further southeastwards ~ 450 km away, another crater chain with the same orientation exists - known as the Abulfeda crater chain (Catena Abulfeda). This feature is believed to have been the result of material from a comet breaking up. Are the two chains related? Depth of Müller ~ 2 km.

Craters

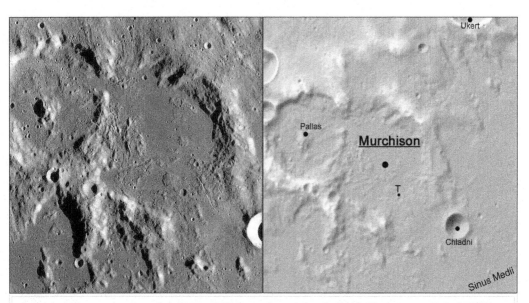

| Murchison | Lat 5.07N | Long 0.21W | 57.83Km |

Sub-craters:

Crater	Lat	Long	Size (km)	Crater	Lat	Long	Size (km)
T	4.44N	0.08E	2.48				

Notes:

Add Info: Murchison lies in an isolated, barely surviving, highlandic region of the Moon; later transformed by formation of the Imbrium Basin to the north. Pallas and Murchison were affected by this event, whose ejecta has partially covered over them. Murchison is bounded by the lava materials in Vaporum to the northeast, Sinus Medii to the southeast, and Sinus Aestuum to the northwest (all considered as large, buried craters); whose history of formations must surely share in the crater's final look. The western sector of Murchison looks clear, where it looks like Pallas has impacted upon its rim and imparted some of its ejecta onto the floor. A small, 6-km-wide 'gap' northwards between both craters may have acted as a link to the lavas seen in each, however, were these lavas at a semi-molten state such that one flowed into the other (the western side of the gap is slightly lower by about 100 metres than in the east)? But what has happened at Murchison's southeast? Where has its rim gone? A wide swade of material exists where it should be - having an approximate height of 0.5 km, however, was its initial demise due to a block of Imbrium's ejecta levelling it out to how we see it today. Or, did the last major flow of lavas in Sinus Medii coming from the east, at one time, breach into Murchison's innards (are the isolated blocks in the crater's floor that look like 'bergs' portions of Murchison's rim?)? Depth ~ 0.9 km.

Craters

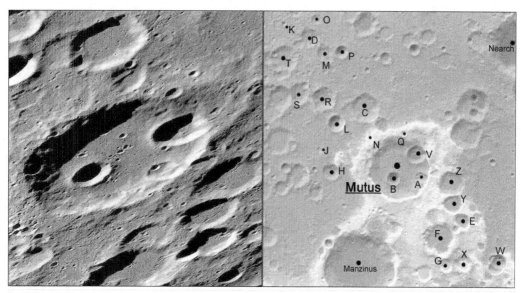

Mutus   Lat 63.65S   Long 29.93E   76.33Km

Sub-craters:

Crater	Lat	Long	Size (km)	Crater	Lat	Long	Size (km)
A	63.95S	31.98E	13.54	B	64.02S	29.56E	15.94
C	61.32S	27.22E	32.88	D	58.5S	23.25E	21.43
E	65.52S	36.0E	22.03	F	66.27S	34.18E	41.67
G	67.24S	34.85E	17.75	H	63.68S	24.11E	20.5
J	62.81S	23.29E	7.15	K	57.89S	21.51E	6.46
L	61.85S	24.81E	19.52	M	59.21S	24.32E	20.31
N	62.44S	27.63E	11.6	O	57.85S	23.86E	11.7
P	59.16S	25.55E	15.16	Q	62.3S	30.37E	6.94
R	60.88S	23.99E	26.64	S	60.63S	22.02E	24.92
T	59.23S	21.26E	31.36	V	63.08S	31.46E	22.0
W	66.81S	40.17E	18.52	X	67.1S	36.62E	20.65
Y	64.87S	34.96E	24.1	Z	64.1S	34.63E	30.25

Notes:

Add Info:

Aerial view of Mutus

Mutus can at times be an awkward crater to find, as so many others nearby, not to mention the countless smaller craters (secondaries of Basins' Nectaris and Orientale) that almost smother this region, creates much confusion. Sub-craters B and V (Orientale secondaries?) act as useful location markers for remembering Mutus (but watch out for crater Pitiscus to the northeast ~ 400 km away), with its two craters on its floor that sometimes may be mistaken for it). Mutus is pre-Nectarian in age (4.6 to 3.92 byo). Depth ~ 3.6 km.

445

Craters

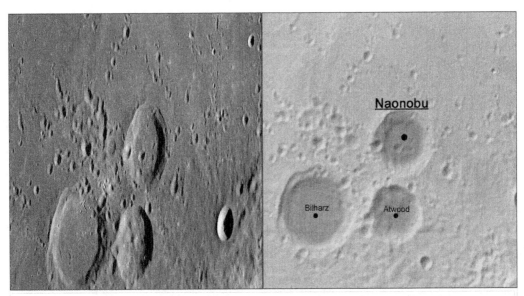

| Naonobu | Lat 4.7S | Long 57.93E | 32.96Km |

Sub-craters: No sub-craters

Notes:

Day-old moon phases

Note: As the Moon goes through various librations throughout the year, suggested times given in text for observations are only approximates.

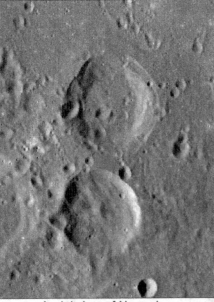

Aerial view of Naonobu

Naonobu and its two neighbours are believed to have formed around the same time. Like the Messier craters that also lie in Mare Fecunditatis (to the west), the trio have almost become as famous for their gouping. While all three look similar, there seems to be a distinction between Naonobu and Bilharz to that of Atwood - the latter having a peak with no terracing, the former having terraces (old though they may be) and no peaks. Naonobu's floor is slightly domed in its centre by about 200 metres, so, perhaps it did at one time have a small peak, but the small, rugby-shaped crater (~5 km in diameter) just off-centre on its floor might have wiped out any signature of said. Note Langrenus's influence to the southeast. D:1.9 km.

Craters

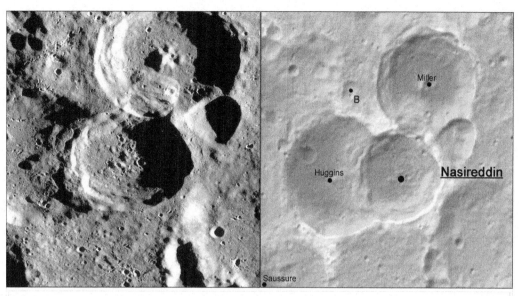

Nasireddin — Lat 41.04S — Long 0.14E — 51.99Km

Sub-craters:

Crater	Lat	Long	Size (km)	Crater	Lat	Long	Size (km)
B	39.47S	1.16W	9.3				

Notes:

Add Info:

Aerial view of Nasireddin

It's very obvious that impact of Nasireddin on top of Huggins to the west has left a definitive signature of ejecta onto its floor. It also looks like similar circumstances occurred between Nasireddin and Miller, too (the clump of material on its floor suggests Nasireddin may be responsible, and so it impacted upon Miller). But consider other impact scenarios: was Miller and Nasireddin a dual impact event (note the relatively sharp-looking ridge between the two)? Or, did Miller impact upon Nasireddin (Nasireddin's floor is some 400 m higher than Miller's is, and might explain its bumpy nature)? D: 3.3 km.

Craters

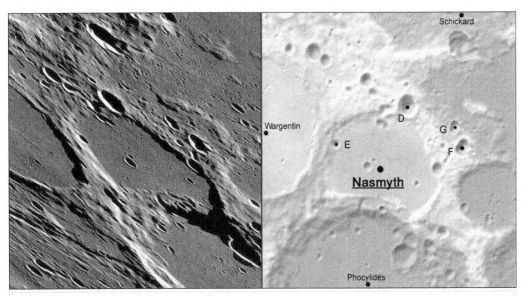

Nasmyth	Lat 50.49S	Long 56.39W				78.37Km	

Sub-craters:

Crater	Lat	Long	Size (km)	Crater	Lat	Long	Size (km)
D	49.23S	55.45W	14.34	E	49.92S	57.79W	6.05
F	50.03S	53.7W	12.05	G	49.64S	53.96W	8.87

Notes:

Add Info:

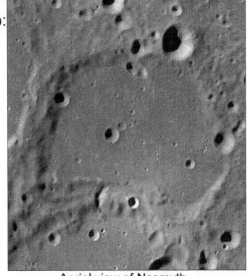

Aerial view of Nasmyth

Notice here how Nasmyth, Phocylides to the south, Wargentin to the west (see also Schickard to the northwest and Schiller to the southeast) all look smooth and flat-like. The reason: possible flooding of lavas within their rims occurred after their formation (the Schiller-Zucchius Basin lies nearby - central coordinates 56S, 45W), which were later covered with ejecta deposits from the Orientale Basin over to the northwest. Nasmyth's floor level is 2 km higher than Phocylides is, but is 1 km lower than Wargentin's. Wargentin lavas may not have over-flowed into Nasmyth, but did Nasmyth's flow into Phocylides's (note the small channel between the two)? Depth 1.1 km.

Craters

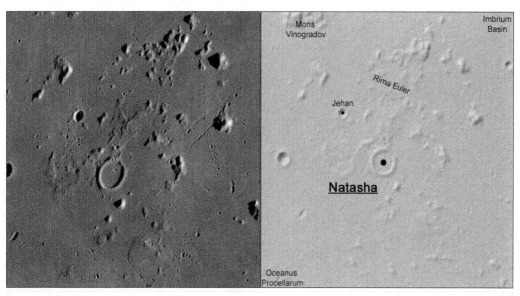

| Natasha | Lat 19.98N | Long 31.16W | 10.98Km |

Sub-craters: No sub-craters

Notes:

Add Info:

Aerial view of Natasha

Natasha (formerly known as Euler P) lies in a region of isolated peaks and mountains possibly connected, at one time, to the much larger set to the southeast (now referred to as Montes Carpatus). The crater hasn't much to offer in terms of significant features, but its bright collar-like rim does show up easily during full moon times. No ejecta pattern at its outer rim sectors is seen, so this is a crater where the series of lavas associated to Mare Imbrium lapped at its shores. Two rille-like features lie closeby: the first, a small, sinuous one to its west just 'kissing' the outer rim; the second, a much broader one (almost looking like a crater itself) to its north that seems to have 'eaten' into Natasha. Rays on its floor are from Copernicus. D: 0.3 km.

Craters

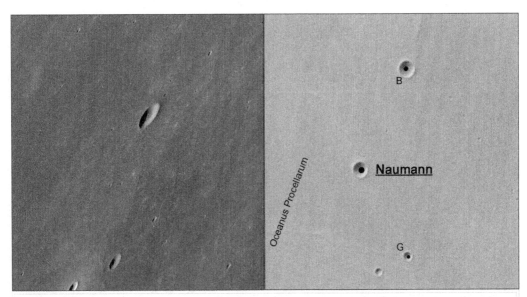

Naumann	Lat 35.38N	Long 62.02W			9.93Km

Sub-craters:

Crater	Lat	Long	Size (km)	Crater	Lat	Long	Size (km)
B	37.46N	60.7W	10.73	G	33.57N	60.73W	6.2

Notes:

Add Info:

Gargantuan Basin

Approximate location of the theorised Gargantuan Basin, which impacted this section of the Moon early on in its history.

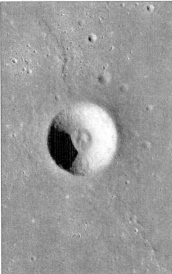

Aerial view of Naumann

Naumann lies in an expanse of mare material related to formation of what today is called Oceanus Procellarum. This sector of the Moon is said to be part of a larger feature usually referred to as the 'Gargantuan Basin' - a 3200 km-wide basin encompassing several of the main mares of the Nearside. The Basin's true existence is a controversial topic; where the mineralogical makeup of such a basin doesn't match up directly in this distribution (other factors come into play as well e.g. topographic details etc.,). One particular feature seen in most basins on the Moon is the tectonic signatures of faulting and the compressional series of wrinkle ridges, which occur when materials (lavas) try to occupy a space smaller than their volumes. Naumann lies on a wrinkle ridge, but note (at low sun times particularly) others nearby; and see how they may relate to the proposed Gargantuan Basin. Depth ~ 1.8 km approx..

Craters

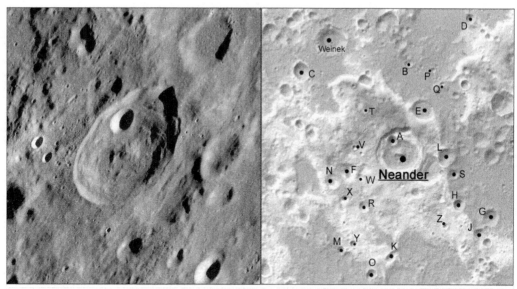

| Neander | Lat 31.35S | Long 39.88E | 49.22Km |

Sub-craters:

Crater	Lat	Long	Size (km)	Crater	Lat	Long	Size (km)
A	30.95S	39.58E	10.7	B	28.26S	40.09E	9.33
C	28.67S	35.95E	19.64	D	26.58S	42.41E	10.75
E	29.91S	40.7E	24.95	F	32.15S	37.78E	23.43
G	33.4S	43.84E	16.66	H	33.08S	42.47E	13.19
J	34.04S	43.46E	13.52	K	35.05S	39.75E	13.09
L	31.43S	41.8E	18.89	M	34.85S	37.66E	8.6
N	32.47S	37.28E	17.01	O	35.66S	38.97E	12.52
P	28.45S	41.07E	5.48	Q	28.91S	41.46E	5.8
R	33.32S	38.54E	11.92	S	32.01S	42.13E	11.8
T	29.95S	38.41E	8.81	V	31.26S	38.08E	5.32
W	32.31S	38.36E	8.65	X	33.05S	37.76E	7.91
Y	34.6S	38.17E	8.21	Z	33.84S	41.96E	6.11

Add Info:

Aerial view of Neander

Neander is quite easy to find if one remembers to see it as lying on an outer clump of material just outside one of the main rings/rims of the Nectaris Basin to the north. The crater would look more circular in form except that its northern sector has been extended slightly by a major slump event; where nearly all of the wall material has slid down into the floor, and sidled up to a well-rounded peak ~ 1.4 km high. Was sub-crater, Neander A, responsible for the event? Or, was it that as Neander formed on an uneven, heighted clump of material at its northwest, most of this was directed towards the southeast, and onto the floor. D: 3.4 km.

Craters

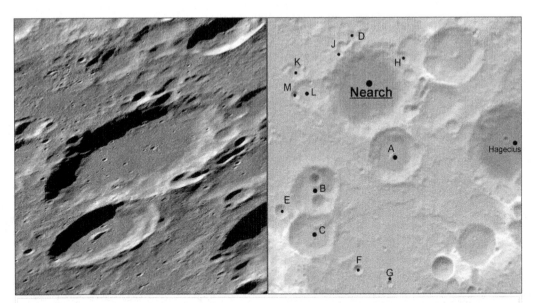

Nearch	Lat 58.58S	Long 39.01E	72.79Km

Sub-craters:

Crater	Lat	Long	Size (km)	Crater	Lat	Long	Size (km)
A	60.2S	40.16E	42.46	B	61.07S	35.91E	40.49
C	62.14S	35.51E	37.58	D	57.17S	38.09E	8.88
E	61.51S	33.95E	10.93	F	63.1S	38.12E	8.24
G	63.46S	39.96E	5.71	H	57.73S	40.58E	9.66
J	57.41S	37.55E	7.93	K	58.03S	35.26E	12.5
L	58.57S	35.6E	23.69	M	58.61S	35.15E	7.51

Notes:

Add Info:

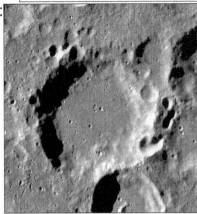

Aerial view of Nearch

Pre-Nectarian in age (4.6 to 3.92 byo), this crater lies in a cluster of other craters similar in appearance, which, at times, will lead to confusion when finding it. Nearch's rim is worn - not only through age as would be expected, but through countless, small-sized impactors that 'chipped' away at every sector. Less obvious is the influence of ejecta by sub-crater, Nearch A, whose impact has imparted partial amounts onto Nearch's floor. This ejecta, together with wall material from Nearch, has 'weathered' down over time on to Nearch's levelled floor - made up of deposits from Nectaris to the north and other main impacts nearby. D: 4.2 km.

Craters

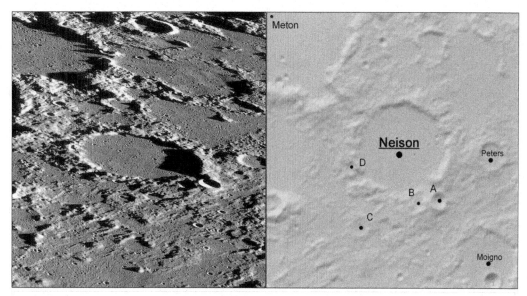

Neison   Lat 68.21N   Long 25.02E   51.03Km

Sub-craters:

Crater	Lat	Long	Size (km)	Crater	Lat	Long	Size (km)
A	67.36N	26.71E	8.93	B	67.32N	25.81E	7.37
C	66.94N	23.03E	9.92	D	67.9N	22.48E	6.64

Notes:

Add Info:

Not to be confused with similar-sounding Nielsen.

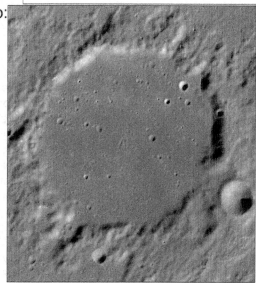

Aerial view of Neison

Lying in an area where nearly all of the craters nearby have been filled with ejecta deposits by the impact that produced the Imbrium Basin to the south, Neison is easily recognised if one remembers to use sub-crater, Neison A, at its southeast as a location marker. A crater of the pre-Nectarian period (4.6 to 3.92 byo), nearly all its rim and outer ejecta deposits hold eroded and gouged signatures that point back to Imbrium's influence. The eastern rim is some 400 metres higher than its western side; which makes for a nice long peak shadow across its floor during low easterly sun times. Depth 1.6 km.

Craters

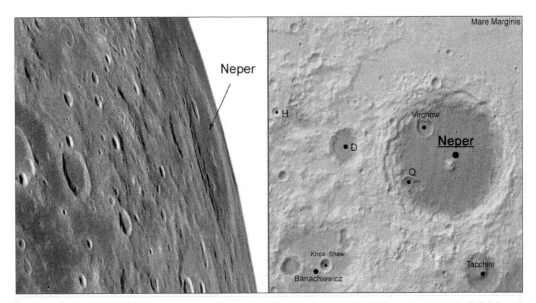

| Neper | Lat 8.76N | Long 84.58E | 144.32Km |

Sub-craters:

Crater	Lat	Long	Size (km)	Crater	Lat	Long	Size (km)
D	9.27N	80.94E	38.21	H	10.41N	78.36E	10.34
Q	8.09N	83.14E	12.88				

Notes:

Add Info:

Aerial view of Neper

Neper is a big crater lying between, and on the ejecta deposits of, two pre-Nectarian-aged mares: Mare Smythii to its south and Mare Marginis to its north. The crater is slightly younger (Nectarian 3.92 to 3.85 byo), its floor is domed and is filled with similar lavas to those of the mares; which undoubtedly were sourced through fractures caused by the initial impact. Neper's peak is some 2.3 km high, and serves as a good location marker for spotting the crater easily whenever it is suitably 'lit' and a favourable libration is in progress. Crater Virchow (formerly Neper G) on the floor shows signs of a shallow-depth impact. Depth of Neper is around 4.5 km.

Craters

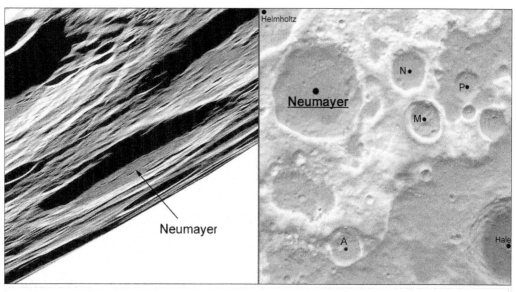

Neumayer   Lat 71.16S   Long 70.92E   84.38Km

Sub-craters:

Crater	Lat	Long	Size (km)	Crater	Lat	Long	Size (km)
A	75.08S	73.88E	34.13	M	71.73S	80.36E	35.44
N	70.54S	78.4E	38.79	P	70.59S	83.49E	23.08

Notes:

Add Info:

Aerial view of Neumayer

Neumayer is a nuisance! Particularly, if you want to see any descent view of it in the eyepiece. Librations, of course, are the best times, but even during such occasions it still proves elusive if the lighting conditions aren't correct (from one or two days before approaching full moon, the crater simply disappears into the bright terrain). The crater lies on ejecta of the outer ring/rim (~ 2500 km in diameter), farside South Pole Aitken Basin (SPA), it has no real distinguishing features to speak of, let alone see, however, there is hint of a chain of small craters across its floor (and mainly across its outer northern sectors) whose source may be that of crater Demonax to its south (~ 220 km away). Depth 3.7 km.

Craters

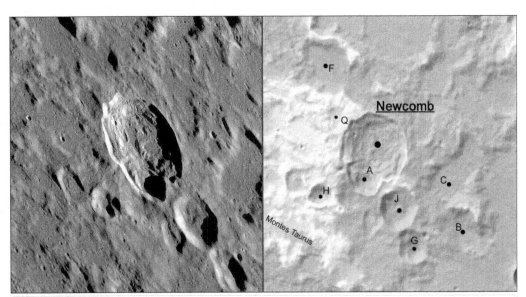

Newcomb   Lat 29.76N Long 43.67E   39.8Km

Sub-craters:

Crater	Lat	Long	Size (km)	Crater	Lat	Long	Size (km)
A	29.22N	43.54E	15.54	B	28.37N	45.53E	23.26
C	29.19N	45.31E	21.77	F	31.42N	42.63E	30.39
G	28.0N	44.53E	16.33	H	28.95N	42.5E	14.9
J	28.72N	44.22E	22.93	Q	30.3N	42.87E	15.66

Notes:

Add Info:

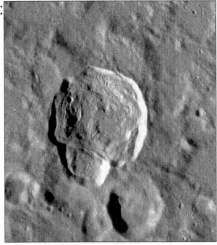

Aerial view of Newcomb

Newcomb is situated on an isolated clump of terrain that is probably original highlands; and not ejecta deposits of, say, Mares' Crisium or Serenitatis to the southeast and west respectively. Obvious to Newcomb is the large, odd crater-like feature (labelled A) attached at its southwest. It looks initially like a major slump event has occurred, where the volume of material wasn't exactly directed towards Newcomb's central floor region, but more to its western, inner rim sector. Impact of sub-crater J to the southeast doesn't look like it caused the slump, as it is an older crater than Newcomb (Eratosthenian in age). D: 2.2 km.

Craters

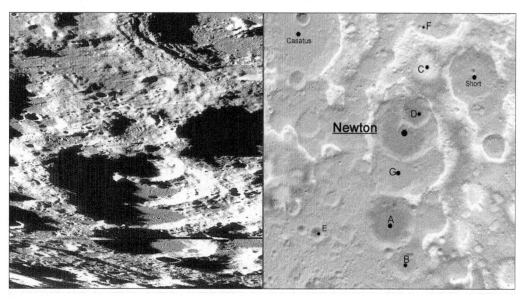

Newton	Lat 76.52S	Long 17.44W					83.85Km

Sub-craters:

Crater	Lat	Long	Size (km)	Crater	Lat	Long	Size (km)
A	80.05S	20.82W	62.26	B	81.45S	16.72W	45.98
C	74.44S	14.23W	35.97	D	76.09S	15.32W	34.82
E	79.86S	37.36W	16.2	F	73.06S	15.14W	6.32
G	78.08S	18.96W	62.47				

Notes:

Add Info:

Aerial view of Newton

It's a good thing that Newton has two distiguishing crater features (impact of Newton D, and Newton's own southern rim lying on Newton G) for finding it, as lack of thereof would cause for a lot of confusion. The crater exhibits more shadows than light whenever in view; not simply because of its location, but that many of the outer deposits (ejecta, highlands or otherwise) at its easterly, westerly sectors have high elevations (east ~ 7 km high, west ~ 5 km). Is that a peak on Newton's central floor, or is it part of ejecta-spat from impact of D? Was D also responsible for the large slumplike feature to its west? D: 5.5 km.

Craters

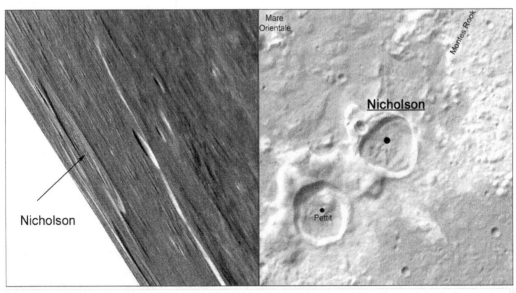

| Nicholson | Lat 26.22S | Long 85.21W | 38.08Km |

Sub-craters: No sub-craters

Notes:

Add Info:

Like Pettit, Nicholson formed on the ejecta deposits from that of the Orientale Basin to their west. Did this relatively 'recent' (Early Imbrian), and possibly weaker, deposit reflect upon how slumping occurred in both craters?

Aerial view of Nicholson

Nicholson lies on one of Orientale's main rings/rims - known as the Montes Rook (the other main ring/rim - Montes Cordillera lies just outside it some 80 km away). At times of favourable libration under suitable lighting conditions, the two rings/rims are obvious, but any detail from Nicholson is extremely hard to see (the big clump/mountain at its outer southern region, and the small 9 km-sized crater on its northwest rim act as good location markers). The crater's floor is virtually non-existant, as nearly all is filled in with slumped deposits from its eastern wall, the clump at its south, and the beginnings of a slump at its northwest (was the small crater re-responsible for initiation?). Nicholson is Eratosthenian in age. D: ~ 3.6 km.

Craters

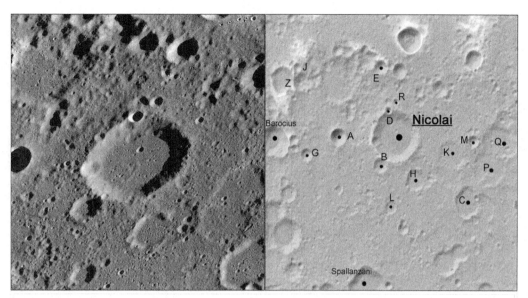

Nicolai		Lat 42.47S	Long 25.87E				40.54Km
Sub-craters:	Crater	Lat	Long	Size (km)	Crater	Lat	Long	Size (km)

Crater	Lat	Long	Size (km)	Crater	Lat	Long	Size (km)
A	42.47S	23.6E	13.71	B	43.21S	25.26E	12.39
C	44.12S	28.9E	25.64	D	41.78S	25.58E	5.75
E	40.68S	25.26E	12.21	G	42.9S	22.32E	10.49
H	43.57S	26.72E	17.66	J	40.65S	22.04E	7.77
K	42.89S	28.09E	22.23	L	44.29S	25.7E	10.79
M	42.52S	29.0E	10.2	P	43.22S	29.63E	28.16
Q	42.53S	30.01E	29.42	R	41.56S	25.87E	5.46
Z	40.96S	21.47E	22.13				

Notes:

Add Info:
Nicolai may also lie between two rings/rims (500, 700 km in diameter) of a proposed basin - the Mutus-Vlacq Basin - whose centre is at 52S, 21E approximately (southeast of crater Baco).

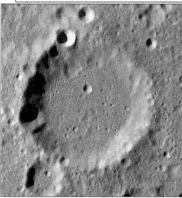
Aerial view of Nicolai

Remember to use sub-crater, Nicolai B, as a location marker for finding this crater (it can get confusing with so many similarly-sized craters around). Nicolai looks relatively fresh for a crater of the pre-Nectarian (4.6 to 3.92 byo), but as it lies in a region where ejecta from the Nectaris Basin (to the northeast) lies, its formation probably borders just before the Nectarian. Nearly all of its rim and inner walls are pocmarked with small impact craters, and while its floor is filled with deposits from the above basin, those from the Imbrium Basin (~ 2600 km to the northwest) fill it, too. Depth ~ 3.6 km.

Craters

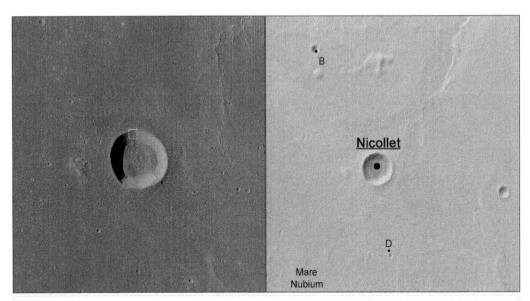

| Nicollet | Lat 21.95S | Long 12.5W | 14.73Km |

Sub-craters:

Crater	Lat	Long	Size (km)	Crater	Lat	Long	Size (km)
B	20.15S	13.57W	4.19	D	23.26S	12.27W	2.23

Notes:

Add Info:

Dark layer deposits at Nicollet's northern inner rim

Nicollet has obviously impacted upon the mare material of Nubium; as during low sun views its ejecta around its outer rim can easily be seen to overlay it. Also during such times, several wrinkle ridges - produced as a result of underlying thrust faults buckling the overlying mare material - are seen to 'converge' at the crater (the ridges were there before Nicollet formed as the ejecta seems to overlie them). The inner wall of Nicollet all around acts as a bright contrast between the outer mare and the inner floor material under any lighting conditions. Depth ~ 2 km approx..

Craters

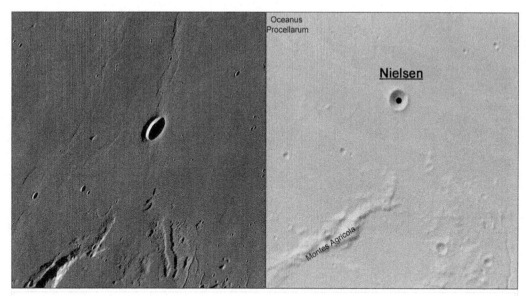

Nielsen   Lat 31.8N   Long 51.76W   9.64Km

Sub-craters: No sub-craters

Notes:

Add Info:

Not to be confused with similar-sounding Neison.

Aerial view of Nielsen

Nielsen looks a bit like crater Wollaston (of similar size and about 130 km away to its east), except its interior seems to have suffered a small slump event at its northeast wall. The crater lies on a wrinkle ridge (caused by underlying thrust faults related to overlying lava developments of Oceanus Procellarum to the west or Mare Imbrium to the east), which shows up wonderfully during low sun views. Like Wollaston, ejecta from the crater can just about be seen, however, more noteworthy is the cluster of small craterlets (all less than a kilometre in size) which are seen on somewhat lighter surface ('gardening' effects) material around the crater. When viewing Nielsen, try also to include Wollaston and Angström (130 km east of Wollaston); as they sometimes can produce a nice contrast in lighting conditions from each. D: 2 km.

Craters

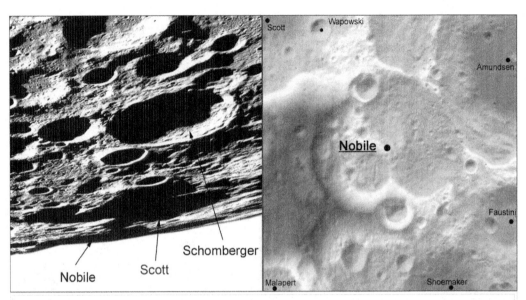

| Nobile | Lat 85.28S | Long 53.27E | 79.27Km |

Sub-craters: No sub-craters

Notes:

Add Info:

Aerial view of Nobile (and surrounds)

There are very few occasions when all the correct lighting conditions, librations etc., are just right to view Nobile. A good time may be during a full moon time when shadows show up, ever-so slightly, in some of the most infamous craters - Amundsen, Scott, Shoemaker, Faustini...nearby at the South Pole. It will never be easy finding Nobile as it is hidden, somewhat, behind mountain terrain and rims of other craters close-by, but a good guide is to firstly find the small bright crater of Wapowski on the southern rim of Scott, and then look just a little to its southwest for a sliver-of-a shadow behind a small mountain that is, in reality, Nobile's northern rim. Use this same location-marker during other favourable times as it is really a challenge to find (be sure to take photos, too). Depth ~ 3.7 km approx..

Craters

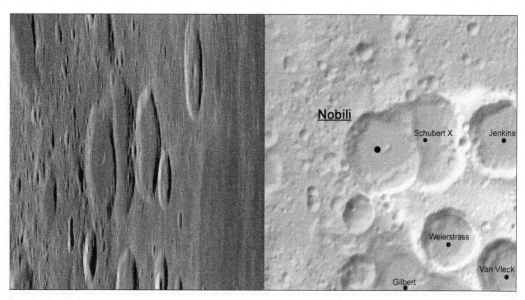

| Nobili | Lat 0.17N | Long 75.95E | 41.79Km |

Sub-craters: No sub-craters

Notes:

Add Info:

Aerial view of Nobili

Not to be confused with crater Nobile in the south, Nobili lies just on the western edge of a hilly clump of material, which is possibly ejecta from the Smythii Basin over to its east. Formerly known as Schubert Y, Nobili has obviously impacted upon Schubert X's western rim, but the ejecta directed onto X's floor was short-lived as lavas sourced from underneath has covered them over partially (note the the signature of another crater at Nobili's northeastern rim, which it has impacted upon also). Nobili's floor is some 700 metres lower than X's, and similar-sourced lavas there have attempted to bury what appears to be the last remains of a peak in the crater (the peak can just about be seen during most lighting conditions). A favourable libration is recommended for viewing Nobili. Depth ~ 3.8 km approx..

Craters

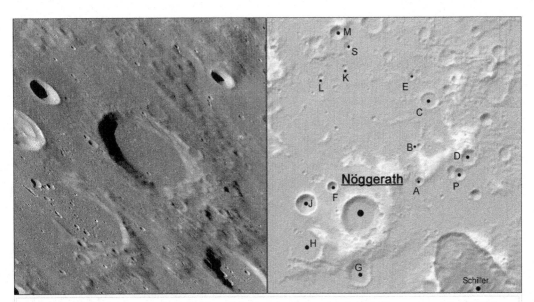

Nöggerath	Lat 48.82S	Long 45.84W	32.12Km

Sub-craters:

Crater	Lat	Long	Size (km)	Crater	Lat	Long	Size (km)
A	47.91S	43.52W	6.85	B	47.02S	43.56W	4.38
C	45.83S	43.2W	13.19	D	47.26S	41.63W	13.94
E	45.22S	43.9W	6.17	F	48.07S	46.96W	9.01
G	50.27S	45.94W	23.15	H	49.57S	48.07W	25.67
J	48.48S	48.01W	17.15	K	44.97S	46.39W	4.35
L	45.2S	47.3W	5.09	M	44.08S	46.68W	10.63
P	47.69S	41.94W	9.93	S	44.45S	46.28W	6.87

Notes:

Add Info:

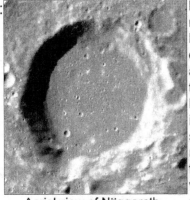

Aerial view of Nöggerath

Nöggerath lies on a clump deposit most likely related to ejecta from the Schiller-Zucchius Basin (whose centre lies some 200 km away due south), and upon the floor of an old, unnamed crater nearly twice its own diameter (note Nöggerath's own ejecta lying on this floor). Having a depth of around 2 kilometres, material in Nöggerath's floor is most likely mare-fill similar to those seen in several other craters nearby (e.g. Schiller to its southeast), and not fluidized ejecta from the Orientale Basin to the west whose signature of striated terrain and secondary impact craters occur in this region.

Craters

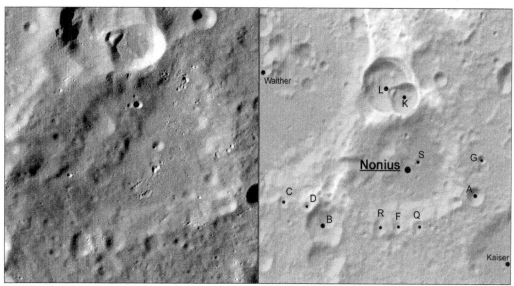

Nonius   Lat 34.9S   Long 3.79E   70.61Km

Sub-craters:

Crater	Lat	Long	Size (km)	Crater	Lat	Long	Size (km)
A	35.39S	5.54E	10.18	B	35.9S	2.03E	23.91
C	35.49S	1.11E	6.32	D	35.57S	1.67E	5.49
F	35.93S	3.74E	6.99	G	34.76S	5.67E	4.83
K	33.73S	3.91E	18.55	L	33.51S	3.52E	30.91
Q	35.93S	4.22E	6.44	R	35.94S	3.3E	8.28
S	34.82S	4.21E	4.05				

Notes:

Add Info: Rays from Tycho over to the west of this crater overlie many craters here, and Nonius didn't escape the onslaught. One or two very small secondary crater chains from that event lie on the floor, but high mag., is required.

It certainly gets confusing, and 'mess-like', when it comes to figuring out how Nonius formed. Initially, it looks like crater Walther to its northwest has imparted a whole lot of its ejecta onto Nonius's floor - making it the younger of the two (possibly on the pre-Nectarian/Nectarian border). However, all this ejecta isn't quite that of Walther's, as subsequent impacts by sub-crater, Nonius L, and then sub-crater, Nonius K, both to the north, have piled additional amounts on top; extending all this debris into Nonius's floor even further. Look also at Nonius's southern rim, and where it meets the eastern rim at nearly right angles to each other. Why does it look that way? Has the set of small impacts (e.g. sub-craters R, F and Q) taken off the round-like appearance of Nonius there? Or is it that clumpy ejecta from the Imbrium Basin (whose centre lies some 2100 km away to the northwest) that fell in this general region covered over, in part, that of Nonius's southern rim? The best times to observe all these features is during low sun times, as in full moon times the crater is hard to see, with only L and K giving hint of its location. Depth is around 3 km approx..

Craters

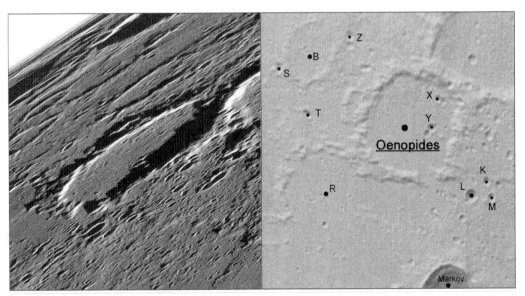

Oenopides   Lat 57.13N   Long 64.2W   73.47Km

Sub-craters:

Crater	Lat	Long	Size (km)	Crater	Lat	Long	Size (km)
B	58.56N	68.93W	36.98	K	55.83N	61.19W	6.39
L	55.55N	61.85W	9.2	M	55.48N	61.02W	6.21
R	55.6N	67.89W	61.99	S	58.19N	70.15W	6.57
T	57.25N	68.81W	7.05	X	57.64N	63.16W	5.15
Y	57.02N	63.48W	6.15	Z	58.96N	67.1W	6.9

Add Info: Notes:

Day-old moon phases

Note: As the Moon goes through various librations throughout the year, suggested times given in text for observations are only approximates.

Aerial view of Oenopides

Like several craters nearby e.g. Babbage or South - both lying eastwards, Oenopides has a 'square-like' look to it. The cause, of course, has to do with the gouging effect of ejecta clumps, produced by formation of the Imbrium Basin to the crater's southeast, 'bevelling' down rim areas perpendicular to its impact's main centre-point. The crater has an easily recognised 'nick'- possibly a small impact created it - at its southeast corner, and its floor holds several, small-sized secondary craters related to Pythagoras at the north. Full moon times, the crater is hard to see, so observe when the terminator isn't far off.

Craters

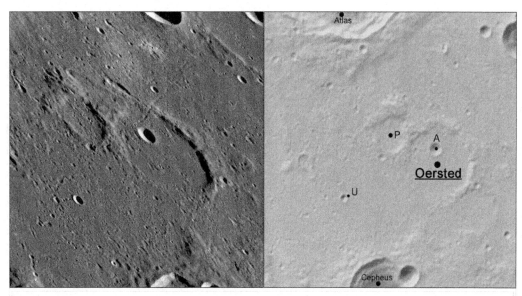

Oersted	Lat 43.09N	Long 47.25E	42.28Km

Sub-craters:

Crater	Lat	Long	Size (km)	Crater	Lat	Long	Size (km)
A	43.35N	47.18E	7.21	P	43.61N	45.91E	20.9
U	42.41N	44.72E	4.09				

Notes:

Add Info:

Aerial view of Oersted

Clementine data show that Oersted lies just within the outer sector of a buried basin ~ 400 km in diameter, centred on crater Shuckberg's (over to the east) southeastern rim. A bigger perspective view will show the extent of the basin's surrounds, whose lava-filling event (3.74 to 3.62 byo) has partially buried not only Oersted but several other (Oersted P, Chevallier) unnamed craters (try spot them) nearby. Oersted was later subjected to another lighter dusting of ejecta from craters Atlas and Cepheus (Eratosthenian and Upr. Imbrian in age respectively), and its northeastern rim was affected by a chain of small craters (what was there source?). Depth 1km.

Craters

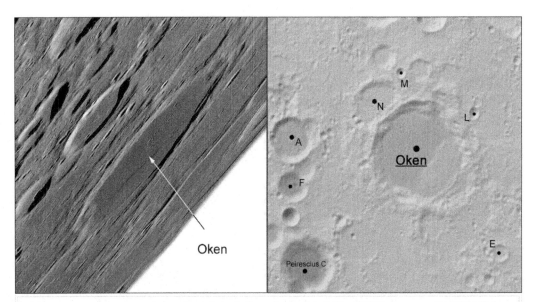

Oken

| Oken | Lat 43.76S | Long 76.09E | 78.67Km |

Sub-craters:

Crater	Lat	Long	Size (km)	Crater	Lat	Long	Size (km)
A	43.34S	71.41E	35.45	E	46.22S	79.25E	13.21
F	44.49S	71.38E	24.63	L	42.79S	78.14E	9.25
M	41.8S	75.48E	8.03	N	42.6S	74.67E	38.23

Notes:

Add Info:

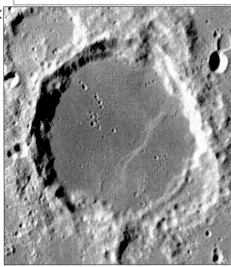

Aerial view of Oken

Oken lies in a relatively dark sector of the moon's southeast, where the Australe Basin lavas have flooded practically most of the regions and features therein. Good librations, of course, are recommended, however, even during such occassions, it can still be hard to identify as several other craters of similar size may lead to confusion. The crater has a wonderful ridge (or is it a scarp?) running across its floor - the southeastern side (platform) of which is some 500 metres higher up. Did the, what looks like a, major slump at its southeast rim give cause to produce the ridge-scarp feature? Depth ~ 2.6 km.

Craters

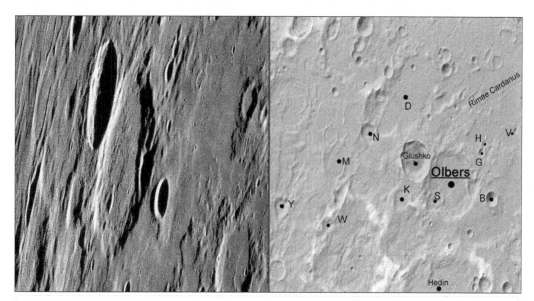

Olbers Lat 7.3N Long 76.14W 73.02Km

Sub-craters:

Crater	Lat	Long	Size (km)	Crater	Lat	Long	Size (km)
B	6.84N	74.21W	16.36	D	10.23N	78.03W	100.03
G	8.4N	74.67W	9.01	H	8.69N	74.55W	7.46
K	6.81N	78.28W	23.77	M	8.0N	81.18W	39.78
N	9.03N	79.82W	21.1	S	6.83N	76.68W	16.23
V	9.13N	73.12W	6.39	W	5.89N	81.61W	18.28
Y	6.46N	83.73W	20.16				

Notes:

Add Info:

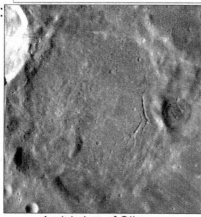

Aerial view of Olbers

Low sun, terminator times (a day or two before full or new moon) are best for observing Olbers (other times it gets hard to see). The crater has a depth of about 2 km, its floor is slightly hummocky to convexed, and several fractures occur mainly in its eastern sector (good librations show them easily). Crater Glushko (Copernican in age ~ 1.1 byo to present) over to its west has just 'clipped' Olbers's northeastern rim. Its ejecta may be responsible for the above-mentioned hummocky, convexed floor, and possibly why no fractures are seen in the western sector as they may have been buried by this material. Note Glushko's rays.

Craters

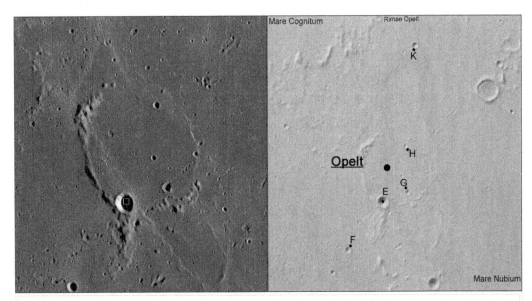

Opelt	Lat 16.32S	Long 17.64W	48.73Km

Sub-craters:

Crater	Lat	Long	Size (km)	Crater	Lat	Long	Size (km)
E	17.04S	17.91W	7.38	F	18.09S	18.8W	4.01
G	16.85S	17.29W	3.73	H	15.82S	17.34W	2.71
K	13.62S	17.09W	4.19				

Notes:

Add Info:

Close-up showing boulder trails in Opelt E

Opelt lies in a 'splotchy' region of Mare Nubium where basaltic activity and magma eruptions - from Early Imbrian to Late Eratosthenian in age - gave rise to those signatures we see today. Their effects led to partial flooding of Opelt (particularly in the east), followed by tectonic 'crumpling' at its western sector where a wrinkle ridge crosses the crater (note, in general, the NW-SE trend of other ridges, and faults - e.g. Rupes Recta over to its east, in Opelt's surrounds). The bright wall material of sub-crater Opelt E acts as a good location marker during full moon times for finding Opelt. Depth ~ 0.3 km.

Craters

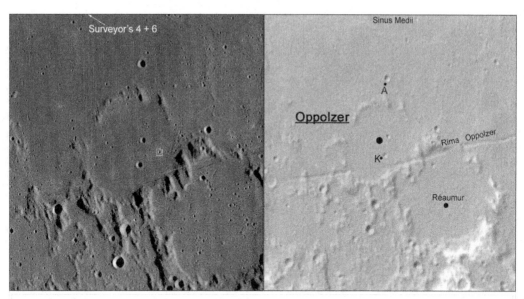

Oppolzer	Lat 1.52S	Long 0.45W	40.87Km

Sub-craters:

Crater	Lat	Long	Size (km)	Crater	Lat	Long	Size (km)
A	0.49S	0.35W	3.31	K	1.72S	0.38W	2.64

Notes:

Add Info:

Surveyor's 4 and 6 landed some 50 km away northwards of Oppolzer in 1967 (17 July and 10 Nov respectively).

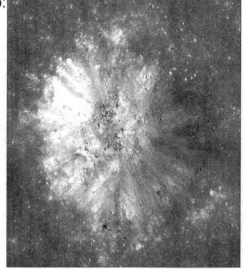

A relatively fresh impact crater on Oppolzer's floor

Oppolzer lies in a trough-like zone; created as a result of it being on the southern edge of Sinus Medii (possibly a buried crater). It also lies within two rings/rims (1700 & 2250 km in diameter) associated to formation of the Imbrium Basin, whose centre is some 1100 km away off to the northwest. This latter event's ejecta has 'grooved' Oppolzer's rim (see how they 'point' back to Imbrium) and several other craters nearby, followed by a series of fluidic lavas that flooded the crater (note possible breach points). A major tectonic, tension event (Rima Oppolzer) occurred later at the crater's southern region, whose extent continues over to Rima Flammarion westwards ~ 240 km away.

Craters

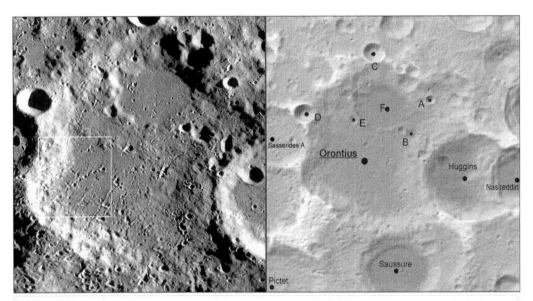

| Orontius | Lat 40.37S | Long 3.96W | 121.02Km |

Sub-craters:

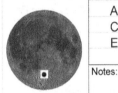

Crater	Lat	Long	Size (km)	Crater	Lat	Long	Size (km)
A	39.12S	2.6W	6.73	B	39.96S	3.16W	8.75
C	37.97S	4.11W	14.52	D	39.41S	6.2W	14.22
E	39.59S	4.82W	7.06	F	39.23S	3.93W	41.41

Notes:

Add Info:

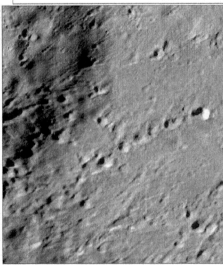

Chain of small impact craters on floor

Upto five relatively big impact craters have changed the shape of Orontius. Huggins to the east is obviously the main culprit, but others likes Sassure, Sasserides A, Orontius F, and an unnamed crater just east of F (on Orontius's northeastern rim) aren't as obvious. Orontius itself has also played the 'impact card', where its effect has wiped out entirely the northern sector of an unnamed crater's rim, whose remaining sectors can just about be seen around Sassure (this crater is just over 100 km in diameter). Tycho to the southwest has 'spat' several, secondary crater chains onto Orontius's floor, while its bright ray material covers the crater entirely. Depth ~ 1.8km approx..

Craters

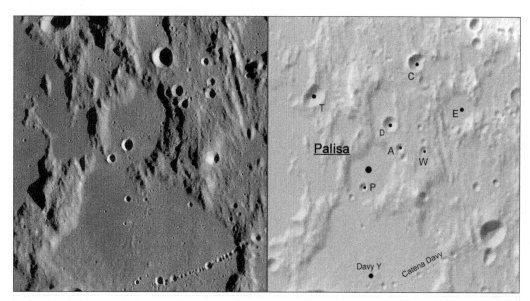

Palisa		Lat 9.47S		Long 7.19W			33.47Km

Sub-craters:	Crater	Lat	Long	Size (km)	Crater	Lat	Long	Size (km)
	A	9.06S	6.74W	4.06	C	7.73S	6.47W	7.87
	D	8.68S	6.91W	7.37	E	8.47S	5.75W	19.94
	P	9.68S	7.37W	4.13	T	8.26S	8.22W	12.08
	W	9.09S	6.33W	4.01				

Notes:

Add Info: Palisa lies in a mid-zoned region of the Moon where: to its west a series of old and younger lavas extend from crater Copernicus (~ 700 km northwestwards away) right down passed the crater to Mare Nubium in the south, while to its east original highlands lunar crust lay subject to huge impacts such as the 153 kilometre-sized crater that is Ptolemaeus. The region has also been at mercy to formation of the Imbrium Basin (centre lies some 1300 km away to the northwest), whose effects has left nearly every pre-existing crater (including Palisa) and feature with a lineated 'stain' – referred usually to as the 'Imbrium Sculpture'. Did huge clumps of ejecta from this major event 'shave-off' Palisa's northwest rim onto its floor (note the 'kink')? Did they stop short as they encountered Palisa's southeast rim, producing what appears to be an abrupt, slump-like deposit feature? The deposits in Palisa's floor (as in Davy Y too) are thus a mix of Imbrium ejecta and lava, which possibly breached not long afterwards into Palisa and Davy Y, as they edged up to their walls (note, as a comparison, the ejecta signatures by crater Davy on Davy Y's southwest floor). A scarp, or is it a crater chain, crosses Palisa's floor in a NW-SE trend, but as to if it is a previous feature or one occurring afterwards is hard to define (was it partially covered?).

Craters

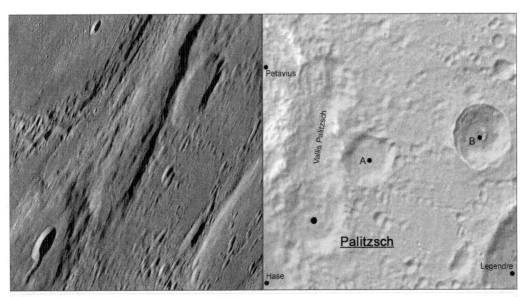

Palitzsch	Lat 28.02S	Long 64.39E	41.87Km

Sub-craters:	Crater	Lat	Long	Size (km)	Crater	Lat	Long	Size (km)
	A	26.96S	65.68E	33.1	B	26.41S	68.39E	37.94

Notes:

Add Info:

Aerial view of Palitzsch

Palitzsch lies east of one of the Moon's most impressive craters - that of Petavius with its distinctive linear-like 'gash' radiating from its centre. Palitzsch is obviosly older than Petavius (Lower Imbrian in age 3.85 to 3.75 byo), and like crater Hase to its south has been subjected to an onslaught of ejecta from the impact. Floor-fractured features occur in Palitzsch (seen in Petavius, too), but these may be due to cooling effects rather than the magmatic uplift effects for Petavius's. The crater makes up the southern end of the Palitzsch Valley, which may be a series of three to four aligned impact craters of similar size.

Craters

Pallas	Lat 5.48N	Long 1.65W	49.51Km

Sub-craters:

Crater	Lat	Long	Size (km)	Crater	Lat	Long	Size (km)
A	5.97N	2.35W	10.05	B	4.2N	2.64W	3.41
C	4.49N	1.12W	5.44	D	2.37N	2.63W	4.1
E	4.03N	1.47W	24.6	F	3.51N	1.39W	18.19
H	4.64N	1.59W	4.82	N	7.02N	0.49E	5.34
V	1.67N	1.59W	2.55	W	3.6N	1.26W	3.47
X	5.14N	3.23W	2.87				

Notes:

Add Info: Pallas is situated in a somewhat isolated clump of material made up of original lunar highlands, which later became subjected to the dynamic, rebound forces associated to formation of the Imbrium Basin; centred 900 km away to the northwest. This major event, with 'loaded' accompanying ejecta, also left its mark in the form of gouged-like striations on several craters and features nearby Pallas (pull back for a bigger perspective view), and altering their shapes forever in the process. Obvious signatures are seen at Pallas's northern rim where three to four of these striations 'point' back to Imbrium, and it looks like most of the rim ended up on Pallas's floor there (was it through actual levelling by ejecta, or slumping effects on Pallas's wall?). Its outer western sector experienced those effects too, leaving an odd, clumpy looking rim composing, possibly, of original rim mixed with Imbrium ejecta. Pallas, of course, has impacted upon Murchison to its east making it slightly younger, and the common border they share has left a breach-point, a series of small rilles, and a rubbly mixture of rims, lavas and possible impact melt. Depth ~ 1.5 km. Peak ~ 1.2 km approx..

Craters

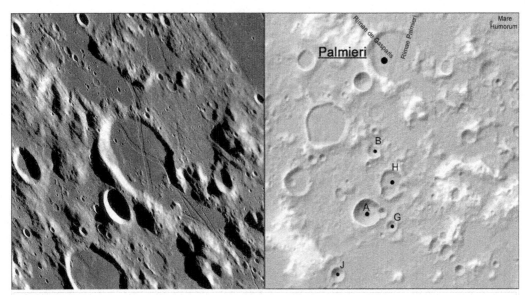

Palmieri		Lat 28.64S	Long 47.8W			39.84Km	

Sub-craters:

Crater	Lat	Long	Size (km)	Crater	Lat	Long	Size (km)
A	32.25S	48.52W	20.85	B	30.84S	48.32W	9.69
E	29.23S	48.63W	14.22	G	32.58S	47.8W	8.88
H	31.56S	47.78W	19.0	J	33.68S	49.42W	10.52

Notes:

Add Info:

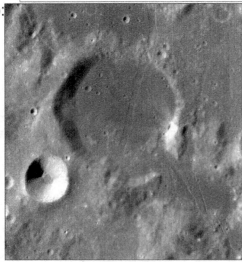

Aerial view of Palmieri

Palmieri lies in a trough-like region between two rings/rims associated with formation of the Humorum Basin to its east. Mare-filling lavas fill the trough, and Palmieri's innards didn't escape as stress fractures, in lieu with cooling effects, in its floor (e.g. Rilles' Palmieri and de Gasparis) testify to possible underlying sources. The rille-cross is easily seen and close-ups show that the Palmieri rille occurred first, followed later by the de Gasparis one (the 'cross' is best seen during low sun times). A small (~ 3.5 km) fresh crater has impacted Palmieri's southeastern rim and wall, leaving a nice ray pattern on its floor. Depth ~ 1 km.

Craters

Parrot   Lat 14.57S   Long 3.29E   70.66Km

Sub-craters:

Crater	Lat	Long	Size (km)	Crater	Lat	Long	Size (km)
A	15.33S	2.08E	20.23	B	13.64S	2.41E	8.33
C	18.57S	1.23E	29.61	D	14.23S	3.6E	22.12
E	16.0S	2.21E	19.79	F	16.12S	1.39E	18.41
G	17.43S	2.54E	28.19	H	17.62S	1.17E	17.62
J	17.03S	1.78E	23.79	K	14.09S	1.58E	38.05
L	18.05S	0.91E	6.37	M	18.0S	1.93E	6.19
N	13.79S	0.43E	4.47	O	16.96S	2.54E	9.41
P	18.74S	2.92E	5.32	Q	15.12S	1.06E	4.71
R	13.53S	3.15E	10.12	S	15.94S	3.55E	9.75
T	15.92S	4.14E	6.84	U	14.02S	4.44E	7.28
V	13.25S	0.8E	27.35	W	13.19S	1.47E	5.03
X	14.53S	1.86E	3.78	Y	13.94S	0.69E	9.79

Notes:

Add Info: Signature of Parrot being an actual crater at all is seen only from hints of its rim, which survived the battering of Imbrium Basin ejecta from the northwest. Huge blocks of the ejecta are presumed to have caused the lineated features (informally referred to as the Imbrium Sculpture) - seen particularly at Parrot's east and south - as they struck sectors of its rim; reducing them down to rubble, melt and dust (see also how small secondary craters e.g. Parrot B and S or R and T produced by the ejecta align up with the lineations). Parrot shares a border with Nectarian-aged (3.92 to 3.85 byo) crater, Albategnius to its north, but as to which formed first is a tough call (does the border show Albategnius's or Parrot's rim, does Albategnius's ejecta overlie Parrot?). Depth 2 km.

Craters

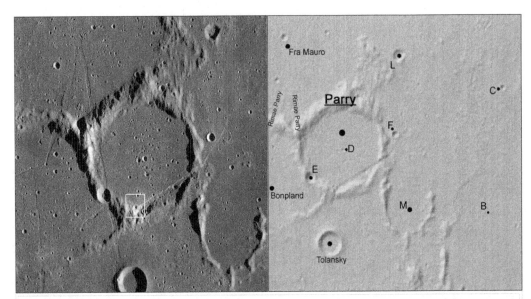

| Parry | Lat 7.88S | Long 15.78W | 47.28Km |

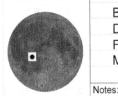

Sub-craters:

Crater	Lat	Long	Size (km)	Crater	Lat	Long	Size (km)
B	8.93S	13.06W	1.14	C	6.86S	12.74W	3.13
D	7.93S	15.73W	2.19	E	8.4S	16.35W	5.71
F	7.66S	14.76W	3.57	L	6.32S	14.7W	6.32
M	8.91S	14.53W	25.22				

Notes:

Add Info:

Apollo 14 landed some 140 km away north of this crater, on 5 Feb 1971 (astronauts: A. Shepard, S. Roosa & E. Mitchell).

Extreme close-up of fresh impact crater on Parry's rim

Though Parry, and its neighbouring crater, Bonpland to its west, lack the more rubbly textured material (Imbrium ejecta) of Fra Mauro's western sectors; all three connect, in a way, through the wonderful rille system that is Rimae Parry. One of these rilles 'cuts' Parry's northwest rim, as well as its floor, which means the feature occurred some time after when the crater formed (sub-crater Parry E wasn't affected, so it is younger still). Southwards, another rille 'cuts' Parry's rim there (shows up wonderfully around last quarter), while a fresh impact crater has left a nice ray display across the floors of Parry and Bonpland.

Craters

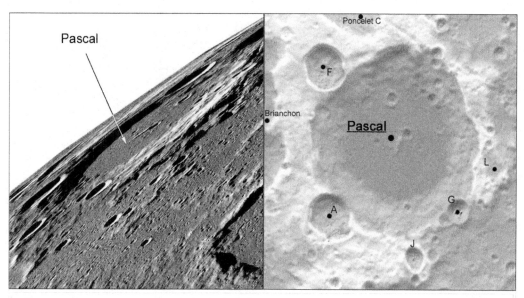

Pascal	Lat 74.36N	Long 70.63W	108.2Km

Sub-craters:

Crater	Lat	Long	Size (km)	Crater	Lat	Long	Size (km)
A	72.92N	75.04W	29.31	F	75.66N	76.21W	27.47
G	73.01N	66.24W	13.65	J	72.18N	69.35W	13.86
L	73.75N	63.56W	18.47				

Notes:

Add Info:

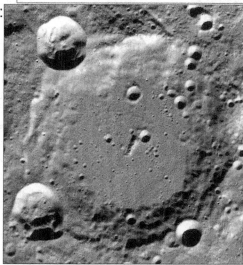

Aerial view of Pascal

Planning for observations of Pascal is an essential requirement to catching any descent view of the crater. Libration times are therefore necessary, however, if the lighting conditions aren't in lieu with them, then try again (a full moon with suitable libration works, too). The crater has all the signs of an old crater - levelled smooth terraces, half-buried peak (~ 700 metres high), numerous impact craters (big and small) on its rim and floor, while striations on its northeastern rim 'point' back to the Imbrium Basin in the southeast whose influence is extent in this general area. Depth ~ 4.6 km.

Craters

| Peary | Lat 88.63N Long 24.4E | 78.75Km |

Sub-craters: No sub-craters

Notes:

Add Info:

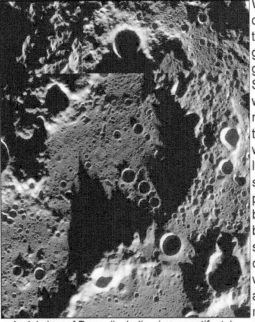

Aerial view of Peary (including image artifacts)

When you start 'hitting' those craters and features on the extreme limb, patience, planning, good librations, and suitable lighting conditions are the norm. Some craters and features, however, will pose serious challenges - Peary is one of them (at times of full moon, waxing and waning also, in lieu with a good libration shows up the crater easily). The crater is aged of the pre-Nectarian Period (4.6 to 3.93 byo), its rim has been battered by countless impactors (big and small secondaries from formation of the Imbrium Basin southwards), while its border, with crater Florey to its southwest, share common ejecta deposits from that same event. Depth 2.7 km.

Craters

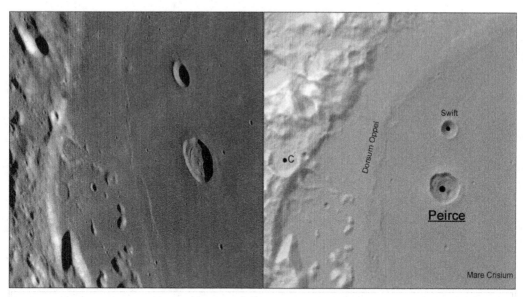

Peirce	Lat 18.26N	Long 53.35E	18.86Km

Sub-craters:	Crater	Lat	Long	Size (km)	Crater	Lat	Long	Size (km)
	C	18.75N	49.89E	19.82				
	Notes:							

Add Info:

Luna 15 impacted in Crisium some 160 km away east of this crater, while southeastwards, Luna 23 and Luna 24 landed some 340 km away, too.

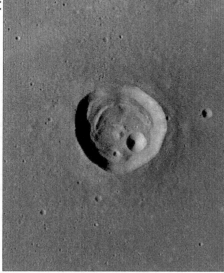

Aerial view of Peirce

Eratosthenian in age (3.15 to 1.1 byo), Peirce lies on somewhat thicker-based lavas (~ 3 km) than those that lie, and lap up, towards Crisium's western rim. Ejecta surrounding Peirce can just about be seen to overlie (mix in with?) the lava plain in the basin; where time, and surface makeup, has left it with a worn down, mottled appearance. Peirce's rim is sharpish all around and is some ~ 500 metres above the plain level. It has two small peaks (the highest is ~ 150 metres) in the centre, and it looks like there were two major slumping events: at the crater's northern rim (firstly), followed (secondly) by another at the rim's northwest. Did the ~ 4-kilometre-sized crater in the floor's southeast set the slumps off? Depth ~ 2 km approx..

Craters

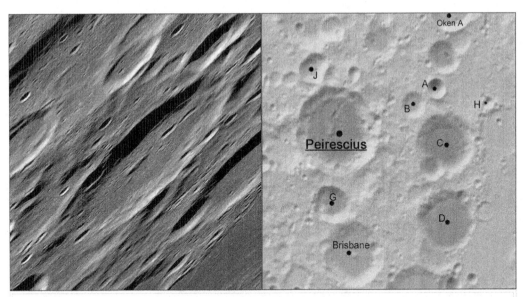

Peirescius	Lat 46.4S	Long 67.8E			61.53Km

Sub-craters:

Crater	Lat	Long	Size (km)	Crater	Lat	Long	Size (km)
A	45.23S	71.24E	14.19	B	45.67S	70.48E	18.84
C	46.47S	71.73E	44.03	D	48.17S	72.11E	42.8
G	48.1S	67.78E	28.09	H	45.4S	73.11E	8.73
J	45.09S	66.75E	17.59				

Notes:

Add Info: The chain that runs from Peirescius to Oken A 'points' to Marinus B in the northeast, but that crater is too old to be the chain's source. Would the chain be concentric secondaries from farside crater, Schrödinger (74.73S, 132.93E)?

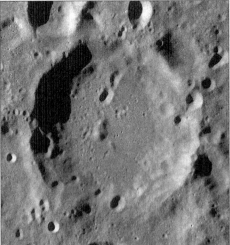

Aerial view of Peirescius

Peirescius can be a hard crater to locate, not because of its limb location but that several similar-sized craters 'in front' and 'behind' it (from an earth perspective) leads to some confusion. Aged of the pre-Nectarian Period (4.6 to 3.92 byo), the crater has been subjected to numerous impacts from basins like Nectaris to the northwest and big craters like Petavius some 600 km away northwards. The crater has two peaks (main one is some 600 metres high), and together with the two small craters (part of a chain that runs upto and beyond Oken A) northeastwards of them, they act as good markers for recognising the crater. Depth 3.7 km.

Craters

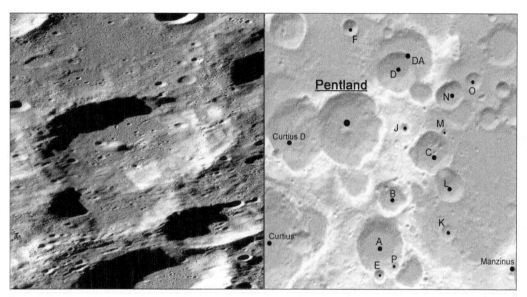

Pentland	Lat 64.57S	Long 11.34E	56.45Km

Sub-craters:

Crater	Lat	Long	Size (km)	Crater	Lat	Long	Size (km)
A	67.32S	13.25E	44.01	B	66.18S	13.93E	28.21
C	65.06S	16.43E	32.13	D	63.22S	13.96E	31.3
DA	62.93S	14.37E	52.4	E	68.09S	13.35E	11.04
F	62.15S	11.31E	12.21	J	64.52S	14.63E	8.49
K	66.92S	17.83E	11.68	L	65.85S	17.77E	22.29
M	64.55S	17.06E	6.24	N	63.65S	17.19E	23.48
O	63.26S	18.51E	13.98	P	67.85S	14.35E	8.11

Notes:

Add Info:

Aerial view of Pentland

Situated in a tight cluster of craters - mostly secondaries from Imbrium, confusion reigns when finding Pentland. The crater has formed on the rim of Curtius D to its west (note ejecta on its floor) and also on an unnamed crater to its south (just west of Pentland B). Having a small peak of around 900 metres high, there also appears to be signs of a possible lobate scarp on the floor. Such scarps occur due to shrinkage of the Moon where the crust is thrust upwards and over another section, resulting in a stair-step like feature (the 'scarp?' extends into Pentland DA and right up to Kinau D in the northeast). Depth ~ 4 km.

Craters

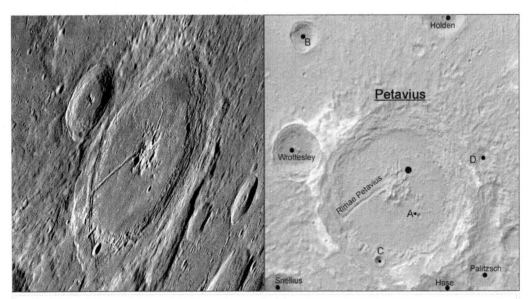

Petavius	Lat 25.39S	Long 60.78E	184.06Km

Sub-craters:

Crater	Lat	Long	Size (km)	Crater	Lat	Long	Size (km)
A	26.15S	61.64E	6.26	B	19.9S	57.0E	31.95
C	27.72S	59.97E	11.19	D	24.04S	64.31E	19.5

Notes:

Add Info:

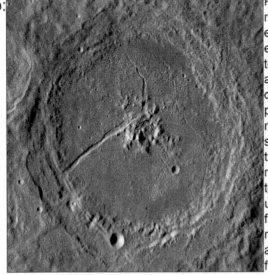

Aerial view of Petavius

Perhaps one of the most favourite craters on the Moon for observing, Petavius offers the viewer everything from a wreath of terraces, several well-defined peaks in its centre, dark mare-like deposits (at 1, 5 and 10-o-clock positions), and several rilles. The main rille at 8-o-clock position is some 4.6 km in width as it 'cuts' the peaks, joining another wide rille starting in the north and down to the southeast where it breaks up into several smaller ones. At full moon times the crater's features get a little too bright (but worth a look), however, a day or two before first quarter, or a day or two after full moon work best. D: 3km.

Craters

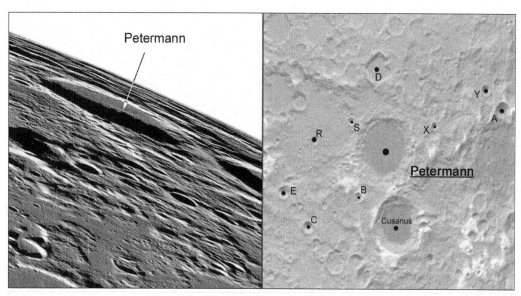

Petermann Lat 74.35N Long 67.89E 76.95Km

Sub-craters:

Crater	Lat	Long	Size (km)	Crater	Lat	Long	Size (km)
A	74.85N	87.63E	18.34	B	72.77N	63.91E	10.34
C	71.51N	57.34E	13.11	D	77.15N	66.52E	33.57
E	72.46N	53.26E	14.53	R	74.78N	55.61E	118.38
S	75.29N	62.04E	9.48	X	75.02N	76.1E	8.95
Y	75.71N	85.69E	12.17				

Notes:

Add Info:

Aerial view of Petermann

Petermann is situated just off one of the rings/rims (1050 km in diameter) associated to Mare Humboltianum (56.92N, 81.54E) in the southeast. Nectarian in age (3.92 to 3.85 byo), the crater's floor is filled with possible Imbrium Basin (~ 1700 km over to the southwest) ejecta, followed by a cluster of small secondary craters in its eastern sector. The rim, and general surrounds, shows signatures of striation effects 'pointing' back to Imbrium, however, one main ejecta valley passing from Petermann to Petermann E has a different orientation, but what was its source. D: 3.3 km.

Craters

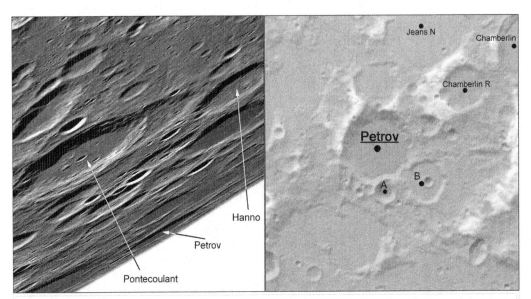

Petrov	Lat 61.36S	Long 88.18E		55.44Km			
Sub-craters:	Crater	Lat	Long	Size (km)	Crater	Lat	Long	Size (km)
	A	62.45S	88.62E	16.96	B	62.2S	90.81E	30.75
	Notes:							

Add Info:

Aerial view of Petrov

Librations are the norm requirements for observing some of the extreme limb-hugging craters. In observing Petrov, however, one would need nearly all librational effects - latitudinal (the 'nodding' libration), longitudinal (the 'wobble' libration), and diurnal libration (where the position of an observer on the Earth allows more of the moon's eastern limb to be seen during moonrise, and more of the western limb during moonset) - to be 'going-on' at one time. The crater lies on the sparse remains of ejecta related to an outer ring-rim system created by formation of the South Pole Aitken Basin on the farside. Depth ~ 2 km approx..

Craters

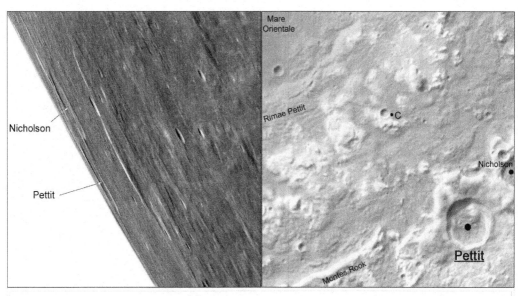

Pettit	Lat 27.52S	Long 86.75W	36.67Km

Sub-craters:	Crater	Lat	Long	Size (km)	Crater	Lat	Long	Size (km)
	C	24.88S	89.07W	7.29				
	Notes:							

Add Info:

Not to be confused with the 5.04 km crater, Petit (formerly called Apollonius W), found at 2.32N, 63.46E.

Aerial view of Pettit

Like its neighbour, Nicholson, to the northeast, Pettit is a crater of the Eratosthenian Period (3.15 to 1.1 byo). Both craters show signs of slump effects in their interiors, however, from the state of Nicholson where it looks like Orientale ejecta - on which both craters formed - has slid down into that crater, would it be possible that Pettit's impact caused the event (implying it as being younger than Nicholson). Pettit will prove somewhat difficult to observe if libration periods are ignored, however, as the western rim section is some 2 km higher than its eastern section, this 'mountain' acts as a good marker for finding the crater. D: 4 km.

Craters

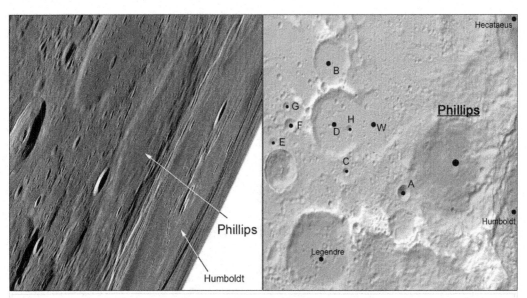

Phillips   Lat 26.57S   Long 75.67E   104.21Km

Sub-craters:

Crater	Lat	Long	Size (km)	Crater	Lat	Long	Size (km)
A	27.15S	73.59E	14.82	B	23.21S	70.59E	41.08
C	26.51S	71.16E	7.52	D	25.04S	70.8E	61.88
E	25.6S	68.12E	10.35	F	25.09S	68.83E	11.93
G	24.55S	68.72E	8.48	H	25.25S	71.31E	7.56
W	25.15S	72.23E	61.65				

Notes:

Add Info:

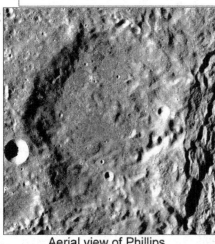

Aerial view of Phillips

Lying to the west of the ever-wonderful crater that is Humboldt, Phillips never really gets a look-in. It's obviously older than Humboldt (Upper, Late Imbrian - 3.75 to 3.2 byo), as ejecta from that major impact event lies within its floor (whose level is some 400 metres above that of Humboldt's). The crater's south-eastern inner rim appears to have a cluster of small craters aligned with orientation towards the northeast. A bigger perspective view, with Humboldt in the picture, shows the cluster may be part of the original chain of craters known as Catena Humboldt. Depth ~ 3.2 km.

Craters

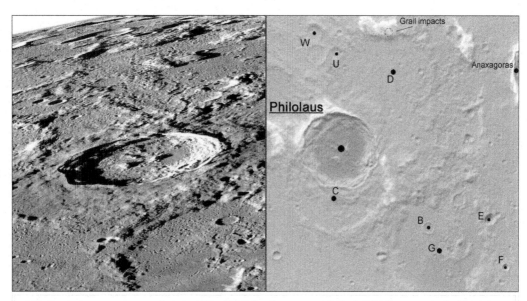

Philolaus   Lat 72.22N   Long 32.88W   71.44Km

Sub-craters:

Crater	Lat	Long	Size (km)	Crater	Lat	Long	Size (km)
B	69.71N	24.47W	10.79	C	71.23N	32.82W	98.14
D	74.37N	27.66W	91.35	E	69.64N	18.79W	10.96
F	68.11N	18.36W	7.1	G	69.18N	23.82W	102.68
U	75.14N	33.21W	12.46	W	75.72N	36.09W	16.04

Notes:

Add Info:

NASA's two spacecrafts (Ebb & Flow) of the GRAIL mission impacted the surface (75.62N, 26.62W) on 17 Dec 2012 - some 115 km away north of this crater.

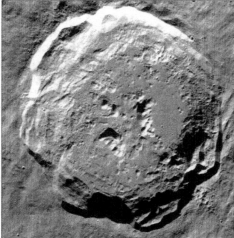

Aerial view of Philolaus

Philolaus lies in an area where much of the terrain is consistant in texture. Ejecta, in various states of consolidation, from the Imbrium Basin event to the south, is to blame, while other later impacts, of a lesser extent, have had their share. The crater has a wonderful series of peaks (highest is the eastern-most one coming in at ~ 1.1 km), sharp-looking terraces (the northwest ones are slightly more prominent - hint, perhaps, of an oblique impact) and rim all around with puddley-looking ejecta at its exterior. The crater has formed upon the old crater that is Philolaus C. Depth ~ 4 km.

Craters

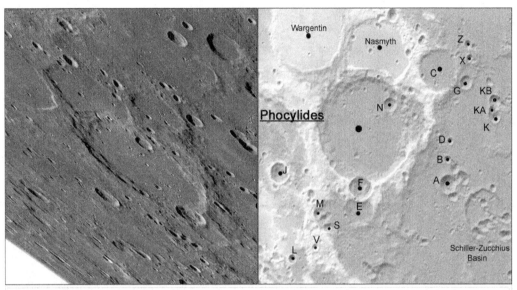

Phocylides Lat 52.79S Long 57.31W 115.18Km

Sub-craters:

Crater	Lat	Long	Size (km)	Crater	Lat	Long	Size (km)
A	54.59S	51.8W	19.24	B	53.79S	51.88W	8.6
C	50.95S	52.74W	46.04	D	53.2S	51.78W	7.48
E	55.42S	57.73W	34.5	F	54.75S	57.58W	28.0
G	51.29S	50.94W	14.18	J	54.05S	62.91W	21.47
K	52.28S	49.05W	14.57	KA	52.12S	49.13W	12.04
KB	51.82S	49.06W	14.54	L	56.83S	62.55W	9.6
M	55.47S	60.58W	10.09	N	52.08S	55.61W	17.04
S	55.93S	59.92W	10.36	V	56.54S	60.87W	8.48
X	50.52S	50.77W	9.07	Z	50.07S	50.92W	9.15

Add Info:
Around 12-day-old moon times (~ waxing gibbous) a wonderful demarcation effect occurs between Phocylides and Nasmyth. While the former's floor is in full light, the latter's is in complete shadow.

Aerial view of Phocylides

Together with its companion, crater Nasmyth, to the north, Phocylides could be considered the 'sole-print' part of this giant shoe-print on the Moon. The crater lies just outside one of the rings/rims of the Schiller-Zucchius Basin to its southeast (centred about 200 km away), it has impacted upon Nasmyth's southern rim, where the elevation between the two has left Phocylides some two kilometres lower. This whole area has been subjected to ejecta from the Orientale Basin over to the northwest (is that what fills the floors of Phocylides, Nasmyth and Wargentin?), where flows of such material between the first two craters may have occurred (note the small channel connecting their rims). Depth ~ 3 km approx..

Craters

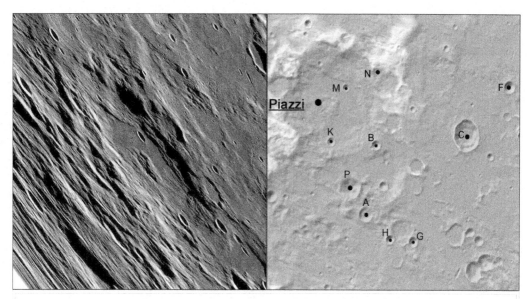

Piazzi		Lat 36.16S	Long 68.01W			102.57Km	

Sub-craters:

Crater	Lat	Long	Size (km)	Crater	Lat	Long	Size (km)
A	39.49S	66.78W	12.96	B	37.5S	66.38W	8.37
C	37.15S	62.75W	28.04	F	35.7S	61.16W	11.03
G	40.23S	64.79W	9.02	H	40.19S	65.76W	7.7
K	37.41S	68.15W	7.26	M	35.88S	67.56W	6.18
N	35.42S	66.21W	15.99	P	38.73S	67.41W	19.87

Notes:

Add Info:

Aerial view of Piazzi

Massive amounts of ejecta created during formation of the Orientale Basin over to Piazzi's northwest is the main reason why the crater is a bit hard to define. The effects would have been worse, however, Piazzi was lucky enough to lie in a zone of avoidance where northeastwards and southwestwards the extent of the ejecta reached well beyond the crater for several hundred kilometres (take in a bigger view of all this area). There's not much to see in Piazzi except the remnants of isolated lava deposits, a broad peak (is it really a peak or just a clump of Orientale ejecta?) and two fresh impacts on its floor. Use Schickard as a marker for finding it. D:1.6 km.

Craters

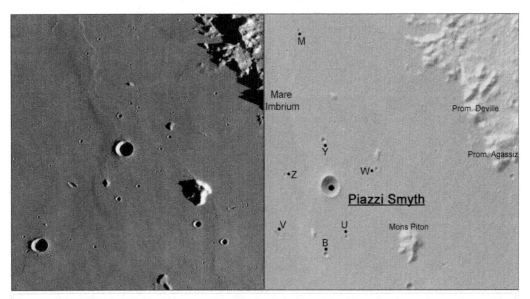

Piazzi Smyth	Lat 41.91N	Long 3.24W				12.96Km

Sub-craters:	Crater	Lat	Long	Size (km)	Crater	Lat	Long	Size (km)
	B	40.53N	3.37W	3.66	M	45.08N	4.24W	2.46
	U	40.88N	2.77W	2.91	V	40.92N	4.78W	7.82
	W	42.25N	1.86W	3.0	Y	42.85N	3.45W	3.73
	Z	42.15N	4.59W	2.55				

Notes:

Add Info:

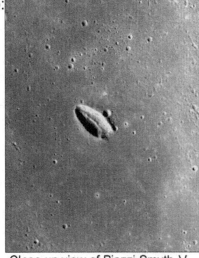

Close-up view of Piazzi Smyth V

Piazzi Smyth is situated on one of the presumed rings/rims (signatured as a well defined ridge) associated to formation of the Imbrium Basin. The crater is similar-looking to its neighbour, crater Kirch, in the southwest, where both are best observed together as an exercise for comparisons in, say, depth, or ejecta displays (during any lighting conditions). Unlike Kirch's floor, which is more flat-bottomed in nature, Piazzi Smyth's has a small mound in its centre, while its wall material seems to be a little brighter, too. Piazzi Smyth V (left) has an unusual shape for a crater, and looks like a miniature version of the much larger crater, Schiller, in the southwest limb of the Moon. What caused V to have such a shape - an impactor striking the surface at a very low angle (< 5°) from the northwest? D:2.5 km.

Craters

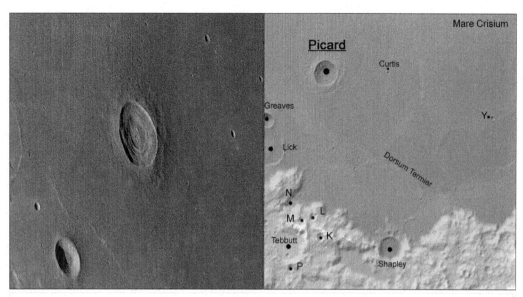

Picard		Lat 14.57N	Long 54.72E				22.35Km
Sub-craters:	Crater	Lat	Long	Size (km)	Crater	Lat	Long	Size (km)

Crater	Lat	Long	Size (km)	Crater	Lat	Long	Size (km)
K	9.73N	54.56E	8.55	L	10.32N	54.31E	7.44
M	10.21N	53.95E	8.19	N	10.52N	53.57E	19.05
P	8.82N	53.62E	7.9	Y	13.18N	60.27E	4.29

Notes:

Add Info:

Barker's Quadrangle: Also known as 'The Trepezium', apparantly, shows up during low and high sun times on the southeast sector of Mare Crisium (Picard Y is said to make up the northwest corner of the Trapezium shape). Initially observed by Robert Barker in 1929, the appearance and disappearance of the feature suggested changes on the Moon.

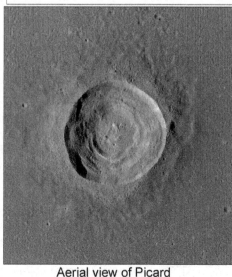

Aerial view of Picard

Like its neighbour northwards, crater Peirce, Picard is very much a show-off during low sun times - around 3-day or 17-day-old moons). Both of their rims are substantially above the level plain of Crisium's lava (Picard's is some 700 metres while Peirce's is about 500 metres) for their sizes; producing nice shadows at times. Picard is just a little bit larger than Peirce, its depth is too, but it's in the ejecta displays from each where the former comes into its own, as it looks fresher (both craters are Eratosthenian in age). Picard is therefore younger, and it's sharp rim to that of Peirce's bares testamont to the fact. Depth 2.3 km.

Craters

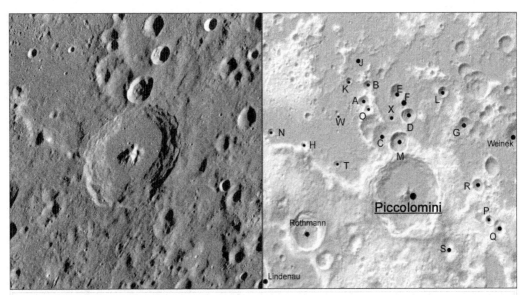

Piccolomini Lat 29.7S Long 32.2E 87.58Km

Sub-craters:

Crater	Lat	Long	Size (km)	Crater	Lat	Long	Size (km)
A	26.44S	30.36E	15.09	B	25.89S	30.52E	11.14
C	27.68S	31.13E	25.34	D	26.95S	32.26E	16.5
E	26.16S	31.78E	18.38	F	26.36S	31.78E	72.22
G	27.27S	34.72E	16.65	H	27.92S	27.68E	7.82
J	25.01S	30.14E	30.01	K	25.72S	29.71E	7.56
L	26.14S	33.75E	12.3	M	27.88S	31.84E	23.13
N	27.37S	26.26E	8.68	O	26.72S	30.54E	10.15
P	30.51S	35.9E	11.42	Q	30.88S	36.38E	13.48
R	29.32S	35.33E	14.72	S	31.64S	34.09E	19.49
T	28.54S	29.07E	7.18	W	26.8S	29.17E	5.52
X	27.03S	31.5E	7.89				

Notes:

Add Info:
As Piccolomini has impacted midway upon a higher section of the Altai Scarp's rim (south of it), was direction of the ejecta more northwards than southwards?

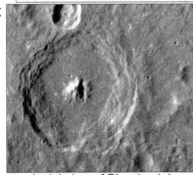

Aerial view of Piccolomini

Like many prominent craters on the Moon known for their sizes and additional features, Piccolomini is recognised more for its unique location on the 'cusp' region of one of the moon's most famous features - that of the Altai Scarp (a ring/rim associated to formation of the Nectaris Basin to the north). The crater is aged in the Upper Imbrian (3.75 to 3.2 byo), it has a wonderful set of peaks (highest is ~ 2.5 km), a wreath of terraces all within (moreso at the south), and ejecta signatures seen predominantly northwards. D: ~ 4.2 km.

Craters

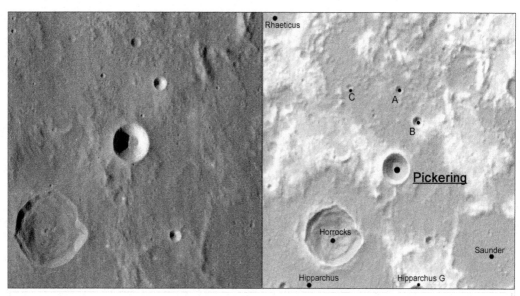

Pickering	Lat 2.88S	Long 6.99E	15.4Km

Sub-craters:

Crater	Lat	Long	Size (km)	Crater	Lat	Long	Size (km)
A	1.58S	7.05E	4.45	B	2.11S	7.39E	5.54
C	1.54S	6.15E	3.32				

Notes:

Add Info: How it looks: The state of a feature e.g. how sharp or smooth a crater or rille ...etc., looks may not always be a signature of age or of geological sequence. Factors like surface make-up in which the feature formed, or proximity to other younger features may also need to be taken into account.

Pickering lies just outside the northeastern rim of crater Hipparchus. The crater looks like it formed on a clump of material that initially one would expect it to be ejecta related to Hipparchus, however, as this whole area underwent an onslaught in volumes of ejecta from formation of the Imbrium Basin over to the northwest (centred some 2300 km away), the 'clump' may have had its origin from that major event rather than from Hipparchus. The reason is that on closeup of Pickering, its rim doesn't seem to have been physically affected like some of the surrounding craters (for example, look at the rims of Hipparchus or Flammarion over to its west). All show striated features that 'point' back to Imbrium - produced by huge blocks of ejecta as they pummelled through higher spots on the rims. However, Pickering's rim is somewhat unaffected. Would Pickering be a secondary impact crater that occurred not long after as Imbrium's ejecta littered the area? It is quite possible because some ejecta from such major events like basins, or large craters, can end up taking high trajectories, which then land several seconds to minutes afterwards when most of the main impact activity has calmed down. Crater Horrocks to the southwest is aged of the Eratosthenian Period (younger than the Imbrian Period), and as its rim looks sharper than Pickering, perhaps, this puts a constraint on the age in which Pickering formed. At full moon times, Pickering's bright walls 'shine'. D: 2.7 km.

Craters

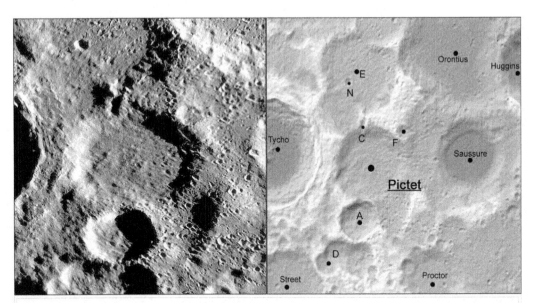

Pictet	Lat 43.56S	Long 7.49W	59.95Km

Sub-craters:

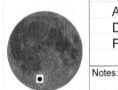

Crater	Lat	Long	Size (km)	Crater	Lat	Long	Size (km)
A	45.01S	7.93W	32.38	C	42.74S	7.8W	6.93
D	46.01S	9.14W	19.98	E	41.45S	7.87W	70.08
F	42.82S	6.32W	10.35	N	41.59S	8.22W	6.41

Notes:

Add Info:

Aerial view of Pictet

Looking at the close-up view of Pictet (left), it seems as if someone with a rough bristle brush swept it right across the entire crater. The brushed-looking material is, of course, ejecta that came from Tycho to its west, as volumes (directed mostly eastwards towards Pictet, as Tycho was an oblique impact) covered the crater and general surrounds (take in a bigger picture view to see the extent). An unusual bright feature - looks like a small cylinder (~5 km long and possibly a slumped feature, or, is it Tycho ejecta?) shows up nicely at the crater's inner southwest rim during easterly to full moon times. D: 1.8 km.

Craters

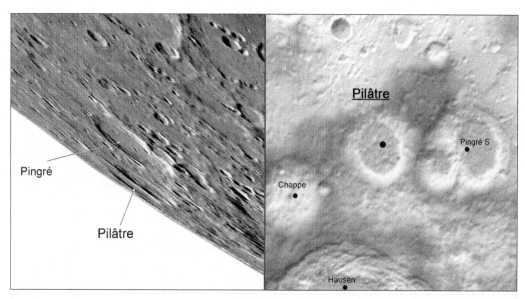

| Pilâtre | Lat 60.17S | Long 86.7W | 64.37Km |

Sub-craters: No sub-craters

Notes:

Add Info:

Aerial view of Pilâtre

You should want to have your wits about you when trying to capture a descent view of Pilâtre. Favourable librations are of course required, however, it can still get confusing as two craters (Pingré S and an unnamed smaller crater behind it) lie in front of Pilâtre - from our Earth view. The wonder of crater Hausen and its peaks will be the obvious location marker for finding Pilâtre, as both lie within a few degrees (about ~ 5 degrees) of each other in latitude. Pilâtre itself has been subject to the ejecta of Hausen (Lower Imbrian in age - 3.75 to 3.2 byo) but as it looks like it lies on other ejecta from the Nectarian-aged Bailly Basin (3.92 - 3.85 byo) over to its east, it, perhaps, puts a constraint on the time it formed.

Craters

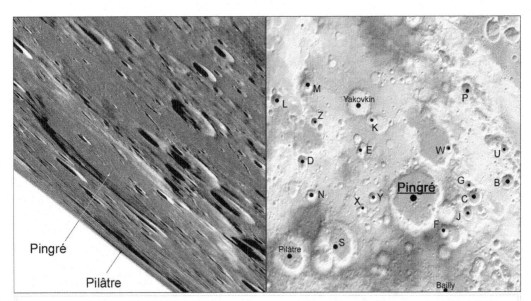

Pingré   Lat 58.64S   Long 73.95W   88.43Km

Sub-craters:

Crater	Lat	Long	Size (km)	Crater	Lat	Long	Size (km)
B	57.56S	65.43W	19.5	C	58.36S	68.46W	22.57
D	56.55S	84.28W	16.58	E	56.31S	78.79W	13.58
F	59.85S	71.32W	18.66	G	57.93S	69.05W	13.53
J	59.05S	68.97W	17.54	K	55.19S	77.91W	16.65
L	53.85S	85.84W	18.36	M	53.49S	83.2W	19.49
N	58.07S	83.91W	18.37	P	53.98S	69.67W	19.67
S	60.33S	82.39W	71.53	U	56.26S	66.0W	15.09
W	56.46S	71.04W	9.06	X	58.79S	79.15W	9.14
Y	58.42S	78.13W	13.38	Z	55.06S	82.81W	12.73

Notes:

Add Info:

Aerial view of Pingré

Pingré doesn't look as old as the Bailly Basin (Nectarian in age 3.92 to 3.85 byo) to it east, however, as there are several secondary crater chains (some less obvious because of their sizes) that 'point' back to crater Hausen (Upper Imbrian 3.75 to 3.2 byo) at its southeast, Pingré must have formed sometime in between. Moreso, assuming that the material on its floor may be ejecta from the Orientale Basin (another Upper Imbrian event like Hausen) ~ 1200 km away to the northwest, it may constrain the age of when Pingré formed that little bit tighter. Depth is around 4 km approx..

Craters

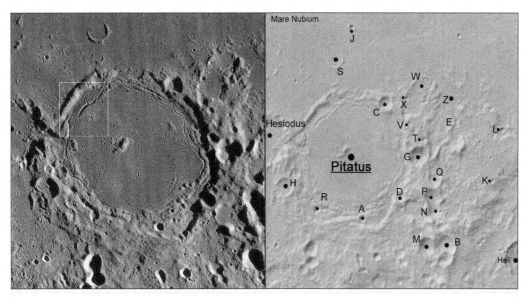

Pitatus	Lat 29.88S	Long 13.53W	100.63Km

Sub-craters:

Crater	Lat	Long	Size (km)	Crater	Lat	Long	Size (km)
A	31.46S	13.23W	5.97	B	32.15S	10.48W	18.3
C	28.5S	12.5W	12.12	D	30.9S	12.0W	9.65
E	28.93S	10.19W	5.79	G	29.83S	11.38W	16.35
H	30.63S	15.77W	14.42	J	26.53S	13.58W	4.69
K	30.42S	8.95W	5.1	L	29.1S	8.68W	4.49
M	32.16S	11.09W	14.19	N	31.2S	10.89W	12.43
P	30.88S	10.93W	14.48	Q	30.44S	10.86W	11.65
R	31.21S	14.73W	6.42	S	27.35S	14.1W	13.07
T	29.41S	11.22W	4.78	V	29.02S	11.71W	4.9
W	27.98S	11.17W	12.83	X	28.5S	11.63W	18.61
Z	28.38S	10.48W	23.47				

Notes:

Add Info:

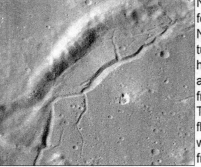

Close-up view of fractures in Pitatus

Nectarian in age (3.92 to 3.85 byo), Pitatus formed on the outer southern ring/rim of the Nubium Basin. The true rim of Pitatus is actually hard to define as nearly every section has been altered through small to large impacts and, of course, the concentric braids of fractures both within the floor and on its rim. The fractures are probably due to uplift of the floor (slightly domed particularly at the north), which undoubtedly acted as source points from where lavas then filled the floor. Peak height is almost the same as its depth ~0.7 km.

Craters

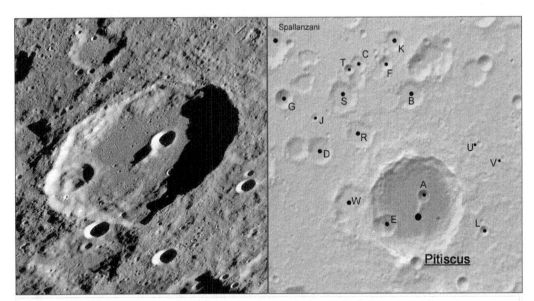

Pitiscus   Lat 50.61S   Long 30.57E   79.85Km

Sub-craters:

Crater	Lat	Long	Size (km)	Crater	Lat	Long	Size (km)
A	50.4S	30.92E	10.15	B	47.79S	30.4E	23.47
C	47.37S	28.27E	14.88	D	49.11S	26.55E	20.49
E	51.12S	29.21E	19.76	F	46.99S	29.42E	11.58
G	47.71S	25.21E	15.18	J	48.25S	26.44E	6.47
K	46.41S	29.81E	15.95	L	51.27S	33.59E	8.19
R	48.75S	28.28E	24.42	S	47.78S	27.63E	23.46
T	47.09S	27.93E	8.11	U	49.05S	33.28E	5.04
V	49.41S	34.31E	4.97	W	50.41S	27.7E	24.03

Notes:

Add Info:

Aerial view of Pitiscus

Two features of Pitiscus can easily be used for locating the crater in this over-cratered region of the Moon: first, is its thick 'band' of ejecta around its exterior, second, its peak 'connected' to sub-crater, Pitiscus A. Several neighbouring craters (e.g. Hommel, Vlacq, Rosenberger, Nearch...etc.,) just southeast of Pitiscus are believed to be pre-Nectarian in age, however, Pitiscus's features look slightly sharper, so it probably is younger - of the Nectarian Period (note it in comparison to Vlacq, which probably is really Nectarian). P:~0.7 km. D:~ 4.6 km.

Craters

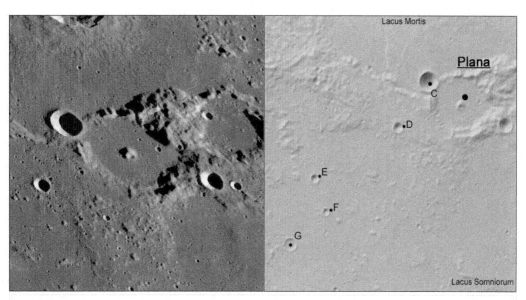

Plana   Lat 42.25N   Long 28.22E   42.97Km

Sub-craters:

Crater	Lat	Long	Size (km)	Crater	Lat	Long	Size (km)
C	42.8N	27.14E	13.69	D	41.76N	26.18E	7.04
E	40.55N	23.59E	5.83	F	39.83N	24.0E	4.79
G	39.06N	22.94E	8.89				

Notes:

Add Info:

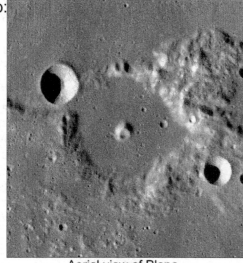

Aerial view of Plana

Plana has obviously formed on the southern sector of the odd-shaped feature known as Lacus Mortis (a possible crater). The crater's rim is somewhat undefined as several clumpy bits - particularly the eastern sectors - overlie, or hide it (just what is the large clump deposit between it and Mason to the east, as it really looks out of place?). Plana has a small peak on its floor, which seems to have a 'dint' on its southern flank - like some crater clipped it. Terminator times (~ 20-day-old moons) show two small arms (ridges?) radiate from the peak to the crater's inner northern rim. D: 1.8 km.

Craters

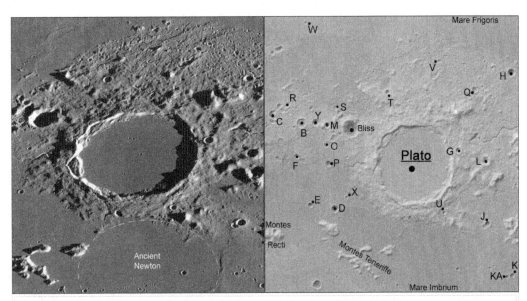

Plato   Lat 51.62N   Long 9.38W   100.68Km

Sub-craters:

Crater	Lat	Long	Size (km)	Crater	Lat	Long	Size (km)
B	53.1N	17.3W	12.45	C	53.37N	19.46W	8.9
D	49.69N	14.57W	9.38	E	49.77N	16.19W	6.63
F	51.72N	17.4W	7.15	G	52.14N	6.28W	7.95
H	55.17N	2.03W	10.69	J	49.05N	4.58W	7.72
K	46.83N	3.26W	6.51	KA	46.79N	3.58W	5.49
L	51.64N	4.4W	10.42	M	53.11N	15.5W	7.92
O	52.31N	15.43W	7.81	P	51.53N	15.21W	8.16
Q	54.61N	4.85W	7.83	R	53.78N	18.46W	6.29
S	53.86N	15.02W	7.14	T	54.58N	11.27W	7.49
U	49.61N	7.39W	5.61	V	55.86N	7.37W	5.78
W	57.24N	17.81W	4.21	X	50.2N	13.82W	4.45
Y	53.18N	16.31W	9.96				

Notes:

Add Info:

At low sun times (easterly or westerly), wonderfully long, peaky shadows - created by high points on the rim - are seen to cross the floor.

Aerial view of Plato

Probably one of the most eye-catching craters on the Moon, Plato never fails to surprise. Aged of the Upper Imbrian (3.75 to 3.2 byo), its dark floor filled with internally-sourced lavas are some 500 metres above that of Imbrium's, and also darker, too (compare their colours). Small craterlets within the floor are a favourite challenge for amateurs - the task to see how many can be seen through an average-sized telescope. A big chunk of the crater's wall, that has broken off at its west, produces nice shadows. D: ~ 2 km.

Craters

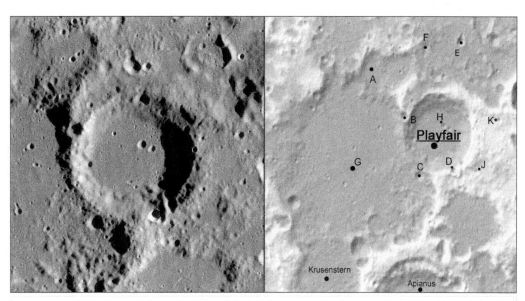

Playfair	Lat 23.56S	Long 8.45E	49.88Km

Sub-craters:

Crater	Lat	Long	Size (km)	Crater	Lat	Long	Size (km)
A	22.3S	6.88E	18.23	B	23.23S	7.59E	5.55
C	24.35S	7.92E	5.95	D	24.27S	8.71E	4.51
E	21.81S	8.89E	5.67	F	21.91S	8.09E	4.76
G	23.99S	6.09E	91.68	H	23.4S	8.49E	3.42
J	24.27S	9.28E	3.53	K	23.29S	9.79E	3.43

Notes:

Add Info: Playfair is situated in an over-cratered region of the Moon where the general makeup of material includes original highlands, and ejecta from both the pre-Nectarian Nubium Basin over to its west and the Nectaris Basin (Nectarian) to its east. The crater obviously formed on the much larger sub-crater, Playfair G, to its west, where signatures of its ejecta can just about be seen on its floor - southwest of Playfair. There is an additional clump of material just outside Playfair's northern rim that almost resembles the older rim of another crater underneath (it shows up quite easily during westerly sun times), however, as to it being an actual rim, or just a clump, is one for further speculation. The walls of Playfair (particularly its eastern and western sides) are pockmarked with numerous small impact craters old and new, while its floor has been levelled by ejecta from the Imbrium Basin (its centre lies some 1800 km away to the northwest), followed by a few small secondaries possibly from that event, too. At full moon times Playfair, like many other prominent craters nearby, simply disappears, however, sub-crater, Playfair D, and another smaller one on its eastern rim, show up bright, fresh impact signatures. Depth is around 3.8 km.

Craters

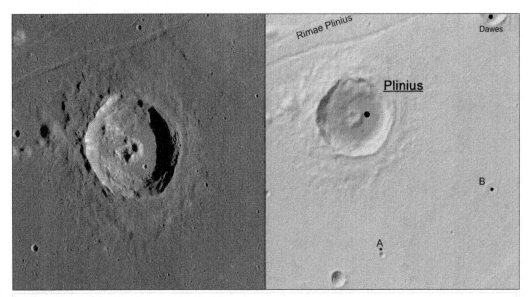

Plinius		Lat 15.36N		Long 23.61E			41.31Km	
Sub-craters:	Crater	Lat	Long	Size (km)	Crater	Lat	Long	Size (km)
	A	12.99N	24.17E	3.45	B	14.09N	26.27E	6.55
	Notes:							

Add Info:

Apollo 17 landed some 250 km away northeastwards from this crater on the 11 Dec 1972 (it was the final mission of the Apollo series to land on the Moon).

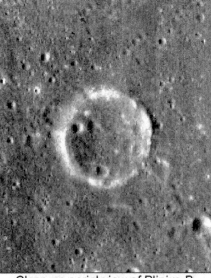

Close-up aerial view of Plinius B

At full moon times Plinius's bright walls, rays and floor signify it as a relatively young-ish crater (Eratosthenian in age 3.15 to 1.1 byo). During terminator times (easterly or westerly) its ejecta - with furrows and ridges - extend out to nearly a crater's diameter all around. At the northern, outer sectors the ejecta obviously overlies the rilles that are of Rimae Plinius, but its occurrence in the southern sector shows the crater formed on lavas associated to formation of the Tranquillitatis Basin (did these lavas breach into the Serenitatis Basin lavas to the north - note the very obvious colour between the two?). Sub-crater, Plinius B (as with Jansen H, or the unnamed crater southeastwards of Jansen R), look as if lavas flowed over them. Depth ~ 3 km.

Craters

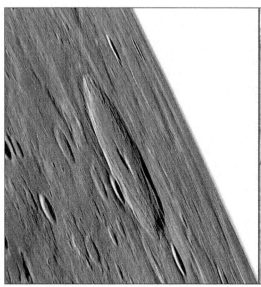
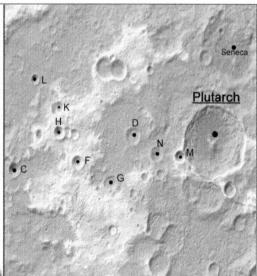

Plutarch   Lat 24.18N   Long 79.05E   69.59Km

Sub-craters:

Crater	Lat	Long	Size (km)	Crater	Lat	Long	Size (km)
C	23.17N	70.96E	13.23	D	24.36N	75.74E	15.8
F	23.51N	73.53E	12.91	G	22.96N	74.85E	16.45
H	24.36N	72.72E	11.97	K	25.08N	72.6E	14.66
L	25.88N	71.64E	9.06	M	23.77N	77.67E	11.71
N	23.83N	76.72E	13.72				

Notes:

Add Info:

Aerial view of Plutarch

Plutarch lies just on the edge of one of Crisium's impact ring/rim system (~1000 km in diameter), and is easily located - given suitable librations - as there are no other craters of similar significance nearby. Aged of the Eratosthenian (3.15 to 1.1 byo), the crater, having formed on uneven ground (highest at the southern sector), led to further unevenness in later effects, such as, terracing and slumping (compare their differences within the south and north). The high terrain at the south produces some nice shadow effects during terminator times, while at full moon it, together with the ~ 2 km-high peak and the unnamed 14 km-sized crater southeast, act as easy markers. D: ~ 4.8 km approx..

Craters

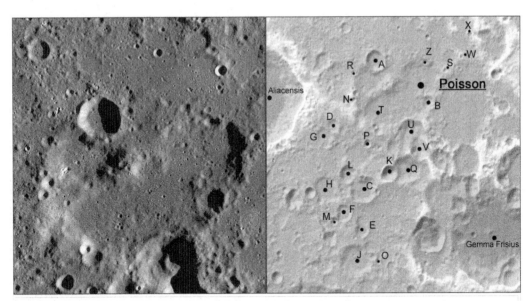

Poisson		Lat 30.34S	Long 10.56E			41.4Km

Sub-craters:

Crater	Lat	Long	Size (km)	Crater	Lat	Long	Size (km)
A	29.72S	9.14E	17.13	B	30.85S	10.88E	11.09
C	33.11S	8.66E	23.35	D	31.44S	7.68E	10.87
E	34.2S	8.57E	12.73	F	33.75S	7.98E	13.75
G	31.71S	7.36E	15.08	H	33.08S	7.37E	20.33
J	35.02S	8.3E	27.45	K	32.73S	9.55E	12.99
L	32.73S	8.15E	14.89	M	33.96S	7.62E	6.6
N	30.75S	8.4E	4.18	O	35.05S	9.12E	4.87
P	31.93S	8.83E	6.84	Q	32.66S	10.13E	24.81
R	30.05S	8.41E	5.23	S	30.01S	11.43E	3.54
T	31.08S	9.2E	24.99	U	31.72S	10.3E	25.54
V	32.1S	10.57E	16.47	W	29.64S	11.98E	3.19
X	29.05S	12.25E	4.93	Z	29.79S	10.75E	3.89

Notes:

Add Info:

Aerial view of Poisson

It's hardly a crater as we expect craters to look. So much geological activity from impacts upon impacts (nearby and afar) to ejecta upon ejecta (less and large volumes) has left the crater battered-to-death, and worn well beyond its age, which must be of the pre-Nectarian (4.6 to 3.92 billions of years ago). Even at terminator times, say, just before first quarter, or around a 20-day-old moon, the crater fails to offer anything in detail. Tycho lies ~ 650 km away southwestwards, but even its rays don't cross the crater. D: 1.7 km.

Craters

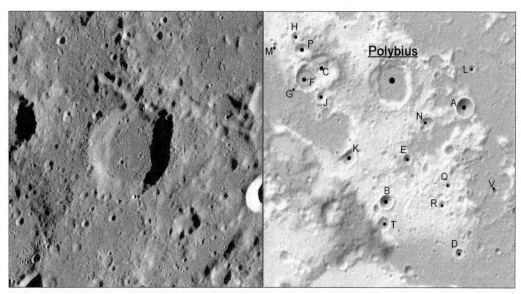

Polybius Lat 22.46S Long 25.63E 40.81Km

Sub-craters:

Crater	Lat	Long	Size (km)	Crater	Lat	Long	Size (km)
A	23.04S	27.98E	16.31	B	25.56S	25.51E	12.04
C	22.08S	23.49E	28.25	D	26.89S	27.9E	8.46
E	24.44S	26.18E	8.18	F	22.27S	23.06E	22.06
G	22.53S	22.7E	4.12	H	21.2S	22.69E	7.97
J	22.8S	23.48E	8.66	K	24.38S	24.35E	12.09
L	22.01S	28.19E	6.15	M	21.4S	22.08E	4.89
N	23.44S	26.77E	13.13	P	21.56S	22.91E	14.92
Q	25.13S	27.5E	5.78	R	25.63S	27.31E	7.03
T	26.14S	25.44E	11.21	V	25.22S	29.04E	5.09

Notes:

Add Info:

Aerial view of Polybius

It looks like Polybius impacted upon ejecta of the Nectaris Basin (to the northeast), so it's therefore younger (but not by too much). Its shape is somewhat odd where at its eastern sector the rim and wall is almost straight, while its western sector follows a more rounded shape. Close-up shows the entire rim experienced numerous small impacts particularly at its northeast and so extending the wall there (that might explain the straightness), while its western sector experienced a slump giving rise to the roundness. Depth ~ 2 km approx..

Craters

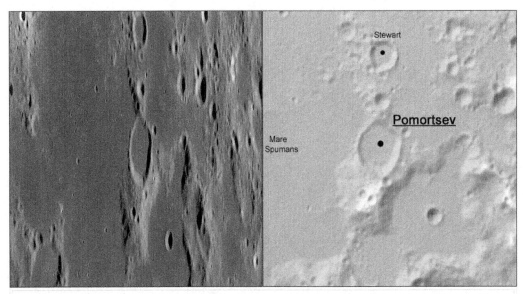

Pomortsev   Lat 0.74N   Long 66.91E   25.51Km

Sub-craters: No sub-craters

Notes:

Add Info:

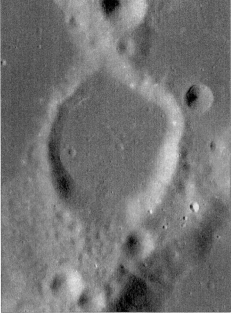

Aerial view of Pomortsev

Pomortsev initially formed in between two rings/rims associated to formation of the Crisum Basin to the north. Later, lavas seeping through fractures in the rock from that main event, filled in and around the lowest spots, but also into the central parts of many other craters as we see in Pomortsev (see, for example, crater Stewart to its north). The level of lava in the crater is the same as that in Mare Spumans to its west, however, the level at the crater's northeastern outer sector is some one kilometre higher up. The crater is quite easy to spot because of its easterly location on Spumans, however, the prominant, 2.5 kilometre-high mountain connected to its south acts as a good marker too. The depth of Pomortsev is at ~ 1.5 km.

Craters

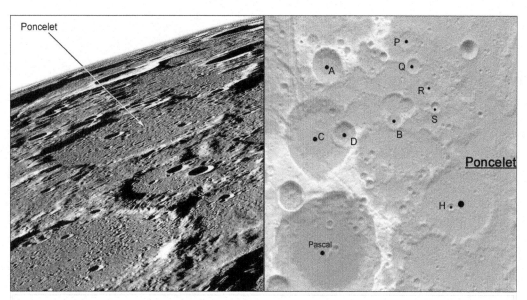

Poncelet Lat 75.91N Long 54.57W 67.57Km

Sub-craters:

Crater	Lat	Long	Size (km)	Crater	Lat	Long	Size (km)
A	79.58N	75.58W	30.96	B	78.35N	63.56W	15.01
C	77.56N	74.5W	67.95	D	77.77N	70.83W	22.75
H	75.82N	55.7W	7.04	P	80.68N	61.75W	16.06
Q	79.93N	60.58W	14.01	R	79.33N	58.09W	9.49
S	78.72N	56.96W	10.05				

Notes:

Add Info:

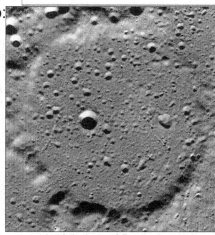

Aerial view of Poncelet

Poncelet can be somewhat confusing to find - not because of its limbward location, but that a lot of the craters and terrain in where it lies have the same texture all around. The 'texture' is believed to be that of volumes of unconsolidated ejecta due to formation of the Imbrium Basin to the south covering the region (and the recipient of thousands of small impactors mottling the surface afterwards). A significant portion of Poncelet's northeastern rim remains covered by this material, while in the opposite southwest direction an 8-kilometre-wide 'chunk' of the rim is missing. Age: Nectarian. Depth ~ 1.2 km approx..

Craters

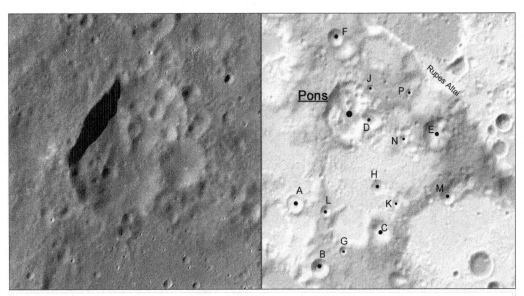

Pons	Lat 25.43S	Long 21.55E	39.7Km

Sub-craters:

Crater	Lat	Long	Size (km)	Crater	Lat	Long	Size (km)
A	27.35S	20.05E	12.12	B	28.75S	20.72E	13.36
C	27.96S	22.33E	17.82	D	25.53S	22.07E	13.72
E	25.84S	23.76E	19.64	F	23.77S	21.15E	11.24
G	28.37S	21.33E	5.87	H	26.99S	22.25E	9.77
J	24.87S	22.05E	5.68	K	27.37S	22.73E	6.82
L	27.52S	20.87E	7.65	M	27.18S	24.07E	10.49
N	25.95S	22.92E	5.76	P	24.94S	23.05E	4.56

Notes:

Add Info: Where some of the highest highland's terrain exist on the Near Side of the Moon, Pons's impactor must surely have struck it - given the state of its northwestern rim that is almost linear to edge-like on this oddly-highted target rock. The crater really has an oval-like shape (its northeast-southwest axis is some seven kilometres longer) to it. However, on closer look, the reason is probably due to slumping being responsible for the extention of the crater at the southwest sector, while several small impacts at the opposite northeast end have 'picked' away at its rim there. This whole area is covered in ejecta from the huge event that was formation of the Nectaris Basin to the northeast, so Pons, like several other craters and features nearby, obviously experienced coverage from huge volumes of that material, being so close. Funnily enough, for such a battered crater amongst numerous other craters of similar size in this general region, it isn't hard to locate (perhaps the small cluster of small impacts on its northeast rim act as a marker, or is it due to its closeness to Rupes Altai?). Depth ~ 2 km.

Craters

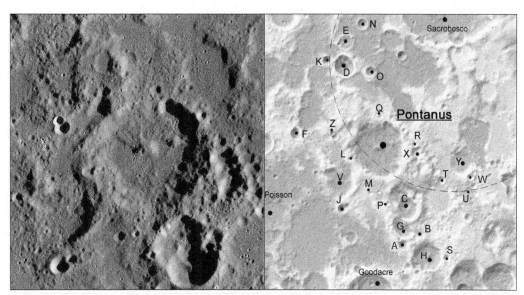

Pontanus — Lat 28.42S   Long 14.36E   55.66Km

Sub-craters:

Crater	Lat	Long	Size (km)	Crater	Lat	Long	Size (km)
A	31.15S	15.24E	9.48	B	30.85S	15.8E	12.12
C	30.12S	15.43E	22.01	D	25.93S	13.22E	20.38
E	25.24S	13.26E	12.82	F	27.88S	11.58E	10.46
G	30.64S	15.24E	23.44	H	31.41S	16.06E	29.89
J	30.09S	13.11E	8.88	K	25.76S	12.65E	8.61
L	28.64S	13.41E	5.85	M	29.65S	14.09E	4.79
N	24.69S	13.83E	9.84	O	26.1S	14.09E	10.09
P	29.94S	14.69E	3.59	Q	27.41S	14.4E	4.7
R	28.2S	15.61E	5.86	S	31.51S	16.77E	6.35
T	29.24S	16.53E	7.08	U	29.48S	17.47E	4.5
V	29.21S	13.07E	33.58	W	29.13S	17.55E	6.76
X	28.46S	15.69E	12.35	Y	28.74S	17.25E	23.37
Z	27.88S	12.81E	4.46				

Notes:

Add Info: Like nearby craters' Wilkins (~ 150 km away to the ESE), Pons (~ 200 km away to the ENE) and Fermat (~ 230 km to the NE), Pontanus may have impacted upon a very old, 225 km-wide crater centred on the outer rim of crater Sacrobosco. Pontanus is therefore younger (bordering on the Nectarian/pre-Nectarian Period, perhaps) as from looking at the state of its rim, walls and inner sectors, time has allowed numerous impacts and ejecta volumes from everywhere to alter its true shape. Ejecta from the Nectaris Basin over to its northeast covers this region, in general, and several impact chains from the Imbrium Basin, centred some 2000 km away to the northwest, marked the terrain (the chain on its outer northeastern rim 'point' back to the basin). Peak height ~ 1.1 km. Depth ~ 1.8 km approx..

Dashed line (in above topography image) represents the 225 km wide crater mentioned in text.

Craters

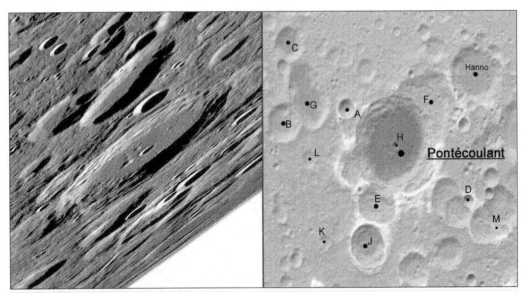

Pontécoulant   Lat 58.78S   Long 66.07E   91.4Km

Sub-craters:

Crater	Lat	Long	Size (km)	Crater	Lat	Long	Size (km)
A	57.69S	62.79E	18.64	B	57.97S	58.54E	37.25
C	55.68S	59.23E	31.71	D	60.22S	71.6E	13.7
E	60.51S	64.95E	43.62	F	57.6S	68.17E	68.68
G	57.52S	60.44E	42.25	H	58.78S	66.17E	7.58
J	61.74S	64.26E	39.54	K	61.57S	60.95E	13.93
L	59.12S	60.09E	17.07	M	60.91S	74.47E	11.0

Notes:

Add Info:

Aerial view of Pontécoulant

For a crater aged of the Nectarian (3.92 to 3.85 byo), Pontécoulant doesn't look at all half bad. The crater obviously formed on sub-crater's, Pontécoulant F, to its northeast, Pontécoulant E, to its southwest (ejecta from Pontécoulant fills its floor) and another unnamed crater just northwest of E. Pontécoulant's terracing isn't even all around, where it looks like a major slump in the southeastern sector has extended the crater's diameter there. The crater lies in an outer trough region related to formation of the Australe Basin to the northeast, while on its floor several clumps may be possible peaks (highest ~ 0.7 km). Depth ~ 5.5 km.

Craters

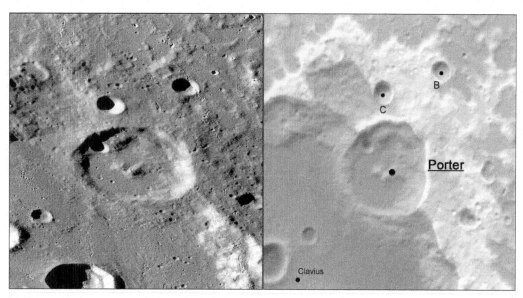

Porter	Lat 56.15S	Long 10.18W	51.46Km

Sub-craters:	Crater	Lat	Long	Size (km)	Crater	Lat	Long	Size (km)
	B	54.47S	8.64W	11.42	C	56.85S	10.46W	11.63
	Notes:							

Add Info:

Aerial view of Porter

Porter obviously formed on the wonder that is Clavius, so it is younger. But look at Porter's rim, in general, and one would think that Clavius's is slightly more fresh-looking than it. Why would this be so? Was it that Porter formed not longer after Clavius did? Was it that Porter's impactor struck 'softer' ground (due to Clavius's initial impact)? Or, is it because volumes of fluidized ejecta from the Orientale Basin (2000 km over to the northwest) is believed to have covered this area, and its features? Slumped material, or is it Orientale ejecta, 'fills' the northeastern sector of the floor, while its highest, ~ 1 km-high peak has a small, fresh impact crater on it.

Craters

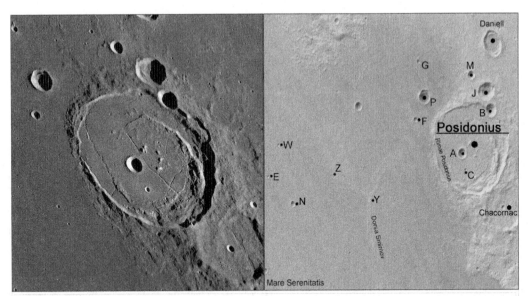

Posidonius   Lat 31.88N   Long 29.99E   95.06Km

Sub-craters:

Crater	Lat	Long	Size (km)	Crater	Lat	Long	Size (km)
A	31.69N	29.52E	11.1	B	33.16N	31.02E	14.05
C	31.13N	29.69E	3.48	E	30.55N	19.7E	3.13
F	32.82N	27.13E	6.04	G	34.79N	27.23E	4.79
J	33.8N	30.79E	22.0	M	34.36N	30.01E	9.32
N	29.7N	21.04E	6.16	P	33.6N	27.58E	14.66
W	31.65N	20.13E	3.01	Y	30.03N	24.91E	2.02
Z	30.75N	22.95E	5.94				

Notes:

Add Info:

The Russian spacecraft, Luna 21, landed some 150 km due south of this crater (see crater Le Monnier).

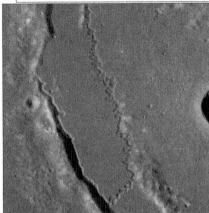

Close-up view of the Posidonius rille

Posidonius is pure eye-candy nearly at any time you can observe it. Its obvious rille system in the left half sector of its floor, the fractures in the middle, and the odd, secondary-looking rim sectioned off its real rim in the east is what makes the crater a 'good catch'. While each feature is in response to some separate, dynamic event e.g. stress, uplift, volcanic...etc., within the crater itself, all must surely have some relation over time to the much bigger event that was the Serenitatis Basin at its west. Low sun views, and high magnifications recommended. D: 1.4 km.

Craters

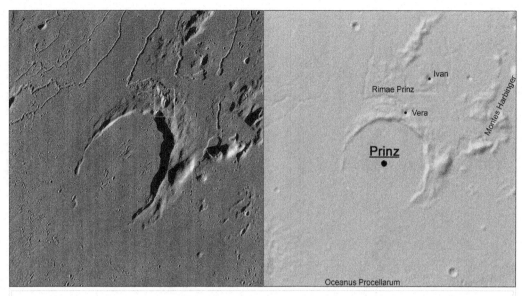

| Prinz | Lat 25.49N | Long 44.14W | 46.13Km |

Sub-craters: No sub-craters

Notes:

Add Info:

Close-up view of one of Prinz's rilles (and Vera)

Sometimes, craters with a portion of their rims covered by lavas usually are found near the edges of Basins - see Sinus Iridum, or Fracastorius for example,(the buried part normally tilts towards the central part of the basins). Prinz does lie near a basin - Imbrium lies to its east, but the buried part is oppositely tilted away from it. Did Prinz initially then form odd-like on a high-ish zone of target rock (terrain is higher north of the crater), or, was its northern sector later uplifted (the series of rilles there may signify such an event). Bright ray material from crater Aristarchus to its west crosses the floor. Depth ~ 1km approx..

Craters

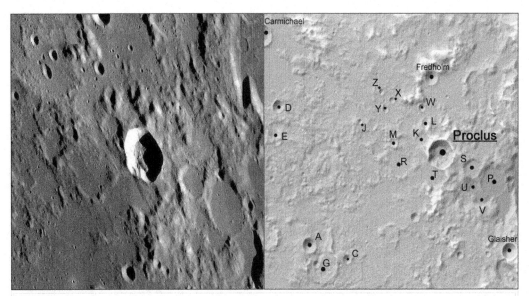

Proclus   Lat 16.09N   Long 46.89E   26.91Km

Sub-craters:

Crater	Lat	Long	Size (km)	Crater	Lat	Long	Size (km)
A	13.32N	42.23E	14.09	C	12.94N	43.58E	9.65
D	17.43N	41.02E	11.74	E	16.58N	40.9E	12.28
G	12.73N	42.7E	32.13	J	17.05N	43.99E	4.83
K	16.52N	46.22E	15.27	L	17.02N	46.35E	10.64
M	16.43N	45.19E	8.45	P	15.34N	48.73E	31.27
R	15.85N	45.45E	28.58	S	15.67N	48.0E	18.46
T	15.39N	46.71E	20.35	U	15.13N	47.99E	13.6
V	14.72N	48.39E	16.53	W	17.47N	46.21E	7.11
X	17.69N	45.13E	5.88	Y	17.42N	44.9E	7.46
Z	17.94N	44.75E	5.51				

Add Info:

Low oblique impacts

O'Neill's Bridge: This infamous apparition can be seen when the terminator (~ 18/19 day-old moon) is just on the eastern side of Proclus P.

Close-up view of melt fractures on Proclus's floor

Notes: The most striking feature of Proclus, when viewed generally through the eyepiece, is its impact ray system. These rays can be seen to cross as far over on to Crisium's floor in the east, however, westwards, a wide swade of the rays is missing. The reason for the odd distribution of the rays is that the impactor, which initially formed the crater, came in at a low angle from the southwest, causing most of the target rock and upper surface material to go eastwards, while a 'zone of avoidance' resulted westwards. Age: Coperncan. Depth ~ 2 km.

Craters

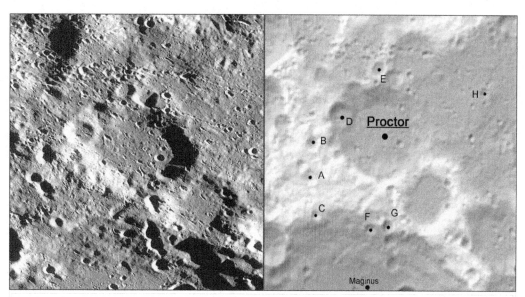

Proctor		Lat 46.43S	Long 5.04W			47.64Km	

Sub-craters:

Crater	Lat	Long	Size (km)	Crater	Lat	Long	Size (km)
A	47.05S	6.79W	7.35	B	46.54S	6.73W	7.01
C	47.61S	6.65W	5.28	D	46.1S	6.06W	11.67
E	45.42S	5.16W	7.73	F	47.83S	5.34W	5.91
G	47.77S	4.9W	7.14	H	45.75S	2.64W	5.69

Notes:

Add Info:

Aerial view of Proctor

Given the worn-looking state of Proctor, its age must surely be in and around that of its larger neighbour, Maginus, to its south (believed to be pre-Nectarian - 4.6 to 3.92 byo). The crater probably formed on Maginus's ejecta, but the demarcation line between the two is difficult to define. Tycho to its northwest (~ 150 km away) has also left its mark(s) across Proctor's floor in the form of several secondary crater chains - each consisting of very small craterlets (high magnification required). However, a more obvious single crater chain just outside Proctor's southern rim sector may be that from the Imbrium Basin ~ 2400 km away to the NW. D: ~2 km.

Craters

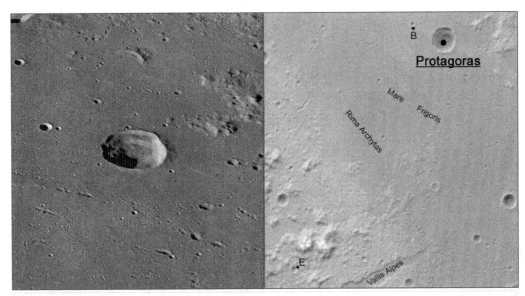

Protagoras	Lat 56.02N	Long 7.34E		21.05Km			
Sub-craters:	Crater	Lat	Long	Size (km)	Crater	Lat	Long	Size (km)
	B	56.43N	5.67E	4.4	E	49.53N	0.54E	5.79
	Notes:						

Add Info:

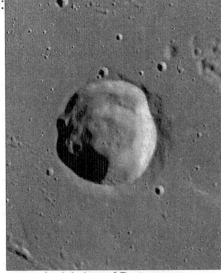

Aerial view of Protagoras

If you look first at crater Archytas - about 80 km away to Protagoras's northwest, you'll notice that while its ejecta can be seen to overlie the lavas of Mare Frigoris, no signatures are seen around Protagoras. The reason, of course, is that while Eratosthenian-aged Archytas formed initially upon the lavas of Frigoris, these same lavas at one time lapped at Protagoras's outer rim, and in the process covered over its ejecta signatures. Frigoris lavas eastwards are generally believed to be older, while in mid Frigoris they are slightly younger. Would this explain why most of the crater's outer rim at its west is missing because series of the younger lavas formed on higher plains there? D: 2.6 km.

Craters

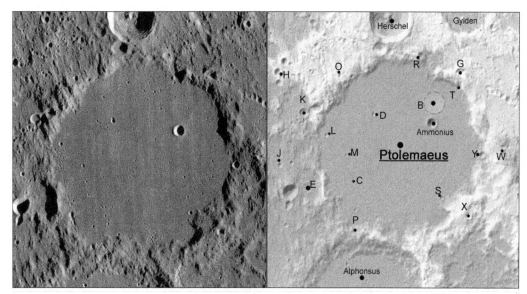

Ptolemaeus   Lat 9.16S   Long 1.84W   153.67Km

Sub-craters:

Crater	Lat	Long	Size (km)	Crater	Lat	Long	Size (km)
B	7.9S	0.8W	18.79	C	10.11S	3.28W	1.92
D	8.29S	2.6W	3.51	E	10.21S	4.52W	28.67
G	7.15S	0.02E	9.42	H	7.13S	5.45W	6.19
J	9.66S	5.4W	4.5	K	8.23S	4.68W	8.16
L	8.85S	4.01W	3.18	M	9.39S	3.42W	2.93
O	7.24S	3.6W	4.05	P	11.4S	3.17W	3.5
R	6.67S	1.17W	6.37	S	10.55S	0.53W	3.32
T	7.49S	0.0W	6.56	W	9.16S	1.33E	4.37
X	11.02S	0.29E	4.91	Y	9.37S	0.7E	5.97

Notes:

Add Info: If there is an infamously-known triad of craters on the Moon, then surely Ptolemaeus, Alphonsus (below Ptolemaeus) and Arzachel (below Alphonsus) is them. The three aren't really related, as there ages differ by periods of millions of years: Ptolemaeus is oldest of the pre-Nectarian Period - 4.6 to 3.93 byo; Alphonsus is Nectarian - 3.93 to 3.85 byo; while Arzachel, the youngest, is of the Lower Imbrian - 3.85 to 3.75 byo. Moreso, each can be subdivided into floor-level heights: Alphonsus's floor is some 1 km below that of Ptolemaeus's while Arzachel's is some 1.5 km below that of Alphonsus's. Ptolemaeus itself may lack the show-off features (peaks, rilles, dark patches, terraces etc.,) of the other two, however, it does have several, shallow, 'saucer-like' craters across most of its floor that the other two don't. Depth of Ptolemaeus ~ 2.5 km.

Saucers: Sub-crater, Ptolemaeus B is the obvious one, however, look below sub-crater M for the next, and above sub-crater D the next again (but there are more).

Craters

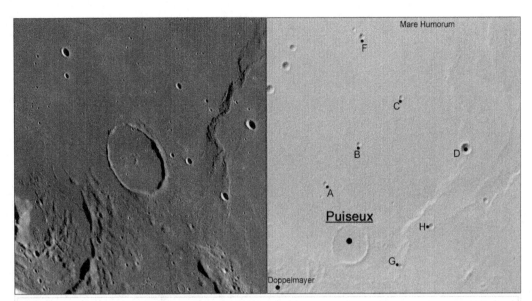

Puiseux	Lat 27.82S	Long 39.19W				24.95Km	

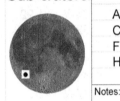

Sub-craters:	Crater	Lat	Long	Size (km)	Crater	Lat	Long	Size (km)
	A	26.53S	39.82W	3.26	B	25.7S	38.97W	3.37
	C	24.7S	37.9W	3.14	D	25.76S	36.23W	7.3
	F	23.42S	38.9W	4.03	G	28.26S	37.87W	3.26
	H	27.4S	37.08W	3.75				

Notes:

Add Info:

The NW-SE trending ridge extends for nearly three crater-diameter's away to the northwest, and close-up views show it may have occurred after Puiseux formed.

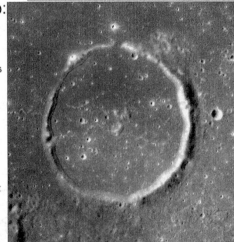

Aerial view of Puiseux

Like its bigger neighbour, Doppelmayer, to its southwest, Puiseux's rim is tilted towards Humorum's centre - because as the central portions of the basin material is presumably thicker and more weighty, it shifted the outer sectors of the basin ever so slightly downwards. Lavas probably filled through fractures inside Puiseux itself, however, a 2-km-wide 'gap' at the northern end of the rim may also have acted as a breach point for Humorum's lavas. This might be why the top half of the floor's lavas appear darker, which may have been stopped by a small NW-SE trending ridge that crosses the floor - an historical hint of geological events.

Craters

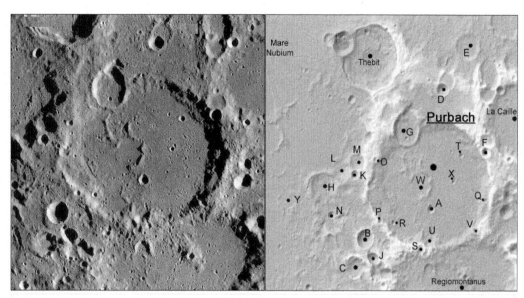

Purbach   Lat 25.15S   Long 2.03W   114.97Km

Sub-craters:

Crater	Lat	Long	Size (km)	Crater	Lat	Long	Size (km)
A	26.17S	1.94W	7.11	B	26.98S	4.26W	15.68
C	27.74S	4.65W	18.24	D	22.88S	1.57W	11.1
E	21.68S	0.73W	23.54	F	24.6S	0.03W	8.89
G	23.94S	2.81W	29.92	H	25.47S	5.69W	28.75
J	27.51S	3.97W	11.02	K	25.2S	4.57W	7.87
L	25.06S	5.05W	16.9	M	24.85S	4.49W	15.24
N	26.3S	5.4W	7.13	O	24.8S	3.86W	5.08
P	26.49S	3.72W	4.73	Q	25.92S	0.04W	4.02
R	26.53S	3.26W	4.24	S	27.26S	2.29W	7.69
T	24.69S	0.92W	4.76	U	26.99S	2.05W	15.37
V	26.75S	0.36W	5.62	W	25.47S	2.29W	20.36
X	25.42S	1.19W	3.47	Y	25.87S	6.93W	15.43

Notes:

Add Info: Like its neighbour, crater Regiomontanus to its south, Purbach appears somewhat 'squished' along its west-east axis. The reason, of course, is that the western sector of the rim may have underwent a major slump, and thus extended that part of the crater. Though it is hard to define, it looks like Purbach impacted upon Regiomontanus - both are of the pre-Nectarianin Period (4.6 to 3.92 byo), but its ejecta on Regiomontanus's floor isn't that obvious as several small impact craters confuse their shared border. Also, did Purbach impact upon another older crater (about 100 km in diameter) seen, less so, just outside its northern rim? Floor material may consist of a mix of ejecta from several craters nearby and from Imbrium (1700 km away to the northwest).

Bright ray material from crater Tycho (to the southwest) crosses Purbach's floor, but is only obvious at its southeastern sector.

Craters

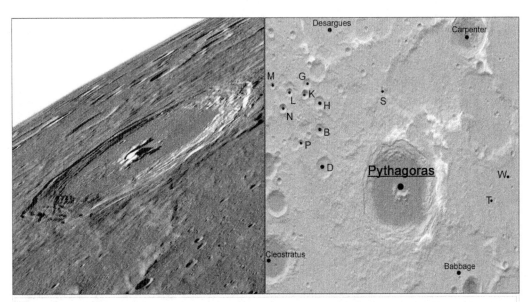

Pythagoras	Lat 63.68N	Long 62.98W	144.55Km

Sub-craters:

Crater	Lat	Long	Size (km)	Crater	Lat	Long	Size (km)
B	66.02N	73.33W	17.3	D	64.51N	72.36W	28.79
G	67.78N	75.79W	14.46	H	67.11N	73.76W	19.05
K	67.35N	75.84W	11.95	L	67.3N	78.03W	11.25
M	67.38N	80.54W	9.25	N	66.59N	78.58W	13.11
P	65.34N	75.6W	9.54	S	67.75N	65.0W	7.09
T	62.5N	51.66W	5.75	W	63.21N	49.35W	4.11

Notes:

Add Info:

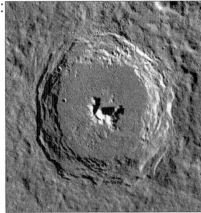

Aerial view of Pythagoras

Named in honour of Pythagoras - he of the 'Pythagorean Theorem', it's a pity that the associated mathematics concerned couldn't have given us a better view of this wonderful crater. The crater could be compared to Copernicus (9.62N, 20.08W), but Pythagoras is much bigger by nearly 50 km, its highest peak is some 2 km over that of Copernicus's, and it is about 1.5 km deeper than it, too. If the two were put alongside each other, Pythagoras would win out at a long shot, even though it is older (Upper Imbrian - 3.75 to 3.2 byo). The crater doesn't have the same 'show-off' ray display as Copernicus has but it does lie in somewhat bright terrain.

Craters

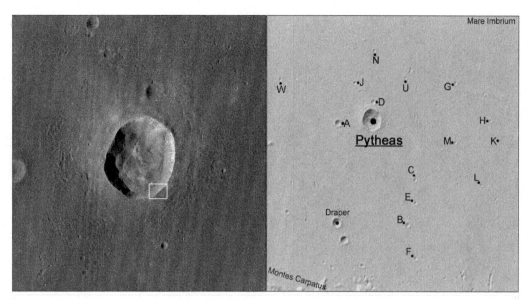

Pytheas   Lat 20.57N   Long 20.59W   18.81Km

Sub-craters:

Crater	Lat	Long	Size (km)	Crater	Lat	Long	Size (km)
A	20.47N	21.74W	5.71	B	17.5N	19.39W	4.15
C	18.81N	19.13W	3.77	D	21.09N	20.54W	4.76
E	18.14N	19.12W	3.62	F	16.55N	19.14W	3.9
G	21.62N	17.73W	3.49	H	20.5N	16.53W	2.59
J	21.63N	21.19W	3.46	K	19.91N	16.18W	2.32
L	18.62N	16.89W	2.61	M	19.9N	17.76W	2.88
N	22.57N	20.5W	3.14	U	21.78N	19.45W	2.94
W	21.71N	23.71W	2.83				

Notes:

Add Info: Herringbone, V-shaped patterns usually are seen around small, secondary impact crater events where the apex of the V (or ridge) 'points' back to the primary crater. They occur more often where the material was ejected from the crater at angles in excess of 60°.

Exposed layers (lava sheets?) on Pytheas's SE wall

Pytheas is Copernican in age (1.1 byo to the Present), and lies in a region of the Imbrium Basin where lavas formed as sheets over millions of years (see left). Ejecta from the crater is easily seen around all its outer rim, but isn't as radial-looking as would be expected (it has a more herringbone look to it). Bright ray material from crater Copernicus to the south covers this region in general, and a patch does appear at Pytheas's northern exterior, however, is it related to the former or the latter? Depth ~ 2.5 km approx..

Craters

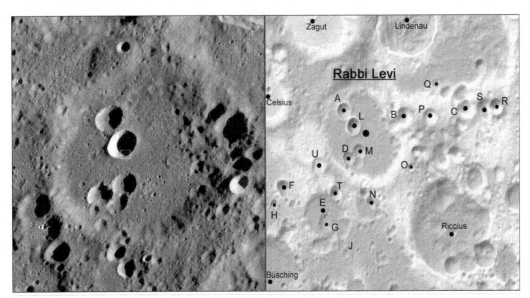

| Rabbi Levi | Lat 34.78S | Long 23.46E | 82.44Km |

Sub-craters:

Crater	Lat	Long	Size (km)	Crater	Lat	Long	Size (km)
A	34.36S	22.66E	11.45	B	34.5S	24.74E	14.13
C	34.29S	26.86E	19.48	D	35.44S	22.83E	10.9
E	36.75S	21.94E	34.7	F	36.12S	20.53E	14.05
G	36.96S	21.94E	11.44	H	36.48S	20.18E	6.73
J	37.65S	22.68E	6.61	L	34.73S	22.99E	11.94
M	35.32S	23.17E	11.32	N	36.51S	23.59E	8.86
O	35.67S	24.98E	5.49	P	34.51S	25.61E	13.29
Q	33.73S	25.79E	7.53	R	34.2S	27.89E	11.92
S	34.27S	27.51E	16.91	T	36.25S	22.35E	10.08
U	35.62S	21.76E	12.29				

Notes:

Add Info: Basin ejecta from Nectaris over to the northeast covers this region, in general, so many of the craters and features may show striated signatures that 'point' back to that major event.

Unusual debris flows in sub-crater, Rabbi Levi L

Rabbi Levi lies between two other craters of similar size (Zagut to its northwest and Riccius to its southeast). All three are pre-Nectarian in age (4.6 to 3.92 byo), however, Rabbi Levi stands out from them all with its quintet of similarly-sized craters on the floor. It looks like a much larger, unnamed crater (about 130 km in diameter) to its northeast may have impacted Rabbi Levi's rim there; imparting ejecta on to its floor in the process. D: 2.5 km.

Craters

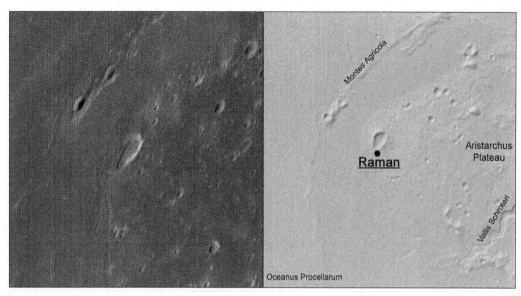

| Raman | Lat 26.96N | Long 55.16W | 10.17Km |

Sub-craters: No sub-craters

Notes:

Add Info:

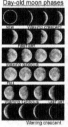

Note: As the Moon goes through various librations throughout the year, suggested times given in text for observations are only approximates.

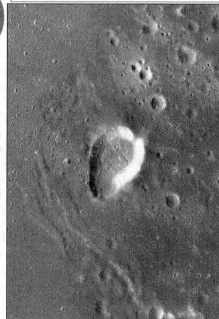

Aerial view of Raman

Raman lies on the western-most corner of the ever-wonderful, volcanic feature that is the Aristarchus Plateau. The crater looks like it was affected by a rille-like feature afterwards; running in a northwest-southeast direction across the rim and floor (others to the southwest of the crater can be seen to parallel it), and an odd 'kink' on its northeastern rim. Raman takes on the chracteristic 'teardrop' shape of a low angle impact crater. Such craters are usually secondaries produced by a block of ejecta derived from a big impact like a basin, which is nearly always found oppositely to the pointy end of the tear. In this case, the tear-end doesn't quite 'point' back to the nearby Imbrium Basin, but more to Sinus Iridum (would this be its source?). That said, as the crater formed on sloped terrain of the Plateau, would that affect derivation?

Craters

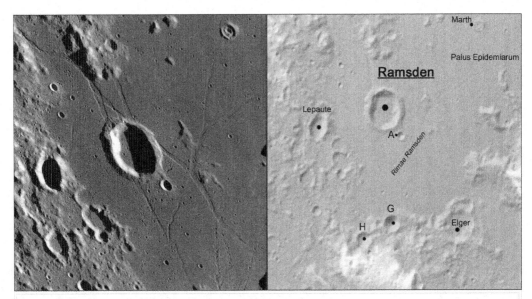

Ramsden	Lat 32.96S	Long 31.87W	25.11Km

Sub-craters:	Crater	Lat	Long	Size (km)	Crater	Lat	Long	Size (km)
	A	33.5S	31.43W	5.26	G	35.35S	31.67W	11.02
	H	35.71S	32.47W	11.43				

Notes:

Add Info:

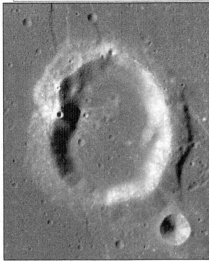

Aerial view of Ramsden

Ramsden is a crater where rilles have affected its final appearance. The rilles weren't there initially when the crater formed, as on close-up views (see left) they 'cut' Ramsden's rim and outer ejecta both at its northern and southern sectors. There are three, but only two are really obvious (the western and eastern ones). The middle one, which at one time might have crossed the floor, may subsequently have been covered over by internal lava fill. Two other features: the odd, mountain-like feature at the crater's inner western rim (a fresh impact crater lies atop it), and the straight 'edge' at the crater's outer southeast. Was uplift underneath Ramsden responsible for all we see? Depth 1.9 km.

Craters

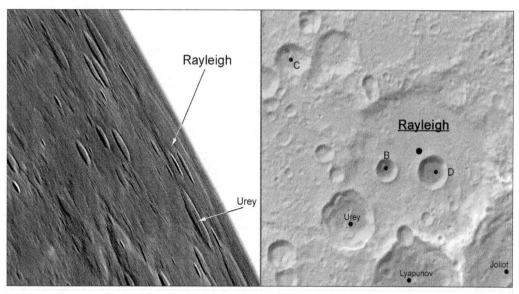

Rayleigh	Lat 29.12N	Long 89.45E	113.77Km

Sub-craters:

Crater	Lat	Long	Size (km)	Crater	Lat	Long	Size (km)
B	29.06N	88.5E	15.48	C	31.44N	85.77E	26.13
D	29.03N	89.79E	22.3				

Notes:

Add Info:

Aerial view of Rayleigh

Rayleigh, of course, will proove the usual challenge for the observer as its limb-hugging location makes for only certain times (librations). The crater isn't as old as pre-Nectarian (4.6 to 3.93 byo) Joliot to its southeast (it looks like it impacted upon its rim), however, given the similar state and appearance of its rim, it may not be far off that age. Level of Rayleigh's floor is some one kilometre above that of crater Joliot, Urey and Lyapunov to its south - all of which 'tapped' into deeper-down lavas that now flood their floors. Both sub-craters, Rayleigh B and D, act as good location markers for finding the crater. Depth is around ~ 1.1 km.

Craters

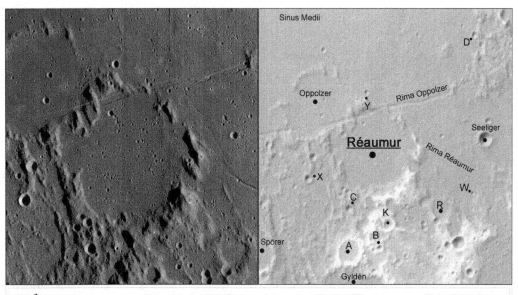

Réaumur	Lat 2.45S	Long 0.73E	51.25Km

Sub-craters:	Crater	Lat	Long	Size (km)	Crater	Lat	Long	Size (km)
	A	4.33S	0.21E	14.95	B	4.26S	0.83E	3.75
	C	3.5S	0.21E	4.27	D	0.24S	2.76E	4.01
	K	3.82S	1.01E	6.94	R	3.51S	2.1E	12.98
	W	3.25S	2.74E	2.58	X	2.92S	0.65W	4.54
	Y	1.3S	0.54E	2.96				

Notes:

Add Info: Is there any portion of Réaumur's rim that hasn't been affected by the dynamic event that was formation of the Imbrium Basin to its northwest (centred some 1100 kilometres away)? A bigger picture of the general area where Réaumur lies is recommended to see Imbrium's effect, but a bigger picture again - taking in Imbrium itself and the crater - will show that the distribution of the huge blocks responsible for the striated look on craters and the surrounding highlands fell a certain distance from the central impact. One would have expected that the initial ejecta from source would have had more affect on regions closer to the event but it doesn't. Why? The ejecta was more amalgamated at first as it was shot out, which then started to break up into smaller units and smaller blocks by the time they reached their further distance from the central impact point. Does this mean also that the striations should show wider, gouged-out 'gaps' closer to the central point event than those further away? Possibly? Have a look! Whatever the results, and reasons, the remaining state of Réaumur still has something to offer the observer: look at the numerous craterlets on its floor, the Oppolzer rille that 'cuts' its northern rim (signature of history of the geological events). D: ~ 0.8 km.

Craters

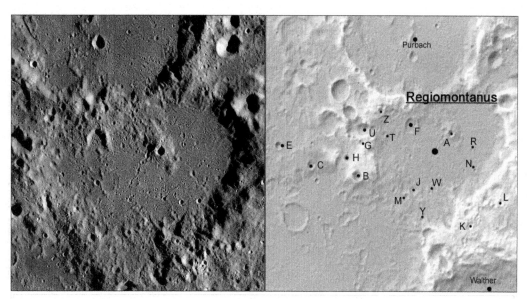

Regiomontanus Lat 28.28S Long 1.09W 126.64Km

Sub-craters:

Crater	Lat	Long	Size (km)	Crater	Lat	Long	Size (km)
A	28.03S	0.7W	5.58	B	29.06S	3.75W	9.64
C	28.77S	5.24W	7.1	E	28.27S	6.26W	6.11
F	27.83S	2.0W	11.08	G	28.34S	3.45W	3.45
H	28.63S	4.08W	5.59	J	29.43S	1.93W	7.87
K	30.26S	0.05W	5.9	L	29.69S	0.98E	5.87
M	29.56S	2.14W	4.39	N	28.91S	0.1E	2.75
R	28.43S	0.03W	3.08	S	28.59S	2.01W	3.26
T	28.28S	2.9W	3.79	U	27.93S	3.53W	11.91
W	29.47S	1.37W	3.32	Y	30.14S	1.63W	4.71
Z	27.47S	2.96W	5.51				

Notes:

Add Info: Before you look at Regiomontanus in close-up, take in a bigger picture view of several other, large-like craters to its north (Purbach, Arzachel, Alphonsus and Ptolemaeus) and to its south (Walther and Stöfler). One would think, from their somewhat close-to-linear alignment, that we were looking at a chain of craters, but they are not as some differ in age by billions of years. World renowned amateur astronomer, Sir Patrick Moore (RIP), in the 1960's proposed* that these huge chains (in this case the 'Ptolemaeus-Walther' chain) were incontrovertible evidence of the volcanic origin of large craters - forming along preferential arcs running north and south. Regiomontanus is pre-Nectarian in age (4.6 to 3.92 byo). Purbach has obviously impacted upon its northern sector, wiping out nearly half its rim. Is the off-centre, one-kilometre-high clump of material on the floor a peak (a small impact crater atop's it), or just Purback ejecta? Depth ~ 3.6 km.

*Survey of the Moon: Ed. 1963. Eyre & Spottiswoode Ltd, (pp. 102-103).

Craters

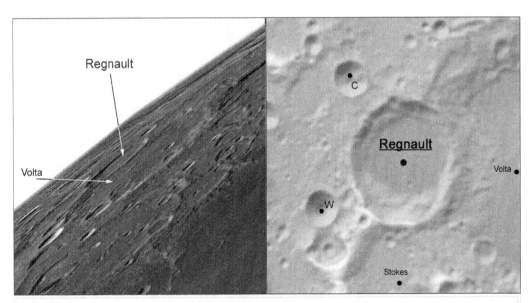

Regnault	Lat 54.04N	Long 87.88W			51.31Km	

Sub-craters:

Crater	Lat	Long	Size (km)	Crater	Lat	Long	Size (km)
C	55.1N	89.16W	13.94	W	53.42N	89.77W	14.18

Notes:

Add Info:

Aerial view of Regnault

Even during the best of librations as required mostly for those craters on the extreme limb, Regnault will prove a challenge. It's not that it doesn't come into view, but more that several large rims associated to craters like Repsold or Volta lying in front of it (from our earth perspective view) can lead to confusion. Any descent view, however, won't give that much detail, and all that stands out is its relatively high southern rim as indication that we are looking at Regnault (Volta helps also in some way as a location marker for finding the crater). The crater is Nectarian in age (3.92 to 3.85 byo). Its impact upon Volta has imparted quite an amount of ejecta onto [its] floor. Depth ~ 2.5 km.

Craters

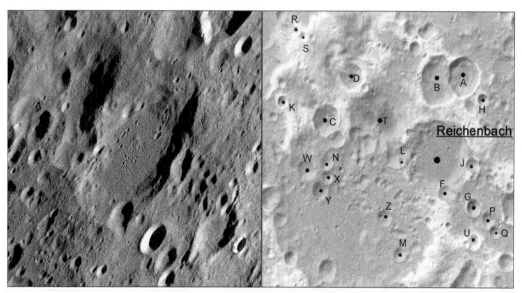

Reichenbach Lat 30.48S Long 47.95E 64.85Km

Sub-craters:

Crater	Lat	Long	Size (km)	Crater	Lat	Long	Size (km)
A	28.28S	48.87E	33.5	B	28.38S	48.08E	42.57
C	29.4S	43.91E	25.37	D	28.22S	44.74E	34.42
F	31.41S	48.27E	13.47	G	31.77S	49.35E	14.55
H	28.88S	49.61E	10.72	J	30.64S	49.25E	13.04
K	28.88S	42.39E	10.72	L	30.59S	46.67E	6.84
M	33.03S	46.54E	11.98	N	30.57S	43.91E	13.33
P	32.08S	49.9E	12.71	Q	32.4S	50.16E	10.61
R	27.0S	42.95E	7.17	S	27.17S	43.2E	9.57
T	29.39S	45.75E	61.5	U	32.69S	49.39E	13.8
W	30.73S	43.18E	19.57	X	30.94S	43.91E	11.12
Y	31.28S	43.66E	15.8	Z	32.07S	45.99E	14.01

Notes:

Add Info:

Aerial view of Reichenbach

Given the state and appearance of Reichenbach crater, it's safe to say it's a crater of age (possibly pre-Nectarian 4.6 to 3.92 byo). It lies in a trough (between a two ring/rim system - 860 km and 1300 km in diameter respectively) related to formation of the Nectaris Basin to its northwest; whose ejecta effect, particularly at the crater's northeastern sector, has partially covered over the rim there (note also the chain of craters to its outer southeast that 'point' back to Nectaris). A lovely cluster of small craterlets (produced by what?) appear nicely on the floor. Depth ~ 3km approx..

Craters

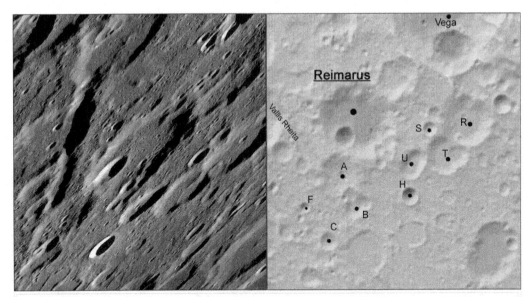

Reimarus   Lat 47.69S   Long 60.42E   47.62Km

Sub-craters:

Crater	Lat	Long	Size (km)	Crater	Lat	Long	Size (km)
A	48.85S	59.93E	29.42	B	49.49S	60.49E	18.85
C	50.19S	59.51E	12.36	F	49.49S	58.71E	6.92
H	49.26S	62.23E	11.17	R	47.65S	64.0E	33.22
S	47.85S	62.83E	9.57	T	48.33S	63.43E	28.08
U	48.57S	62.3E	22.05				

Notes:

Add Info:

Aerial view of Reimarus

One good location-marker for finding crater Reimarus is to use the extraordinary 'gash' of Vallis Rheita as a pointer to it (it's, roughly, the last crater on its end). The gash, of course, is the result of ejecta coming from the Nectaris Basin to the northwest, and while [it], physically, hasn't affected Reimarus itself, there is a less-smaller one just to its east that has. Looking over at pre-Nectarian Vega to its northeast, Reimarus looks older still, however, as Nectaris ejecta covers this region, in general, smaller craters may look older than they actually are. The unnamed, ~ 15 km-sized crater on the floor is some 2.3 km below its level of the floor. D:~3 km.

Craters

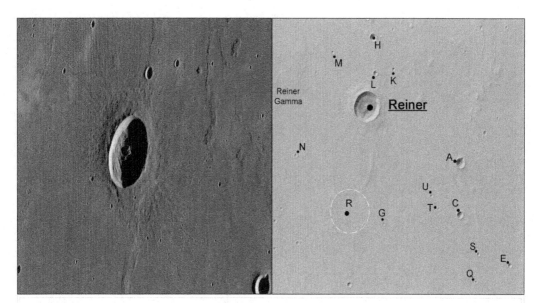

Reiner — Lat 6.92N   Long 54.98W   29.85Km

Sub-craters:

Crater	Lat	Long	Size (km)	Crater	Lat	Long	Size (km)
A	5.13N	51.51W	9.87	C	3.5N	51.52W	7.12
E	1.9N	49.64W	4.46	G	3.23N	54.34W	3.08
H	9.1N	54.76W	7.68	K	8.11N	53.98W	3.13
L	7.97N	54.66W	5.36	M	8.62N	56.25W	3.0
N	5.37N	57.6W	3.67	Q	1.38N	50.92W	2.88
R	3.68N	55.69W	46.08	S	2.26N	50.84W	3.39
T	3.68N	52.32W	2.57	U	4.1N	52.58W	3.24

Add Info:
Reiner Gamma
Several theories have been proposed as to its formation. One suggests unusually strong, localised magnetic fields in the region may have deflected away solar wind particles that are known to darken the lunar surface; while another has shockwaves, produced by a meteoroid impact that occurred on the farside of the Moon, forced lighter material to the surface - giving us the wonderful feature we see today.

Notes:

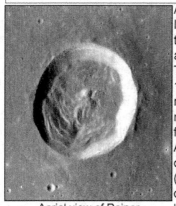
Aerial view of Reiner

As Reiner's rim is some 800 metres above the level of the mare lavas in where it impacted, it therefore produces some wonderful shadows and effects - particularly during low sun times. The crater is aged in the Eratosthenian (3.15 to 1.1 byo) and so it is relatively young, however, not so so as to have left bright ray signatures remaining, as seen in more younger craters like, for example, Kepler or Copernicus to its east. A wonderful series of sharp, slumped deposits occurred at the crater's southern inner sectors (they may have happened very shortly after the crater formed), and its outer ejecta distribution hints Reiner may be a low impact event. D: 3 km.

Craters

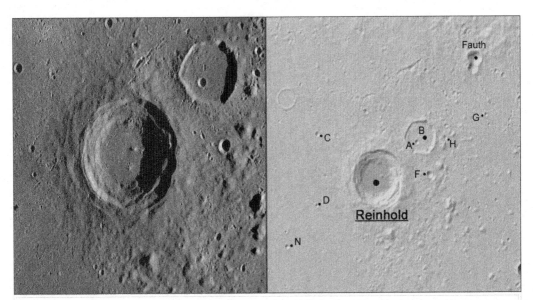

Reinhold	Lat 3.28N	Long 22.86W	43.28Km

Sub-craters:

Crater	Lat	Long	Size (km)	Crater	Lat	Long	Size (km)
A	4.16N	21.79W	3.74	B	4.28N	21.67W	26.16
C	4.36N	24.57W	4.08	D	2.59N	24.6W	2.34
F	3.36N	21.42W	5.57	G	4.83N	19.83W	3.16
H	4.27N	20.98W	3.72	N	1.58N	25.43W	3.83

Notes:

Add Info:

Aerial view of Reinhold

Reinhold, whilst looking almost like a miniatuized version of the ever-wonderful crater Copernicus to its north, it's obvious from the outstart that both are of a different age. The crater is of the Eratosthenian (3.15 to 1.1 byo) Period, it doesn't have the wonderful display of bright rays, its ejecta signature has almost blended into the clump of material it lies upon (possibly Imbrium ejecta?), and it lacks that certain 'oooomphhh', which Copernicus continually shouts. It does, however, have a wonderful series of terraces that Copernicus has, and a few small peaks that lie on its floor (highest is about 350 metres). Full moon times will always see Copnicus 'shine', but you will have to skew your eyes to find Reinhold. Depth ~ 2.7 km.

Craters

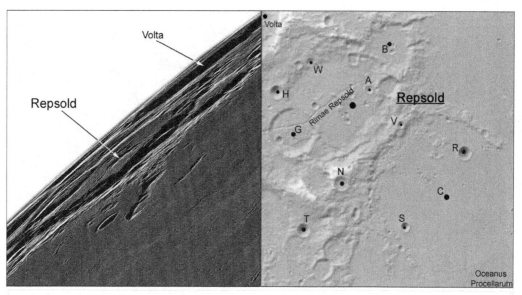

Repsold	Lat 51.31N	Long 78.41W	108.65Km

Sub-craters:

Crater	Lat	Long	Size (km)	Crater	Lat	Long	Size (km)
A	51.84N	76.98W	8.18	B	53.09N	75.92W	45.67
C	48.73N	73.78W	135.09	G	50.55N	80.65W	43.98
H	51.69N	81.48W	12.62	N	49.06N	78.29W	14.54
R	49.86N	72.4W	12.49	S	47.75N	75.33W	8.65
T	47.64N	79.95W	12.66	V	50.77N	75.41W	7.14
W	52.6N	79.88W	9.64				

Notes:

Add Info:

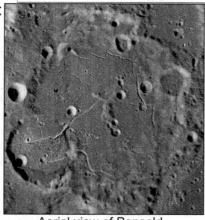

Aerial view of Repsold

It's a pity that Repsold lies where it does, as from an overhead perspective (left), it really is a crater worth listing amongst some of the best floor-fractured craters (for example, Posidonius, Gassendi, Petavius and more) on the Nearside. The fractures are of course responses to underlying stresses in the crater (the floor is very comvexed), and it's a wonder that more of its centre isn't flooded by lavas sourced deep down (Repsold's floor may be very thick). Both the jutting arms of sub-crater, Repsold C, and Repsold's own peak (~ 800 m), act as good markers for noting the crater's location. Depth is around 1.9 km approx..

Craters

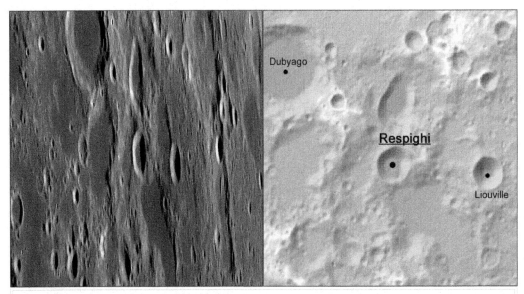

| Respighi | Lat 2.86N | Long 71.9E | 17.88Km |

Sub-craters: No sub-craters

Notes:

Add Info:

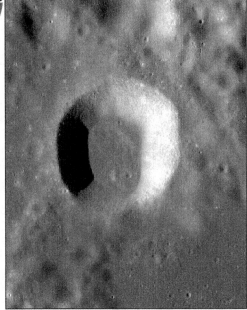

Aerial view of Respighi

Respighi lies in a region where the terrain takes on a more mottled look - not of craters but of puddles of lava deposits. The reason, of course, is that two major Basins - Crisium to the north and Smythii to the southeast - 'churned' up the terrain so much so that fractures in the crust allowed for deep down lavas to seek the surface. Respighi itself is relatively fresh-looking - it is very similar-looking to Liouville over to its east, but it really hasn't much to offer in terms of notable features. Both craters align back nicely towards the Tranquillitatis Basin to the west - a suggestion that they might be secondaries from that event, but they can't be as [that] basin is just too old. What then might be their source?

Craters

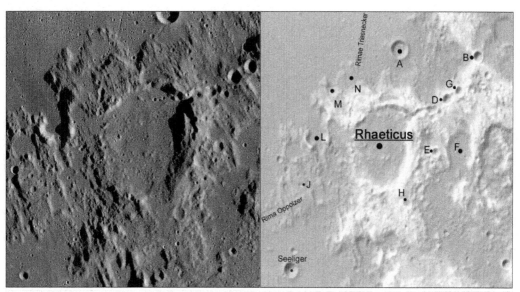

Rhaeticus   Lat 0.03N   Long 4.92E   44.43Km

Sub-craters:

Crater	Lat	Long	Size (km)	Crater	Lat	Long	Size (km)
A	1.73N	5.18E	10.5	B	1.6N	6.8E	6.29
D	0.84N	6.17E	6.07	E	0.06S	5.93E	4.96
F	0.06S	6.44E	18.42	G	1.03S	6.41E	5.52
H	1.03S	5.34E	5.56	J	0.2N	3.18E	2.77
L	0.2N	3.48E	13.78	M	1.19N	3.81E	7.33
N	1.19N	4.21E	12.38				

Notes:

Add Info: It's obvious from looking at Rhaeticus, along with several other craters nearby, that something major has affected the whole region. The answer, of course, is Imbrium - this huge basin that formed nearly 4 billions years ago, and affecting nearly every feature for hundreds and hundreds of kilometres away from its explosive point. Rhaeticus seems to have born extensive abuse, as tons and tons of Imbrium ejecta not only scoured into high points of its rim, but plopped volumes into its centre (look at the crater's northern inner rim - the ejecta there must have simply 'spilled' over into the crater, still in its fluidized state). As a result, all we really see of Rhaeticus's originality is the eastern sector of its rim, which has a height above the floor of around 2 kilometres. The western sector of the rim is virtually non-existant as it blends into local terrain and [said] ejecta. Look at the chain of craters just outside Rhaeticus's northeastern rim (one actually 'cuts' the rim), and follow their line over to Rima Oppolzer in the west. Are they related? A similar-like relationship is suggested at Rhaeticus's north where a branch of Rimae Triesnecker (~ 130 km away) seems to follow a north-south trending line that 'connects' to a smaller chain at the crater's outer northwest.

Craters

Rheita	Lat 37.1S	Long 47.17E	70.81Km

Sub-craters:

Crater	Lat	Long	Size (km)	Crater	Lat	Long	Size (km)
A	38.05S	50.05E	11.46	B	39.11S	52.71E	19.36
C	35.17S	44.21E	9.69	D	39.05S	50.09E	5.51
E	34.24S	48.92E	67.72	F	35.38S	48.41E	13.49
G	40.61S	54.3E	13.88	H	39.81S	51.66E	6.93
L	37.69S	52.99E	10.74	M	35.35S	50.05E	25.92
N	35.08S	49.41E	7.32	P	37.95S	44.44E	10.32

Notes:

Add Info:

Aerial view of Rheita

The long valley-like feature to Rheita's west was created by ejecta from the Nectaris Basin to the northwest. Rheita later impacted the valley (officially known as Vallis Rheita), and so it must be younger than a time when the Basin formed (3.95 to 3.85 billions of years ago). But looking at the state of its rim and inner walls, and it's clear that Rheita isn't that far off of being similar in age (compare it to crater Metius to its southwest, which is aged of the Nectarian). Rheita has a depth of over four kilometres in parts, so it really was a deep-down type impact event, however, material on its floor doesn't look lava-like. The crater has also impacted on an unnamed crater to its east whose floor level is some 2 km higher. D:1.2 km.

Craters

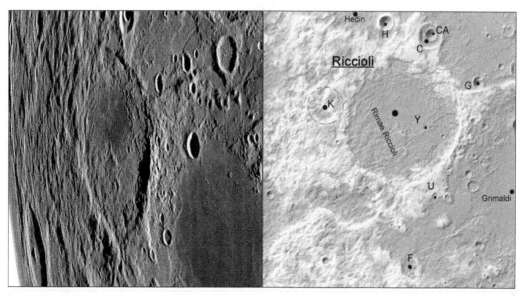

Riccioli	Lat 2.9S	Long 74.42W	155.66Km

Sub-craters:

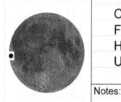

Crater	Lat	Long	Size (km)	Crater	Lat	Long	Size (km)
C	0.58N	73.2W	31.55	CA	0.64N	73.15W	14.13
F	8.73S	73.89W	30.27	G	1.27S	71.15W	14.97
H	1.12N	75.04W	17.46	K	2.32S	77.63W	40.76
U	5.79S	72.95W	8.39	Y	3.04S	73.35W	6.73

Notes:

Add Info:

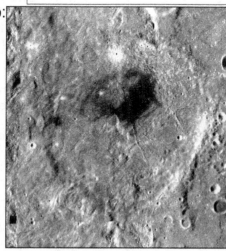

Aerial view of Riccioli

If you happen to lie some ~ 750 kilometres away from one of the biggest impact basins on the Moon, then chances are your existing features will receive huge volumes of ejecta from that event. The basin, of course, is Orientale, and its effect has produced a lineated look to Riccioli as clumps and rocks scoured into high points (rim, peaks...etc.,) of the crater, and increasing its floor level by several hundred metres upwards. Look over at Grimaldi's flooded lava floor to its southeast - would Riccioli's floor initially have looked like it before the onslaught of Orientale's ejecta (the floor subsequently underwent serious fracturing). Depth of Riccioli is around ~ 2.8 km.

Craters

Riccius	Lat 37.02S	Long 26.43E					71.79Km

Sub-craters:

Crater	Lat	Long	Size(km)	Crater	Lat	Long	Size(km)
A	35.89S	27.21E	23.51	B	37.62S	27.7E	21.32
C	36.24S	28.77E	21.81	D	40.4S	28.87E	15.43
E	39.96S	26.38E	21.41	G	38.56S	24.44E	13.37
H	35.36S	26.07E	17.86	J	40.76S	26.0E	12.61
K	39.17S	25.62E	5.23	L	41.55S	26.83E	6.55
M	37.9S	26.5E	13.68	N	41.21S	27.57E	12.89
O	36.19S	27.68E	8.74	P	35.73S	28.04E	10.86
R	41.44S	30.66E	5.51	S	37.15S	26.47E	9.49
T	36.34S	24.97E	5.87	W	38.95S	25.17E	19.26
X	38.84S	26.59E	10.2	Y	35.85S	29.11E	8.9

Notes:

Add Info:

Aerial view of Riccius

Straight away you can see that Riccius is a crater of age (it is pre-Nectarian 4.6 to 3.92 byo). That left it open to subsequent coverage of ejecta from formation of the Nectaris Basin to its northeast (centred some ~ 700 km away). As a consequence, all that we really see of Riccius's original formation is in its western rim, as practically every other portion, including its floor, is totally altered through impacts, infill...etc. Both craters, Rabbi Levi and Zagut (also pre-Nectarian), to its northwest are similar in size to Riccius, and in a way, the trio make for a good method for noting/remembering each. D: ~ 1.7 km.

540

Craters

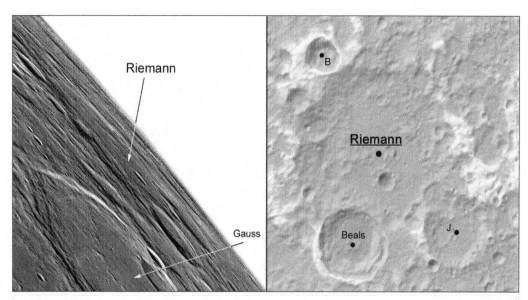

Riemann		Lat 39.38N	Long 87.18E				117.85Km

Sub-craters:

Crater	Lat	Long	Size (km)	Crater	Lat	Long	Size (km)
B	41.41N	85.56E	26.26	J	37.42N	89.79E	48.54

Notes:

Add Info:

Aerial view of Riemann

There's something very odd, at first, to the general appearance of Riemann - said to be Nectarian - 3.92 to 3.85 byo in age. It looks way older than its neighbour, Gauss, (also Nectarian) to its west, and looks as if it is covered over by material that Gauss escaped. What would this material be, and what was its source? It can't be ejecta from the Humboltianum Basin to its north as the crater lies on an outer ring/rim of it, so is Gauss's ejecta the culprit? There are striated features between the two craters that 'point' back towards the Crisium Basin in the southwest, so could that also be a candidate? If so, the sequence of events might have went like this: Humboltianum formed first, then Riemanm upon its ejecta, then Crisium ejecta covered Riemann, and Gauss then not long afterwards. Depth ~ 3 km.

Craters

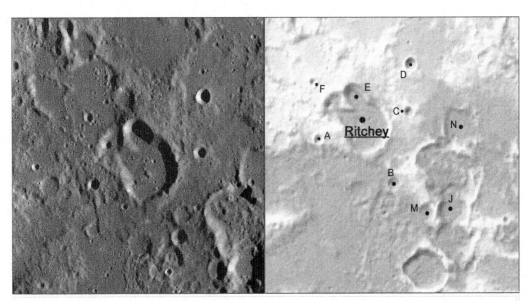

Ritchey	Lat 11.13S	Long 8.48E	24.07Km

Sub-craters:

Crater	Lat	Long	Size (km)	Crater	Lat	Long	Size (km)
A	11.34S	7.74E	4.68	B	11.98S	8.93E	6.03
C	10.95S	9.18E	5.18	D	10.26S	9.21E	5.84
E	10.72S	8.31E	12.37	F	10.52S	7.65E	3.68
J	12.41S	9.88E	15.65	M	12.39S	9.47E	8.2
N	11.16S	10.0E	16.86				

Notes:

Add Info: Ritchey is situated in an area where extensive ejecta deposits from the Nectaris Basin (centred some ~ 800 kilometres over to its east) has fallen, but also from the Imbrium Basin (centred some ~ 150 kilometres to its northwest). Of course, we could go down the Formations' route here, too, when viewing Ritchey; mentioning both the Descartes and Cayley Formations (a fluidized ejecta form of Imbrium) - the latter being more lighter-toned and smoothed. The crater itself has an odd shape, almost resembling an outline of 'Pooh Bear' - the fictional bear created by A. A. Milne. Of course, the shape is due to several impacts upon Ritchey, which essentially has wiped out all of its northwest sector, and extended its west (Pooh's right ear). The left ear was obviously created by a single impact, however, it gets very confusing as to figuring out how the rightmost one formed. Is it made up of two coallesced impact craters, or is it just a single impact crater whose northwest sector slumped down afterwards, then later covered by the basin ejecta (Pooh's left ear is obviously younger than his right). Peak is 400 metres high, and depth of crater is 1.3 kilometres.

Craters

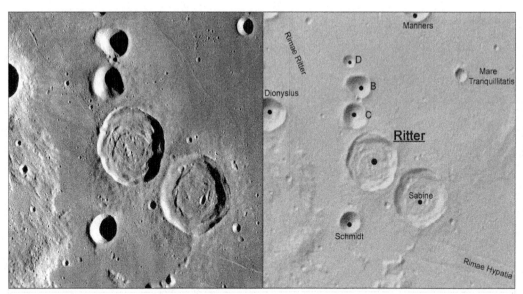

Ritter	Lat 1.96N	Long 19.17E	29.52Km

Sub-craters:

Crater	Lat	Long	Size (km)	Crater	Lat	Long	Size (km)
B	3.25N	18.93E	14.2	C	2.75N	18.86E	12.68
D	3.65N	18.74E	6.58				

Notes:

Add Info: Ritter, along with its southeastern partner, Sabine, joins those elite groupings of craters where one unavoidably cannot, should not, be viewed without the other. The two make for a nice pair in the eyepiece, acting as a good exercise for the observer in comparing similar-looking, similar-sized craters lying close together. Both formed probably in the Upper Imbrian Period (3.75 to 3.2 byo), both show ejecta deposits on lavas of Mare Tranquilitatis, and each has features not lacking in the other. The features include slumping and terraces, concentric rilles to fractures, hint of peaks, and relatively sharp rims all around. Close inspection of each, however, shows there's a slight difference between all, with Ritter's being slightly sharper than Sabine's (look at Sabine's northwestern rim - it looks like some of Ritter's ejecta may have landed upon it - 'softening' its appearance). The outer rim area - a sort of a highway channel 'connecting' both craters - is some 1100 metres higher than the level of the mare, and, as expected, there's a slight 'dip' of both craters - towards its central region and also due southeastwards. The craters under most lighting conditions always show similar colouring to the mare, while at full moon times their rims and floor features display as lighter (dark, and bright, rays, from crater Dionysius to the northwest just about kiss Ritter's rim). Depth of each crater is around ~ 1.3 km.

Craters

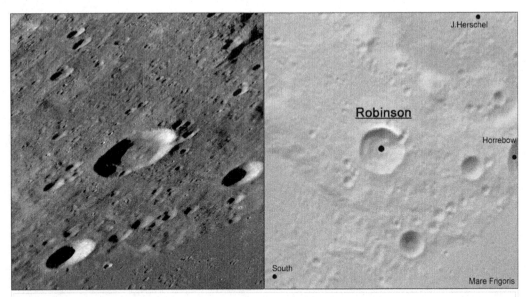

| Robinson | Lat 59.06N | Long 46.03W | 24.09Km |

Sub-craters: No sub-craters

Notes:

Add Info:

Aerial view of Robinson

Robinson lies on a 70-kilometre-wide 'tongue' of highlands betw- crater J. Herschel to its northeast and crater South to its southwest. Both of these craters (aged of the pre-Nectarian Period - 4.6 to 3.92 byo), as well as the tongue, have been covered over with what is believed to be Imbrium Basin ejecta from the southeast, however, Robinson doesn't show such characteristics as it impacted upon all. Robinson is thus younger, its floor is relatively flat, and slumping features appear at it southwest sector. A small 'gash' at its northeastern rim suggests a small impact crater 'clipped' it there, however, this feature could also be the result of Robinson impacting upon a small, pre-existing crater (the feature looks somewhat odd). Depth 2.5 km.

Craters

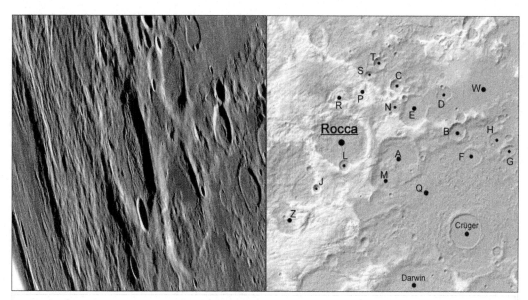

Rocca		Lat 12.89S	Long 72.89W			84.06Km	

Sub-craters:

Crater	Lat	Long	Size (km)	Crater	Lat	Long	Size (km)
A	13.76S	70.32W	64.25	B	12.58S	67.46W	23.43
C	10.7S	70.33W	17.47	D	11.02S	68.09W	23.35
E	11.75S	69.59W	43.2	F	13.63S	66.72W	26.56
G	13.32S	65.02W	22.7	H	12.92S	65.53W	24.52
J	14.97S	74.03W	12.76	L	14.05S	72.74W	17.1
M	14.57S	70.89W	44.13	N	11.66S	70.33W	24.2
P	11.12S	71.89W	31.82	Q	15.29S	69.14W	58.51
R	11.17S	73.13W	34.82	S	10.25S	71.64W	10.37
T	9.76S	71.14W	15.94	W	10.26S	66.52W	104.65
Z	16.23S	75.54W	54.36				

Notes:

Add Info:

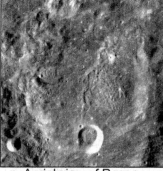

Aerial view of Rocca

Not to be confused with crater Rocco (found at Lat 28.91N, Long 45.0W, and east of crater Krieger), Rocca is a crater of the Nectarian Period - 3.92 to 3.85 byo). Lying some ~ 600 km away northeastwards from the Orientale Basin's centre, the crater's floor and rim has been covered by ejecta from that major event. Volumes of the ejecta distributed unequally across the floor - its eastern half level is some ~ 500 m higher than its west (it may also have covered internal lava deposits there, too, similar to those in Rocca A). Use Rocca L as a marker for finding the crater.

Craters

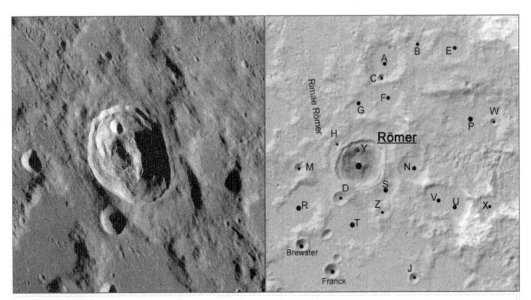

Römer	Lat 25.43N	Long 36.41E	43.7Km

Sub-craters:

Crater	Lat	Long	Size (km)	Crater	Lat	Long	Size (km)
A	28.1N	37.13E	34.82	B	28.5N	38.11E	22.51
C	27.66N	37.06E	7.24	D	24.51N	35.81E	12.08
E	28.48N	39.18E	30.76	F	27.19N	37.25E	23.16
G	26.82N	36.24E	14.67	H	25.95N	35.72E	6.49
J	22.38N	37.95E	8.54	M	25.29N	34.61E	9.14
N	25.31N	37.97E	25.54	P	26.54N	39.81E	59.37
R	24.15N	34.45E	52.57	S	24.91N	36.79E	43.03
T	23.82N	36.17E	44.65	U	24.28N	39.15E	27.23
V	24.43N	38.64E	26.71	W	26.44N	40.41E	6.07
X	24.29N	40.12E	23.92	Y	25.76N	36.31E	6.17
Z	24.13N	37.05E	11.6				

Notes:

Add Info: Luna 21, or Lunik 21, landed some 160 km away west from this crater on the 15 Jan 1973. It deployed a rover, 'Lunokhod 2', afterwards onto the surface for research purposes.

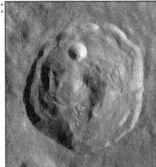

Aerial view of Römer

The reason why Römer looks the way it does is because of the extensive slumping on nearly every part of its rim and walls. The featured effect is almost stair-like; where several 'stepped' slumpings meet several others - it gets very complex (see, for example, the southern sector in the left image). Did such slumpings happen through seperate event hundreds of years apart, or, are we looking at a single event (was impact of Römer Y responsible?)? The slumped deposits reached in as far as Römer's 1.7-km-high peak - so much so as to 'round' its top. Depth of Römer ~ 3.5 km.

Craters

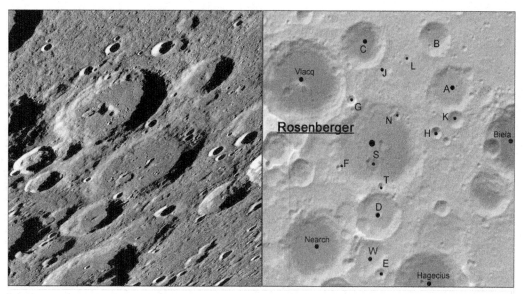

Rosenberger   Lat 55.49S   Long 43.15E   91.65Km

Sub-craters:

Crater	Lat	Long	Size (km)	Crater	Lat	Long	Size (km)
A	53.65S	47.17E	48.27	B	52.06S	46.21E	33.76
C	52.28S	42.2E	46.67	D	57.6S	43.04E	46.58
E	59.43S	43.15E	11.14	F	56.09S	40.6E	5.84
G	53.99S	41.45E	9.81	H	54.98S	46.47E	11.89
J	53.04S	43.26E	20.54	K	54.53S	47.58E	18.03
L	52.69S	44.67E	7.91	N	54.48S	44.13E	8.25
S	56.04S	42.69E	14.04	T	56.71S	43.22E	7.92
W	58.86S	42.65E	30.82				

Notes:

Add Info:

Aerial view of Rosenberger

It's quite easy to get lost here in this corner of the Moon where numerous, similarly-sized craters dominate the view. However, 'groupings' of craters can be seen (this doesn't imply a relationship between their impacts), which act as a good way for remembering where one is when looking at a specific feature or target. Rosenberger lies in a group that includes Piticus, Vlacq, Hommel, Nearch, Hagecius and Biela over to its east - all of which are 'oldies' - ranging from the Nectarian to the pre-Nectarian Period). View Rosenberger at low sun times, as during full moons it simply is 'gone'. Depth ~ 3 km.

Craters

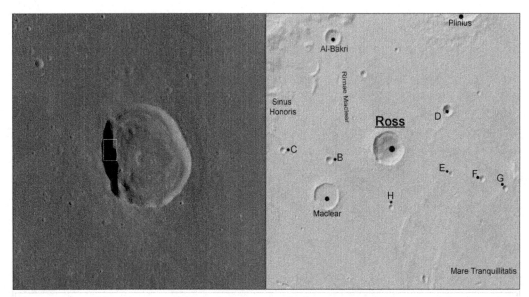

Ross	Lat 11.67N	Long 21.74E		24.49Km			
Sub-craters:	Crater	Lat	Long	Size (km)	Crater	Lat	Long	Size (km)
	B	11.38N	20.17E	5.62	C	11.62N	18.93E	4.75
	D	12.57N	23.32E	8.41	E	11.04N	23.41E	4.2
	F	10.91N	24.23E	4.56	G	10.68N	24.86E	4.55
	H	10.24N	21.81E	4.17				
	Notes:							

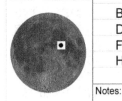

Add Info:

Ranger 6 impacted some 75 km away southwards from this crater on the 2 Feb 1964.

Flow-like signatures on the western wall of Ross

Like crater Plinius to its northeast, Ross is of the Eratosthenian Period (3.15 to 1.1 byo). The two make for a good exercise in comparing how the rims, ejecta and floors in each developed as a consequence to the sizes of the initial impactors that produced them. Obviously, Plinius's impactor must have been slightly bigger than Ross's, however, the resultant features in both (peaks, terracing, crenulation of rim...etc.,) all look similar on a scale to the respective sizes of each crater. Such exercises don't always apply, as other factors like target rock, angle of approach and impactor make-up will determine crater size and features formed. But as we're looking here in where the two formed on Tranquillitatis lavas, each showing no signatures of obliqueness (note ejecta distribution), could we imply impactor make-up was the same. D: 1.9 km.

Craters

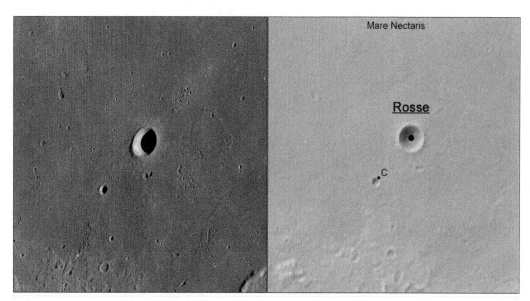

Rosse	Lat 17.95S	Long 34.98E			11.43Km		
Sub-craters:	Crater	Lat	Long	Size (km)	Crater	Lat	Long	Size (km)
	C	18.55S	34.43E	3.92				
	Notes:							

Add Info:

Not to be confused with similar-sounding crater, Ross, found at 11.67N, 21.74E.

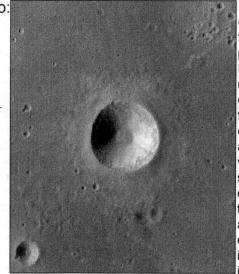

Aerial view of Rosse

Rosse looks quite alone on the lavas of Mare Nectaris. Given its relatively fresh-looking appearance, the crater probably formed in the Eratosthenian Period (3.15 to 1.1 byo). With it's almost perfect circular shape, its evenly-distributed ejecta, and its bowl-shaped floor, the impactor that produced the crater most likely came in at a high angle to the mare surface. There is a very obvious ray off to its northeast to suggest an oblique impact event, however, this ray is most likely one related to formation of Tycho crater ~1400 km away in the opposite direction. Viewed at any time, Rosse always makes it presence known - at low sun times its floor is usually filled in shadow, while full moon times it 'shines'. D:2.4km.

Craters

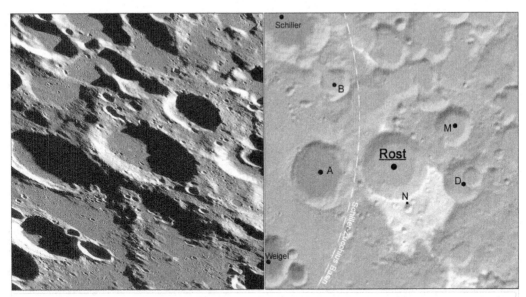

Rost   Lat 56.42S   Long 33.84W   46.85Km

Sub-craters:

Crater	Lat	Long	Size (km)	Crater	Lat	Long	Size (km)
A	56.62S	36.84W	44.65	B	54.68S	36.3W	20.25
D	56.63S	31.0W	30.55	M	55.5S	31.52W	26.77
N	57.29S	33.22W	5.98				

Notes:

Add Info:

Aerial view of Rost

Rost is quite easy to find in this overpopulated cratered region of the Moon when one remembers that Schiller's long axis sort of physically 'points' to it. Rost lies just outside the main 335-km diameter-wide ring/rim of the Schiller-Zucchius Basin (pre-Nectarian in age), and on its ejecta that together overlaid original highlands (the area due south of Rost's rim is nearly 3 km higher than the basin floor level). As a consequence of this higher terrain on which Rost lies, it hasn't 'dug' down deep enough to possibly 'tap' into the mare materials (that may have happened with craters like Schiller or Rost A). Smoother materials in Rost (as too with several other craters nearby) may be Orientale ejecta.

Craters

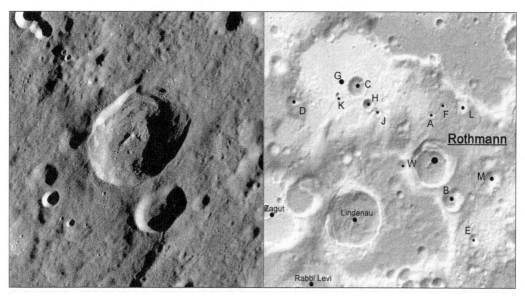

Rothmann	Lat 30.81S	Long 27.7E	41.67Km

Sub-craters:

Crater	Lat	Long	Size (km)	Crater	Lat	Long	Size (km)
A	29.46S	27.59E	7.35	B	31.82S	28.42E	21.28
C	28.64S	25.01E	17.87	D	28.97S	22.78E	12.87
E	32.94S	29.19E	9.64	F	29.19S	28.01E	6.47
G	28.39S	24.39E	92.01	H	29.16S	25.4E	11.15
J	29.39S	25.73E	6.85	K	28.86S	24.36E	5.2
L	29.22S	28.67E	13.19	M	31.24S	29.84E	16.42
W	30.87S	26.57E	10.36				

Notes:

Add Info:

Aerial view of Rothmann

Like Lindenau over to its southwest, Rothmann seems to have struck upon an old unnamed crater some 130 km in diameter. Both craters look relatively the same (Lindenau is of the Late Imbrian Period - 3.75 to 3.2 byo), but is Rothmann just that little bit younger? Major slumping in the crater occurred nearly in every sector of its walls, but it's at the east is where it is predominantly obvious, where it reaches as far into the centre to meet the odd peak (~ 900 metres high). The crater formed on ejecta deposits of the Nectaris Basin over to its northeast, and lies just 60 km away from its main ring/rim (Rupes Altai). Depth ~ 4 km.

Craters

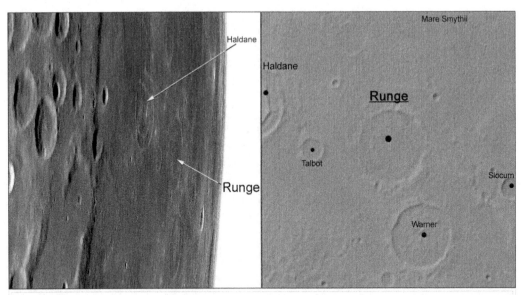

Runge   Lat 2.43S   Long 86.81E   38.89Km

Sub-craters: No sub-craters

Notes:

Add Info:

Aerial view of Runge

Craters and features on the extreme limbs of the Moon can be hard enough to observe - even when they have significant high rims or other terrain characteristics that can cast shadows to 'outline' their presence. However, when such features are half-buried-to-death by lavas - as is here with crater Runge in the centre of Mare Smythii, then it's an additional challenge for the observer to put up with. Runge's rim is just about 200 metres above the level of the lavas, so any shadows from it will be hard to detect. Like several other craters nearby, for example, Warner to the south, Haldane to the west, Runge shows additional, inner ring-like features within the main rim. Are they signatures of complex crater dynamics as they formed on Smythii's floor? Librations are required for viewing Runge (also try just after Full Moon).

Craters

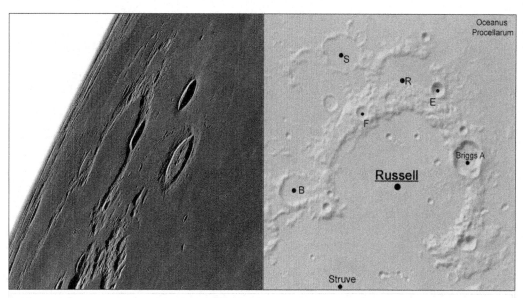

| Russell | Lat 26.51N | Long 75.55W | 103.37Km |

Sub-craters:

Crater	Lat	Long	Size (km)	Crater	Lat	Long	Size (km)
B	26.39N	78.3W	20.49	E	28.6N	74.58W	10.03
F	28.09N	76.57W	9.4	R	28.92N	75.43W	40.38
S	29.34N	77.1W	24.76				

Notes:

Add Info:

Aerial view of Russell

You just can't view Russell without noting its two other bigger neighbours - Struve to its south and Eddington further to its southeast. All have been flooded by lavas similar to those found in the vast plains of Oceanus Procellarum (note the breach points in all three craters), and further bombardment by ejecta from the Orientale Basin (due southwestwards some 1400 km away). An obvious 'gash' by this ejecta is seen on Russell's outer northwestern rim, however, less-obvious signatures are found on its southeastern rim. All three craters are old, however, which formed first (is Russell the youngest?)? Librations are required to get any descent view, or photograph, of the crater. Depth ~ 0.9 km.

Craters

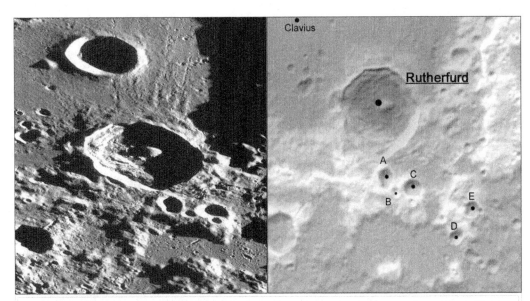

Rutherfurd   Lat 61.15S   Long 12.25W   49.98Km

Sub-craters:

Crater	Lat	Long	Size (km)	Crater	Lat	Long	Size (km)
A	62.31S	11.99W	10.45	B	62.57S	11.58W	6.27
C	62.54S	10.9W	13.34	D	63.32S	8.96W	9.02
E	62.83S	8.38W	8.78				

Notes:

Add Info:

Aerial view of Rutherfurd

Rutherfurd should equally be considered as interesting, and eye-catching, as crater Clavius on which it has impacted upon. Copernican in age (1.1 byo to the Present), it's got a sharp, crenulated rim, several impressive terraces (particularly at its south), an off-centred peak (~ 1 km high), lava deposits (or is it impact-melt?) in parts of its floor (its level is about a kilometre below Clavius's floor level), and a very odd display of ejecta on its outer northeastern sector (the ejecta looks like it produced a series of closely-packed craters resembling chains). Observing the show-offs, like Clavius and others, can at times distract us away from those 'less showie'. However, imagine Clavius without Rutherfurd, and it's obvious why it shouldn't be ignored.

Craters

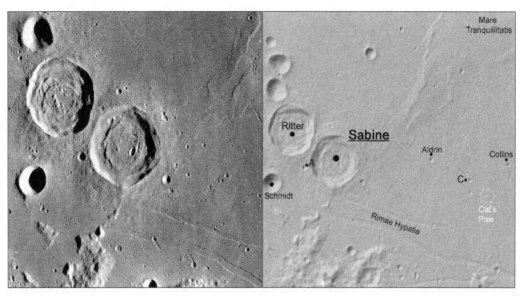

Sabine	Lat 1.38N	Long 20.07E		29.75Km			
Sub-craters:	Crater	Lat	Long	Size (km)	Crater	Lat	Long	Size (km)
	A	1.24N	19.46E	3.65	C	1.02N	22.92E	2.84
	Notes:							

Add Info:

Missions Apollo 11 and Surveyor 5 landed some 100 km away from this crater, as too with Ranger 8 that impacted the surface on 20 Feb 1965.

Dark deposits on Sabine's floor

Both Sabine and its neighbour, Ritter to its northwest, look almost similar - each showing old terraces, concentric ridges and rilles, and hummocky floors - a state of their respective features suggest Sabine the older of the two. Sabine's northern floor (left) shows several small craterlets to have a dark 'splotch' around them (they look almost chain-like having a NW-SE trend). Usually, such dark-halos are produced by the impactor excavating dark deposits to the surface, however, in this case the trending characteristic suggest something very odd (see an obvious dark-haloed crater (unnamed) outside Sabine's SE rim.

Craters

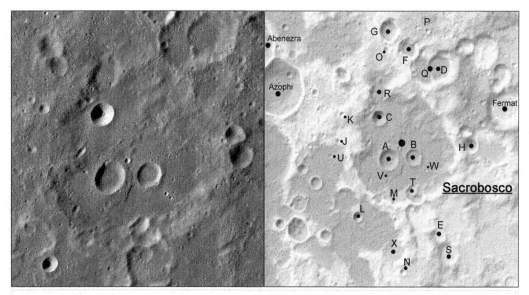

Sacrobosco   Lat 23.75S   Long 16.64E   97.67Km

Sub-craters:

Crater	Lat	Long	Size (km)	Crater	Lat	Long	Size (km)
A	24.1S	16.12E	16.49	B	24.04S	16.89E	13.22
C	23.0S	15.82E	12.55	D	21.67S	17.8E	26.52
E	26.11S	17.69E	12.18	F	21.16S	16.73E	17.56
G	20.7S	16.15E	18.11	H	23.72S	18.68E	12.08
J	23.61S	14.53E	4.8	K	22.9S	14.69E	5.54
L	25.61S	15.12E	8.42	M	25.3S	16.27E	6.93
N	27.09S	16.56E	5.73	O	21.19S	16.05E	4.84
P	20.7S	17.32E	4.57	Q	21.66S	17.39E	36.37
R	22.32S	15.78E	20.55	S	26.6S	17.98E	19.6
T	24.94S	16.83E	11.54	U	23.99S	14.3E	3.69
V	24.56S	16.13E	3.61	W	24.38S	17.28E	2.27
X	26.63S	16.27E	23.27				

Notes:

*A similar feature also appears across crater Scheiner's floor (60.35S, 27.81W).

Add Info: Pre-Nectarian in age (4.6 to 3.92 byo), Sacrobosco lies in a zone where formation of the Nectaris Basin - centred some 500 km away to its northeast, deposited fluidized and blocky ejecta across its rim and floor (at low sun times striated features can be seen to 'point' back to the Basin). The crater looks odd when considered a single impact event, as looking over to its western sector, in general, it appears like two 40-kilometre-sized craters (sub-crater, Sacrobosco C lies on the northern one while sub-crater, Sacrobosco A, lies on the southern one) wiped out Sacrobosco's original rim there. Significant deposits (Nectaris ejecta and rim slumping?) from Sacrobosco's northeast meets ejecta at A that together, literally, halves the crater into a NE-SW ridge-like feature* across the floor (it drops down by ~ 500 m on the eastern side). Is this further evidence for the two 40 km craters proposed?

Craters

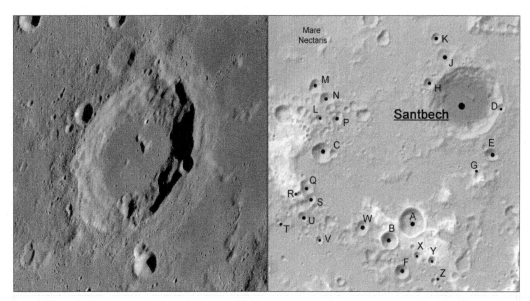

Santbech   Lat 20.99S   Long 44.06E   62.24Km

Sub-craters:

Crater	Lat	Long	Size (km)	Crater	Lat	Long	Size (km)
A	24.27S	42.29E	24.69	B	24.74S	41.57E	14.89
C	22.29S	39.51E	17.34	D	21.14S	45.14E	7.44
E	22.42S	44.84E	12.94	F	25.61S	41.94E	12.56
G	22.96S	44.39E	4.72	H	20.47S	42.88E	8.96
J	19.74S	43.39E	14.06	K	19.23S	43.12E	9.29
L	21.37S	39.46E	7.16	M	20.47S	39.36E	13.06
N	20.83S	39.66E	10.72	P	21.42S	40.0E	9.24
Q	23.22S	38.96E	12.71	R	23.41S	38.75E	5.5
S	23.53S	39.08E	10.02	T	24.13S	38.04E	4.68
U	24.03S	38.83E	8.79	V	24.68S	39.33E	6.81
W	24.38S	40.69E	14.28	X	25.18S	42.46E	7.77
Y	25.32S	42.94E	7.8	Z	25.88S	42.98E	5.37

Add Info:

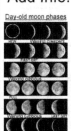

Day-old moon phases

Note: As the Moon goes through various librations throughout the year, suggested times given in text for observations are only approximates.

Notes:

Though it's not very obvious initially, Santbech lies on one of the main rings/rims (about ~ 600 kilometres in diameter) associated to formation of the Nectaris Basin - centred some 300 km away to the northwest. Later, upwelled lavas of that major event, and of others over to its northeast that 'connect' to Mare Fecunditatis eventually, seem to have filled in the lower terrain around the crater; leaving a well-defined, thick ejecta band around Santbech that blends in seamlessly to them. Lavas sourced internally in its floor also seem to have occurred; covering most of its central, once ring-like(?) peak ~ 0.6 km high. During full moon times, these floor lava signatures are seen in its western sector. Depth of Santbech ~ 3.5 km.

Craters

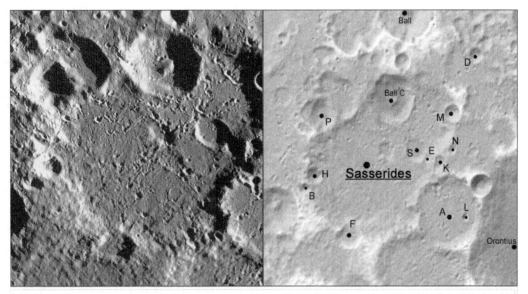

Sasserides	Lat 39.28S	Long 9.44W	81.74Km

Sub-cratersː

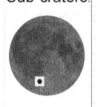

Crater	Lat	Long	Size (km)	Crater	Lat	Long	Size (km)
A	39.98S	7.11W	45.48	B	39.52S	11.23W	8.17
D	36.78S	6.46W	10.28	E	38.96S	7.72W	8.14
F	40.52S	9.96W	15.07	H	39.27S	10.94W	10.92
K	38.97S	7.39W	7.99	L	40.09S	6.63W	4.97
M	37.98S	7.06W	11.34	N	38.72S	7.06W	6.03
P	38.04S	10.81W	21.57	S	38.75S	8.08W	14.14

Notes:

Add Info:

Surveyor 7 landed some 70 km away due southwest from this crater on 10 Jan 1968

Aerial view of Sasserides

Immediately obvious when looking at crater Sasserides for the first time is that it's been through the impact wars. Practically, all of the northeastern half of its rim has been obliterated by major impact craters - Ball C, Sasserides A and an unnamed crater similar in size to Sasserides A (Ball C has impacted upon its northwestern sector). Of course, the instant feature we see of Sasserides (and other craters nearby) is the effect of Tycho's ejecta. Lying some 150 km away due southwards, Tycho not only coated Sasserides's floor with the material, but it also 'marked' it for life with small, pit-like secondaries. Depth ~ 1. 8 km.

Craters

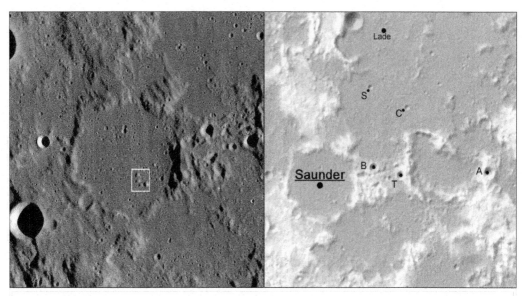

Saunder	Lat 4.26S	Long 8.72E		44.38Km		

Sub-craters:	Crater	Lat	Long	Size (km)	Crater	Lat	Long	Size (km)
	A	4.01S	12.26E	7.42	B	3.93S	9.81E	5.55
	C	2.77S	10.52E	3.3	S	2.38S	9.74E	3.17
	T	4.07S	10.39E	4.91				

Notes:

Add Info:

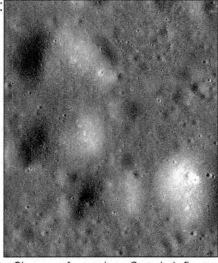

Close-up of mounds on Saunder's floor

Like most craters in this region (take in a wider view), Saunder has been subject to the Imbrium ejecta wars (note the 'groove-like' appearance on all of their rims that 'point' back to its direction. Infilling of Saunder's floor is probably a fluidized form of Imbrium ejecta, which levelled it almost to the rim (depth of the crater is at 0.6 km). Southeast on the crater's floor, an odd series of 'mounds' appear during low sun times. Are they volcanically-related, or, simply just the last vestiges of a small, half-buried impact crater whose rim has been altered by said Imbrium ejecta? Note, image (left) is a mosaic - the righmost 'mound' was taken under different lighting conditions to the other three.

Craters

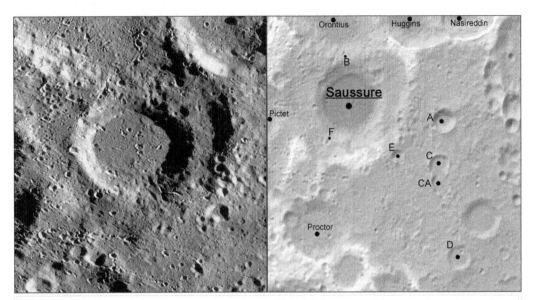

Saussure	Lat 43.38S	Long 3.88W		54.56Km			

Sub-craters:

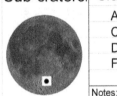

Crater	Lat	Long	Size (km)	Crater	Lat	Long	Size (km)
A	43.82S	0.54W	18.4	B	42.27S	4.02W	3.95
C	44.86S	0.68W	15.6	CA	45.3S	0.59W	18.59
D	47.01S	0.15E	18.18	E	44.67S	2.16W	10.72
F	44.34S	4.64W	4.37				

Notes:

Add Info:

Close-up of ejecta effects on Saussure's floor

In this region of the Moon where so many similarly-sized, similarly-aged craters can lead to confusion, oddly enough, Saussure is quite easy to locate. What makes it so is its relationship to having impacted centrally upon an old, unnamed crater over 100 kilometres in diameter (its most northern rim had subsequently been 'wiped' out by Orontius to the north). Saussure's floor level is some two kilometres below the 'level' of the unnamed crater, and ~ 500 metres below that of Orontius's. Both Saussure and Orontius are joined at the hip (their rims and ejecta...etc.,), but which formed last – Saussure?

Craters

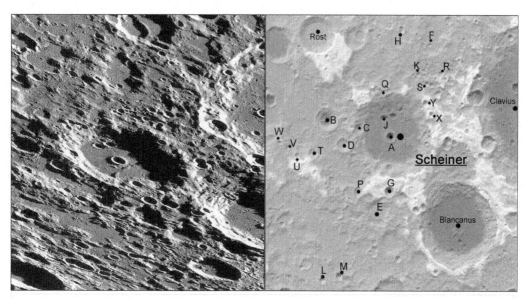

Scheiner   Lat 60.35S   Long 27.81W   110.07Km

Sub-craters:

Crater	Lat	Long	Size (km)	Crater	Lat	Long	Size (km)
A	60.49S	28.28W	12.05	B	59.7S	33.57W	29.1
C	60.13S	30.93W	12.14	D	60.72S	32.3W	16.73
E	63.22S	29.33W	24.4	F	56.89S	25.26W	6.11
G	62.59S	28.39W	13.47	H	56.35S	27.46W	8.16
J	59.76S	28.83W	11.39	K	58.05S	26.08W	7.36
L	65.87S	35.22W	9.09	M	65.77S	33.45W	10.39
P	62.56S	31.16W	12.37	Q	58.79S	28.9W	8.14
R	57.99S	24.27W	7.88	S	58.39S	25.33W	6.77
T	60.96S	34.95W	12.19	U	60.96S	36.31W	6.68
V	60.61S	37.02W	6.68	W	60.37S	37.66W	6.82
X	59.58S	24.86W	7.35	Y	59.13S	25.39W	8.19

Notes:

Add Info:

At full moon times, a bright ray from Tycho (to the north-east) 'clips' at Scheiner's outer north-western sector.

Aerial view of Scheiner

Scheiner almost looks like a smaller version of its famous neighbour, Clavius, to its east. It's got a group of hard-to-ignore craters on its floor (are the top two secondaries from the Orientale Basin centred 1800 km away westwards?), an odd ridge-like feature running across the bottom half (is it ejecta from Clavius?), and its inner walls, rim and outer ejecta are just as significantly noticeable under the best of lighting conditions (low to high sun view times). Clavius (Nectarian in age - 3.93 to 3.85 byo) looks like it is older, but not by much. Depth 3.5 km.

Craters

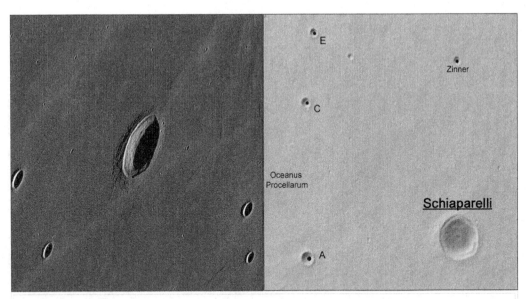

Schiaparelli	Lat 23.38N	Long 58.82W			24.2Km

Sub-craters:

Crater	Lat	Long	Size (km)	Crater	Lat	Long	Size (km)
A	22.97N	62.1W	7.58	C	25.83N	62.22W	5.66
E	27.12N	62.07W	4.88				

Notes:

Add Info:

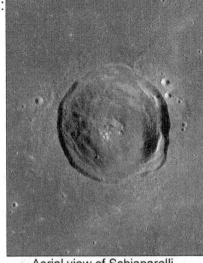

Aerial view of Schiaparelli

Like most young-ish craters, for example, crater Reiner ~ 500 km due southwards, or crater Lambert ~ 1000 km eastwards, that have impacted in extensive lava plains on the lunar surface, Schiaparelli's has taken on the same, low albedo featured look of its target. Of course, this isn't always the case (see, for example, Seleucus ~ 250 km westwards, or Lichtenberg ~ 350 km northwestwards), however, such darkish, and young-ish craters may give signatures as to the depth of lavas, or, say something about their make-up in a particular zone. Schiaparelli has a relatively sharp rim, its outer ejecta is extensive more westwards than at its east. It has a small, 250 metre-high peak, and the crater formed on a wrinkle ridge to its north. Depth ~ 2.5 km.

Craters

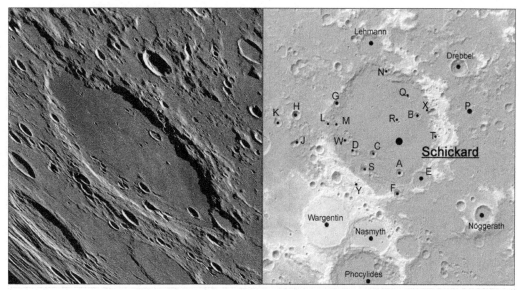

Schickard Lat 44.38S Long 55.11W 212.18Km

Sub-craters:

Crater	Lat	Long	Size (km)	Crater	Lat	Long	Size (km)
A	46.87S	53.71W	14.45	B	43.74S	52.25W	13.89
C	45.86S	56.0W	13.34	D	45.77S	57.72W	9.33
E	47.28S	51.67W	32.95	F	48.0S	53.86W	14.32
G	43.04S	58.98W	12.19	H	43.52S	62.34W	16.1
J	45.01S	62.44W	12.54	K	43.86S	63.9W	13.67
L	44.12S	59.8W	8.59	M	44.19S	58.98W	8.74
N	41.3S	54.69W	6.51	P	42.91S	48.5W	94.39
O	42.74S	53.05W	5.53	R	44.13S	53.82W	5.23
S	46.66S	56.8W	16.01	T	44.81S	50.37W	4.86
W	45.12S	58.19W	8.12	X	43.55S	51.26W	7.78
Y	47.44S	57.6W	4.87				

Add Info: There's no mistaking crater Schickard in the eyepiece. The contrast of lighter material (Orientale sourced) on its floor against the darker material (interior lava flooding) is what, firstly, catches your eye, and then, secondly, its thick band of ejecta all around its exterior (shows up easily around low sun times). The crater is an 'oldie' of the pre-Nectarian Period (4.6 to 3.93 byo), its floor is slightly convexed (observe this effect when the terminator is actually on the floor), and its rim has been altered by numerous impacts of all sizes. Note the cluster of craters on the floor's southwest (Orientale secondaries?), and the tiny, fresh impact crater just below sub-craters R and B. Depth is around 3 km approx..

Aerial view of Schickard

Craters

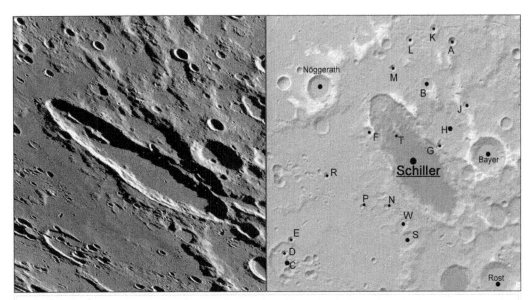

Schiller Lat 51.72S Long 39.78W 179.36Km

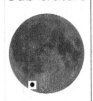

Sub-craters:

Crater	Lat	Long	Size (km)	Crater	Lat	Long	Size (km)
A	47.18S	37.65W	10.9	B	48.87S	39.15W	17.2
C	55.37S	49.31W	45.04	D	55.08S	49.36W	9.4
E	54.59S	48.8W	7.93	F	50.66S	42.99W	12.74
G	51.24S	38.3W	8.72	H	50.6S	37.74W	72.44
J	49.64S	36.69W	9.16	K	46.7S	38.77W	10.25
L	47.12S	40.23W	10.25	M	48.22S	41.26W	8.76
N	53.63S	41.98W	6.24	P	53.58S	43.66W	6.03
R	52.25S	45.87W	7.87	S	54.95S	40.42W	21.22
T	50.77S	41.29W	6.1	W	54.47S	40.97W	15.57

Notes:

Add Info:

Aerial view of Schiller

Schiller has to take the biscuit for being the weirest-looking crater on the Nearside. What would have caused it to look this way? Is it the result of several, large-sized impact craters (produced by break-up of a comet?) 'coalesced' together? Was it created by a single impactor coming in at a very low angle to the surface (if so, from what direction northwest or southeast?)? Or, is it a combination of both type events? Adding to the confusion is the odd-looking, linear-like feature at the floor's northwestern sector (again, what type of event, or events, would have produced it?). Schiller is an oddity for certain, but a wonderful one. D: 2.5 km.

Craters

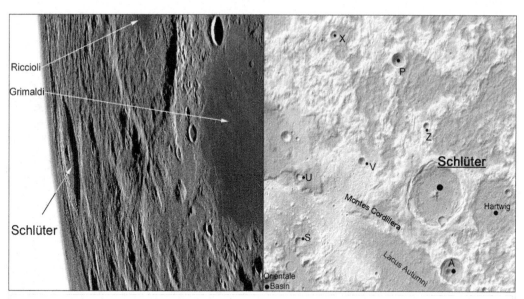

Schlüter   Lat 5.93S   Long 83.39W   87.55Km

Sub-craters:

Crater	Lat	Long	Size (km)	Crater	Lat	Long	Size (km)
A	9.22S	82.57W	36.75	P	0.05N	85.21W	20.78
S	7.9S	90.04W	13.13	U	5.06S	90.01W	9.83
V	4.41S	86.97W	12.1	X	1.17N	88.3W	12.95
Z	2.87S	83.85W	10.35				

Notes:

Add Info:

Lunar Orbiter 5 (the last in the series) impacted some 100 km away north of this crater on 31 Jan 1968.

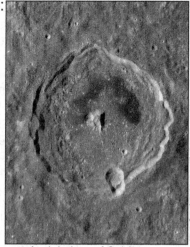

Aerial view of Schlüter

Schlüter formed on outer terrain (usually referred to as the Montes Cordillera whenever in view) of the Orientale Basin - centred just 500 kilometres away to the southwest. It's a pity that Schlüter lies on the limb - dictating we have to rely on favourable librations to glean any good detail from it. Most obvious, is its peak (~ 2.3 km high), its crenulated rim and terraces (all sharp-looking), lava-floods in its northern floor, and an impact crater just within its southern rim (which at close-up shows a sectioned-off slump of the rim may have afterwards slid down in the small crater's floor). At full moon times, the dark part of the floor acts as a good marker for noting its location. The crater is of the Upper Imbrian Period (3.75 to 3.2 byo). Depth ~ 3 km approx..

Craters

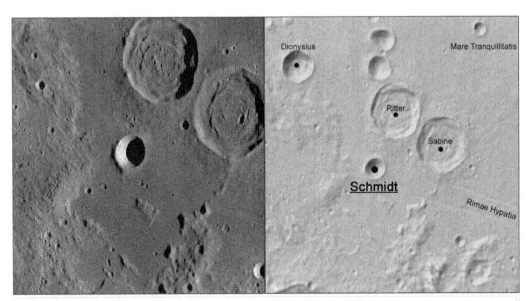

| Schmidt | Lat 0.96N | Long 18.78E | 11.13Km |

Sub-craters: No sub-craters

Notes: er

Add Info:

Aerial view of Schmidt

Judging from the state of Schmidt's rim, it may be as old (perhaps, Upper Imbrian - 3.75 to 3.2 byo) as crater Ritter to its north. A closer look shows the crater lies just on the edge of Ritter's ejecta, but as it's quite hard to see Schmidt's own ejecta anywhere around its outer rim, would it be safe to say that it is the older (not by much) of the two? Certainly, at full moons times, the crater is a tad brighter than Ritter (Sabine, too); which might then change our minds into thinking it is younger. However, as Schmidt has impacted closer to brighter, highland's material westwards where the thickness of Tranquillitatis's lavas may be thinned more slightly than at Ritter's location, would this explain the above 'tad' difference, and thus re-affirm it is older?

Craters

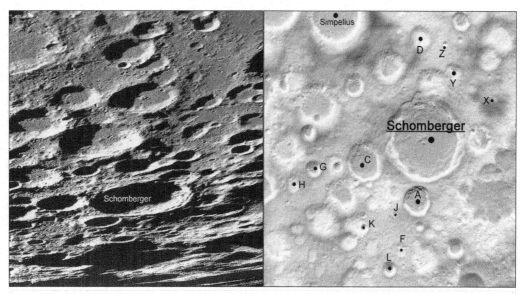

Schomberger   Lat 76.64S   Long 24.69E   85.8Km

Sub-craters:

Crater	Lat	Long	Size (km)	Crater	Lat	Long	Size (km)
A	78.61S	23.52E	29.44	C	77.17S	15.45E	40.73
D	73.38S	24.22E	23.68	F	80.18S	20.17E	10.65
G	77.01S	7.43E	17.15	H	77.26S	3.6E	15.75
J	78.98S	19.31E	7.4	K	79.28S	13.56E	8.79
L	80.69S	17.27E	15.98	X	75.13S	34.41E	8.53
Y	74.44S	28.54E	18.9	Z	73.58S	27.29E	6.17

Notes:

Add Info:

Aerial view of Schomberger

There's no easy way, or marker, that will make Schomberger stick out in the mind for remembering its location. It's a crater of the Early Imbrian Period (3.85 to 3.75 byo), it's got a crenulated rim with significant terracing (slumping effects), it's got several peaks (highest is about ~ 1 km-high), but signatures of its ejecta aren't quite obvious as they are 'lost' to the jumble of other smaller craters nearby. Favourable librations aren't always required, but when they do occur, they bring the crater into a better view where all of the above become apparent. Depth ~ 5.0 km.

Craters

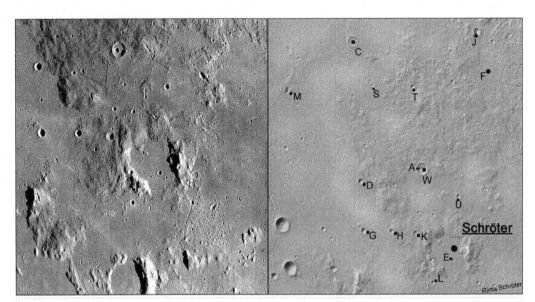

Schröter Lat 2.74N   Long 6.98W   36.67Km

Sub-craters:

Crater	Lat	Long	Size (km)	Crater	Lat	Long	Size (km)
A	4.81N	7.8W	3.68	C	8.26N	9.79W	8.38
D	4.49N	9.53W	4.92	E	2.36N	6.83W	2.85
F	7.41N	5.89W	29.2	G	3.17N	9.44W	5.2
H	3.16N	8.63W	4.38	J	8.51N	6.12W	6.46
K	3.09N	7.93W	5.22	L	1.77N	7.43W	3.34
M	6.94N	11.68W	5.3	S	7.07N	9.2W	2.51
T	7.03N	8.01W	3.96	U	4.08N	6.68W	3.81
W	4.83N	7.74W	9.5				

Notes:

Add Info: About the only original portion of crater Schröter that can be seen today is at its western sector (you can just make out its inner rim wall). Every other portion has been either obliterated by small impacts that 'picked' away at the rim, or affected by the most prominent effect of ejecta from the Imbrium Basin - centred some 900 km away to the northwest. This latter event seems to have 'dolloped' a substantial amount of ejecta onto Schröter's northern rim; burying nearly half of the crater in the process (see the extent of it to the crater's west). As to what happened at the crater's south is anyone's guess, where a 'gap' of nearly 10 kilometre's wide feeds into lavas that extend right up to Sinus Aestuum in the north. The crater lies in one of the largest pyroclastic deposits (LPDs) on the moon, which formed (in this region, anyway) through processes related to extensional stresses of the surface. These allowed volumes of ash and black crystalline beads (e.g. high-titanium ilmenite) to cover the area through fire-fountains from larger vents. Depth of Schröter is around 3 km approx..

Large LPD regions on the Nearside Moon include: Taurus-Littrow, Sinus Aestuum, Vaporum, and Rimae Bode (sometimes referred to as the 'blue' members of the LPDs).

Craters

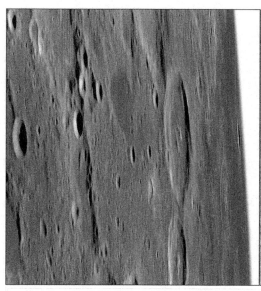
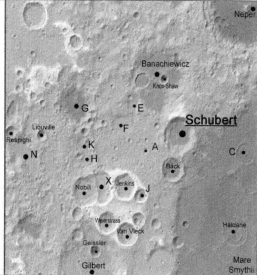

Schubert	Lat 2.78N	Long 81.01E				51.94Km	

Sub-craters:

Crater	Lat	Long	Size (km)	Crater	Lat	Long	Size (km)
A	2.14N	79.34E	1.88	C	1.94N	84.64E	30.72
E	4.25N	78.43E	27.62	F	3.29N	77.87E	34.45
G	4.19N	75.34E	56.38	H	1.64N	76.03E	27.66
J	0.01S	78.94E	22.34	K	2.41N	75.88E	30.17
N	1.79N	72.65E	65.13	X	0.31N	76.75E	49.18

Notes:

Add Info:

Aerial view of Schubert

Schubert formed on a zone between where two rings/rims (~ 400 km to ~ 600 km in diameter respectively), related to Smythii Basin to its southeast exist. Smythii is pre-Nectarian in age (4.6 to 3,93 byo), so Schubert, as too with crater Back to its south, must be younger, but by how much? Both craters are probably Nectarian in age (3.92 to 3.85 byo), but as to which formed first is somewhat confusing because neither show any signatures of the other's ejecta on the other's floor (did the two form simultaneously?) Favourable librations are required best for observing, and use both the dark mare of Smythii and the fresh impact display of bright material from sub-crater, Schubert A, for finding the crater.

Craters

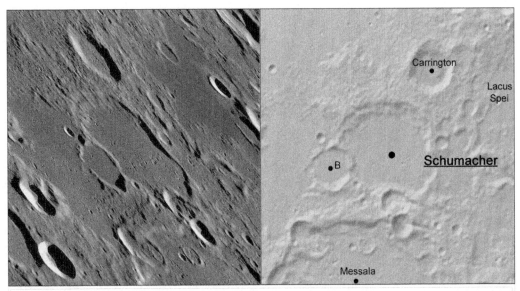

| Schumacher | Lat 42.42N | Long 60.81E | 61.31Km |

Sub-craters:

Crater	Lat	Long	Size (km)	Crater	Lat	Long	Size (km)
B	42.18N	59.35E	24.17				

Notes:

Add Info:

Not to be confused with similar-sounding, crater Shoemaker (88.41S, 45.91E) towards the South Pole region.

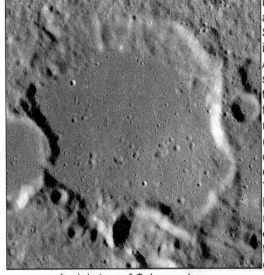

Aerial view of Schumacher

Like Lacus Spei over to its east and crater Messala to its south, Schumacher has experienced internal mare flooding of sorts. As the Humboltianum Basin lies some ~ 600 km away to the northeast, the floodings are attributed to it; where a depressed zone between two outer rings/rims (~ 1000 km and ~ 1400 km in diameter respectively) from that event allowed the mare infilling. The crater's southeastern floor level is ~ 200 m higher than at its northwest (probably of Crisium's ejecta to the south), while its rim there, too, looks as if it has been covered over after possible slumping effects occurred. D: ~1.8km.

Craters

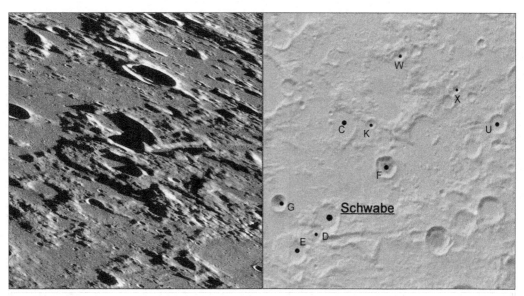

Schwabe	Lat 65.1N	Long 45.48E		25.47Km

Sub-craters:

Crater	Lat	Long	Size (km)	Crater	Lat	Long	Size (km)
C	67.74N	46.95E	29.3	D	64.68N	44.78E	15.29
E	64.14N	43.43E	18.78	F	66.45N	50.15E	19.71
G	65.5N	42.01E	14.88	K	67.7N	49.23E	8.85
U	66.57N	57.06E	17.44	W	69.62N	52.52E	9.05
X	68.33N	56.6E	7.34				

Notes:

Add Info:

Gash source: Imbrium is favoured over Humboltianum, because the feature 'line's up' more to it. Line-up towards the Humboltianum direction falls midway between the Basin's centre and crater Bel'kovich (61.53N, 90.15E), off to its northeast.

Aerial view of Schwabe

It's quite easy to find Schwabe - when one remembers to 'link' it to the obvious 'gash' just kissing its southern rim. The gash doesn't have an official title, which it certainly deserves, but as to what it is and what was its source is a puzzle (is it a secondary crater chain event from the Imbrium Basin (~ 1400 km away westwards), or of the Humboltianum Basin (~ 550 km away eastwards?). Schwabe's southern rim may have received partial covering of small amounts of ejecta from one of the 'leading' secondaries, but look also at sub-crater, Schwabe D's hummocky floor - is that secondary ejecta too? D: 1.3 km.

Craters

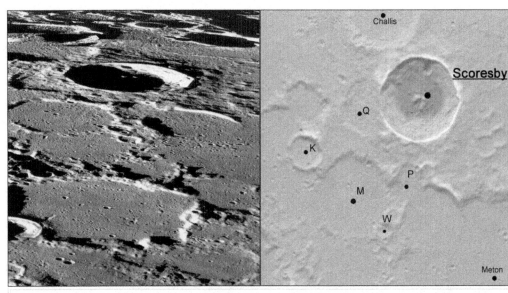

Scoresby	Lat 77.73N	Long 14.13E		54.93Km			

Sub-craters:

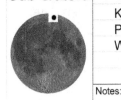

Crater	Lat	Long	Size (km)	Crater	Lat	Long	Size (km)
K	76.35N	2.95E	22.2	M	75.57N	8.15E	56.52
P	75.89N	12.78E	22.98	Q	77.48N	8.81E	38.25
W	74.38N	11.14E	10.4				

Notes:

Add Info:

Aerial view of Scoresby

Scoresby would be a nice crater to be viewing in the eyepiece - if only it was placed favourbly. Eratosthenian in age (3.15 to 1.1 billions of years ago), the crater has a sharp rim all around, a wonderful series of peaks in its centre (highest, on the west, is about 1.2 kilometres-high), and several slumpings that are worth noting - look at the northern ones, which almost look ripple-like (did the small crater just within the rim there set off the slumps?). Its location dictates good libration times are required, but also suitable lighting conditions to 'see' some of the above details. Depth ~ 4 km.

Craters

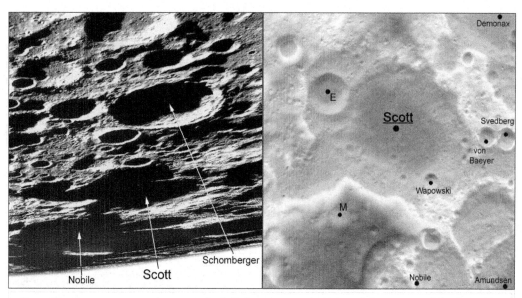

Scott Lat 83.35S Long 48.52E 107.82Km

Sub-craters:

Crater	Lat	Long	Size (km)	Crater	Lat	Long	Size (km)
E	81.17S	35.7E	29.22	M	83.86S	34.74E	17.85

Notes:

Add Info:

Aerial view of Scott

If you've prepared well, that is, you've researched that an upcoming, favouable libration will bring Scott (and other craters of high latitudes at the South Pole) over the next day or so into view, be sure also to have a camera at hand. Yes, you might get away swapping back and forth between the eyepiece and your lunar atlas for reference, but the extreme foreshortening effect on craters at these latitudes, not to mention shadows, intervening high points like mountains or rims etc., will almost certainly lead to confusion. Photographing the region allows you to view the features at leisure, and time to figure out what rim, shadow or mountain that is. Of course, nothing can beat the actual view, but sometimes technology other than the eye is an advantage. D:~6 km.

Craters

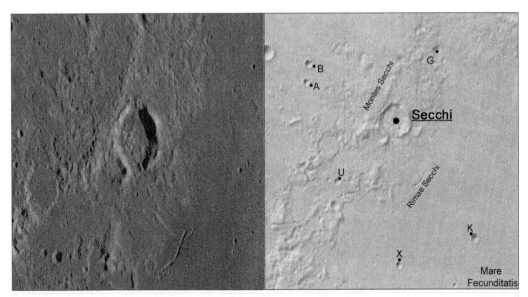

Secchi	Lat 2.4N	Long 43.56E				22.13Km

Sub-craters:	Crater	Lat	Long	Size (km)	Crater	Lat	Long	Size (km)
	A	3.26N	41.47E	4.85	B	3.66N	41.52E	5.16
	G	3.94N	44.57E	6.69	K	0.17S	45.5E	6.01
	U	1.08N	42.19E	4.93	X	0.77S	43.67E	4.69
	Notes:							

Add Info:

Aerial view of Secchi

Secchi is situated in an area where two systems of rings/rims related to basins' Tranquillitatis to the northwest and Fecunditatis to the southeast produced the Secchi Mountains (Montes Secchi). The crater is an odd one of sorts with its thin western rim contrasting against the more fatter one at the east. East is also higher too by about 700 metres, which during low sun times casts long shadows than those from other parts of the rim. The floor also is unusual where its centre is slightly convexed upwards by about 200 metres as hummocky material takes over at its west, while being smoother at its east (mare material?). Given all of the above, is Secchi slightly tilted towards Tranquillitatis than Fecunditatis? Depth ~ 1.4 km.

Craters

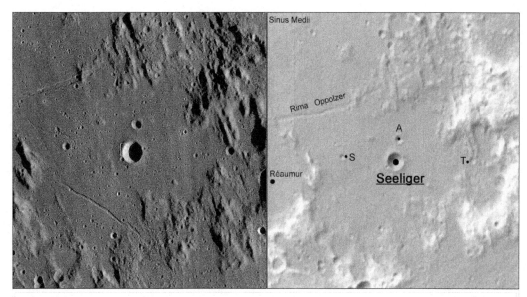

Seeliger	Lat 2.22S	Long 3.0E			8.28Km

Sub-craters:	Crater	Lat	Long	Size (km)	Crater	Lat	Long	Size (km)
	A	1.88S	3.04E	3.78	S	2.14S	2.1E	2.88
	T	2.2S	4.31E	3.32				

Notes:

Add Info:

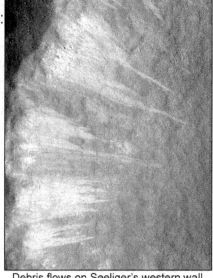

Debris flows on Seeliger's western wall

Though is isn't very obvious unless you are viewing Seeliger and its general surrounds under higher sun times, the crater lies in more lighter coloured terrain than that to its northwest in Sinus Medii. This area is some ~ 500 metres higher than at the Sinus, and may have something to do with ejecta make-up from the Imbrium Basin - centred some 1000 km off to the northwest. Seeliger, then, has got to be younger than the basin, and if you look further at the striated effects on craters and mountains nearby produced by Imbrium, the crater doesn't show any similar-like signatures. Two small craters (each ~ 1.5 km in diameter) attached on either side of its easterly and westerly rims will prove a challenge to see in the smaller-sized telescopes. Depth ~ 1.8 km.

Craters

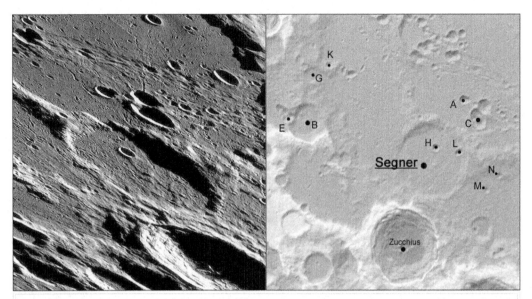

Segner	Lat 58.96S	Long 48.68W			67.84Km		

Sub-craters:

Crater	Lat	Long	Size (km)	Crater	Lat	Long	Size (km)
A	57.27S	47.03W	8.51	B	57.74S	55.94W	36.07
C	57.8S	46.2W	18.74	E	57.48S	57.0W	11.19
G	56.39S	55.57W	12.48	H	58.64S	48.61W	7.06
K	56.17S	54.46W	6.95	L	58.76S	47.31W	6.17
M	59.78S	45.49W	5.32	N	59.34S	44.74W	5.12

Notes:

Add Info:

Aerial view of Segner

Segner formed in a trough zone between two impact rings/rims associated to formation of the Schiller-Zucchius Basin. As a result, it wiped out a portion of one of its rings (the ~150 kilometres in diameter one), leaving the crater looking like it has two 'horn-like' appendages at its east and northwest. As Schiller-Zucchius is of the pre-Nectarian Period (4.6 to 3.93 byo), Segner has to be younger, but from the state of its worn rim (and 'horns') its age mustn't be far off the same mark. Younger crater Zucchius (Copernican in age - 1.1 byo to the Present) to its south has 'spat' a load of its ejecta onto its floor (see left), which show up nicely as rivulets under low sun times. Depth 1.5 km.

Craters

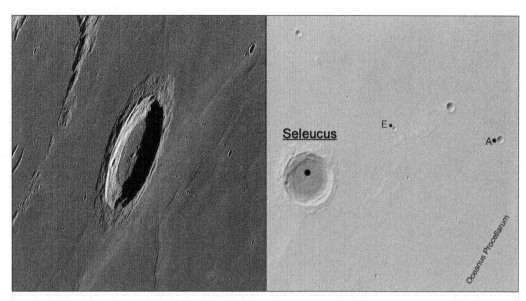

Seleucus	Lat 21.09N	Long 66.66W	45.01Km

Sub-craters:

Crater	Lat	Long	Size (km)	Crater	Lat	Long	Size (km)
A	22.04N	60.52W	6.36	E	22.43N	63.92W	3.99

Notes:

Add Info:

What is the source of the bright ray material off to the crater's southeast, that further extends beyond the crater's northeast? It 'points' back to crater Glushko in the southwest, but is that really its source?

Aerial view of Seleucus

Seleucus almost looks like its two similar-looking neighbours - crater Krafft and crater Cardanus (view all three in a single show under terminator times - a day or so before full moon or new moon), down to its southwest. Obvious in all three is the wonderful, wide band of ejecta around their exteriors, and while each is Upper Imbrian in age - 3.75 to 3.2 byo, Seleucus, with its slightly sharper features, is probably the youngest. Its features include a series of terraces, rilles (on the eastern sector of its floor) and possible faults (on the floor's western sector). It's got a small peak of around 350 metres high. D:~ 3 km.

Craters

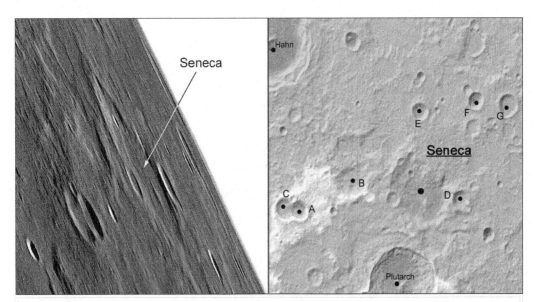

Seneca	Lat 26.71N	Long 79.81E	47.57Km

Sub-craters:

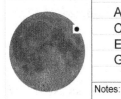

Crater	Lat	Long	Size (km)	Crater	Lat	Long	Size (km)
A	26.28N	75.07E	18.35	B	27.17N	77.01E	29.47
C	26.32N	74.48E	21.34	D	26.71N	81.29E	17.55
E	29.32N	79.71E	17.11	F	29.63N	81.93E	16.41
G	29.48N	83.23E	20.66				

Notes:

Add Info:

Aerial view of Seneca

Given its location, viewing Seneca is best done around a day or so after new moon or full moon, as shadows within its rim and floor define its feattures (use crater Plutarch to its south also as a good location marker for it). Looking at the aerial view (left), the crater's size can really be defined from its ~600 metres-high central peak (if we take it to be an actual peak at all), and not from the additional 'ear-like' appendage on its northwest, which looks like an older crater (~ 45 km in diameter) that Seneca impacted upon. Like most of its surrounding area, Seneca is covered by ejecta from the Crisium Basin to its southwest.

Craters

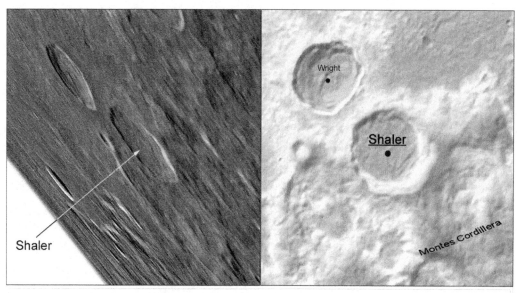

| Shaler | Lat 32.89S | Long 85.27W | 48.47Km |

Sub-craters: No sub-craters

Notes:

Add Info:

Aerial view of Shaler

There's a reason why Shaler's southeastern rim and wall look somewhat extended in that direction - it's because the initial impactor that produced the crater struck upon uneven ground (the southeastern rim is some ~ 2.3 kilometres higher than its northwestern rim), which later then led to slumping effects. The uneven ground is ejecta of the Orientale Basin (centred ~ 450 kilometres off to the northwest), so formation of Shaler has to be younger than some 3.75 to 3.2 byo - the Late Imbrian Period (Wright, to the northwest, is younger again). The higher side can sometimes hide the crater's interior details, so favourable libration times are best for good views. D: ~ 2.5 km.

Craters

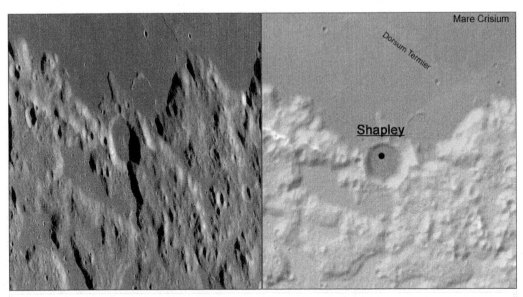

| Shapley | Lat 9.35N | Long 56.83E | 24.82Km |

Sub-craters: No sub-craters

Notes:

Add Info:

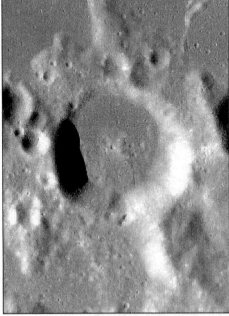

Aerial view of Shapley

Shapely has formed on an inner ring/rim (it's about ~ 500 km in diameter) associated to ejecta from creation of the Crisium Basin to its north. Given the look of the ejecta in general surrounds, the crater's impactor must initially have struck odd-highted ground, but not so to have affected its force that caused it to tap deep down into lava resources similar to those as in Crisium (the floor level of Shapley is some 300 metres below the level of the mare). The crater shows signatures of it having being tilted in towards Crisium's centre as the basin and its maria settled over time (where its southern rim is some ~ 1 km higher than at its north, the floor also drops by about 200 metres to the centre). Depth is around ~ 4.0 km.

Craters

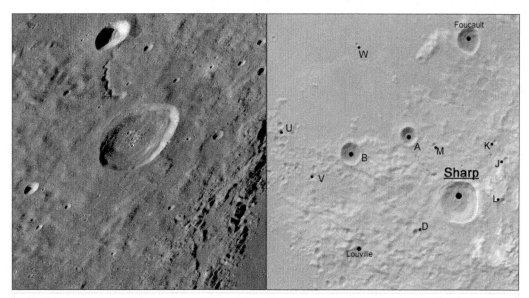

Sharp		Lat 45.75N	Long 40.22W				37.61Km

Sub-craters:

Crater	Lat	Long	Size (km)	Crater	Lat	Long	Size (km)
A	47.62N	42.67W	17.26	B	47.01N	45.35W	20.62
D	44.83N	42.17W	7.07	J	46.96N	38.06W	5.47
K	47.5N	38.56W	4.43	L	45.78N	38.23W	4.96
M	47.39N	41.46W	4.75	U	47.41N	48.68W	6.33
V	46.23N	47.06W	6.64	W	50.2N	45.44W	3.92

Notes:

Add Info:

Aerial view of Sharp

Sharp lies on outer ejecta deposits of the wonderful crater that today is called Sinus Iridum (Bay Of Rainbows). If you were to compare its rim height at its southeast to that at its northwest, a drop of ~ 200m would be returned. This initially might suggest a slight tilting of the crater away from Iridum's centre. However, it isn't really (the floor is fairly level overall), but its whole northwestern rim seems to have undergone a sort of backwards 'slump' away from Sharp's centre. An obvious, deep-looking rille (it's not Rima Sharp) off to its northwest may have drained lavas in the same direction, where Sharp A now lies. Did the rille eventually affect Sharp's 'slumped' northwestern rim?). Depth ~ 4.0 km approx..

Craters

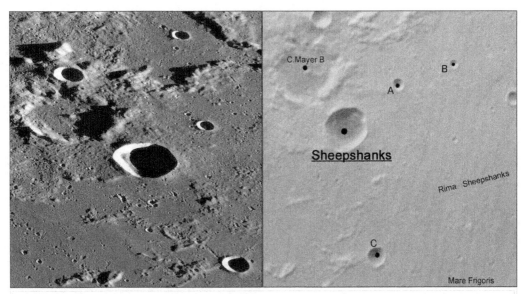

Sheepshanks	Lat 59.24N	Long 17.04E		23.67Km

Sub-craters:

Crater	Lat	Long	Size (km)	Crater	Lat	Long	Size (km)
A	60.0N	18.96E	6.65	B	60.32N	21.09E	4.77
C	57.03N	18.09E	10.39				

Notes:

Add Info:

Aerial view of Sheepshanks

Sheepshanks almost looks like its neighbour, Archytas, over to its west. Sheepshanks, however, is slightly older (Sheepshanks is of the Upper Imbrian Period - 3.75 to 3.2 byo, while Archytas is of the Eratosthenian - 3.15 to 1.1 byo). The crater formed on textured ejecta from the Imbrium Basin (to the southwest), but signature of its own ejecta on it can be hard to see (low sun times show it sparingly). Several slump events within the crater's innards have altered the crater's shape over its lifetime, producing some odd shadows to be cast (left). Usually, at full moon times, many craters 'disappear', however, brighter material on Sheepshanks's northern wall makes it stand out somewhat. Depth 3.9 km.

Craters

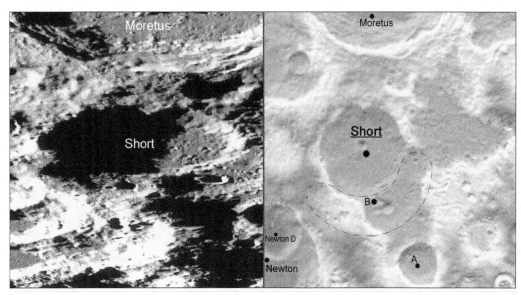

Short	Lat 74.54S	Long 7.68W	67.94Km

Sub-craters:

Crater	Lat	Long	Size (km)	Crater	Lat	Long	Size (km)
A	76.95S	1.28W	32.49	B	75.89S	5.17W	89.84

Notes:

Add Info:

Aerial view of Short

For a crater situated in the high latitudes of the southern limb, and, in an overly complex region of the Moon where many other craters 'group', Short isn't that hard to locate when you mark it down as being below its more obvious neighbour, crater Moretus. Short has impacted upon an old crater - sub crater, Short B (~ 90 km in diameter), leaving the level of its floor nearly a kilometre below the level of B. A large clump of material overlies the crater's southwest rim and floor, but what is it, and where did it originate from (it also seems to have filled Newton D to the southwest)? Is it just ejecta from Moretus?

Craters

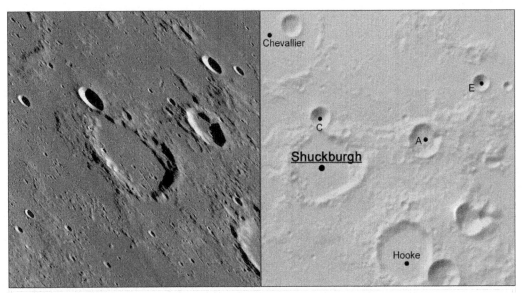

Shuckburgh   Lat 42.65N   Long 52.71E   37.73Km

Sub-craters:

Crater	Lat	Long	Size (km)	Crater	Lat	Long	Size (km)
A	43.08N	55.42E	18.08	C	43.54N	52.72E	11.58
E	44.05N	57.0E	9.22				

Notes:

Add Info:

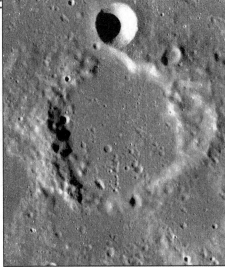

Aerial view of Shuckburgh

The additional, 'ear-like appendage' on Shuckburgh's eastern side is what makes this crater easy to find. But what is the appendage - its level is ~ 400 metres above the level of Shuckburgh's floor level? Is it an old crater that Shuckburgh impacted upon, is it some kind of slump event, or is it an oblique impact signature from formation of the crater? If that's a puzzle, then look at the crater's western side too, where a 1.2 km-high clump-of-a-deposit seems to have buried Shuckburgh's rim there. Shuckburgh is said to be centred generally in a very old basin whose size of ~ 400 km in diameter extends from Cepheus at its west to Schumacher at its east. Depth of Shuckburgh ~ 1.3 km.

Craters

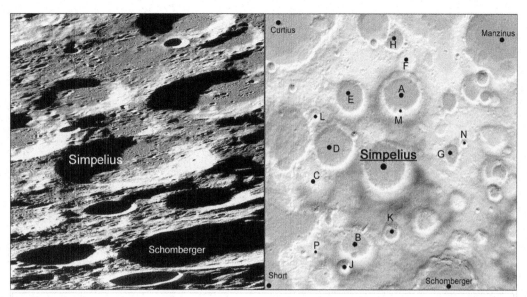

Simpelius   Lat 72.61S   Long 14.74E   68.89Km

Sub-craters:

Crater	Lat	Long	Size (km)	Crater	Lat	Long	Size (km)
A	69.9S	16.1E	57.54	B	75.17S	10.1E	50.16
C	72.64S	5.38E	48.79	D	71.56S	8.29E	55.93
E	70.0S	10.6E	42.96	F	68.69S	16.73E	27.08
G	71.78S	22.76E	22.32	H	67.94S	15.34E	28.68
J	75.86S	8.11E	17.54	K	74.71S	15.35E	22.43
L	70.44S	6.53E	16.05	M	70.52S	16.33E	6.82
N	71.36S	24.09E	9.38	P	75.29S	3.98E	5.66

Notes:

Add Info:

Aerial view of Simpelius

Crater confusion territory again, however, a useful location-marker for remembering Simpelius's position is to use sub-craters' Simpelius A, E and D that sort of encircle it. The crater having an approximate depth reaching nearly six kilometres in parts dictates shadows rule during most times, but, given suitable librations, and lighting, say, around Waxing Gibbous or Waning Gibbous times, details within the crater can be gleaned. The crater has a small peak of around 250 kilometres high (not really obvious in observations), and like several other craters nearby, Simpelius's southern wall is broader than at its northern wall. Why?

Craters

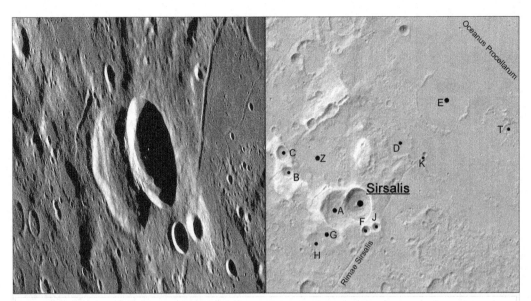

Sirsalis   Lat 12.49S   Long 60.51W   44.17Km

Sub-craters:

Crater	Lat	Long	Size (km)	Crater	Lat	Long	Size (km)
A	12.67S	61.4W	48.46	B	11.11S	63.77W	14.85
C	10.35S	64.0W	22.42	D	9.98S	58.7W	34.86
E	8.16S	56.7W	70.29	F	13.56S	60.15W	13.16
G	13.85S	61.71W	24.24	H	14.02S	62.5W	25.49
J	13.42S	59.69W	11.86	K	10.39S	57.47W	7.04
T	9.28S	53.59W	14.98	Z	10.37S	62.04W	90.32

Notes:

Add Info:

Aerial view of Sirsalis

Sirsalis is Eratosthenian in age (3.15 to 1.1 byo), and lies on a mix of material from original highlands and ejecta from Basins like Grimaldi to the northwest and Orientale to the southwest. Sirsalis has obviously impacted upon sub-crater, Sirsalis A, whose floor level is nearly a kilometre above that of Sirsalis. The central peak is just over a kilometre in height, and it meets an interesting series of slump-like 'wavelets' that cover nearly half the floor in the southeast. Fresh impact crater, Sirsalis F, has dusted brightly the crater, while Rimae Sirsalis to the southeast is said to be the longest on the Moon (405 km). Depth 4.7 km approx..

Craters

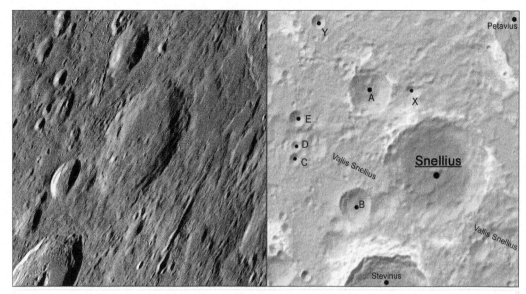

Snellius   Lat 29.33S   Long 55.7E   85.98Km

Sub-craters:

Crater	Lat	Long	Size (km)	Crater	Lat	Long	Size (km)
A	27.43S	53.68E	34.51	B	30.16S	53.21E	25.25
C	29.02S	51.41E	8.65	D	28.71S	51.46E	8.13
E	28.05S	51.5E	11.86	X	27.41S	54.95E	9.11
Y	25.74S	52.21E	11.1				

Notes:

Add Info:

Aerial view of Snellius

Snellius really didn't have a chance in escaping an onslaught of ejecta: firstly, by Petavius (Late Imbrian - 3.75 to 3.2 byo) to its northeast, and then later by Stevinus (Copernican - 1.1 byo to Present) to its southwest. The crater almost looks like similar-sized, crater Hase, over to its east, and so could probably be assigned to a Nectarian/pre-Nectarian age - 4.6 to 3.85 byo). The crater is criss-crossed with bright ray material from both craters' Furnerius A to its southeast and Stevinus A to its southwest. The brightness can, at times, drown out detail in the crater, but close-to-terminator times (around waxing crescent or after full moon) are good.

Craters

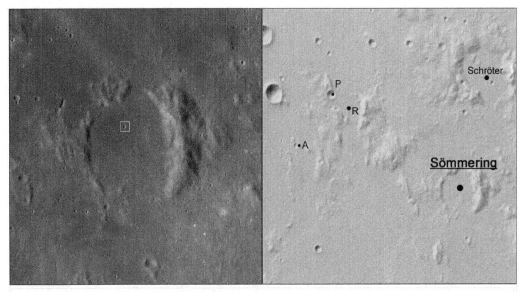

Sömmering	Lat 0.19N	Long 7.53W	27.97Km

Sub-craters:

Crater	Lat	Long	Size (km)	Crater	Lat	Long	Size (km)
A	1.1N	11.09W	2.72	P	2.14N	10.33W	5.91
R	1.88N	9.87W	16.56				

Notes:

Add Info:

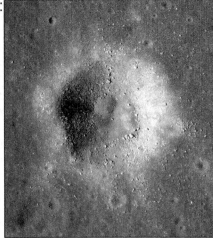

Extreme close-up of an impact crater (with boulders) on Sömmering's floor

Surveyors' 4 and 5 impacted, and landed respectively, not some 160 km away east from this crater.

Like its neighbouring crater, Schröter, to its north, Sömmering has been subjected to flooding of lavas (~ 2.5 byo) through breach-points. These breaches may have been produced by blocks of ejecta from the Imbrium Basin (centred ~ 1300 km away northwards) oblitering their rims at such points. Sömmering's eastern rim is some 800 metres higher than at its west, and as the floor level drops by about ~ 200 metres towards that same direction, did the crater experience some kind of 'tilting' at one time? A small ~ 9 km-sized crater (left) on its floor can just about be made out. But, it doesn't appear fresh, as mass-wasting (soil erosion or morphology degradation) signatures suggest otherwise. D: 1.3 km.

Craters

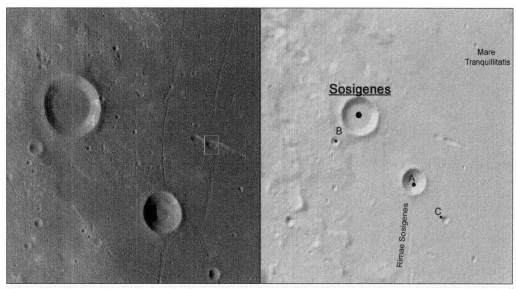

Sosigenes	Lat 8.7N	Long 17.6E	16.99Km

Sub-craters:

Crater	Lat	Long	Size (km)	Crater	Lat	Long	Size (km)
A	7.76N	18.47E	11.29	B	8.33N	17.21E	3.74
C	7.22N	18.97E	3.31				

Notes:

Add Info:

More on Gaseous Features:
Stooke, P.J. (2012). Lunar Meniscus Hollows
http://www.lpi.usra.edu/meetings/lpsc2012/pdf/1011.pdf

Close-up view of a gaseous feature east of Sosigenes

Sosigenes seems to lie more on ejecta from the Imbrium Basin (centred ~ 1200 km away northwestwards) rather than on the ejecta remains from one of the rings/ rims of the Tranquillitatis Basin (to east). The floor is very smooth suggesting it might be internally-sourced with lavas similar to those in/of Tranquillitatis. Its wall shows equal thickness, equally-sloped all within (note the two bright debris flows on the eastern wall). Extreme close-up of the crater's floor shows unusual features that are believed to be signatures of volcanic gas escape (left, shows a more extreme example found east of the crater). These odd-like features exist in places all over the Moon, but many are found between craters' Ross, Arago and Sosigenes.

Craters

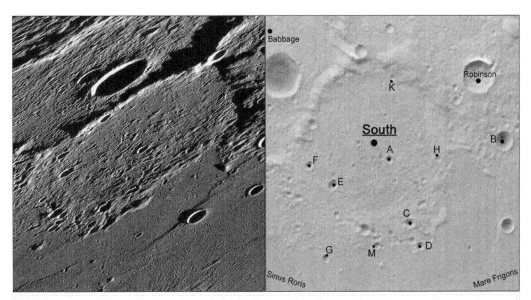

South	Lat 57.58N	Long 50.94W	119.04Km

Sub-craters:

Crater	Lat	Long	Size (km)	Crater	Lat	Long	Size (km)
A	57.28N	50.41W	6.28	B	57.54N	45.05W	14
C	55.81N	49.57W	7.7	D	55.27N	49.13W	6.16
E	56.7N	53.0W	8.11	F	57.13N	54.09W	6.52
G	55.0N	53.25W	5.68	H	57.25N	48.18W	5.51
K	59.17N	50.19W	4.42	M	55.31N	51.02W	5.72

Notes:

Add Info: What is the source of the small chain of craters on South's southeastern floor? Alignment doesn't exactly 'point' to Sinus Iridum! So, would it be in the other direction, towards crater Pascal (74,36N, 70.63W), in the northwest?

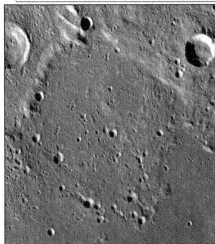

Aerial view of South

Looking at the aerial image (left) of the crater, South, one thing becomes obviously clear - that of its 'squarish' shape. Why does it look that way? The same also goes for South's neighbouring craters to its northwest - that of, Babbage and Oenopides, which also appear 'squarish'. Would ejecta from the Imbrium Basin to the southeast be the cause? Huge blocks striking at circular rim areas non-parallel or perpendicular to the main source of Imbrium's impact might have been affected more. Certainly, the 'pummelling' effect of the ejecta is seen on all portions of the remaining rims, but is the above theory correct for the 'squarishness'? Depth ~ 1 km approx..

Craters

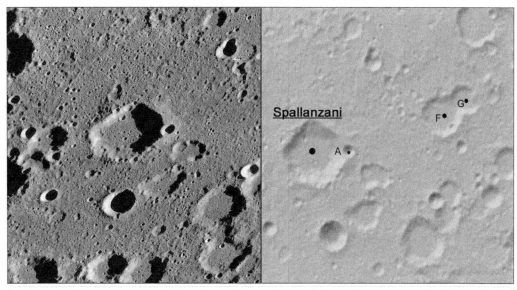

Spallanzani	Lat 46.38S	Long 24.73E		30.86Km	

Sub-craters:

Crater	Lat	Long	Size (km)	Crater	Lat	Long	Size (km)
A	46.32S	25.62E	6.65	D	46.19S	28.57E	6.29
F	45.67S	28.03E	21.99	G	45.42S	28.53E	15.04

Notes:

Add Info:

Aerial view of Spallanzani

Spallanzani lies in a textured region having a 'pitted' look to it. The 'pits' are due to numerous, small impact craters covering the area, which us-usually implies older units exist ('the older an area is, the greater the number of craters found'). Some of the 'pits' look more worn than others, so Spallanzani impacted initially upon an older unit (the ejecta of the Nectaris Basin - centred some 900 km away to the northwest). This later experienced further coverage by ejecta from other basins and craters, near and far. Sub-crater, Spallanzani A, attached to its eastern rim, acts as a good marker for finding/remembering Spallanzani's location. Depth is around 2.3 km.

Craters

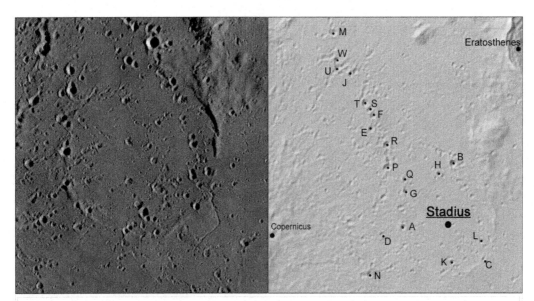

Stadius		Lat 10.48N	Long 13.77W				68.48Km

Sub-craters:

Crater	Lat	Long	Size (km)	Crater	Lat	Long	Size (km)
A	10.44N	14.84W	4.27	B	11.84N	13.62W	5.6
C	9.74N	12.9W	2.82	D	10.29N	15.35W	3.51
E	12.61N	15.63W	4.17	F	13.02N	15.67W	4.34
G	11.24N	14.77W	4.22	H	11.62N	13.98W	3.37
J	13.82N	16.1W	4.83	K	9.67N	13.66W	4.04
L	10.12N	12.96W	2.87	M	14.73N	16.57W	6.18
N	9.4N	15.65W	4.5	P	11.75N	15.23W	5.66
Q	11.48N	14.81W	2.84	R	12.25N	15.26W	5.21
S	12.9N	15.58W	4.33	T	13.18N	15.75W	5.84
U	13.97N	16.45W	4.92	W	14.12N	16.48W	4.3

Notes:

Add Info:
Pit craters: Why would Copernicus be their source? Most have a characteristic herringbone feature, where their apexes point towards Copernicus, while their ridges point away from it.

Apex Ridge

Close-up of small impacts

Stadius is more famous not for the crater itself, but for the hundreds of 'pits' that lie within and around it. The 'pits' are, of course, small impact craters (all under 7 km or so across in diameter), which, presumably, were produced by impact of Copernicus to its southwest. Their distribution wasn't exactly even, as from a bigger picture view some, for example, like those from sub-crater Stadius P up to Stadius M, clustered in a concentric formation from Copernicus (signature of its impact and associated impact dynamics). Stadius was partially buried (its original rim appears in the NE, SE and SW) by lavas before the Copernicus impact onslaught. Note the rille off its southeast rim, and the dark deposits at its north. D: 0.6 km.

Craters

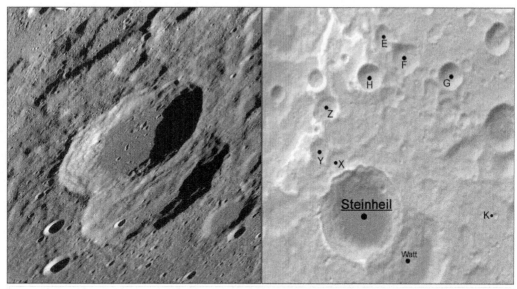

Steinheil   Lat 48.71S   Long 46.66E   63.28Km

Sub-craters:

Crater	Lat	Long	Size (km)	Crater	Lat	Long	Size (km)
E	44.88S	47.49E	14.13	F	45.34S	48.3E	21.26
G	45.7S	49.99E	18.58	H	45.77S	46.97E	19.17
K	48.63S	51.97E	4.32	X	47.6S	45.76E	17.19
Y	47.35S	45.06E	15.59	Z	46.41S	45.34E	21.36

Notes:

Add Info:

Aerial view of Steinheil

Steinheil has impacted upon ejecta of Janssen (pre-Nectarian in age) to its northwest, but most obviously on its neighbour, Watt (again, pre-Nectarian). This latter event 'dolloped' a huge volume of material onto Watt's floor; covering nearly half of it in the process, and raising its level in parts to some ~ 800 metres above that of the level of Steinheil. Steinheil is assigned to the Nectarian Period (3.92 to 3.85 byo), it's got a wreath of terraced walls (thinner at the southeast, possibly because of impact on Watt), and its floor is filled with external material from several impact events near and far. Depth ~ 5 km.

Craters

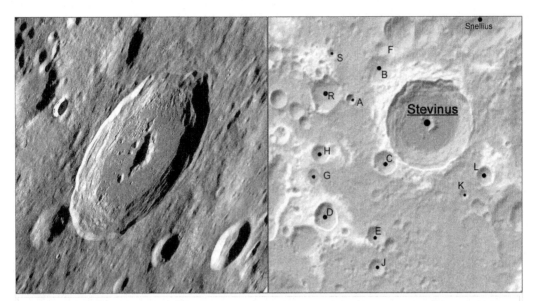

Stevinus	Lat 32.49S	Long 54.14E			71.54Km

Sub-craters:

Crater	Lat	Long	Size (km)	Crater	Lat	Long	Size (km)
A	31.86S	51.65E	7.87	B	31.12S	52.55E	20.44
C	33.46S	52.71E	19.25	D	34.76S	50.74E	21.69
E	35.29S	52.36E	15.74	F	30.61S	52.62E	9.49
G	33.77S	50.37E	11.24	H	33.19S	50.62E	15.63
J	36.08S	52.36E	13.17	K	34.26S	55.27E	8.15
L	33.78S	55.99E	14.83	R	31.69S	50.87E	25.14
S	30.69S	51.07E	7.68				

Notes:

Add Info:

Aerial view of Stevinus

In this southeastern sector of the Moon, five main, big craters with similar features - terraces, bright walls, peaks etc., - can all be viewed at low power in the eyepiece. Formed in a U-like configuration, they include: Theophilus and Piccolomini on the left part of the U, Langrenus and Petavius on the right part, and then Stevinus is on the U's bottom. It is the youngest of them all (Copernican 1.1 byo to the Present), but also the smallest (some 16 km under Piccolomini). The crater has a wonderfully, sharp-looking peak (nearly 2 km high), and fresh crater, Stevinus A, off to its west has brightly dusted it and surrounds.

Craters

| Stiborius | Lat 34.49S | Long 31.99E | 43.76Km |

Sub-craters:

Crater	Lat	Long	Size (km)	Crater	Lat	Long	Size (km)
A	36.91S	35.54E	32.41	B	37.39S	33.5E	8.93
C	33.94S	33.25E	20.49	D	33.35S	35.57E	15.67
E	34.73S	34.03E	14.23	F	35.8S	32.39E	8.04
G	37.33S	35.78E	8.28	J	36.07S	35.57E	9.68
K	35.57S	34.53E	14.68	L	34.96S	33.35E	8.05
M	35.46S	32.69E	6.71	N	36.36S	32.84E	8.7
P	33.22S	33.89E	5.81				

Notes:

Add Info:

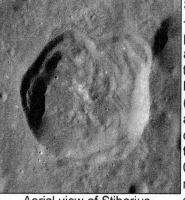

Aerial view of Stiborius

Stiborius is aged of the Late Imbrian Period - 3.75 to 3.2 byo (a good exercise is to compare it to Late Imbrian-aged crater Lindenau, and Eratosthenian-aged crater Rothmann, to to its northwest). Stiborius lies in a trough-zone between two rings/rims of the Nectaris Basin (centred some 600 km away to the northeast), and seems to have impacted on an unnamed crater nearly twice its own diameter. The crater has a series of peaks (the highest is some 0.8 km high) on its floor, and major slumps at nearly every sector of its walls have occurred, except for at its southwest (why?). D: 3.8 km.

Craters

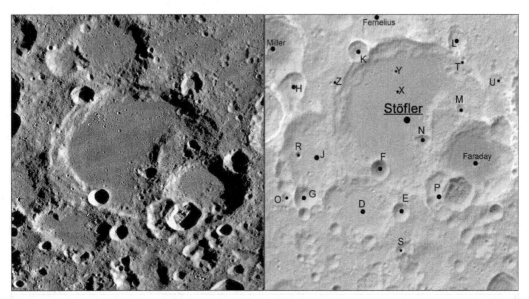

Stöfler   Lat 41.24S   Long 5.93E   129.87Km

Sub-craters:

Crater	Lat	Long	Size (km)	Crater	Lat	Long	Size (km)
D	43.71S	4.29E	53.24	E	43.86S	5.75E	15.92
F	42.7S	4.92E	17.01	G	43.4S	1.91E	19.5
H	40.37S	1.62E	22.86	J	42.11S	2.49E	64.19
K	39.47S	4.14E	18.49	L	39.19S	7.84E	17.29
M	41.02S	8.09E	9.42	N	41.91S	6.61E	14.19
O	43.37S	1.19E	9.17	P	43.38S	7.33E	40.15
R	42.25S	1.73E	5.85	S	44.89S	5.71E	6.64
T	39.69S	8.1E	4.68	U	40.17S	9.58E	5.06
X	40.58S	5.5E	2.69	Y	40.03S	5.43E	3.22
Z	40.33S	3.08E	3.5				

Notes:

Add Info:

Aerial view of Stöfler

Physically, nearly all of Stöfler's southeast rim and floor of the crater has been lost to several major impacts. The crater has been around for a long time (it is aged of the pre-Nectarian Period 4.6 to 3.92 byo), and so it's not so surprising that some part inevitably came under the influence of ejecta from basins near and far. The crater doesn't show any sign of a peak - expected for such a big crater, and so may lie underneath the smooth material in its floor (note the large clump off Faraday's northwest, and on Stöfler's floor - basin ejecta, or an old crater rim?). D: 3 km.

Craters

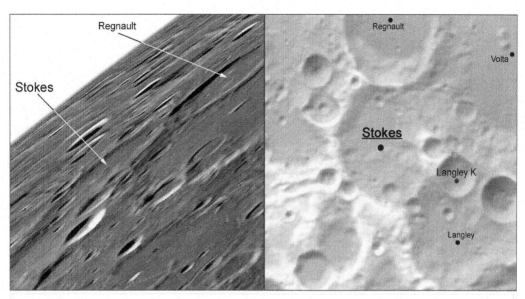

| Stokes | Lat 52.36N Long 88.11W | 53.85Km |

Sub-craters: No sub-craters

Notes:

Add Info:

Aerial view of Stokes

Practically all of Stokes's eastern side is missing or hidden due to impact ejecta coverage from both crater Volta, to its northeast, and from Langley, to its southeast. A 'spit' of a deposit lies on Stokes's floor (see left), but as to its source it's impossible to say (is it the ejecta of crater Langley K, to the east, or is it from the small unnamed crater northwest of K?). All this impact activity has left Stokes's floor level higher up than the floor levels of the above-mentioned craters, as it blends in seamlessly with its well-worn, remaining rim on the westerns side. Stokes is old - of the pre-Nectarian Period (4.6 to 3.92 byo), and has a depth of around ~ 1.8 km approx..

Craters

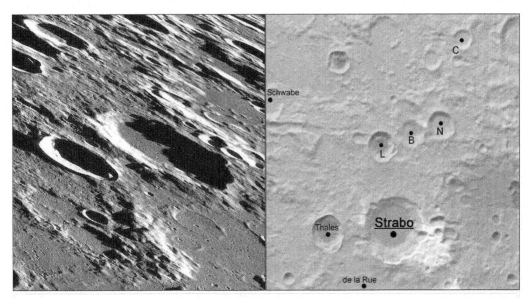

Strabo	Lat 61.94N	Long 54.42E	54.72Km

Sub-craters:

Crater	Lat	Long	Size (km)	Crater	Lat	Long	Size (km)
B	64.57N	55.48E	23.26	C	67.1N	59.43E	17.5
L	64.21N	53.48E	24.8	N	64.77N	57.53E	24.91

Notes:

Add Info:

Aerial view of Strabo

Strabo lies just outside one of the rings/rims (the 650 km in diameter one) associated to formation of the Humboltianum Basin - centred 450 km away to its northeast. Strabo is Nectarian in age (3.92 to 3.85 byo), and while it obviously formed on de la Rue's northern rim, the impactor also struck an odd-highted clump of ejecta of the basin (did this impact scenario affect how Strabo's southeast walls developed some time afterwards?). Nearby, fresh impact crater, Thales, to its west, has imparted bright ray material across this whole area, but direction of it seems to be less towards Strabo (view it around full moon times). Depth ~ 4.5 km.

Craters

Street   Lat 46.58S   Long 10.74W   58.52Km

Sub-craters:

Crater	Lat	Long	Size (km)	Crater	Lat	Long	Size (km)
A	47.02S	9.08W	16.76	B	47.05S	12.15W	12.86
C	48.42S	15.44W	13.1	D	48.88S	12.73W	10.05
E	47.48S	11.89W	12.1	F	48.25S	16.72W	7.37
G	46.69S	15.13W	10.25	H	48.38S	12.25W	27.76
J	48.74S	13.7W	6.11	K	47.56S	13.18W	8.11
L	50.74S	13.61W	7.83	M	47.82S	14.58W	48.19
N	48.29S	10.34W	5.0	P	45.63S	11.98W	5.58
R	49.16S	14.61W	4.9	S	48.99S	14.83W	3.61
T	49.29S	15.21W	8.74				

Notes:

Add Info:

Aerial view of Street

When you lie not some ~ 100 kilometres away from one of the most eye-catching, brightest impact craters on the Moon, that is Tycho, then it's a sure thing you're going to be affected by it. You would think that lying so close, Tycho's bright ray material would be more denser in its coverage and appearance, on Street, however, it isn't, as it's darker. The darker material may be due to deeper ejecta deposits of Tycho (the closer ejecta is to a rim, the deeper and older it is), whose depth usually reaches down to 1/3 the transient crater diameter. Observe the effect around full moon times. Depth ~ 2.7 km.

Craters

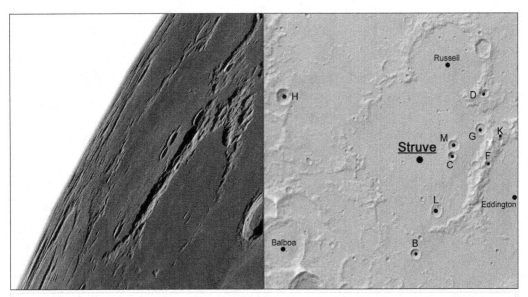

Struve		Lat 23.41N	Long 76.65W			164.34Km	

Sub-craters:	Crater	Lat	Long	Size(km)	Crater	Lat	Long	Size(km)
	B	18.98N	77.14W	13.37	C	22.86N	75.4W	11.03
	D	25.34N	73.76W	10.06	F	22.48N	73.71W	8.26
	G	23.9N	74.0W	14.28	H	25.18N	83.36W	21.17
	K	23.46N	73.07W	6.49	L	20.68N	76.15W	15.04
	M	23.28N	75.3W	14.94				

Notes:

Add Info:

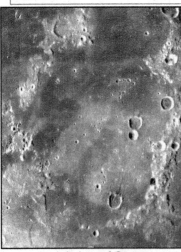

Aerial view of Struve

Struve, Russell and Eddington while impressive in the eyepiece, are craters that obviously have been subjected to major floodings. Their rims really aren't so obvious, as more it is their ejecta and rubbly remains that are left from all. Judging how each relates to the other in terms of their ejecta and rims, it looks like Struve is the oldest of all three (Eddington's ejecta seems to overlie Struve's eastern floor more, while Russell looks like it impacted on Struve's ejecta - see the portion east of sub-crater, Struve D). The bright material on Struve's floor may be Orientale in origin, but as the floor is somewhat convexed (more at its eastern sector), other factors, for example like non-flooding of that sector, might also be responsible. Depth of Struve ~ 2.3 km.

Craters

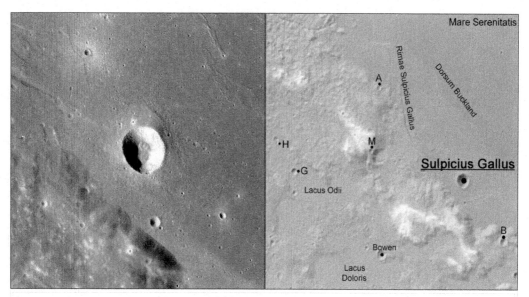

Sulpicius Gallus   Lat 19.63N   Long 11.68E   11.61Km

Sub-craters:	Crater	Lat	Long	Size (km)	Crater	Lat	Long	Size (km)
	A	22.08N	8.93E	3.99	B	17.98N	12.97E	6.43
	G	19.8N	6.31E	5.4	H	20.55N	5.74E	3.42
	M	20.4N	8.72E	4.34				

Notes:

Add Info: Sulpicius Gallus lies just within the main ring/rim (approximately 675 kilometres in diameter), which was produced as a result in formation of the Serenitatis Basin. Copernican in age (1.1 billions of years old to the Present), the crater seems to have impacted close to, if not upon, the southeastern end of one of the rilles related to Rimae Sulpicius Gallus (does its ejecta cover it?), and also in a zone where a series of lava flows occurred (note the difference in their albedos). These local lavas (Titaniuim-rich) are older than those found in the central mare (Titanium-poor), and probably extruded, in part, through the above-mentioned rilles. The crater also lies on a bench feature that is some 200 metres above the main level of the mare lava fills, where a wrinkle ridge (Dorsum Buckland), and a change in the lava's albedo, marks its edge and boundary. The ridge is one of several seen around the entire basin (take in a wider view), and is a feature produced in response to gravitational re-adjustment of the mare (Buckland's formation may also have responded to smaller, pre-mare impact basins that underlie the lava fills). Sulpicius Gallus really hasn't much to offer in terms of detail, however, low to high sun views will show all of the above features. Depth of Sulpicius Gallus ~ 2.0 km approx..

Craters

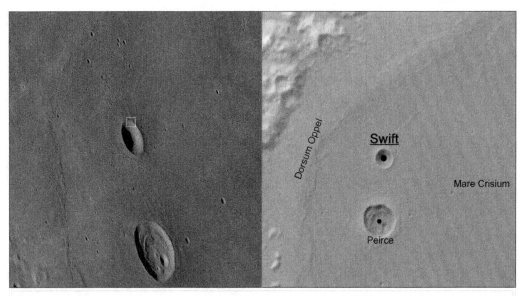

Swift	Lat 19.35N	Long 53.44E	10.06Km

Sub-craters: No sub-craters

Notes:

Add Info:

Close-up view of an impact crater on Swift's northeastern (sloped) wall

Swift just can't be viewed without noticing its southerly neighbour, Peirce, in the eyepiece, too. The two lie on mare deposits of Crisium, and each at low sun times displays their mottled-like ejecta - more obvious at Peirce than at Swift. Swift's rim is slightly less sharp as Peirce's, so it probably is of an older age (perhaps, Late Imbrian - 3.75 to 3.2 byo, Peirce is Eratosthenian - 3.15 to 1.1 byo in age). Swift appears to have impacted near a small ridge that veers off to its outer, eastern rim, but look also just west of the rille where a small depression meets Swift's northern rim (the two features show up easily during low sun times). Coincidentally, a small crater (left) lies on the rim's inner wall here, whose impactor, striking a sloped target wall, must have lead to some interestng impact dynamics. D: ~ 2.0 km.

Craters

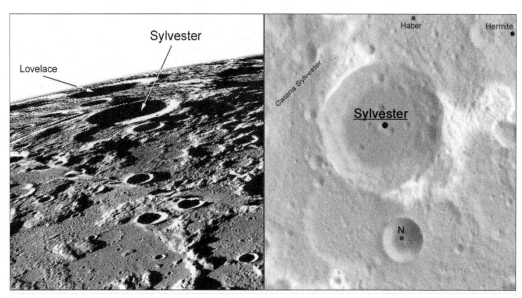

Sylvester	Lat 82.65N	Long 81.22W		59.28Km

Sub-craters:	Crater	Lat	Long	Size (km)	Crater	Lat	Long	Size (km)
	N	82.41N	68.68W	19.98				

Notes:

Add Info:

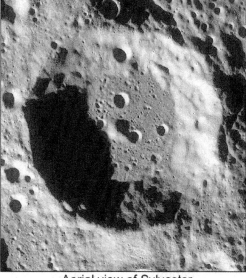

Aerial view of Sylvester

Sylvester lies just some 220 kilometres away from the lunar North Pole. As a result, any observations attempted will undoubtedly involve views of the crater's interior filled partially, or entirely, with shadows that hide details. The perspective problem (for us earth viewers) doesn't help either, however, favourable librations, and good lighting conditions do show hints of its rim, walls and central peak (~ 700 metres). The crater formed on outer ejecta deposits of crater Hermite (pre-Nectarian in age 4.6 to 3.92 byo), off to its northeast, but also on Haber, to its north, and an unnamed crater to its south. Sylvester is Nectarian in age. Depth 3.5 km.

Craters

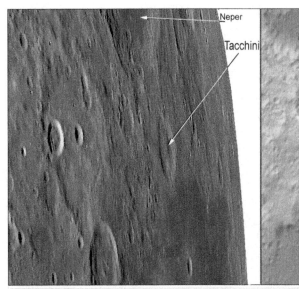
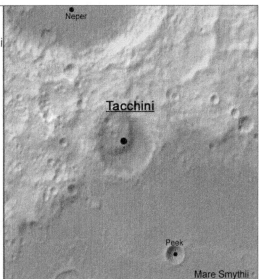

| Tacchini | Lat 5.08N | Long 85.82E | 42.58Km |

Sub-craters: No sub-craters

Notes:

Add Info:

Note: As the Moon goes through various librations throughout the year, suggested times given in text for observations are only approximates.

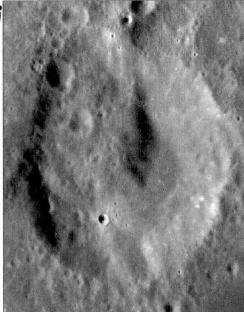

Aerial view of Tacchini

Tacchini impacted on an outer, main ring/rim (~ 370 km in diameter) associated to formation of the Smythii Basin to its south. The crater looks relatively sharp but not so sharp as that of crater Neper to its north, which is Nectarian (3.92 to 3.85 byo) in age. If this is the case, then Tacchini may have impacted upon ejecta of Neper, and the obvious clump of material seen to cover the top half of the floor isn't ejecta from Neper, but a major slumped feature (Tacchini's northern rim is over 2.5 km higher than its southern rim, so the slump event certainly had a huge volume of material to play with). Tacchini will prove a challenge to see - even during good libration times. Depth is around ~ 3.5 km.

Craters

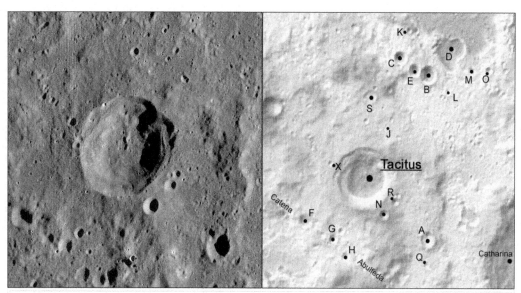

Tacitus   Lat 16.2S   Long 18.95E   39.81Km

Sub-craters:

Add Info:

Crater	Lat	Long	Size (km)	Crater	Lat	Long	Size (km)
A	17.47S	20.46E	10.21	B	14.05S	20.45E	12.17
C	13.7S	19.79E	9.01	D	13.52S	20.92E	22.59
E	13.95S	20.13E	8.78	F	17.09S	17.55E	9.84
G	17.46S	18.23E	6.12	H	17.83S	18.52E	5.82
J	14.99S	19.59E	2.74	K	13.12S	19.99E	3.54
L	14.4S	20.9E	4.06	M	13.97S	21.48E	5.85
N	16.95S	19.43E	6.79	O	13.94S	21.89E	4.66
Q	18.0S	20.46E	4.26	R	16.7S	19.69E	4.98
S	14.5S	19.09E	8.9	X	15.89S	18.17E	3.83

Notes:

Tacitus has formed on the main ring/rim (usually referred to as the Altai Scarp) produced as a result by formation of the Nectaris Basin - centred some 450 kilometres to its east. The crater looks relatively sharp in comparison to Nectarian-aged crater, Catharina, over to its east, but not so much as to assign it close to the boundary between the Nectarian Period and the Early Imbrian Period (3.85 to 3.15 byo). The crater has paid its toll over that time, where almost all of the floor has been covered by minor to major slumping deposits (moreso at its northeastern walls), which sidled up, eventually, to a small off-centered peak some ~ 500 metres high (the peak actually has a small, 0.5 kilometre-sized impact crater atop it). Several small craters surround Tacitus's outer rim (at its south and at its northwest respectively) that suggest an alignment of sorts. The former group aligns towards Orientale in the southwest, while the latter to Imbrium in the northwest - are they secondaries of those events? Depth of Tacitus is around ~ 2.8 km approx..

Craters

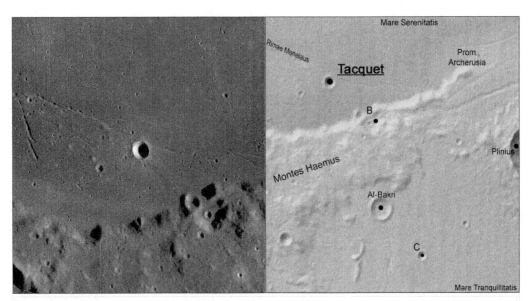

Tacquet		Lat 16.64N	Long 19.2E			6.43Km	
Sub-craters:	Crater	Lat	Long	Size (km)	Crater	Lat	Long	Size (km)
	B	15.94N	20.07E	15.36	C	13.47N	21.08E	4.98
	Notes:							

Add Info: As observers of the countless lunar features all over the Moon, we usually tend to like/view those showing significant details through aspects of their geology and formation. Some will obviously show you 'in-your-face' details about themselves, however, others, like Tacquet, require a little more time to divulge the varied series of events that led (or may have led) to what we see today. Looking at the crater itself: the initial interpretation is that it has formed on a portion of Serenitatis's lavas. Its small size suggests the impactor that produced it, too, was small. Its perfectly, circular shape, and also its bowl-like depression, says the impactor more than likely came in at a relatively high angle. Its brightish material that surrounds its exterior suggests the crater is somewhat fresh, but its rim isn't very sharp. Ejecta from Tacquet seems to cover, or at least blends into, a section of a rille associated to those of Rimae Menelaus, however, a high-resolution close-up shot suggests that the rille in question may have formed afterwards (a section of it appears to sub-divide the ejecta). Easterly, low sun views shows up a small ridge at its southeast, too, but as to say if the ridge is an old one or a recent one relative to Tacquet's formation, it isn't as clear in high-rez views. When the crater is in full light, say, around/close to full moon, its walls 'shine' (the bright, outer material less so), and contrasts nicely against the local dark mare lavas.

Craters

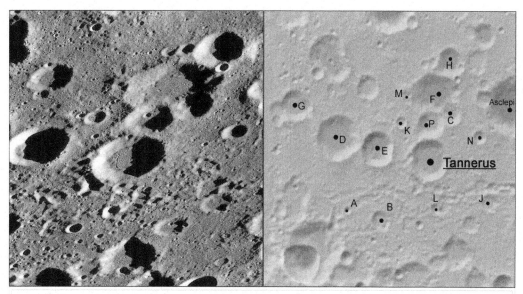

Tannerus　　　　Lat 56.44S　　Long 21.92E　　　　28.07Km

Sub-craters:

Crater	Lat	Long	Size (km)	Crater	Lat	Long	Size (km)
A	57.56S	18.23E	5.41	B	57.79S	19.73E	13.98
C	55.42S	22.8E	15.61	D	55.93S	17.93E	31.64
E	56.2S	19.64E	24.95	F	54.97S	22.04E	35.68
G	55.15S	16.17E	21.55	H	54.25S	22.74E	20.23
J	57.35S	24.71E	12.72	K	55.63S	20.68E	8.33
L	57.54S	22.2E	8.54	M	55.0S	20.91E	6.16
N	55.91S	24.12E	9.76	P	55.66S	21.94E	19.19

Notes:

Add Info:

Aerial view of Tannerus

In this over-populated cratered region of the Moon, Tannerus is quite easy to locate, when one remembers to note its alignment with sub-craters, Tannerus E and D, as useful markers. The three are all relatively similar in size, similar in age - from the worn-down look to their rims, and each show similar-type materials (Imbrium ejecta?) in their floors. As the alighnment is towards the Humorum Basin (Nectarian in age 3.93 to 3.85 byo) off to the crater's northwest, are all three secondaries from that event? Note the two rilles on Tannerus's floor, and ejecta effects (from Orientale?) on its western rim and exterior.

Craters

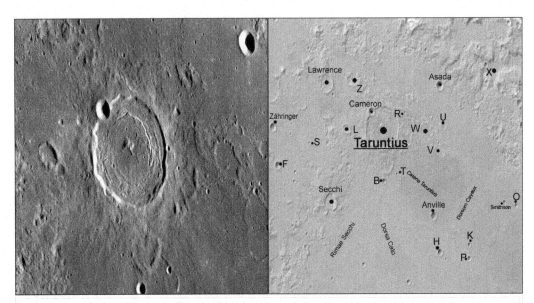

Taruntius	Lat 5.5N	Long 46.54E	57.32Km

Sub-craters:

Crater	Lat	Long	Size (km)	Crater	Lat	Long	Size (km)
B	3.29N	46.67E	7.32	F	3.92N	40.51E	10.16
H	0.33N	49.88E	8.59	K	0.65N	51.57E	5.65
L	5.46N	44.52E	14.26	O	2.23N	54.33E	6.0
P	0.06N	51.58E	6.78	R	6.14N	47.85E	4.76
S	4.85N	42.35E	4.84	T	3.6N	47.4E	4.28
U	5.54N	50.13E	8.69	V	4.45N	49.83E	19.83
W	5.44N	49.05E	16.82	X	7.86N	52.94E	22.63
Z	7.56N	44.9E	18.78				

Notes:

Add Info:

Aerial view of Taruntius

Aged of the Eratosthenian (3.15 to 1.1 byo), Taruntius lies on sparse deposits between two basins - Tranquillitatis to its northwest and Fecunditatis to its south (level of the latter to the former drops by some 1.5 km in a NW-SE trend). Taruntius's wonderful series of rilles and fractures concentric on the floor are what catches the eye at first, followed next by its central peak (~ 0.9 km), its relatively bright material around its exterior, and finally, hints of dark ash deposits in the northwestern part of its floor (around Cameron, too). Its ejecta shows up nicely during low sun times. Depth ~ 1.4 km.

Craters

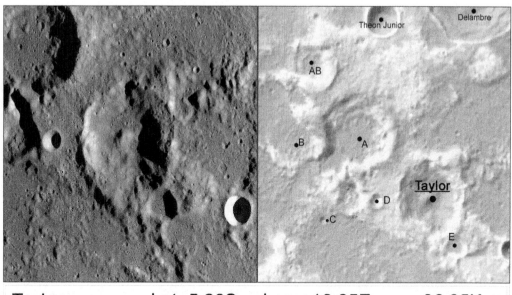

Taylor	Lat 5.28S	Long 16.65E	36.35Km

Sub-craters:	Crater	Lat	Long	Size (km)	Crater	Lat	Long	Size (km)
	A	4.23S	15.35E	39.3	AB	3.2S	14.61E	22.97
	B	4.34S	14.29E	28.85	C	5.65S	14.75E	4.2
	D	5.34S	15.69E	7.58	E	6.06S	17.11E	12.1

Notes:

Add Info: Taylor lies on highlands terrain, which extends over to Mare Tranquillitatis and Sinus Asperitatis ~ some 300 kilometres away eastwards. The crater is an odd one, especially when looking at its shape, as it looks relatively sharp-ish at its western side, but more worn at it east. Why would this be so, and why does the crater look squished (compare it to its more circular sub-crater neighbour, Taylor A), to its northwest? The reason might be ejecta from the Imbrium Basin (it lies centred some ~ 1500 kilometres away to the northwest), where a particular clump may have struck Taylor's southern rim, and thus extending it in that direction. Two other big clumps also lie just outside the crater's northeast and southeast rim respectively, however, while the former may actually be Imbrium ejecta that 'settled' (still in its molten state) somewhat over Taylor's rim (outline of the rim is seen), the latter clump appears older that Taylor may have impacted upon. The crater has a small, rounded peak (~ 1.4 kilometres high), while hints of its floor (partially filled with another form of Imbrium ejecta, perhaps, fluidized?) are seen only within the northeastern sector. With a depth of around 3.0 kilometres, terminator times, say, a day before first quarter, or around waning gibbous, are best for observing, as during full moon times the crater is difficult to find.

Craters

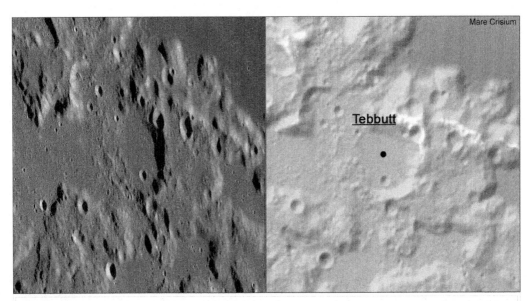

Tebbutt   Lat 9.46N   Long 53.52E   33.99Km

Sub-craters: No sub-craters

Notes:

Add Info:

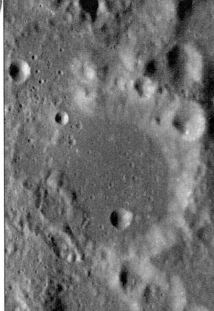

Aerial view of Tebbutt

Tebbutt lies just some ~ 50 kilometres away southwards from the main ring/rim associated to formation of the Crisium Basin. Interpretated as pre-basin rock uplifted during Crisium's formation, the crater has been subjected to both flooding and pummelling by small impacts over its lifetime. Practically all of its western rim is gone possibly due to flooding (and breaching?), and its floor is now lava-filled with said. However, where did the lavas come from - Fecunditatis to its southwest or Crisium to its northeast (Fecunditatis's mare level is some 1.5 kilometres above that of Crisium's). Whatever the source, these lavas may later have been covered over with a thin layer of ejecta from the Imbrium Basin (centred some ~ 2000 km away to the northwest). The small impact crater (~ 4 km) on its floor might have been obliquely formed. D: 0.6 km.

Craters

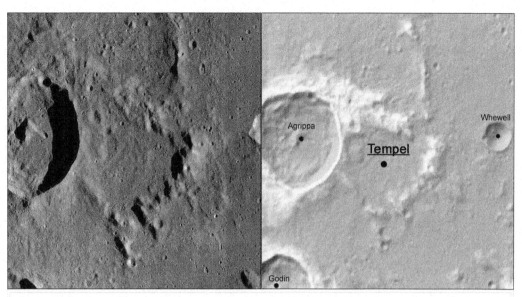

| Tempel | Lat 3.76N | Long 11.86E | 43.19Km |

Sub-craters: No sub-craters

Notes:

Add Info: Is Tempel a crater at all? Barely perceptible - even at the best of times when the terminator's effect of shadows and light aren't far off - the crater (?) just doesn't say 'I'm a crater' as we're used to seeing them. Appearance of a rim is seen as constructs of clumpy, broken-up mountainy material everywhere, except at its northwest, which gives more a squarish look to Tempel than the usual circular form. Of course, the reason for all this is due to two major dynamic events: that of Agrippa's formation which totaly wiped out Tempel's northwestern rim, and, earlier before that, formation of the Imbrium Basin (centred some ~1200 kilometres away to the northwest). The latter may have involved huge volumes of ejecta material striking not only the higher parts of Tempel's rim, but also its floor and walls etc.,; which all must have mixed together (note the odd clumps and distribution thereof to give you an idea of what happened) while being blasted downrange from the central impact point. That occurred some 3.8 billions ago, however, a billion of years later, the former event of Agrippa's effects then splashed its ejecta across the remaining portions of the crater and floor; leaving the 'mess that is the crater (?) as we see it today (Tempel's floor level is actually some ~ 1.5 km higher than Agrippa's level). Imbrium's effects are obviously seen as striations on Tempel's rim (pull back your view to see the effect more on nearby craters and general terrain), however, another series of smaller striations cross at right angles to these, which 'point' back to crater Godin (Copernican in age 1.1 byo to the Present) to its southwest. If the terminator times prove difficult for observing the crater, then full moon times will be worse (Agrippa and Godin 'shine'), as it is impossible to see. D: ~0.7 km.

Craters

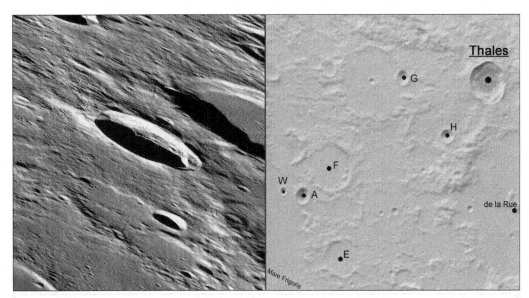

Thales	Lat 61.74N	Long 50.27E	30.75Km

Sub-craters:

Crater	Lat	Long	Size (km)	Crater	Lat	Long	Size (km)
A	58.56N	40.89E	13.1	E	57.21N	43.23E	29.41
F	59.37N	42.08E	37.02	G	61.75N	45.54E	11.46
H	60.4N	48.13E	10.49	W	58.59N	39.9E	6.06

Notes:

Add Info:

Aerial view of Thales

Copernican in age (1.1 billions of years ago to the Present), Thales, is as expected, fresh-looking, sharp, and, even at this location, it is an eye-catcher at most times. At full moon times, it's bright rays splash predominantly southwestwards towards us, while at other times two long arms of mountainy material (one going northwestwards, the other southwestwards), follow the rays's direction. The first arm may be ejecta from Mare Humboltianum (~ 500 km away to the northeast), while the other arm is that of de la Rue's northwestern rim. Thales's southwestwards rays suggests the crater might be oblique in origin, however, did the arms have anything to do with a preferred direction of these rays (does the crater, like Tycho, have a dark, exterior annulus?).

Craters

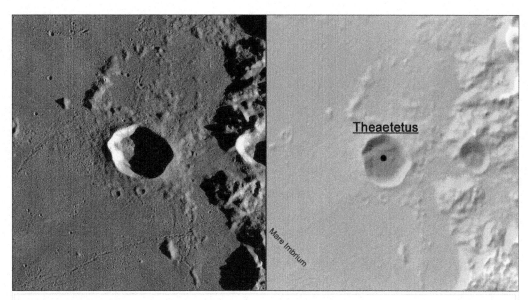

Theaetetus   Lat 37.01N   Long 6.06E   24.59Km

Sub-craters: No sub-craters

Notes:

Add Info:

Aerial view of Theaetetus

Theaetetus seems to have formed not only on the lavas of Imbrium (to its west), but also on sparse ejecta deposits of it, too (Theaetetus's own ejecta seems to blend in to it). The crater, therefore, has to be younger than Imbrium, however, as ray material from early, Eratosthenian-aged (3.15 to 1.1 byo) crater Aristillus, to its southwest, seems to overlie the ejecta, did Theaetetus form then at some time in between? Aerial view (left) of the crater shows three main slump events occurred (most particularly at its northwest rim and wall) over the intervening time; partially filling/mixing into material (from Aristillus, or other nearby impacts?) on its floor. Low sun views shows up the ejecta nicely, while at full moon its walls appear bright against the somewhat dark floor material. D: 2.8 km.

Craters

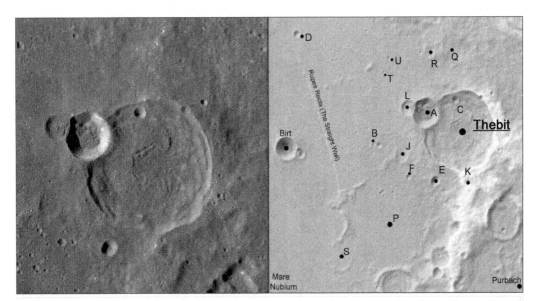

Thebit   Lat 22.01S   Long 4.02W   54.64Km

Sub-craters:

Crater	Lat	Long	Size (km)	Crater	Lat	Long	Size (km)
A	21.58S	4.93W	19.91	B	22.29S	6.28W	3.58
C	21.27S	4.12W	5.35	D	19.79S	8.26W	5.0
E	23.12S	4.65W	6.93	F	23.04S	5.37W	3.57
J	22.55S	5.53W	9.59	K	23.14S	3.79W	4.61
L	21.47S	5.37W	10.64	P	24.06S	5.63W	77.09
Q	20.08S	4.24W	15.86	R	20.2S	4.8W	8.38
S	24.88S	7.22W	17.27	T	20.67S	6.0W	2.15
U	20.35S	5.87W	3.75				

Notes:

Add Info:

Ranger 9 impacted some 250 kilometres away northwards from this crater, while the upper stage of Luna 5 impacted some 320 kilometres away southwards.

Close-up of Thebit A and L

Upper Imbrian in age (3.75 to 3.2 billions of years ago), Thebit not only formed on a main ring/rim (~ 715 km in diameter) of the Nubium Basin, to its west, but it also clipped the north-eastern rim of an older crater that was Thebit P (Thebit's ejecta is seen to partially cover it). The crater's rim is relatively sharp for its age, it has several terraces due to slumping events on nearly all its walls, and an obvious fracture concentric to its floor centrally. The crater has a gentle slope in towards Nubium's centre (its eastern rim is ~ 1 km above that of its western rim), and sub-crater, Thebit A (left) seems like it impacted between both Thebit and Thebit L.

Craters

Theon Jnr.	Lat 2.41S	Long 15.79E		17.61Km			
Sub-craters:	Crater	Lat	Long	Size (km)	Crater	Lat	Long	Size (km)
	B	2.19S	13.27E	6.33	C	2.38S	14.63E	3.83
	Notes:						

Add Info:

Try get a close-up view of sub-crater, Theon Jnr. B, over to the west of this crater, which shows a wonderful scarp-like feature on its western wall (a relatively high power may be required).

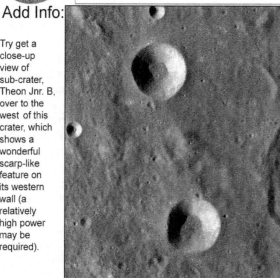

Aerial view of Theon Jnr. (bottom) with Theon Snr. (top)

No, you're not 'seeing double' as you view Theon Junior with Theon Senior to its north, when both are centred as a single shot in the eyepiece. Each is similar in size, similar in depth (around 3.5 kilometres approximately), and both have bright walls which 'shine' - especially during full moon times. The craters don't appear to have been affected by ejecta from the Imbrium Basin (to the northwest), which produced a striated look across nearby craters and on the general terrain, so both craters may be younger if that is the case. Theon Junior doesn't have much to offer in terms of detail, except for what looks like a very small slump occurred at its southwest wall. At low sun times, the crater is usually shadow-filled (also in Theon Snr.).

Craters

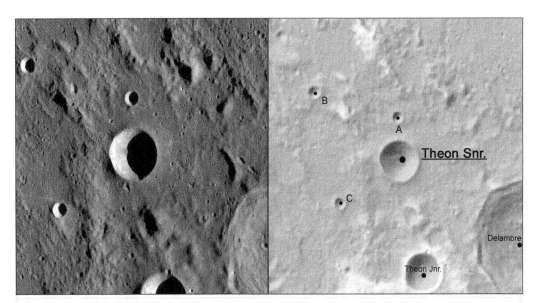

| Theon Snr. | Lat 0.81S | Long 15.42E | 18.02Km |

Sub-craters:

Crater	Lat	Long	Size (km)	Crater	Lat	Long	Size (km)
A	0.2S	15.39E	5.21	B	0.17N	14.12E	5.66
C	1.42S	14.51E	5.25				

Notes:

Add Info:

Aerial view of Theon Senior

Both Theon Senior and Theon Junior do look similar in age (post-Imbrian?). But does Theon Senior look just that little bit sharper? This latter point might suggest that Theon Senior then is younger, however, this may not always be true in comparing such-like craters that happen to be close together, and look the same. Other factors like their relative ejecta, shape, target rock, proximity to other craters (and their possible effects)...etc., must always be in mind. However, the visual evidence alone of the feature or crater itself being scrutinised can serve as a reliable marker. Observe both Theons' under all lighting conditions to see any differences, and note how the above factors might lead to allegiance of one over the other.

Craters

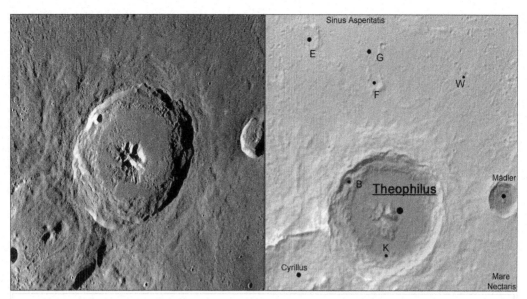

Theophilus	Lat 11.45S	Long 26.28E	98.59Km

Sub-craters:

Crater	Lat	Long	Size (km)	Crater	Lat	Long	Size (km)
B	10.59S	25.23E	8.2	E	6.87S	23.98E	22.25
F	8.02S	25.98E	12.22	G	7.21S	25.77E	19.2
K	12.6S	26.29E	5.52	W	7.79S	28.61E	3.59

Notes:

Add Info:

Aerial view of Theophilus

Theophilus is like a slightly smaller version of Langrenus (to its east ~ 1000 km away), but only better. Both are aged of the Eratosthenian (3.15 to 1.1 byo), each have a wonerful series of terraces, peaks, impact melt signatures and crenulated rims. If the two, however, were put together, then Theophilus would win hands down - simply because it looks sharper, has more impressive peaks (highest is ~ 2.8 km, as opposed to Langrenus's at ~ 2.5 km), and its outer ejecta, during low sun times, puts on a better show than Langrenus's does. Observations of the crater at any suitably-lit time will always prove rewarding. D: ~ 4.2 km.

Craters

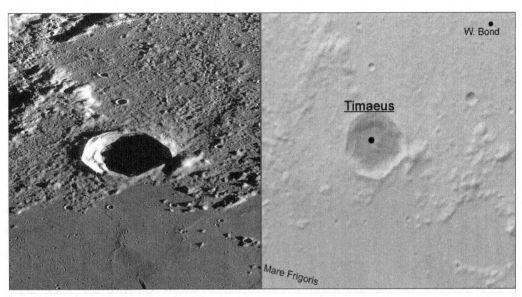

| Timaeus | Lat 62.91N | Long 0.55W | 32.81Km |

Sub-craters: No sub-craters

Notes:

Add Info:

Aerial view of Timaeus

Timaeus almost looks like is southwest neighbour, Archytas, which at times can lead to confusion. Both are aged of the Eratosthenian (3.15 to 1.1 byo), each has a relatively sharp rim, their peaks are about the same height (~ 600 metres high), and the two lie on a contact zone between rubbly material to their north and lavas to their south (Timaeus has also impacted upon W. Bond's southwestern rim). Unlike Archytas, however, Timaeus has a nice set of rilles concentric to its peak, but they can be hard to see at times, as light and shadows don't suit their general orientations (also, slumped material has covered them partially). At times of full moon, bright ray material from Epigenes A, to its north, can just about be seen to cross Timaeus's eastern and western rim. Depth ~ 3.0 km.

Craters

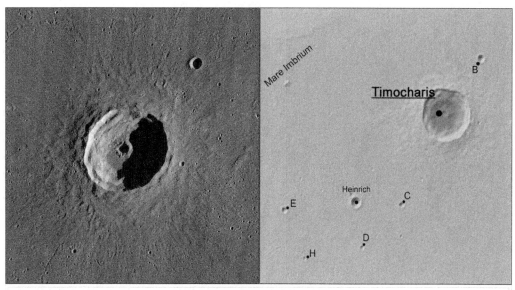

| Timocharis | Lat 26.72N | Long 13.1W | | 34.14Km | |

Sub-craters:

Crater	Lat	Long	Size (km)	Crater	Lat	Long	Size (km)
B	27.88N	12.18W	5.14	C	24.81N	14.21W	3.42
D	23.9N	15.19W	3.05	E	24.64N	17.14W	3.78
H	23.65N	16.62W	2.24				

Notes:

Add Info:

Aerial view of Timocharis

Eratosthenian in age (3.15 to 1.1 byo), Timocharis looks like a sligtly sharper version of Lambert (also Eratosthenian) over to its west. Obvious to the crater, particularly at low sun times, is its wonderful ejecta equally distributed around the crater, while at full moon times bright ray material unequally distributed is seen at its east. The crater has a relatively sharp rim (its terraces are sharp, too), with slumped material on the floor where it meets its ~ 500 metre-high peak, which appears to have been impacted upon (a 4.5 km diameter crater remains). Note the additional 'lip' outside the crater's eastern rim.

Craters

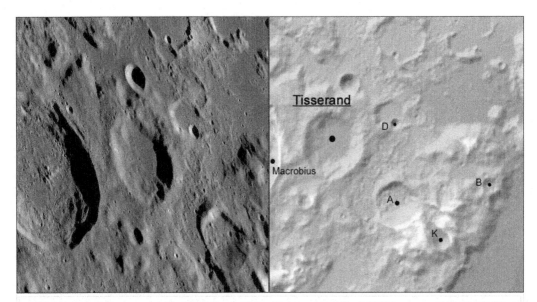

Tisserand	Lat 21.41N	Long 48.17E				34.63Km	

Sub-craters:

Crater	Lat	Long	Size (km)	Crater	Lat	Long	Size (km)
A	20.36N	49.48E	24.53	B	20.65N	51.37E	8.72
D	21.76N	49.47E	5.67	K	19.74N	50.33E	13.0

Notes:

Add Info:

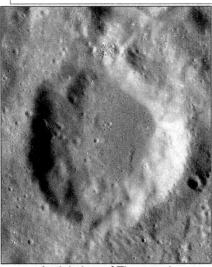

Aerial view of Tisserand

Like its younger neighbour, Macrobius (Lower Imbrian 3.85 to 3.75 byo), over to its west, Tisserand (Nectarian in age - 3.92 to 3.85 byo) lies just outside an outer ring/rim associated to formation of the Crisium Basin to its southeast. Ejecta from Macrobius overlies most of Tisserand's western rim and floor (the eastern portion of the floor appears to have undergone some internal flooding), while its eastern rim, particularly the northeastern part, has been affected by either a later slumping event, or that the impactor initially responsible for the crater struck on odd-highted terrain there. Tisserand's outer rim (and ejecta) at its southeast ends abruptly where its meets other material (is it Crisium ejecta?).

Craters

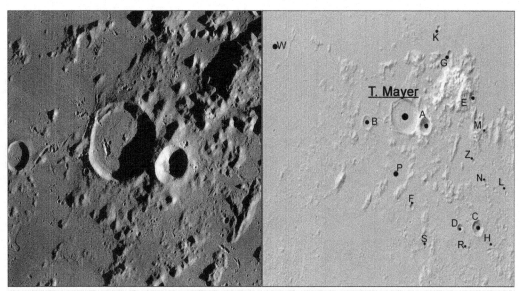

T. Mayer		Lat 15.54N	Long 29.17W		33.15Km	

Sub-craters:

Crater	Lat	Long	Size (km)	Crater	Lat	Long	Size (km)
A	15.24N	28.3W	16.16	B	15.35N	30.91W	12.14
C	12.21N	26.0W	14.85	D	12.23N	26.8W	7.64
E	16.04N	26.19W	8.42	F	12.87N	28.94W	5.09
G	17.35N	27.17W	6.97	H	11.68N	25.44W	4.65
K	18.12N	27.66W	4.82	L	13.2N	24.72W	4.03
M	14.91N	25.81W	5.11	N	13.56N	25.61W	4.26
P	14.02N	29.58W	37.03	R	11.63N	26.39W	4.23
S	11.66N	28.34W	2.85	W	17.57N	34.96W	33.47
Z	14.18N	26.15W	4.34				

Notes:

Add Info: The 'T' stands for Tobias - a German astronomer of the eighteenth century. The crater lies on ejecta deposits (usually referred to as Montes Carpatus) that were shot out in every direction during formation of the Imbrium Basin to its northeast. The crater is obviously younger than the Imbrium Basin event that occurred just over 3.8 billions of years ago, but by how much is it older? Certainly, it has to be older than the Copernican Period (1.1 byo to the Present) as ray material from Copernicus to its southeast fell on its floor and outer rim (easily observable around full moon times), so T. Mayer has to be of an age between both - perhaps it is of the Eratosthenian (as an exercise, compare it to crater Eratosthenes over to its east - some ~ 500 km away). Sub-crater, T. Mayer A, has also imparted its ejecta onto T. Mayer's floor, which again during full moon appears as dark as the mare to its west. A small slump has occurred at the crater's northwestern rim. A large 'gash' on the floor's western sector suggest a small, but thick rille, while just off-centre a peak (if it is one?) is seen. Depth of T. Mayer is around ~ 3 km approx..

Craters

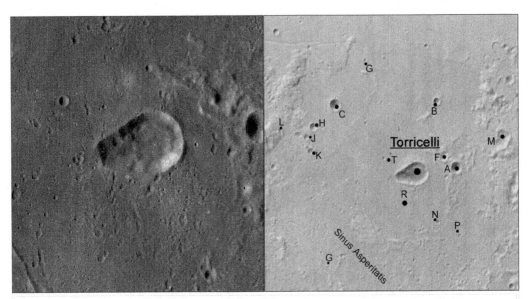

Torricelli		Lat 4.72S		Long 28.4E			30.87Km

Sub-craters:

Crater	Lat	Long	Size (km)	Crater	Lat	Long	Size (km)
A	4.54S	29.78E	10.45	B	2.63S	29.17E	6.84
C	2.71S	26.01E	10.42	F	4.22S	29.37E	7.43
G	1.44S	26.94E	3.23	H	3.35S	25.31E	7.23
J	3.64S	25.09E	5.25	K	4.04S	25.22E	5.74
L	3.47S	24.27E	3.96	M	3.62S	31.27E	12.08
N	6.11S	29.2E	4.02	P	6.5S	29.89E	3.74
R	5.24S	28.14E	87.64	T	4.26S	27.51E	3.57

Add Info: Torricelli joins that elite group of craters where their shapes are more asymmetric than the symmetric (circular) norm that we're used to seeing. Such craters can occur through several factors, for example: produced by impactors coming in at low angles (usually around 15 degrees and less) to the surface; breakup of the impactor (as several closely-landing pieces cause an elongated crater); and slumping in a preferred section of a crater's rim and wall. The source of original impact can sometimes be inferred from such asymmetric craters by noting the orientation of their longer axes, and where the 'pointy' section dictates, approximately, the downrange side. For example, craters' Alphonsus B (13.26S, 0.2W) and Abulfeda D (21.76S, 9.65E), southwest of Torricelli, both may have their origin from the Nectaris Basin. With Torricelli, the orientation suggests the origin might be the Fecunditatis Basin over to its east, but that is an old basin of the pre-Nectarian Period (4.6 to 3.92 byo) and Torricelli looks way too fresh. There doesn't seem to be any other impact crater or basin that may have sourced Torricelli, so was this asymmetric crater created from one of the other above factors (several small, separate slumps have occurred at the crater's western wall, which might favour that type of event). Depth of Torricelli is around 2.5 km.

Craters

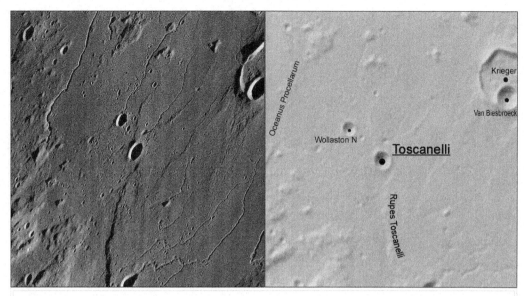

| Toscanelli | Lat 27.96N | Long 47.61W | 7.05Km |

Sub-craters: No sub-craters

Notes:

Add Info:

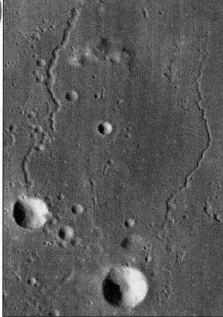

Rille system north of Toscanelli

Toscanelli lies in an interesting region of the Moon, where rilles to wridges rule. The slope of the general terrain runs gently northwards; for example, from Toscanelli to the end of the small, unnamed rillle above it, a drop of about six hundreds metres occurs over a 40 kilometre length. The rille, like most others nearby, looks like a meandering river (originally of a river of lava), which is presumed to have flowed northwards. Toscanelli seems to have formed on an old wrinkle ridge (old because it doesn't appear to have affected the crater), which also interrupted the Rupes Toscanelli fault (not scarp). The crater really doesn't have much to offer in terms of detail, but its relationship to the above features makes for a good observational exercise. Depth ~ 1.5 km approx..

Craters

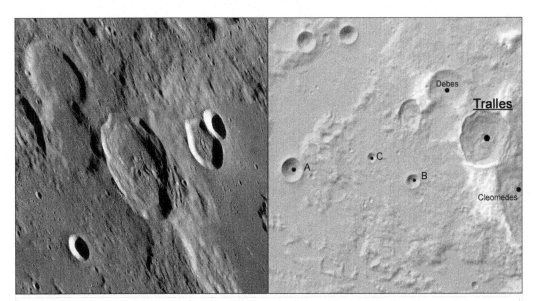

Tralles	Lat 28.32N	Long 52.85E				44.16Km	

Sub-craters:

Crater	Lat	Long	Size (km)	Crater	Lat	Long	Size (km)
A	27.42N	47.03E	17.18	B	27.26N	50.66E	11.15
C	27.8N	49.4E	7.34				

Notes:

Add Info:

Aerial view of Tralles

Both Tralles and its larger neighbour, Cleomedes off to its right, formed in a trough-like zone between two rings/rims of the Crisium Basin to the southeast. Tralles's impactor initially struck Cleomedes's rim and outer ejecta, which in turn 'dolloped' a huge volume of its own ejecta back onto Cleomedes's northwestern floor. Tralles has to be younger than Nectarian-aged (3.92 to 3.85 byo) Cleomedes, however, comparing its relatively sharp rim to, say, Eratosthenian-aged (3.15 to 1.1 byo) Newcomb over to its west, the crater most likely formed sometime between those ages. Tralles's floor level is some ~ 700 metres above Cleomedes, and the highest of peaks (?) is ~ 500 metres.

Craters

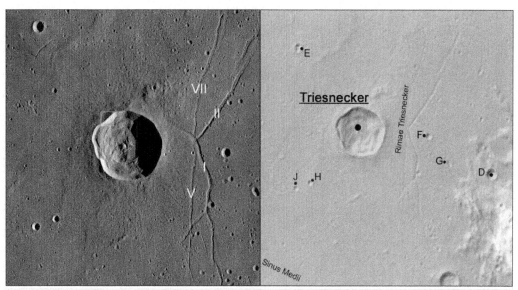

Triesnecker	Lat 4.18N	Long 3.6E	24.97Km

Sub-craters:

Crater	Lat	Long	Size (km)	Crater	Lat	Long	Size (km)
D	3.48N	5.95E	5.85	E	5.55N	2.48E	4.24
F	4.1N	4.82E	3.24	G	3.66N	5.19E	3.52
H	3.34N	2.7E	2.55	J	3.25N	2.45E	2.92

Notes:

Add Info:

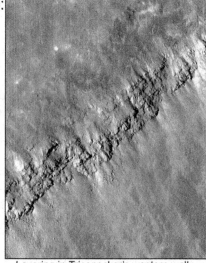

Layering in Triesnecker's western wall

Try hard as you might, it's near impossible to view Triesnecker without noting the wonderful rilles (Rimae Triesnecker) off to its east. Roman numerals were once assigned to the rille system, however, these are no longer used - a pity, as it would help in referencing today. Rille I above, which bends in towards Triesnecker's rim, is the only one that was partially buried by ejecta from the crater. Triesnecker is Copernican in age (1.1 byo to the Present), its interior walls have suffered several major slump events exposing some wonderful layers (left) that most likely represent separate lava flows on which Triesnecker now lies. The floor of Triesnecker only appears at it eastern sector, where it meets two small peaks (highest ~250 metres). D: 2.8 km.

Craters

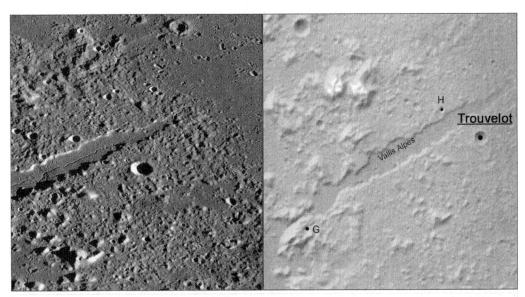

Trouvelot   Lat 49.37N   Long 5.79E   8.5Km

Sub-craters:

Crater	Lat	Long	Size (km)	Crater	Lat	Long	Size (km)
G	47.45N	0.24E	3.76	H	49.92N	4.51E	4.4

Notes:

Add Info:

Vallis Alpes
Nearly 2.5 km wide, its eastern side, which is approximately 300 m higher than its western side, produces a cliff with a relatively shallow slope of about 10 degrees. As none of the structures through which the fault passes show any apparent lateral off-set, movement along the fault is believed to have been predominantly vertical.

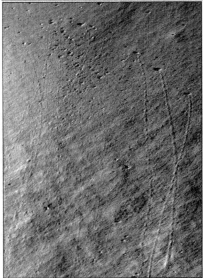
Boulder trails on Trouvelot's floor

You'd hardly notice this crater - simply because of its close proximity to the wonder of Vallis Alpes to its west. The crater lies on hummocky terrain believed to be composed of blocky clumps of ejecta from the Imbrium Basin to its southwest. The crater then is younger than Imbrium, which formed some four billion years ago, however, close-up of its rim shows a worn appearance (would Trouvelot's original impactor have been a highly-lofted clump of Imbrium ejecta that fell not long afterwards?). Further, close-up (left) of the crater shows some wonderful trails by boulders that rolled down Trouvelot's southern wall, and whose momentum finally slowed down as they met the level of the floor. Depth ~1 km approx..

Craters

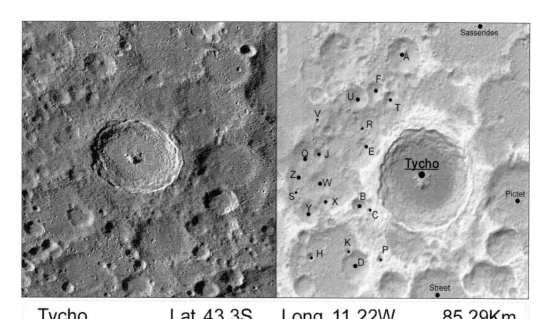

| Tycho | Lat 43.3S | Long 11.22W | 85.29Km |

Sub-craters:

Crater	Lat	Long	Size (km)	Crater	Lat	Long	Size (km)
A	39.94S	12.07W	28.96	B	43.99S	13.92W	13.64
C	44.12S	13.46W	7.17	D	45.58S	14.07W	25.97
E	42.34S	13.66W	12.87	F	40.91S	13.21W	16.57
H	45.29S	15.92W	7.59	J	42.58S	15.42W	10.66
K	45.18S	14.38W	6.4	P	45.44S	13.06W	7.08
Q	42.5S	15.99W	20.29	R	41.91S	13.68W	4.5
S	43.47S	16.3W	3.37	T	41.15S	12.62W	14.22
U	41.08S	13.91W	19.88	V	41.72S	15.43W	3.84
W	43.3S	15.38W	20.69	X	43.84S	15.25W	11.84
Y	44.12S	15.93W	21.74	Z	43.23S	16.35W	23.35

Add Info:

Melt fractures on Tycho's floor

If ever there was a crater that could be assigned 'promoter for the wonder of craters on the Moon', Tycho is certainly a contender. Not because of its crenulated rim and series of terraces. Not because of its peaks (highest is ~ 2.7 kilometres high) and the impact-melt on its floor (left). Not because of the annulus of darkish material surrounding its outer rim (note it at full moon times), or, the wonder of its bright rays that extend in nearly every direction (just look at the one off to its east that runs for nearly ~ 1500 km over to Mare Nectaris). But, simply, because Tycho is the most-likely crater to impress you when initially observing the Moon for the first time. Depth of Tycho ~ 4.8 km approx..

Craters

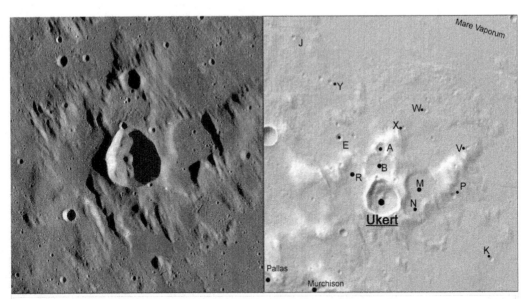

Ukert   Lat 7.71N   Long 1.37E   21.71Km

Sub-craters:

Crater	Lat	Long	Size (km)	Crater	Lat	Long	Size (km)
A	8.72N	1.35E	9.03	B	8.35N	1.28E	20.25
E	8.96N	0.4E	4.73	J	11.03N	0.61W	2.86
K	6.47N	3.76E	3.21	M	7.99N	2.18E	20.43
N	7.58N	2.01E	17.13	P	7.77N	2.93E	4.22
R	7.93N	0.69E	18.26	V	8.73N	3.24E	2.25
W	9.51N	2.33E	2.65	X	9.16N	1.85E	2.52
Y	10.11N	0.21E	3.19				

Notes:

Add Info: Ukert lies between two lava plains - that of Mare Vaporum (pre-Nectarian in age 4.6 to 3.92 billions of years old) to its north and Sinus Medii (again, pre-Nectarian) to its south. The crater formed initially on highlandic material that was subsequently pummelled-to-death by huge ejecta blocks created by the Imbrium Basin (centred some ~ 850 kilometres off to the northwest), giving a groove-like apppearance to the general terrain. Obvious, also, is the crater's sharp-ish rim that suggests it is younger than the three craters it impacted on, that is, sub-crater B, M and N (and possibly another crater at its western outer rim), however, sparse hints of its ejecta is really only seen to overlie Ukert B. The shape of Ukert is due to two main slump events - the first occurring at the crater's southeast wall, but the second, a major event, at nearly all of the crater's western wall (did the small ~ 4 km-sized crater on its northern rim set off the slump?). As a consequence, all of Ukert's floor remains buried by their deposits, which sidled up to a rounded peak (~ 0.5 km). Depth ~ 2.8 km.

Surveyors 4 and 6 impacted and landed respectively, some 150 kilometres due south of this crater.

Craters

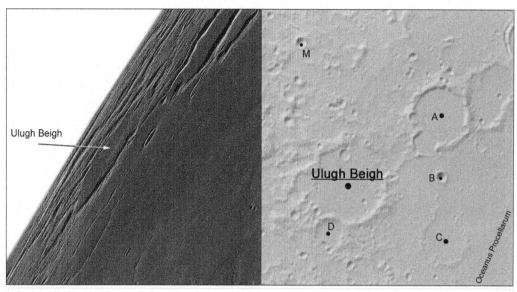

Ulugh Beigh	Lat 32.67N	Long 81.96W			57.04Km

Sub-craters:

Crater	Lat	Long	Size (km)	Crater	Lat	Long	Size (km)
A	34.13N	79.34W	41.18	B	32.79N	79.3W	7.59
C	31.4N	79.24W	33.6	D	31.54N	82.51W	21.47
M	35.7N	83.46W	8.03				

Notes:

Add Info:

Day-old moon phases

Note: As the Moon goes through various librations throughout the year, suggested times given in text for observations are only approximates.

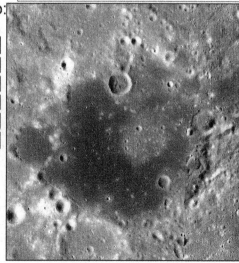

Aerial view of Ulugh Beigh

There are quite a few craters here on the northwestern limb that can lead to confusion - so many of similar size, of similar appearance (a good indicator for Ulugh Beigh is the dark lavas on its western floor, which contrasts against the brighter patch on its east). The lavas on the floor are most likely interior-sourced, but as there also appears to be a low, possible breach section at the crater's northeastern rim, did the lavas of Procellarum fill through there? Of course, favourable librations are the best times for observing, but you'll also need some shadowing going on at its rim (say, times approaching a full or new moon). Depth ~ 0.6 km.

Craters

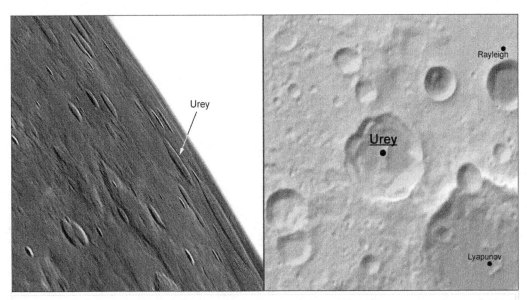

Urey	Lat 27.93N Long 87.43E	39.29Km

Sub-craters: No sub-craters

Notes:

Add Info:

Aerial view of Urey

In the very high longitudes, Urey, from the outstart, will prove to be a challenge - even for the serious observer (suitable librations and good lighting conditions are obviosly the best times). The crater lies on the southeastern rim of Rayleigh, which formed in the pre-Nectarian Period (4.6 to 3.92 byo), so Urey has to be younger (it isn't as young, however, as crater Lyapunov to its southeast). Several major slumps have occurred on Urey's walls (particularly at its south), which seems to have reached in so far to Urey's once peak(?), forming a long mountain-like feature across the floor (note also the small crater ~ 4 km in diameter, terminating the feature). Use the unnamed crater on Lyapunov's northern rim as a location marker for Urey. D: ~ 3.5km.

Craters

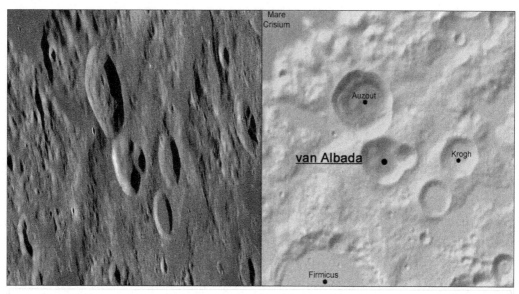

| van Albada | Lat 9.36N | Long 64.35E | 22.92Km |

Sub-craters: No sub-craters

Notes:

Add Info:

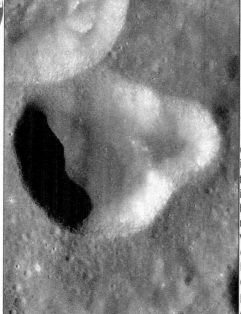

Aerial view of van Albada

The additional extension on to van Albada's eastern sector makes for easy finding of the crater. The extention, of course, is due to a small impact crater forming on van Albada's original eastern rim (ejecta of the small crater lies on van Albada's eastern floor). The question one has to ask is: 'was the impactor that produced the small crater initially part of a much larger impactor that came in at a low angle (< 15 degrees) from the west?'. It may not be, but what if it was? Then the impactor would have broken up on impact, a separate piece might have richocheted downrange, leaving an oblong crater behind. Van Albada's northern rim has been clipped by formation of Auzout (ejecta lies on van Albada's floor) to the north; putting a constraint on its age as Auzout is of the Eratosthenian Period (3.15 to 1.1 byo). Depth ~ 3.5km.

Craters

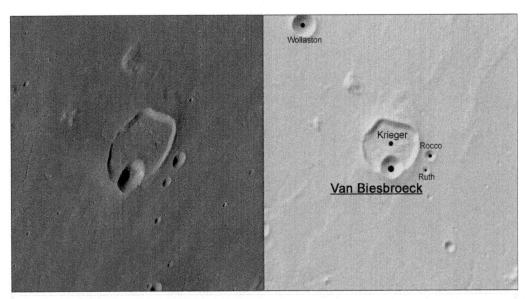

Van Biesbroeck Lat 28.77N Long 45.59W   9.08Km

Sub-craters: No sub-craters

Notes:

Add Info:

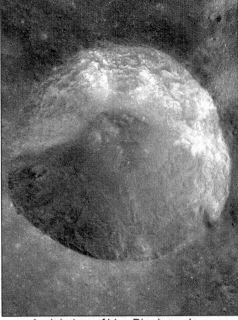

Aerial view of Van Biesbroeck

Van Biesbroeck has obviously impacted upon the southern rim of Krieger. The impactor that produced the crater must have intially struck on odd-highted ground; where Krieger's own rim was some 0.8 kilometres above the level of its floor. As a result, was ejecta from Van Biesbroeck distributed more towards Krieger's central floor area (see the clumpy material there) than being directed around the crater equally. Certainly, the impact forces involved in impact cratering will demolish nearly everything in their path, however, with some impacts, the topography, for example, as with Krieger's higher southern rim and wall, may play a role in ejecta distribution and direction. Depth ~ 1.6 km approx..

Craters

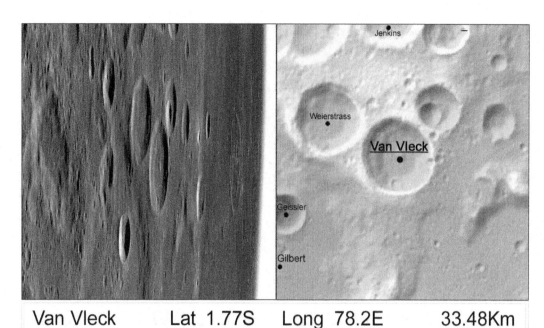

Van Vleck   Lat 1.77S   Long 78.2E   33.48Km

Sub-craters: No sub-craters

Notes:

Add Info:

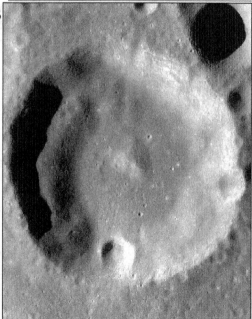

Aerial view of Van Vleck

Van Vleck lies on the ejecta deposits of Mare Smythii over to its east. Together with Weierstrass, the two look very similar in appearance, and it looks like both impacted on Gilbert's northeastern rim. Van Vleck does look that little bit sharper than its neighbour, but it's a hard call to see any of its ejecta on Weierstrass's floor; whose level is just over a kilometre above Van Vleck's. The crater has a small rounded peak ~ 0.5 kilometres in height, while the western side of its floor has a major deposit resembling another rim. It isn't a rim, of course, so is it a slump deposit, or, a rebound deposit off of Van Vleck's western wall after the crater formed. Depth of the crater is around 3.5 km approx..

Craters

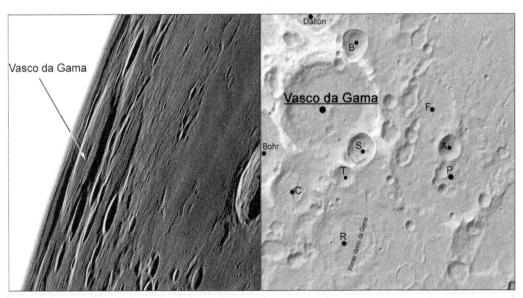

Vasco da Gama Lat 13.78N Long 83.94W 93.52Km

Sub-craters:

Crater	Lat	Long	Size (km)	Crater	Lat	Long	Size (km)
A	12.63N	80.07W	22.03	B	15.72N	83.12W	25.06
C	11.48N	85.07W	46.83	F	13.86N	80.78W	55.78
P	12.0N	80.34W	102.54	R	9.91N	83.51W	60.9
S	12.63N	82.94W	30.29	T	11.87N	83.5W	19.94

Notes:

Add Info:

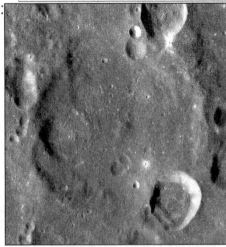

Aerial view of Vasco da Gama

Like several craters and other features in the general area, Vasco da Gama has been subjected to coverage by ejecta deposits from the Orientale Basin to its southwest (centred some 1000 km away). The crater is probably of the pre-Nectarian Period (4.6 to 92 byo) given the state of its appearance, and the numerous impacts - new and old that affected its rim. Vasco da Gama's floor is slightly convexed upwards, it is also tilted slightly towards Oceanus Procellarum in the east, and it has a ~ 1 km-high peak in its centre. Good libration times and suitable lighting conditions will bring out the best details in the crater. Depth ~ 3.0 km.

Craters

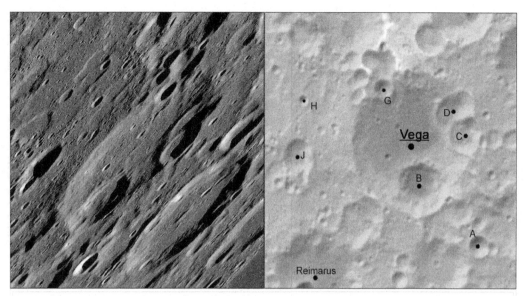

| Vega | Lat 45.41S | Long 63.27E | 73.51Km |

Sub-craters:

Crater	Lat	Long	Size (km)	Crater	Lat	Long	Size (km)
A	47.25S	65.15E	12.45	B	46.23S	63.5E	31.17
C	45.27S	64.72E	20.14	D	44.89S	64.47E	25.01
G	44.43S	62.46E	11.52	H	44.59S	60.27E	6.12
J	45.56S	60.04E	18.64				

Notes:

Add Info:

Aerial view of Vega

Vega is situated midway approximately between two basins - Australe Basin to its southeast and Nectaris Basin to its northwest. The crater is as old (pre-Nectarian 4.6 - 3.92 byo) as the former, but has been affected by the latter where ejecta from it overlies Vega. Practically, all of the crater's eastern rim, and floor, has been affected by three good-sized impact events - see sub-craters' Vega B, C and D, the latter two having interrupted a chain of secondaries from Nectaris (they parallel the oddly-aligned Rheita Valley crater chain feature over to the west). Depth of Vega is around ~ 3.5 km approx..

Craters

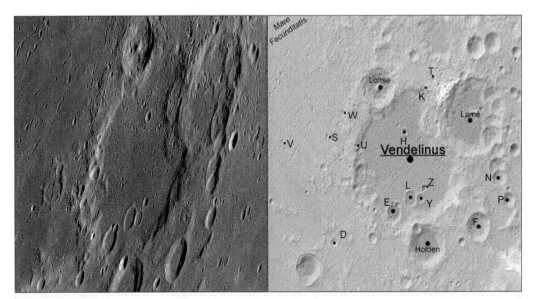

| Vendelinus | Lat 16.46S | Long 61.55E | 141.21Km |

Sub-craters:

Crater	Lat	Long	Size (km)	Crater	Lat	Long	Size (km)
D	19.05S	58.24E	10.02	E	18.08S	61.02E	19.46
F	18.49S	64.92E	32.02	H	15.3S	61.5E	7.98
K	13.82S	62.44E	8.92	L	17.56S	61.79E	17.29
N	16.82S	65.87E	17.32	P	17.56S	66.33E	16.45
S	15.4S	57.98E	5.81	T	13.49S	62.8E	5.65
U	15.91S	58.73E	5.39	V	15.55S	55.92E	5.77
W	14.59S	58.7E	4.93	Y	17.58S	62.24E	10.94
Z	17.2S	62.41E	7.38				

Notes:

Add Info:

Aerial view of Vendelinus

You can just about make out Vendelinus's rim and floor at most times (close to terminator times are best). The problem seeing it is mainly due to its low depth (~ 2.0 km), its rim which has been affected by numerous impacts (note the three major ones of Holden, Lamé and Lohse), and its floor that is flooded (the crater lies in an intersection zone between two ring/rim systems; that of the Balmer-Kapteyn Basin to its southeast and the Fecunditatis Basin to its northwest). Rays from Eratosthenian-aged (3.15 to 1.1 byo) crater Langrenus to its north overlie its floor, but they're not so bright to be obvious (they have an odd alignment, too).

Craters

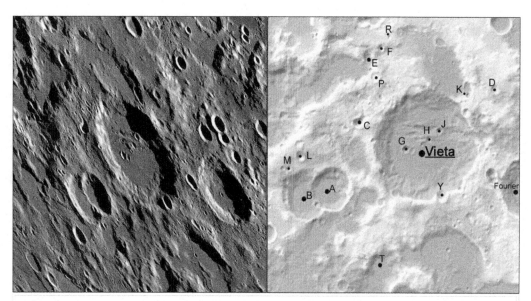

Vieta		Lat 29.31S	Long 56.53W			87.16Km	
Sub-craters:	Crater	Lat	Long	Size (km)	Crater	Lat	Long	Size (km)

Crater	Lat	Long	Size (km)	Crater	Lat	Long	Size (km)
A	30.35S	59.47W	34.14	B	30.49S	60.33W	39.32
C	28.68S	58.56W	12.2	D	27.84S	54.3W	7.77
E	26.95S	58.25W	9.81	F	26.79S	57.87W	6.19
G	29.37S	57.12W	6.32	H	29.1S	56.39W	5.09
J	28.89S	56.06W	5.47	K	28.0S	55.15W	4.81
L	29.5S	60.42W	7.31	M	29.79S	60.8W	4.74
P	27.52S	58.05W	8.28	R	26.52S	57.61W	3.3
T	32.39S	57.93W	28.29	Y	30.57S	55.96W	10.47

Notes:

Add Info:

Aerial view of Vieta

When you lie not some ~ 1000 kilometres away from one of the most impressive impact basins on the Moon, that is, Orientale Basin over to its west, then it's a sure thing you're going to be affected in some way or other. The 'gashes', then, across Vieta's floor is a consequence of that major event, which most likely involved a cluster of large chunks of ejecta gouging into Vieta's floor as they broke up on landing (see the effects at crater Fourier to its east, too). Vieta is of the Nectarian Period (3.92 to 3.85 byo). Bright material from the unnamed crater on its northeastern rim stands out easily, as too with rays from Byrgius A, to the northwest.

Craters

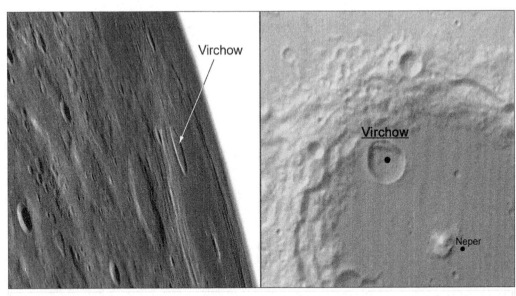

| Virchow | Lat 9.88N | Long 83.77E | 18.83Km |

Sub-craters: No sub-craters

Notes:

Add Info:

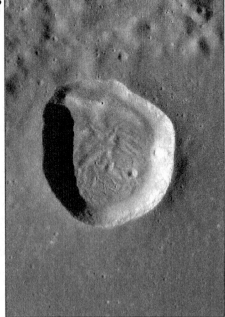

Aerial view of Virchow

Virchow's impactor initially struck on the lava plain of Nectarian-aged crater, Neper, however, as the crater grew, its effect inevitably reached into a portion of Neper's northern inner wall. As a consequence, ejecta south of Neper must be predominantly of deeper deposits overlying the plains, while north it must be a mixture including both Neper's own wall and terraced material. Another consequence, perhaps, is the major slump event at Virchow's northern rim - did the initial impact produce a weaker zone there? The slumped deposits cover nearly half of Virchow's floor, which appears to have been previously fractured. It isn't hard to spot as a crater once in view during a favourable libration, as it contrasts nicely against Neper's floor. Virchow's floor level is some ~1 km below that of Neper's. D: 1.8 km.

Craters

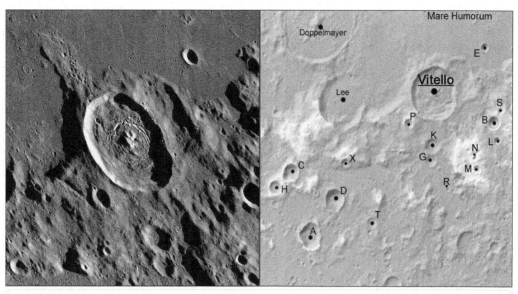

Vitello	Lat 30.42S	Long 37.55W	42.51Km

Sub-craters:

Crater	Lat	Long	Size (km)	Crater	Lat	Long	Size (km)
A	34.11S	41.97W	23.1	B	31.17S	35.45W	10.93
C	32.43S	42.58W	16.5	D	33.17S	41.06W	16.88
E	29.18S	35.82W	7.34	G	32.2S	37.67W	9.48
H	32.8S	43.13W	11.63	K	31.8S	37.63W	12.89
L	31.63S	35.32W	6.16	M	32.39S	36.02W	6.18
N	32.12S	36.09W	4.76	P	31.2S	38.43W	8.26
R	32.95S	37.06W	3.19	S	30.84S	35.23W	5.81
T	33.81S	39.76W	8.98	X	32.22S	40.68W	7.31

Notes:

Add Info:

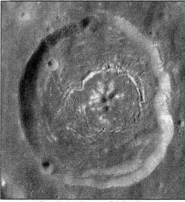

Aerial view of Vitello

Vitello is almost like a smaller version of the wonderful crater, Gassendi, to its northwest, just ~ 380 kilometres away. Both lie on the edge of Mare Humorum and its main ring/rim (~ 420 kilometres in diameter) system, and both have wonderful series of fractures and ring peaks in their floors. But there's a slight difference - Vitello seems to be just that little bit sharper, fresher, and possibly younger (Gassendi is of the Nectarian Period - 3.95 to 3.85 byo). Moreover, as it formed on higher ground, Vitello's floor level is some ~ 500 metres above Gassendi's (and the mare floor, too). Depth ~ 1.7 kilometres.

Craters

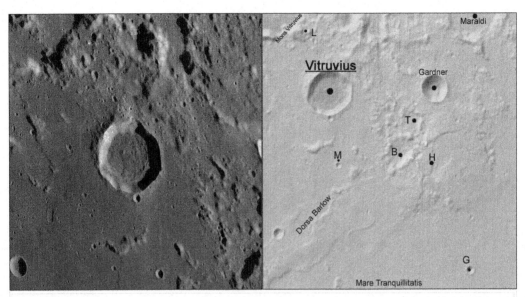

Vitruvius	Lat 17.66N	Long 31.28E				30.94Km	

Sub-craters:

Crater	Lat	Long	Size (km)	Crater	Lat	Long	Size (km)
B	16.37N	32.96E	17.82	G	13.89N	34.61E	4.99
H	16.36N	33.8E	22.36	L	18.94N	30.68E	5.58
M	16.13N	31.51E	3.66	T	17.06N	33.24E	13.95

Notes:

Add Info:

Apollo 17 landed some 80 km away north of this crater on 11 Dec 1972. It was the final mission in the Apollo series, and the astronauts involved were E. Cernan, R. Evans and H. Schmitt.

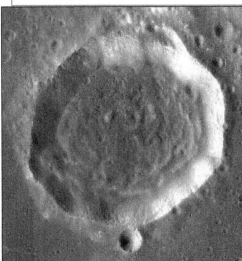

Aerial view of Vitruvius

Vitruvius almost escaped forming entirely within an outer ring/rim of the proposed Tranquillitatis Basin, as it just about kisses hummocky terrain to its north. As a consequence, ejecta from the crater lies outside its northern rim only, while the remaining outer sectors of the crater have been covered by lavas that lapped nearly right up to the rim (see the southern sector particularly). Vitruvius's interior has un-undergone changes as well (it's as dark as the mare), where slumping of material at both its eastern and southwestern walls have 'ovaled' the crater in shape (see left). Vitruvius has a depth of 2.0 km approx..

Craters

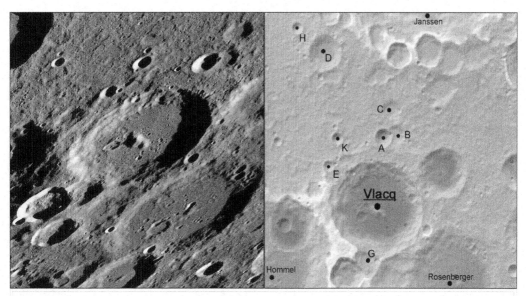

Vlacq	Lat 53.39S	Long 38.69E	89.21Km

Sub-craters:	Crater	Lat	Long	Size (km)	Crater	Lat	Long	Size (km)
	A	51.28S	39.01E	16.56	B	51.15S	39.76E	17.19
	C	50.45S	39.41E	18.89	D	48.73S	36.17E	31.93
	E	52.09S	36.15E	10.44	G	55.0S	38.01E	27.11
	H	47.95S	34.88E	11.07	K	51.29S	36.66E	11.49

Notes:

Add Info:

Aerial view of Vlacq

Vlacq and crater Hommel, over to its southwest, are both believed to be of the pre-Nectarian Period 4.6 to 3.92 byo. However, closer inspection of its rim and general features, its terraces and peaks etc.,, shows Vlacq to be slightly sharper in appearance, and so it probably formed after Hommel (it's not quite clear if Vlacq's ejecta lies within Hommel's eastern floor). Both craters put on impressive 'shadow shows' during low sun times, but high sun times, as at full moon, it becomes extremely hard to locate them (Vlacq's bright-ish peak may help?). Depth of Vlacq is around 3.8 km.

Craters

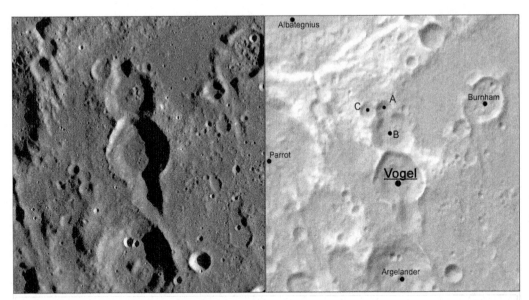

Vogel	Lat 15.11S	Long 5.83E				26.3Km

Sub-craters:	Crater	Lat	Long	Size (km)	Crater	Lat	Long	Size (km)
	A	14.06S	5.57E	8.85	B	14.42S	5.69E	21.74
	C	14.08S	5.3E	8.57				

Notes:

Add Info: The main eye-catching, characteristic-related feature when viewing Vogel is the wide 'gash' attached at its south. Initial observations point to the gash as being a lineated feature produced by a huge chunk of ejecta from the Imbrium Basin (centred some 1500 kilometres away to the northwest) as it gouged into the target terrain. If you take in a larger view of the area, several other similar-like, lineated features can be seen on the crater rims of, say, Ptolemaeus, Alphonsus and others nearby, which do 'point' back to the basin. But look again at Vogel's gash in this larger view and you'll see it really doesn't have the same 'pointy' orientation to Imbrium as the others do. Moreover, as crater Argelander to the south of Vogel seems to have impacted/interrupted the gash, then it can't be Imbrium in origin as Argelander is of an older age (it is of the Nectarian Period - 3.92 to 3.85 byo). So, what is this 'gash', if indeed that is what it is at all. Could it be just an old crater that Vogel impacted upon, and that the final settlement of material and ejecta between the two produced the feature seen (see a similar type of relationship between sub-craters' Vogel A and B)? Whatever the sequence of events for the two, they make for some nice shadow interchanges at low sun times. Vogel itself is generally plain in terms of details. It seems to have impacted upon B's southern rim. A major slump has occurred at its western wall. Hint of a possible, off-centred peak is seen (~250 metres high) on the floor's northeast. Depth of Vogel ~ 2.5 km approx..

Craters

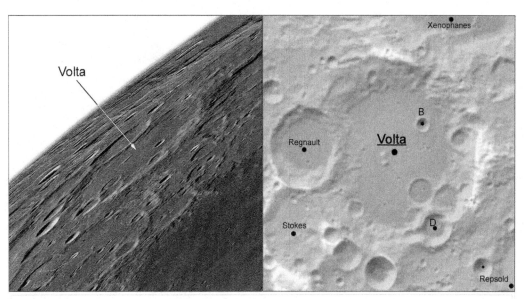

Volta	Lat 53.9N	Long 84.77W	117.15Km

Sub-craters:	Crater	Lat	Long	Size (km)	Crater	Lat	Long	Size (km)
	B	54.53N	83.53W	8.29	D	52.49N	83.09W	19.53
	Notes:							

Add Info:

Aerial view of Volta

Volta isn't lost in this northwest sector of the Moon where some of the largest craters (like Repsold and Repsold C, Xenophanes and Gerard Q Outer) can be observed. All are old-looking (possibly of the pre-Nectarian Period - 4.6 to 3.92 byo), but, unfortunately, all also require favourable librations and good lighting conditions. Volta's floor holds a host of features: it is slightly convexed, it has branching rilles, Regnault's ejecta lies on its western sector, a small peak (~800 metres high) at its centre, and several later impact craters show additional features, too (see the crater above Volta D). Depth of Volta ~ 3.5 km approx..

Craters

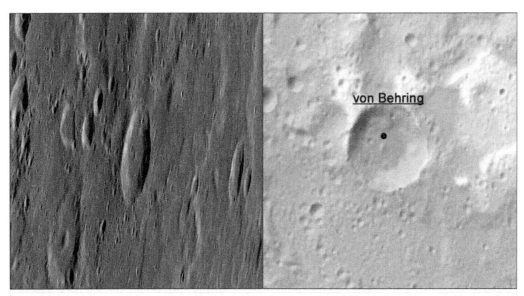

von Behring   Lat 7.75S   Long 71.72E   37.65Km

Sub-craters: No sub-craters

Notes:

Add Info:

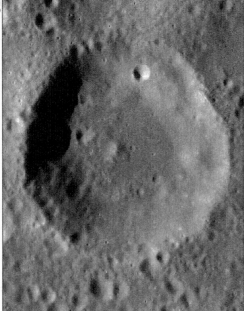

Aerial view of von Behring

Von Behring looks like it formed on the outer ejecta deposits of the Smythii Basin (centred some 500 kilometres away east). The crater also lies within the proposed Balmer-Kapteyn Basin (centred 200 kilometres to the south), whose underlying remnants may be responsible for the darker portion seen in von Behring's eastern floor. The crater has a small peak (~ 500 metres high), nearly all its rim has been pitted by numerous small impacts, while its eastern wall has slumped (did the small, 3.5 km-sized crater on its northern wall set off the slump, or is it an old slump?). The crater is easily found under most lighting conditions (at full moon times, its dark-ish floor, and a dark patch at its northeast helps). D: 3.8 km.

Craters

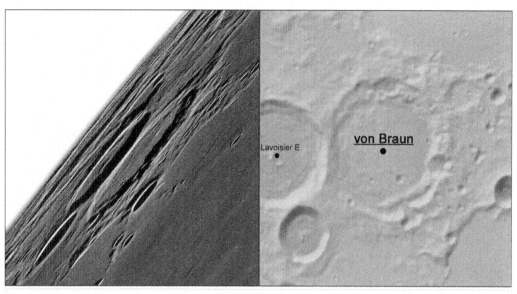

von Braun   Lat 41.04N Long 78.08W   61.83Km

Sub-craters: No sub-craters

Notes:

Add Info:

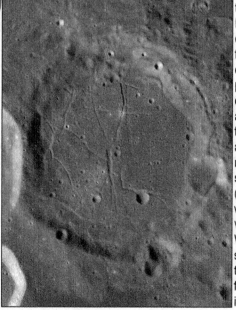

Aerial view of von Braun

It's a pity that we can't see this crater better due to its location, as von Braun has some wonderful rilles to offer the viewer. They criss-cross the floor in every direction, and are believed to have formed due to magma ponding below under pressure. Its effect caused the floor to dome upwards, leading to fracturing of the surface (von Braun's convexed floor is almost level with Oceanus Procellarum to its east). Close-up of the rilles show some reached in to the walls (for example, see the western walls), where they eventually disappear as walls weathered/slumped over time. What has happened at the eastern side of the crater, particularly within the rim there, where a long rille-like feature hugs nearly its full length? Is it due to subsidance, leading then to scarp? Depth of von Braun ~1.5 km.

Craters

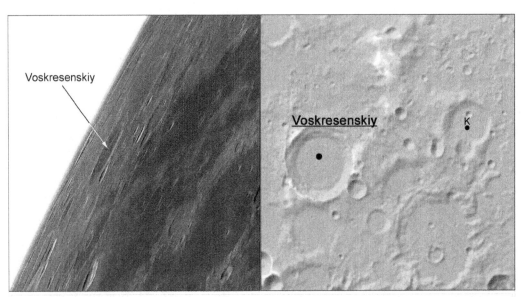

Voskresenskiy   Lat 27.91N   Long 88.12W   49.37Km

Sub-craters:

Crater	Lat	Long	Size (km)	Crater	Lat	Long	Size (km)
K	28.74N	84.24W	36.08				

Notes:

Add Info:

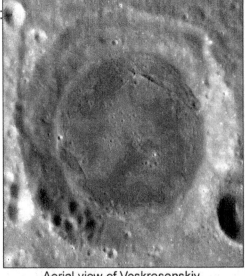

Aerial view of Voskresenskiy

Voskresenskiy lies just within the 'mean limb' zone of observation (that is, where features can just about be seen before the limb itself - even during the best of libration times - prevents any further viewing advantages). It will pose a problem, therefore, to get any descent view of detail of the crater's innards and floor, which has been flooded internally by lavas (the aerial view, left, shows some circumferential fractures that possibly were where the lavas extruded onto the floor). The crater is possibly of the Nectarian Period 3.92 to 3.85 byo). The chain of craters on its western rim 'point' to the Orientale Basin. Depth 2.5 km.

Craters

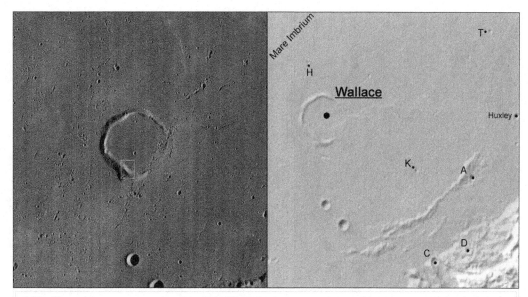

Wallace	Lat 20.26N	Long 8.75W		25.71Km		

Sub-craters:	Crater	Lat	Long	Size (km)	Crater	Lat	Long	Size (km)
	A	19.18N	5.6W	3.71	C	17.65N	6.42W	5.0
	D	17.84N	5.71W	4.01	H	21.29N	9.08W	2.06
	K	19.31N	6.79W	2.4	T	21.83N	5.16W	2.44
	Notes:							

Add Info:

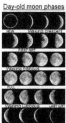

Note: As the Moon goes through various librations throughout the year, suggested times given in text for observations are only approximates.

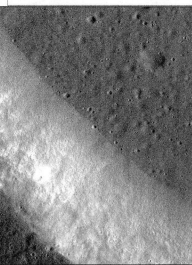

Close-up of Wallace's southwest wall and floor

It's obvious that Wallace and its floor has been flooded by lavas in Mare Imbrium (the breach point occurred at its eastern side). What's not so obvious, however, is that Wallace over time has experienced two separate flow events - the lavas west of the crater are oldest ~ 3.5 byo than those at its east ~ 3.4 byo (it's possible to see an albedo change on the surface during suitable lighting conditions - the east lavas are slightly darker). The floor has a slight slope (from the east side to the west side it drops by ~ 50 metres) towards Imbrium's centre, and bright ray material from Copernicus to its southwest cover mainly the southeastern sector of the crater (the chain of small craters at its southeast may be from crater Eratosthenes. Depth 0.3 km.

Craters

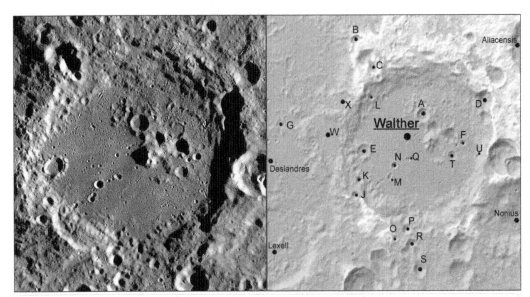

Walther	Lat 33.25S	Long 0.62E	134.23Km

Sub-craters:

Add Info:

Crater	Lat	Long	Size (km)	Crater	Lat	Long	Size (km)
A	32.43S	0.75E	11.21	B	30.48S	1.43W	9.02
C	31.2S	0.85W	12.68	D	32.02S	2.84E	16.46
E	33.35S	1.27W	11.74	F	33.15S	2.09E	5.35
G	32.59S	4.02W	7.7	J	34.5S	1.58W	6.39
K	34.09S	1.41W	6.83	L	31.96S	0.98W	5.69
M	34.06S	0.4W	4.26	N	33.74S	0.22W	5.57
O	35.63S	0.22W	5.05	P	35.4S	0.18E	8.13
Q	33.57S	0.24E	4.55	R	35.81S	0.34E	9.99
S	36.47S	0.62E	11.28	T	33.48S	1.74E	7.63
U	33.5S	2.61E	4.34	W	32.83S	2.57W	31.98
X	32.16S	1.96W	11.55				

Notes:

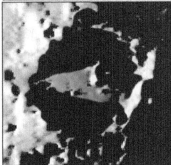

Crater ray of light in Walther

Reference to Walther in modern-day books, articles and general media sometimes entitles the crater as 'Walter' (history of the confusion is quite interesting, but 1.26 km-sized Walter, found at 28.04N, 33.81W, today, is a separate, officially-named crater). Walther has impacted on Deslandres (pre-Nectarian in age 4.6 to 3.92 byo) to its west. It has a ~ 2.5 km-heighted peak - joined to more hummocky terrain at its northeast (Imbrium ejecta possibly makes up the smoother material on the floor's western sector). Depth of Walther ~ 4.0 km approx..

Craters

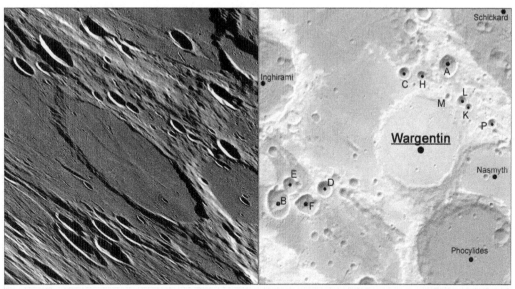

Wargentin   Lat 49.53S   Long 60.44W   84.69Km

Sub-craters:

Crater	Lat	Long	Size (km)	Crater	Lat	Long	Size (km)
A	47.09S	59.2W	20.68	B	51.34S	67.74W	17.03
C	47.42S	61.34W	12.48	D	51.03S	65.3W	16.11
E	50.91S	67.14W	16.0	F	51.51S	66.26W	19.82
H	47.47S	60.48W	9.68	K	48.31S	58.0W	8.26
L	48.12S	58.36W	11.52	M	48.05S	59.09W	6.71
P	48.72S	56.74W	9.33				

Notes:

Add Info:

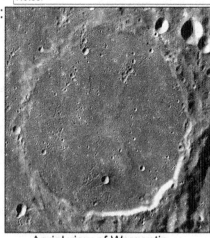

Aerial view of Wargentin

Wargentin must surely be a contender for the most 'filled-to-the-brim' crater on the Moon. The southeast sector of its floor is some 300 metres higher than at its northwest, while levels with Nasmyth and Phocylides, to its east, are 0.8 and 2.8 kilometres below respectively. But what is the source of the lava fill (it isn't as dark as that seen in Nasmyth's or Schickard's floors)? Is it internally derived, is it related to the Schiller-Zucchius Basin (lies centred some ~ 350 km to Wargentin), or is it a huge dollop of ejecta from The Orientale Basin (centred ~ 1200 km to the northwest). At low sun times note several ridges on its floor. Depth ~ 0.3 km.

Craters

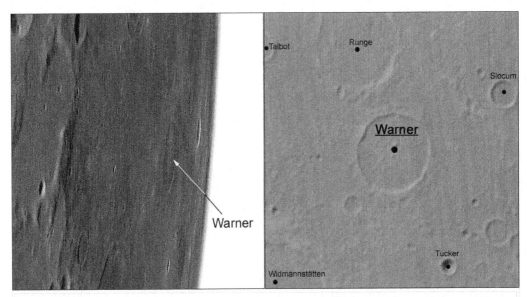

| Warner | Lat 3.98S | Long 87.35E | 34.51Km |

Sub-craters: No sub-craters

Notes:

Add Info:

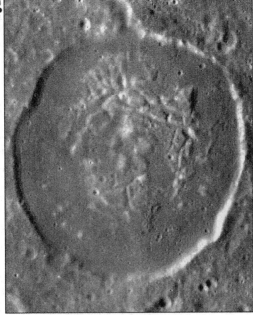

Aerial view of Warner

Warner formed just off centre in the limb-hugging mare that is Smythii. It joins several other craters (ten in all) each over 30 kilometres in diameter; where everyone of them has been flooded. Oddly with most of these craters, their interiors show additional features resembling a series of repeating rims and ridges concentric to the craters (see, for example, Runge and Haldane). The features are most likely due to some form of dynamic uplift of their centres. Warner's floor is no different as it is convexed, and portions of it rise above the mare floor by over a hundred metres. It will prove a difficult crater to catch, so depend on suitable librations and lighting conditions. D: ~ 0.3 km.

Craters

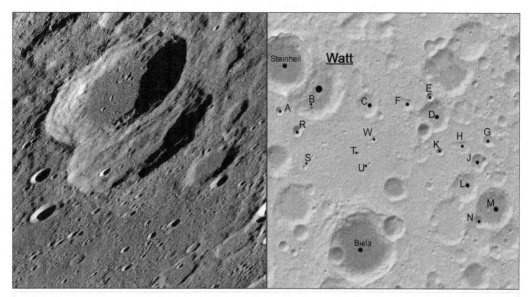

Watt	Lat 49.6S	Long 48.48E	66.54Km

Sub-craters:

Crater	Lat	Long	Size(km)	Crater	Lat	Long	Size(km)
A	50.35S	46.41E	9.23	B	50.28S	48.3E	5.22
C	50.11S	51.48E	24.81	D	50.35S	55.46E	34.54
E	49.7S	55.13E	9.58	F	50.88S	54.95E	16.37
G	50.95S	58.83E	13.36	H	51.29S	57.31E	16.59
J	51.74S	58.42E	18.83	K	51.53S	56.0E	8.48
L	52.65S	57.75E	32.3	M	53.2S	59.76E	42.66
N	53.73S	58.92E	13.01	R	51.06S	47.42E	11.66
S	52.29S	47.86E	5.7	T	51.78S	51.14E	4.23
U	52.14S	51.76E	4.7	W	51.28S	51.98E	6.17

Notes:

Add Info:

Not to be confused with crater Watts (8.84N, 46.13E)

Aerial view of Watt

Steinheil deperately tries to steal the limelight from Watt, as it stands out more than it when viewed in the eyepiece. But Watt is the more interesting really (see left), as the ejecta from which Steinheil has deposited onto its floor gives us some idea as to how such dynamic events can change the appearance of craters that happen to lie in their neighbourhood (Watt's floor is nearly entirely covered by the ejecta). Watt is obviously the oldest of the two - Steinheil is of the Nectarian Period 3.92 to 3.85 byo, and Watt isn't far off that age either (perhaps on the pre-Nectarian border). Depth ~ 3.5 km.

Craters

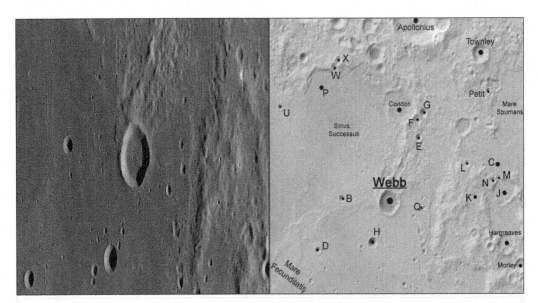

Webb	Lat 0.98S	Long 60.0E	21.41Km

Sub-craters:

Crater	Lat	Long	Size (km)	Crater	Lat	Long	Size (km)
B	0.85S	58.37E	6.45	C	0.15N	63.83E	34.34
D	2.38S	57.52E	6.53	E	0.94N	61.05E	6.82
F	1.47N	61.0E	9.54	G	1.67N	61.22E	9.07
H	2.15S	59.44E	9.98	J	0.62S	63.99E	26.53
K	0.85S	62.96E	18.61	L	0.16N	62.73E	6.3
M	0.26S	63.81E	5.83	N	0.32S	63.62E	5.86
P	2.39N	57.69E	37.86	Q	1.13S	61.24E	4.28
U	1.85N	56.23E	5.73	W	3.02N	58.12E	7.75
X	3.22N	58.25E	7.98				

Notes:

Add Info:

Aerial view of Webb

It looks like Webb formed more on ejecta deposits from the possible crater underlying Sinus Successus (126.65 kilometres in diameter) than from the Basin that is of Fecunditatis to its west. Its own ejecta deposits can just about be seen southwards of the crater where it 'melds' into the mare, while to its northeast it isn't as obvious as a slump on its rim there has taken place. Is the unnamed rille on the crater's outer northeastern rim one that occurred later or did Webb form on it (a hard call). Webb's floor under high sun times appears as dark as the mare's (internal flooding?). D:~2 km.

Craters

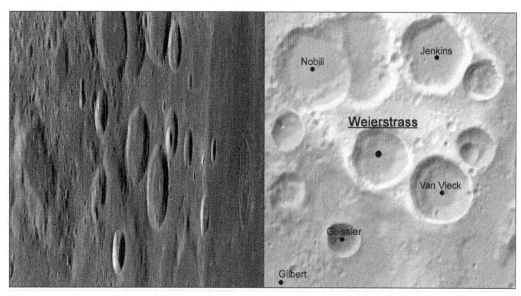

| Weierstrass | Lat 1.26S | Long 77.15E | 31.32Km |

Sub-craters: No sub-craters

Notes:

Add Info:

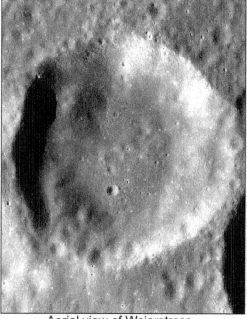

Aerial view of Weierstrass

Weierstrass has impacted upon the northeastern rim of Gilbert to its southwest - depositing ejecta onto its floor. Both craters are of the pre-Nectarian Period (4.6 to 3.92 byo), so the ejecta signatures have been well lost to time. However, as crater Geissler has also imparted deeper deposits on top of Weierstrass's ejecta again, the mix of materials between both craters must make for interesting geology (was Weierstrass's southwestern rim, which is lower, affected by Geissler's ejecta blasting a portion of it away?). Floor levels between all three craters has Weierstrass's being highest. Van Vleck's is a kilometre lower, while Gilbert's is at 1.5 km lower again. Depth of Weierstrass ~ 2.6 km.

Craters

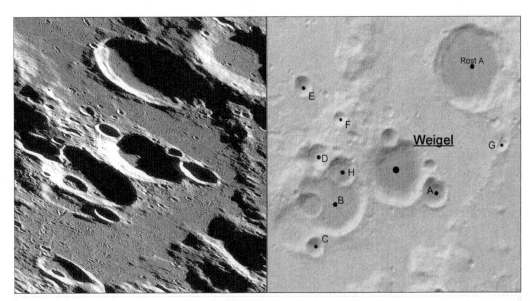

Weigel   Lat 58.39S   Long 39.34W   34.85Km

Sub-craters:

Crater	Lat	Long	Size (km)	Crater	Lat	Long	Size (km)
A	58.71S	37.98W	15.19	B	58.82S	41.53W	38.46
C	59.61S	42.12W	9.8	D	58.15S	42.0W	14.08
E	56.95S	42.39W	10.19	F	57.5S	41.2W	7.8
G	57.86S	35.73W	6.39	H	58.37S	41.12W	16.52

Notes:

Add Info:

Aerial view of Weigel

Weigel, along with several of its titled sub-craters, has impacted in a trough zone between two rings/rims created by formation of the Schiller-Zucchius Basin (centred ~ 120 km away to the northwest). Weigel as well as Weigel B, which lie so close together and are so similar in appearance, suggest the two may have formed simultaneously; each possibly of the Nectarian Period (3.92 to 3.85 byo). Weigel's impactor struck more in the trough (and mare of Schiller-Zucchius) zone than did B, allowing it to tap deeper deposits (B's floor level is some 800 metres above Weigel's). Depth of Weigel ~ 2.5 km.

Craters

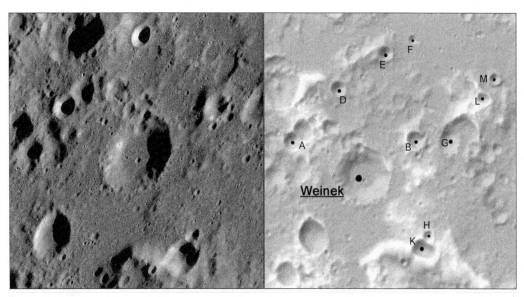

Weinek	Lat 27.57S	Long 37.06E	32.01Km

Sub-craters:

Crater	Lat	Long	Size (km)	Crater	Lat	Long	Size (km)
A	26.95S	35.52E	9.41	B	26.95S	38.22E	10.13
D	26.01S	36.54E	8.93	E	25.39S	37.58E	8.59
F	25.14S	38.16E	4.68	G	26.97S	38.97E	14.82
H	28.66S	38.51E	6.18	K	28.96S	38.43E	16.62
L	26.19S	39.71E	8.29	M	25.83S	39.96E	6.41

Notes:

Add Info:

Aerial view of Weinek

One thing you'll notice about Weinek when viewing it generally amongst its neighbouring craters is that of the thickness of its crater walls. Numerous small impacts have pockmarked the now slumped walls that reach so far into the crater's floor to have covered over half of it. The crater is obviously an old one, however, is isn't as old as the Nectaris Basin (less than 4 billions years old, and lying ~ 400 kilometres away northwards); where it lies between two outer rings/rims in a trough created by that major event. Weinek's floor is probably a mix of local crater ejecta (of Piccolomini westwards?). D: 3.5km.

Craters

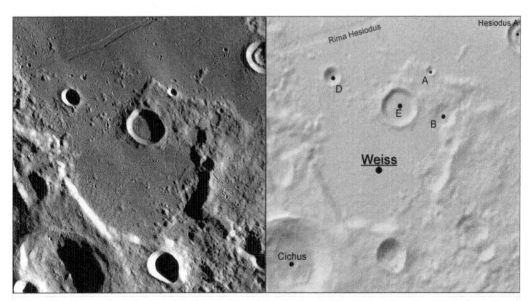

Weiss   Lat 31.76S   Long 19.59W   66.56Km

Sub-craters:

Crater	Lat	Long	Size (km)	Crater	Lat	Long	Size (km)
A	30.63S	18.68W	3.94	B	31.24S	18.43W	9.83
D	30.7S	20.42W	7.83	E	31.13S	19.24W	16.47

Notes:

Add Info:

Aerial view of Weiss (with Tycho's rays)

Weiss's one outstanding feature has got to be its long, straight-edged wall at its southwest. Normally, we would expect crater walls to be curved following their diameter, but why does Weiss's wall of length approximately ~ 28 kilometres here look the way it does (Weiss's southern wall has a similar, linear feel, too, except for a small 'kink' midway?). Is Weiss one of those so-called, Polygonal features and craters (e.g. Lacus Mortis), where slumping along particular parts of their rims and walls follow pre-existing fractures within the target rock? Whatever the cause, such features may hold signatures to underlying structural properties in the lunar crust. D: ~0.5 km.

Craters

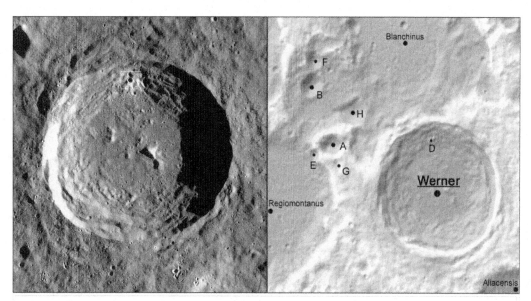

Werner   Lat 28.03S   Long 3.29E   70.59Km

Sub-craters:

Crater	Lat	Long	Size (km)	Crater	Lat	Long	Size (km)
A	27.24S	1.05E	14.45	B	26.21S	0.68E	13.27
D	27.12S	3.17E	1.95	E	27.39S	0.72E	6.68
F	25.79S	0.73E	9.47	G	27.59S	1.2E	8.4
H	26.68S	1.5E	16.02				

Notes:

Add Info:

Close-up of Werner D's rays (on Werner's northern wall)

Not to be confused with crater Warner on the east side of the Moon (3.98S, 87.35E), Werner faces us nearly head-on, and is a crater of the Eratosthenian Period (3.15 to 1.1 byo). Ejecta signatures from Werner can be seen on the floors of several craters nearby (see Blanchinus or Aliacensis, for example), it has a small, off-centered peak (one kilometre high), and developed terraces still sharp-looking. Sub-crater, Werner D, steals the limelight, literally, as its bright rays (left) 'shine' when Werner is in suitable light. Depth ~ 4.5 km.

Craters

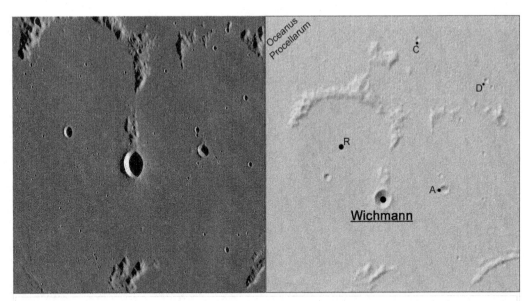

Wichmann	Lat 7.54S	Long 38.13W	9.77Km

Sub-craters:

Crater	Lat	Long	Size (km)	Crater	Lat	Long	Size (km)
A	7.37S	36.92W	5.13	B	7.14S	39.22W	4.36
C	4.67S	37.45W	2.63	D	5.41S	36.12W	2.66
R	6.64S	38.95W	64.13				

Notes:

Add Info:

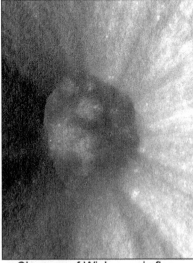

Close-up of Wichmann's floor

From a wider perspective view of this area in where Wichmann lies, note how several relatively large craters (e.g. Wichmann R to Flampsteed P in the northwest, or the unnamed crater southeastwards of Wichmann itself) show the extent of lava plains which have flooded the region. Wichmann may lie on the rim (the northwestern rim of a possible ~ 100 km-sized crater centred approximately east of Wichmann A) of one of these craters. Wichmann is younger than the lavas, hints of its ejecta can just about be seen to overlie them. The crater shows some bright ejecta material around its exterior when in suitable lighting conditions (full moon times especially), while close terminator periods shows the crater lies actually on a slight rise, or hilly swell. Depth of Wichmann ~ 2.0 km.

Craters

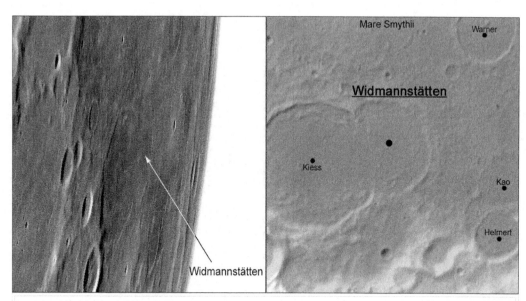

Widmannstätten Lat 6.09S   Long 85.43E   52.88Km

Sub-craters: No sub-craters

Notes:

Add Info:

Aerial view of Widmannstätten

The question to be asked when trying to view those craters on the extreme limb regions of the Moon is: 'Why should you?'. It's going to be a task, if not a nuisance, to start with initially. You're going to have to depend mainly on libration times that may not always 'fit in' with weather conditions (hence, opportunities and plans lost). Patience will play a major role as you flit back and forth between the eyepiece; judging all the time if you actually are viewing the intended, awkward target. And, finally, what am I going to gain by viewing such craters as I'm not really seeing any decent detail at all. The answer: 'Because of the challenge, the experience, the unexpected from capturing a feature - elusive'.

Craters

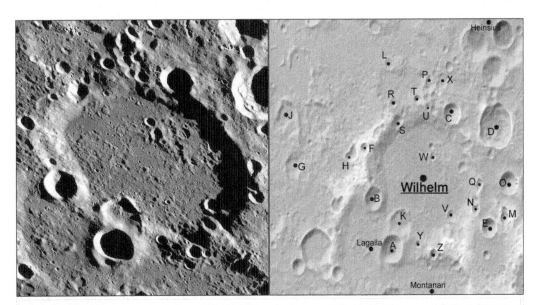

Wilhelm   Lat 43.21S   Long 20.94W   100.83Km

Sub-craters:

Crater	Lat	Long	Size (km)	Crater	Lat	Long	Size (km)
A	44.7S	22.11W	19.74	B	43.54S	22.85W	16.06
C	41.6S	19.61W	15.49	D	41.88S	17.85W	31.43
E	44.19S	18.01W	13.63	F	42.38S	23.16W	8.49
G	42.7S	26.12W	16.8	H	42.55S	23.81W	7.05
J	41.5S	26.25W	18.93	K	44.13S	21.76W	19.72
L	40.48S	22.13W	8.15	M	43.96S	17.37W	8.51
N	43.78S	18.58W	8.42	O	43.16S	17.26W	16.67
P	40.91S	20.52W	11.59	Q	43.2S	18.44W	7.06
R	41.37S	21.97W	7.06	S	41.82S	21.83W	9.79
T	41.29S	21.0W	6.68	U	41.47S	20.58W	5.0
V	43.93S	19.62W	7.51	W	42.6S	20.38W	5.67
X	40.86S	20.01W	10.39	Y	44.6S	20.98W	5.42
Z	44.83S	20.34W	7.7				

Add Info:

Aerial view of Wilhelm

Though all three craters - Wilhelm, Lagalla and Montanari - are craters of age, Wilhelm looks to be the youngest. It has impacted upon the rims of Lagalla and Montanari, but signature of its ejecta on their floors has been lost to time, and to subsequent coverage by ejecta from younger impacts craters - Wilhelm A, B, K, Y and Z - which border all three. Wilhelm's floor is smooth to hummocky in parts, and rays from Tycho to its east (~ 200 kilometres away) are seen easily under most suitable lighting conditions. Depth ~ 3.3km.

Craters

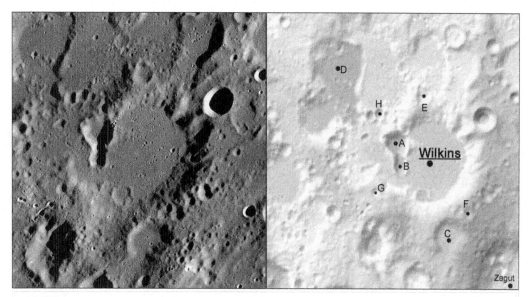

Wilkins	Lat 29.58S Long 19.58E	59.44Km

Sub-craters:

Crater	Lat	Long	Size (km)	Crater	Lat	Long	Size (km)
A	29.17S	18.87E	13.56	B	29.55S	18.92E	7.55
C	30.7S	19.99E	17.99	D	27.87S	17.63E	30.22
E	28.37S	19.42E	9.0	F	30.35S	20.39E	5.86
G	29.99S	18.41E	4.64	H	28.67S	18.49E	5.7

Notes:

Add Info:

Aerial view of Wilkins

Like craters Zagut and Rabbi Levi to the southeast, Wilkins seems to be covered by ejecta from the Nectaris Basin (centred some ~ 600 km away to the northeast). Its floor is relatively flat - filled, possibly, with ejecta deposits from the Imbrium Basin (centred some ~ 2100 km away to the northwest). The distinctive feature to Wilkins, however, has to be the keyhole-like shape of the crater duo that is of craters' Wilkins A and B (they act as a good marker for finding Wilkins). A's floor level is ~ 200 m below B's - are we looking at two craters produ- by two separate impactors, or two by by a single, oblique angled impactor?

Craters

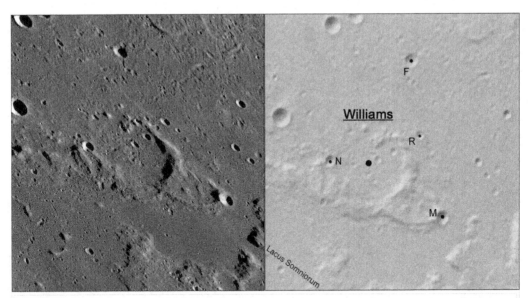

Williams	Lat 42.02N	Long 37.31E					36.36Km
Sub-craters:	Crater	Lat	Long	Size (km)	Crater	Lat	Long	Size (km)
	F	43.6N	38.18E	5.81	M	41.28N	38.83E	6.13
	N	42.13N	36.44E	4.73	R	42.5N	38.37E	3.82
Notes:								

Add Info:

Aerial view of Williams

You would never think Williams to be an actual crater, as its appears more like just an additional chunk of terrain adjoining to the southern sector of Lacus Mortis to the west. The confusion, of course, lies with the missing portion of Williams's northern rim, but also with the odd-looking 'tongue' of material off its southeastern rim, which extends right over to sub-crater, Williams M. The floor certainly has undergone some sort of flooding by lavas, but the question is were they sourced from the north, or internally (note the two squiggly features on its floor hinting at signatures of old lava channels). Depth ~ 0.8 km.

Craters

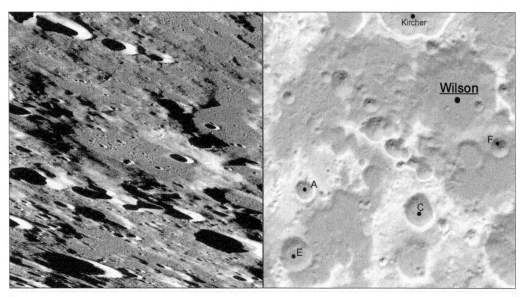

Wilson   Lat 69.33S Long 42.83W   66.57Km

Sub-craters:

Crater	Lat	Long	Size (km)	Crater	Lat	Long	Size (km)
A	71.24S	53.78W	15.4	C	71.96S	45.38W	25.47
E	72.54S	55.35W	24.26	F	70.46S	39.59W	13.19

Notes:

Add Info:

Aerial view of Wilson

There's only about a hundred metres in level between Wilson's floor and Kircher's to the north. Wilson, however, is oldest (possibly of the pre-Nectarian Period 4.6 to 3.92 byo as Kircher is possibly an age of the Nectarian 3.92 to 3.85 byo). While no hint of Kircher's ejecta is seen on WIlson's floor, the reason most likely is due to both having been covered by later ejecta (perhaps, fluidized) from the Orientale Basin - centred some 1800 km away to the northwest. Wilson at high sun times can be a bit hard to see, but other suitable lighting conditions are okay. Depth of Wilson ~ 3.8 km.

Craters

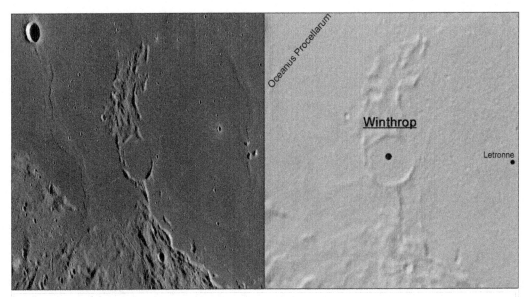

| Winthrop | Lat 10.76S | Long 44.46W | 17.25Km |

Sub-craters: No sub-craters

Notes:

Add Info:

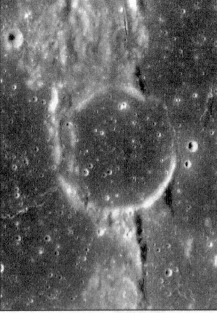

Aerial view of Winthrop

Judging from the state of Winthrop's rim in comparison to Letronne's rim, the two craters are probably of similar age (Lower Imbrian 3.85 to 3.75 byo). Both craters in their formation would have produced respective ejecta, however, while Letronne's is obviously seen nearly all at its south, Winthrop's was nearly all lost due to subsequent flooding of lavas coming from every direction (east and west particularly). Letronne's northern rim was buried because structural shifts in the region tilted the crater towards Oceanus Procellarum in the north, however, Winthrop also may have underwent similar circumstances as it, too, is tilted (it tilts by about 200m across its diameter) in to Letronne's centre). Would it be safe to say that flooding of Winthrop came predominantly from the east? Depth ~0.4 km.

Craters

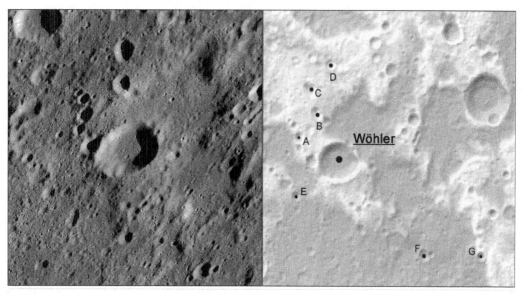

Wöhler	Lat 38.25S	Long 31.35E	28.07Km

Sub-craters:

Crater	Lat	Long	Size (km)	Crater	Lat	Long	Size (km)
A	37.72S	30.27E	7.06	B	37.25S	30.77E	9.83
C	36.76S	30.58E	11.54	D	36.25S	31.14E	7.39
E	38.94S	30.12E	5.79	F	40.16S	33.84E	8.38
G	40.14S	35.53E	6.45				

Notes:

Add Info:

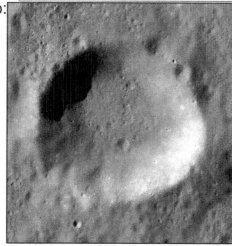

Aerial view of Wöhler

Wöhler isn't as worn-looking as pre-Nectarian aged (4.6 to 3.92 byo) crater Riccius, to its northwest, nor as fresh-looking as crater Stiborius, to its north, which lies on ejecta from the Nectaris Basin to its northeast. The crater formed, then, sometime in between, but as to if it lies on the Nectaris ejecta, or, was covered by it later on, it's a bit unclear (Wöhler's, southwestern wall is broader, and is less steep, than its wall in the opposite, northeastern direction - does it say something about impact time?). Rays from Tycho, centred some 900 kilometres away to the west, sparsely overlies the crater. Depth ~ 2.5 km.

Craters

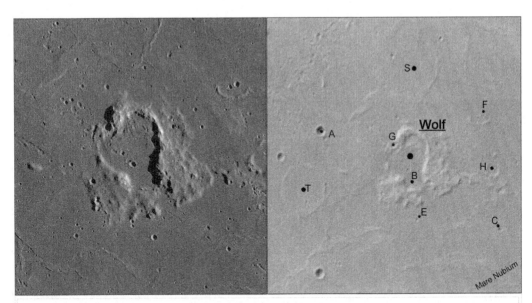

Wolf		Lat 22.79S	Long 16.63W			25.74Km	

Sub-craters:

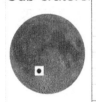

Crater	Lat	Long	Size (km)	Crater	Lat	Long	Size (km)
A	22.29S	18.49W	5.59	B	23.18S	16.51W	15.11
C	24.13S	14.55W	2.65	E	23.95S	16.39W	2.31
F	22.02S	14.96W	2.34	G	22.6S	16.86W	5.7
H	23.01S	14.73W	7.75	S	21.09S	16.37W	29.84
T	23.39S	18.9W	27.1				

Notes:

Add Info:

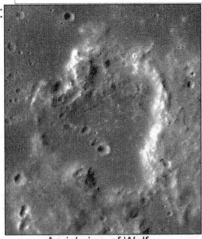

Aerial view of Wolf

Just exactly where does Wolf begin and end as a crater? We can see hint of a rim both at its northeast and southwest sectors, but the sectors at its northwest and southeast need for some clarification. We might suggest the southeast sector was created by an impact crater adjoining on to Wolf's original crater form, but as to the northwestern sector, it's a bit of a puzzle as there appears to be an additional clump or deposit that fills nearly all of the crater's top half sector of the floor (is it ejecta from some nearby basin - Imbrium?). Whatever is it, it certainly seems to have formed on Wolf's rim there. Depth of Wolf ~ 0.8 km.

Craters

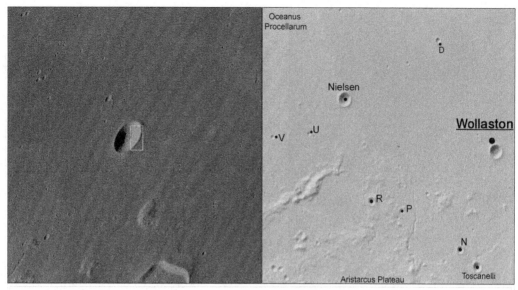

Wollaston	Lat 30.6N	Long 46.98W	9.64Km

Sub-craters:

Crater	Lat	Long	Size (km)	Crater	Lat	Long	Size (km)
D	33.16N	48.79W	4.32	N	28.36N	48.14W	5.52
P	29.26N	49.94W	4.0	R	29.47N	50.87W	5.76
U	30.99N	52.87W	2.87	V	30.89N	53.98W	3.16

Notes:

Add Info:

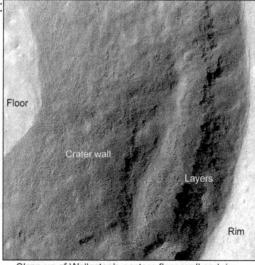

Close-up of Wollaston's eastern floor, wall and rim

At low sun times Wollaston, and its mottled ejecta display around its outer rim, makes for some nice shadow effects. The crater has impacted onto lava plains associated with Oceanus Procellarum, but these lavas are younger (ranging from 1.5 to 3.0 billions of years old) than those found generally further east and west away from the crater. A geologist climbing down Wollaston's wall (left) would spend a most wonderous time taking samples of the many layers of lava; each older than the one above it (according to the Law of Superposition). Depth ~ 2.0 km.

Craters

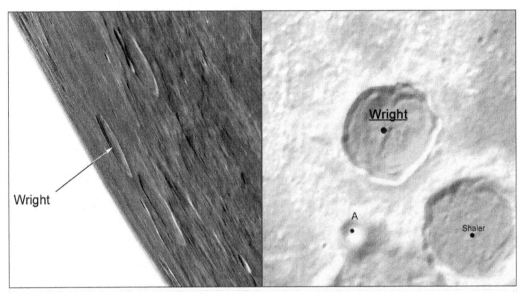

| Wright | Lat 31.55S | Long 86.74W | 40.16Km |

Sub-craters:

Crater	Lat	Long	Size (km)	Crater	Lat	Long	Size (km)
A	32.82S	87.31W	11.33				

Notes:

Add Info:

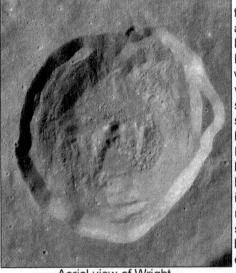

Aerial view of Wright

Wright has impacted between the two main rings/rims (officially known as Montes Rook and Montes Cordillera respectively) of the Orientale Basin - centred 400 km northwestwards from the crater. Wright looks very sharp all around - sharp rims, sharp terraces (created by several slumps) - to suggest that it is a relatively fresh impact crater (possibly Eratosthenian in age - 3.15 to 1.1 byo). Shaler, to its southeast, also looks fresh, however, ejecta deposits on its northwestern rim and exterior may be those from Wright, and so it may have formed some time before Wright. Wright's limb location dictates librations are required, and a little patience, too. Depth ~ 4.0km.

Craters

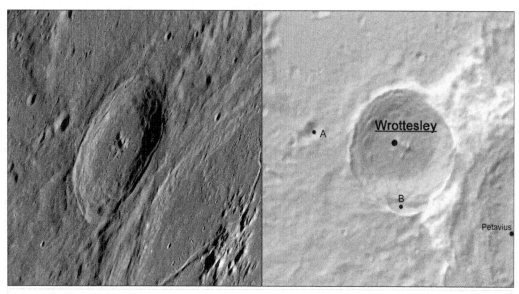

Wrottesley	Lat 23.9S	Long 56.62E	58.38Km

Sub-craters:	Crater	Lat	Long	Size (km)	Crater	Lat	Long	Size (km)
	A	23.57S	54.93E	9.43	B	24.97S	56.87E	7.81

Notes:

Add Info:

Aerial view of Wrottesley

When you lie nearby to one of the most impressive-looking craters - Petavius - on the moon's eastern side, it's a sure thing you're not going to get that 'much-of-a-look-in'. It's a pity, really, as Wrottesley has some nice features: from terraces to peaks, to hummocky floor - all of which produce some nice shadow effects, particularly during low sun times. Wrottesley has obviously impacted upon Petavius's northwestern rim, however, signatures of its ejecta on Petavius's rim, or its floor, are hard to see (they're there, but worn). Wrottesely does, however, lie on Patavius's ejecta, and also on similar, but sparse deposits from the Nectaris Basin to the northwest.

Craters

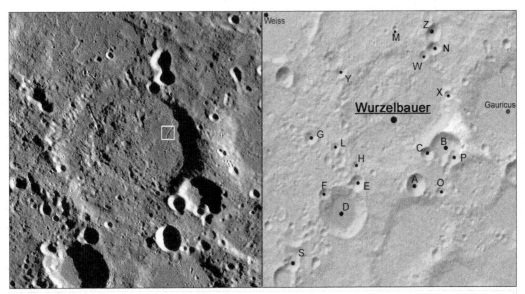

Wurzelbauer   Lat 34.04S   Long 16.06W   86.77Km

Sub-craters:

Crater	Lat	Long	Size (km)	Crater	Lat	Long	Size (km)
A	35.76S	15.42W	16.25	B	34.91S	14.54W	23.86
C	35.04S	15.1W	9.61	D	36.36S	17.7W	38.83
E	35.7S	17.24W	10.74	F	35.95S	18.25W	9.02
G	34.62S	18.63W	10.71	H	35.3S	17.29W	6.03
L	34.87S	17.89W	7.3	M	32.14S	16.06W	4.26
N	32.58S	14.92W	11.42	O	35.91S	14.66W	7.85
P	35.14S	14.28W	8.83	S	37.52S	19.32W	11.7
W	32.76S	15.2W	7.35	X	33.67S	14.46W	6.46
Y	33.13S	17.68W	8.95	Z	32.23S	14.94W	13.83

Notes:

Add Info:

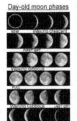

Day-old moon phases

Waning crescent

Note: As the Moon goes through various librations throughout the year, suggested times given in text for observations are only approximates.

Close-up of rille on Wurzelbauer's eastern floor

The charateristic feature to crater Wurzelbauer has to be the rubbly clump of material on its floor. Is it ejecta of some sort from the Nubium Basin to its north? Is it volcanically-related, say, from an underlying source (the crater's floor is convexed upwards)? Is it another crater within Wurzelbauer (note how it looks separated from Wurzelbauer's northwest inner rim and wall)? Whatever it is, fracturing of the clump has occurred across several parts of it. Depth ~ 1.8 km.

Craters

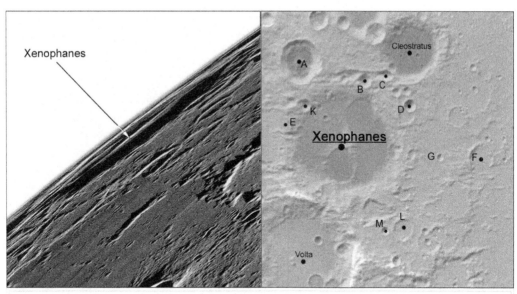

Xenophanes   Lat 57.49N   Long 82.01W   117.57Km

Sub-craters:

Crater	Lat	Long	Size (km)	Crater	Lat	Long	Size (km)
A	60.07N	84.83W	43.18	B	59.48N	80.56W	14.42
C	59.58N	78.95W	9.85	D	58.62N	77.63W	12.32
E	58.06N	85.74W	15.66	F	56.69N	73.46W	27.8
G	56.94N	75.99W	7.28	K	58.7N	84.56W	14.1
L	54.81N	78.54W	22.41	M	54.8N	79.62W	9.08

Add Info: Notes:

Aerial view of Xenophanes

Xenophanes must surely be the best candidate for 'the crater with the most obvious striations/gouges on nearly every part of its rim'. The features are believed to be the result of ejecta produced during the Imbrium Basin impact centred some ~ 1500 km away to the southeast, which 'bombed' their way into all the high points of Xenophanes (including what appears to be its peak attached to the unnamed crater in its centre). Deposits on Xenophanes's floor haven't been affected; implying they are later (are they fluidized, non-melted ejecta from Imbrium?). Depth of Xenophanes is around 2.8 km.

Craters

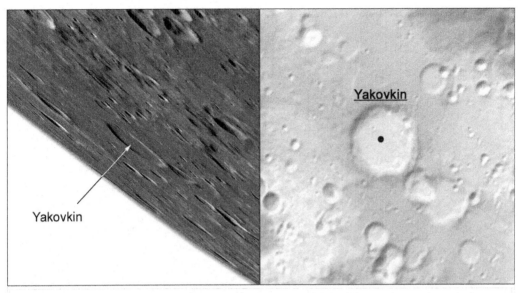

Yakovkin   Lat 54.42S   Long 78.93W   35.92Km

Sub-craters: No sub-craters

Notes:

Add Info:

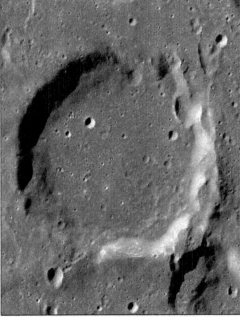

Aerial view of Yakovkin

Yakovkin is situated just within a main ~ 600 km diameter ring/rim associated to formation of the Mendel-Rydberg Basin (centred some ~ 330 km to the northwest). It also lies some ~ 1200 km away from the Orientale Basin, again to the northwest, whose ejecta may overlie Yakovkin (some striated features on its northeastern rim suggests this might be so). The crater's floor depth is just under 1.6 km, is very smooth, and may have underwent some sort of internal flooding, before Orientale's ejecta draped over it. Obviously, having a high latitude as well as a high longitude designation, observations are limited to suitable libration periods. So prepare well, use an atlas, use a camera, and be patient - for the rewards.

Craters

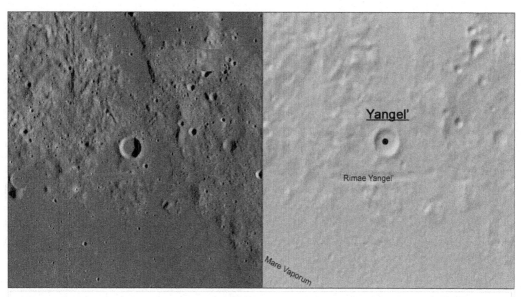

Yangel' Lat 16.96N Long 4.69E 7.87Km

Sub-craters: No sub-craters

Notes:

Add Info:

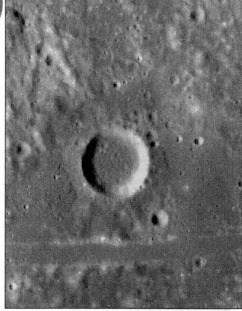

Aerial view of Yangel'

Yangel' formed in an area where the outer ejecta deposits from two main basins lie - those of the Imbrium Basin off to its northwest (at some 800 km centred), and the Serenitatis Basin off to its northeast (at some 500 km centred). The crater also lies on the edge of Mare Vaporum, whose lavas most likely flowed northwards up to Yangel''s rim and surrounds. The crater's floor material takes on a similar albedo characteristic to that of the mare (note it at full moon times) suggesting possible internal flooding, leaving its floor level some 200m below that of the Mare. Rimae Yangel' to its south formed later, as it obviously 'cuts' into the terrain (ejecta) there. The depth of Yangel' is around 0.6 km.

Craters

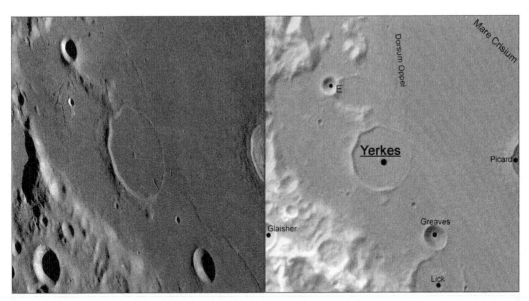

Yerkes	Lat 14.6N	Long 51.7E	34.94Km

Sub-craters:	Crater	Lat	Long	Size (km)	Crater	Lat	Long	Size (km)
	E	15.9N	50.67E	9.91				
	Notes:							

Add Info:

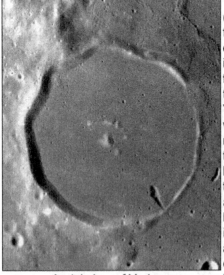

Aerial view of Yerkes

Yerkes lies within one of the rings/rims (~ 555 km in diameter) associated to formation of the Crisium Basin to its east. Aged of the Upper Imbrian Period (3.75 to 3.2 byo), the crater seems to have struck on a clump of material that possibly is ejecta from Crisium. Yerkes's own ejecta is seen only at its western exterior, as that to its east is missing due to the crater having been tilted from structural settlement of the basin (later lavas in Crisium covered the eastern ejecta). Yerkes's floor has also undergone some sort of flooding, but as to if it was internal or external is unclear (the floor is convexed, and about 250 m above the Mare's). Highest of Yerkes's peak is around 250 m high, while the crater's depth is 0.6 km.

Craters

| Young | Lat 41.54S | Long 50.98E | 71.44Km |

Sub-craters:

Crater	Lat	Long	Size (km)	Crater	Lat	Long	Size (km)
A	41.15S	51.18E	11.76	B	41.0S	50.74E	7.56
C	41.56S	48.24E	29.58	D	43.49S	51.73E	45.11
F	44.85S	51.67E	19.56	R	42.43S	55.39E	9.45
S	43.36S	53.88E	9.98				

Notes:

Add Info:

Aerial view of Young

Young is pre-Nectarian in age (4.6 to 3.92 byo). Nearly all of its western sector is missing (you can just about see a portion of the remaining rim there), due to the series of secondary craters believed to have formed by impactors (ejecta) coming from the Nectaris Basin - some ~ 890 km away to its northwest. The series of craters produced what is today known as the Rheita Valley - over 500 km length and as wide as 30 km in parts. Note how the valley north of Young doesn't quite align with the valley south of Young (are they two separate crater chains?). And why don't both exactly 'point' back to Nectaris's centre? D: 1.5 km.

Craters

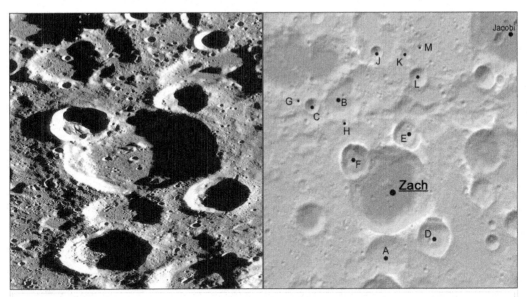

Zach   Lat 60.92S   Long 5.25E   68.54Km

Sub-craters:

Crater	Lat	Long	Size (km)	Crater	Lat	Long	Size (km)
A	62.61S	4.99E	35.33	B	58.65S	2.78E	30.19
C	58.68S	1.17E	12.15	D	62.12S	7.85E	28.94
E	59.54S	6.21E	24.19	F	60.16S	3.3E	28.4
G	58.54S	0.46E	5.51	H	59.17S	2.88E	5.89
J	57.41S	4.67E	10.5	K	57.48S	6.11E	8.35
L	58.05S	6.79E	15.59	M	57.28S	6.87E	4.83

Notes:

Add Info:

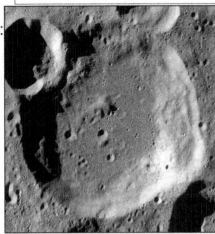

Aerial view of Zach

Lying in a region where similar-sized, similar-aged craters are the norm as Zach, it's quite easy to get lost here. Fortunately, Zach's sub-craters, D, E, and F, act as good markers for noting and remembering the crater's location. It's clear from the ejecta signatures on Zach's northern floor from E and F that they are younger than Zach, but as to Zach D it is another call (perhaps it is older). The material on Zach's floor is relatively smooth - possibly from some young basin (Orientale?), while its peak is some 800 m high. Tycho is ~ 600 km away to the northwest, but its rays aren't as obvious as eleswhere. Depth 4.5 km.

Craters

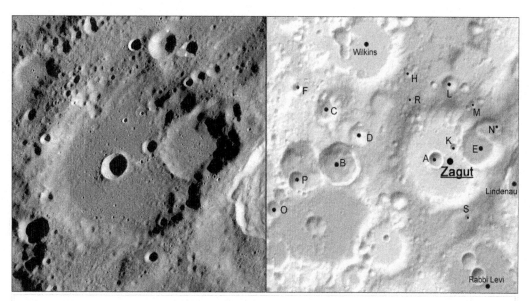

Zagut	Lat 31.94S	Long 21.89E	78.92Km

Sub-craters:

Crater	Lat	Long	Size (km)	Crater	Lat	Long	Size (km)
A	32.05S	21.64E	11.4	B	32.1S	18.71E	30.36
C	30.91S	18.38E	19.78	D	31.46S	19.28E	15.06
E	31.73S	23.04E	32.84	F	30.32S	17.51E	8.12
H	30.02S	20.76E	8.01	K	31.75S	22.17E	6.36
L	30.3S	22.02E	11.99	M	30.75S	22.72E	5.45
N	31.21S	23.45E	7.96	O	33.08S	16.65E	11.41
P	32.47S	17.4E	13.7	R	30.65S	20.82E	4.37
S	33.38S	22.63E	5.73				

Notes:

Add Info:

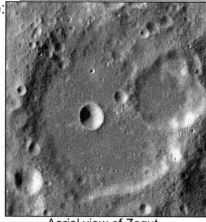

Aerial view of Zagut

Initial observations of Zagut would seem that the crater has formed on outer ejecta deposits from the Nectaris Basin (centred ~ 600 km away to the northeast). But, the opposite is true, as Zagut formed before Nectaris (therefore, pre-Nectarian in age - 4.6 to 3.92 byo), as portions of its rim and walls (see the northeastern sectors in the image, left) show striated effects that most likely were produced by ejecta from the Basin. Sub-crater, Zagut E, which has impacted onto Zagut's floor, looks as old as Zagut, but as it doesn't show the above striations, it must be younger than Nectaris, but not by much. Depth of Zagut ~ 3.5 km.

Craters

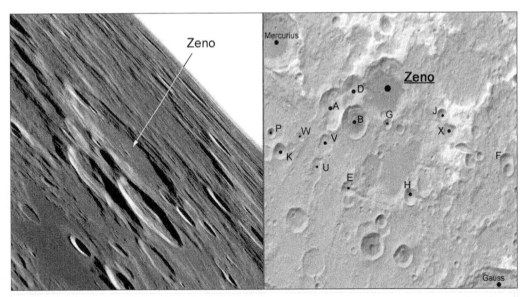

| Zeno | Lat 45.15N | Long 72.98E | 66.78Km |

Sub-craters:

Crater	Lat	Long	Size (km)	Crater	Lat	Long	Size (km)
A	44.43N	69.69E	43.98	B	44.01N	71.09E	35.78
D	45.14N	70.96E	29.06	E	41.71N	70.79E	18.07
F	42.41N	80.02E	18.43	G	43.93N	73.04E	11.44
H	41.42N	74.35E	17.47	J	44.19N	76.33E	13.93
K	42.83N	66.73E	19.12	P	43.43N	66.14E	11.23
U	42.44N	68.91E	14.39	V	43.17N	69.36E	23.84
W	43.39N	67.85E	11.94	X	43.57N	76.73E	18.69

Notes:

Add Info:

Aerial view of Zeno

Though it's not that difficult to observe at most times for its limbward location, Zeno lies just outside one of the rings/rims (~ 650 km in diameter), and ejecta, associated with formation of the Humboltianum Basin (centred ~ 400 km to the northeast). The crater is Nectarian in age (3.92 to 3.85 byo), but appears to have impacted on an even older, unnamed crater (actually, within its northern rim and floor), that is over twice Zeno's own diameter (note it around low sun times). Zeno's peak(?) is some 450 m high, and is connected by a 'spit' of land that forms into a valley over to sub-crater, Zeno D. Depth ~ 4.5 km.

Craters

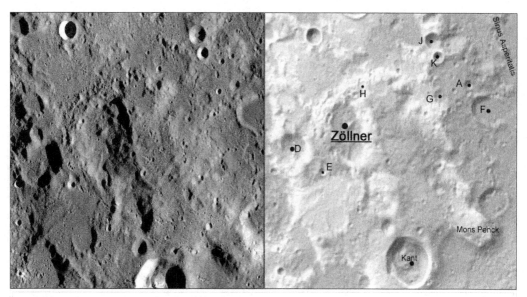

Zöllner	Lat 7.97S	Long 18.9E	47.69Km

Sub-craters:

Crater	Lat	Long	Size (km)	Crater	Lat	Long	Size (km)
A	7.08S	21.48E	6.15	D	8.39S	17.66E	24.36
E	8.82S	18.28E	4.7	F	7.58S	21.99E	27.0
G	7.32S	20.87E	9.51	H	7.09S	19.13E	6.59
J	6.2S	20.7E	10.36	K	6.53S	20.82E	7.2

Notes:

Add Info:

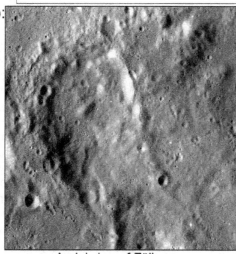

Aerial view of Zöllner

The most distinguishing feature to Zöllner has got to be the large deposit on its east, and which nearly covers half of the crater. But where did it come from, what was its source? Judging by the extent and size of the deposit it may be ejecta from some major event, like a basin or large crater. The Nectaris Basin to the southeast is obviously a candidate, but other basins, like Fecunditatis, Tranquillitatis, or Imbrium may also be responsible. Whatever the source, its top is some 5 km above the level of Sinus Asperitatis's level over to the east, and ~ 1.8 km above Zöllner's western rim. Depth 2.0 km.

Craters

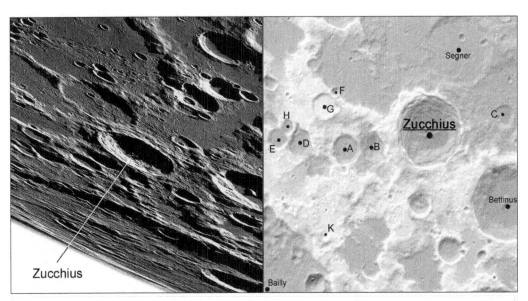

Zucchius	Lat 61.38S	Long 50.65W		63.18Km			

Sub-craters:

Crater	Lat	Long	Size (km)	Crater	Lat	Long	Size (km)
A	61.81S	56.27W	27.55	B	61.84S	54.37W	24.25
C	60.88S	45.72W	21.31	D	61.38S	59.14W	21.08
E	61.24S	60.69W	20.59	F	60.11S	56.67W	8.93
G	60.48S	57.4W	24.05	H	60.94S	59.95W	14.29
K	64.27S	58.3W	10.48				

Notes:

Add Info:

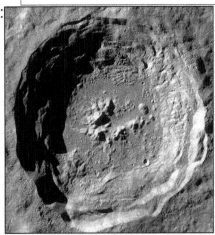

Aerial view of Zucchius

While Zucchius obviously shows all the signatures of a young crater (it is of the Copernican Period - 1.1 byo to the Present), it lies in an old region where two major impacts occurred - that of Bailly (Nectarian) to its southwest, and of the Schiller-Zucchius Basin (pre-Nectarian) to its northeast. It's a pity that the crater lies where it does, as Zucchius has a wonderful series of sharp terraces, impact melt both on the floor and interstitial too on the terraces, and peaks to die for (the highest is over a kilometre high). Zucchius, with its two neighbours over to its east (Bettinus and Kircher), makes for an easy find in this southern 'trio'.

Craters

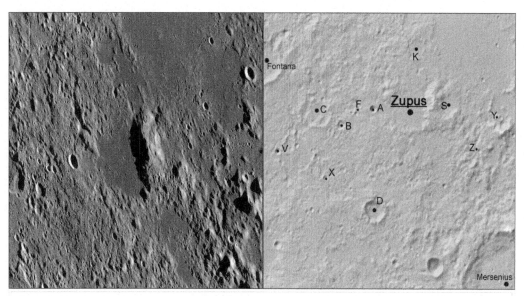

Zupus	Lat 17.18S	Long 52.37W	35.29Km

Sub-craters:

Crater	Lat	Long	Size (km)	Crater	Lat	Long	Size (km)
A	17.22S	53.57W	6.2	B	17.61S	54.52W	6.36
C	17.33S	55.18W	20.62	D	19.7S	53.54W	17.77
F	17.33S	54.07W	3.47	K	15.81S	52.28W	14.75
S	16.96S	51.34W	24.84	V	18.22S	56.43W	5.57
X	18.89S	55.0W	4.91	Y	17.43S	49.73W	3.72
Z	18.24S	50.34W	3.52				

Notes:

Add Info:

Aerial view of Zupus

To 'see' Zupus as a crater is simply a 'no-no' (just where is it?)! Hint of a rim can be seen at its southern end, but that could just be an odd clump like many others in the area. Instead, Zupus should really be studied more for its two obvious features: its dark floor, and the big mountain to its east. The first is most likely due to some lava flooding of sorts, which connects to the much larger deposits of Oceanus Procellarum in the north. The second may be due to some old remnant portion of an outer ring/rim associated to the Humorum Basin (centred 400 km to the southeast). Nice long shadows form from this mountain (around 12-day or 26-day old moons).

Features

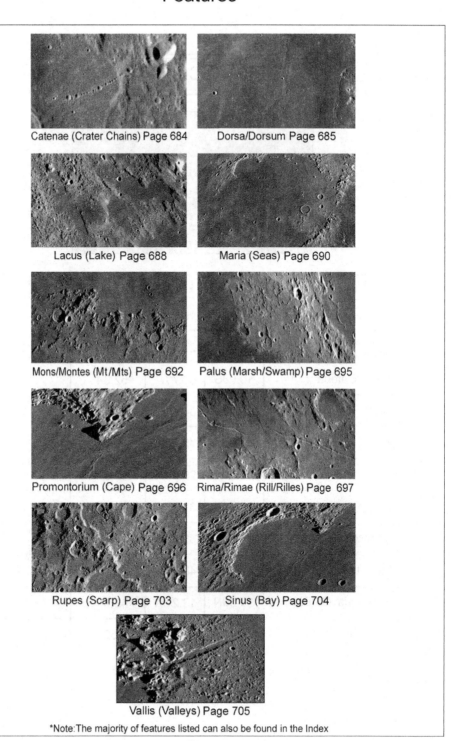

Catenae (Crater Chains) Page 684

Dorsa/Dorsum Page 685

Lacus (Lake) Page 688

Maria (Seas) Page 690

Mons/Montes (Mt/Mts) Page 692

Palus (Marsh/Swamp) Page 695

Promontorium (Cape) Page 696

Rima/Rimae (Rill/Rilles) Page 697

Rupes (Scarp) Page 703

Sinus (Bay) Page 704

Vallis (Valleys) Page 705

*Note: The majority of features listed can also be found in the Index

Catenae

CATENAE (Crater Chains)

FEATURE	LAT	LONG	SIZE (km)	NOTES
Abulfeda, Catena	16.59S	16.7E	209.97	
Brigitte, Catena	18.5N	27.49E	7.65	
Davy, Catena	10.98S	6.27W	52.34	
Humboldt, Catena	21.98S	84.7E	162.29	
Krafft, Catena	14.91N	72.25W	55.11	
Littrow, Catena	22.23N	29.61E	10.3	
Pierre, Catena	19.76N	31.86W	9.44	
Sylvester, Catena	79.99N	83.12W	139	
Taruntius, Catena	3.04N	48.71E	69.24	
Timocharis, Catena	29.09N	13.21W	48.37	
Yuri, Catena	24.41N	30.38E	4.52	

Dorsa/Dorsum

DORSA/DORSUM (Ridge/Ridges)

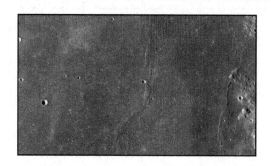

FEATURE	LAT	LONG	SIZE (km)	NOTES
Aldrovandi, Dorsa	23.61N	28.65E	126.81	
Andrusov, Dorsa	1.56S	56.77E	80.98	
Argand, Dorsa	28.29N	40.34W	91.94	
Barlow, Dorsa	14.04N	30.57E	112.57	
Burnet, Dorsa	26.18N	56.78W	194.99	
Cato, Dorsa	0.21N	47.7E	128.64	
Dana, Dorsa	2.26N	89.6E	82.31	
Ewing, Dorsa	11.06S	38.31W	261.57	
Geikie, Dorsa	4.21S	52.83E	218.35	
Harker, Dorsa	13.79N	63.65E	213.14	
Lister, Dorsa	19.76N	23.52E	180.09	
Mawson, Dorsa	7.77S	52.48E	142.71	
Rubey, Dorsa	9.88S	42.36W	100.64	

Dorsa/Dorsum

FEATURE	LAT	LONG	SIZE (km)	NOTES
Smirnov, Dorsa	26.41N	25.53E	221.78	
Sorby, Dorsa	18.65N	13.7E	76.03	
Stille, Dorsa	26.91N	18.85W	66.32	
Tetyaev, Dorsa	20.02N	64.06E	187.53	
Whiston, Dorsa	29.77N	56.96W	138.93	
Arduino, Dorsum	24.77N	36.27W	99.73	
Azara, Dorsum	26.86N	19.17E	103.2	
Bucher, Dorsum	30.76N	39.55W	84.65	
Buckland, Dorsum	19.43N	14.3E	369.13	
Cayeux, Dorsum	0.76N	51.22E	95.14	
Cloos, Dorsum	1.15N	90.41E	103.09	
Cushman, Dorsum	1.42N	49.19E	85.65	
Gast, Dorsum	24.38N	8.71E	64.87	
Grabau, Dorsum	29.76N	14.19W	123.69	
Guettard, Dorsum	9.92S	18.26W	40.46	
Heim, Dorsum	32.2N	29.83W	146.79	
Higazy, Dorsum	27.93N	17.47W	63.1	
Nicol, Dorsum	18.32N	22.66E	43.74	
Niggli, Dorsum	29.01N	52.28W	47.75	
Oppel, Dorsum	19.31N	52.09E	297.62	

Dorsa/Dorsum

FEATURE	LAT	LONG	SIZE (km)	NOTES
Owen, Dorsum	25.14N	11.09E	33.47	
Scilla, Dorsum	32.34N	60.0W	107.52	
Termier, Dorsum	11.63N	57.15E	89.65	
Thera, Dorsum	24.4N	31.42W	7.25	
Von Cotta, Dorsum	23.6N	11.95E	183.06	
Zirkel, Dorsum	29.55N	24.82W	195.22	

Lacus

LACUS (Lake)

FEATURE	LAT	LONG	SIZE (km)	NOTES
Aestatis, Lacus	14.83S	68.57W	86.39	
Autumni, Lacus	11.81S	83.17W	195.65	
Bonitatis, Lacus	23.18N	44.32E	122.1	
Doloris, Lacus	16.8N	8.61E	102.9	
Excellentiae, Lacus	35.65S	43.58W	197.74	
Felicitatis, Lacus	18.52N	5.36E	98.48	
Gaudii, Lacus	16.33N	12.27E	88.54	
Hiemalis, Lacus	15.01N	13.97E	48.04	
Lenitatis, Lacus	14.32N	12.05E	78.25	
Mortis, Lacus	45.13N	27.32E	158.78	
Odii, Lacus	19.22N	7.27E	72.68	
Perseverantiae, Lacus	7.84N	61.93E	70.64	
Somniorum, Lacus	37.56N	30.8E	424.76	

Lacus

FEATURE	LAT	LONG	SIZE (km)	NOTES
Spei, Lacus	43.46N	65.2E	76.67	
Temporis, Lacus	46.77N	56.21E	205.3	
Timoris, Lacus	39.42S	27.95W	153.65	
Veris, Lacus	16.48S	85.91W	382.88	

Maria

MARIA (Seas)

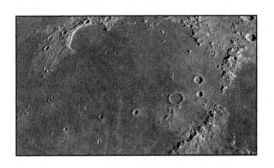

FEATURE	LAT	LONG	SIZE (km)	NOTES
Anguis, Mare	22.43N	67.58E	145.99	
Australe, Mare	47.77S	91.99E	996.84	
Cognitum, Mare	10.53S	22.31W	350.01	
Crisium, Mare	16.18N	59.1E	555.92	
Fecunditatis, Mare	7.83S	53.67E	840.35	
Frigoris, Mare	57.59N	0.01W	1446.41	
Humboldtianum, Mare	56.92N	81.54E	230.78	
Humorum, Mare	24.48S	38.57W	419.67	
Imbrium, Mare	34.72N	14.91W	1145.53	
Insularum, Mare	7.79N	30.64W	511.93	
Marginis, Mare	12.7N	86.52E	357.63	
Nectaris, Mare	15.19S	34.6E	339.39	
Nubium, Mare	20.59S	17.29W	714.5	

Maria

FEATURE	LAT	LONG	SIZE (km)	NOTES
Orientale, Mare	19.87S	94.67W	294.16	
Serenitatis, Mare	27.29N	18.36E	674.28	
Smythii, Mare	1.71S	87.05E	373.97	
Spumans, Mare	1.3N	65.3E	143.13	
Tranquillitatis, Mare	8.35N	30.83E	875.75	
Undarum, Mare	7.49N	68.66E	244.84	
Vaporum, Mare	13.2N	4.09E	242.46	

Mons/Montes

MONS/MONTES (Mountain/Mountains)

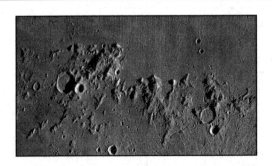

FEATURE	LAT	LONG	SIZE (km)	NOTES
Agnes, Mons	18.66N	5.34E	0	
Ampère, Mons	19.32N	3.71W	29.96	
Argaeus, Mons	19.33N	29.01E	61.48	
Bradley, Mons	21.73N	0.38E	76.49	
Delisle, Mons	29.42N	35.79W	32.42	
Esam, Mons	14.61N	35.71E	7.92	
Gruithuisen Delta, Mons	36.07N	39.59W	27.24	
Gruithuisen Gamma, Mons	36.56N	40.72W	19.65	
Hadley, Mons	26.69N	4.12E	26.4	
Hadley Delta, Mons	25.72N	3.71E	17.24	
Hansteen, Mons	12.19S	50.21W	30.65	
Herodotus, Mons	27.5N	52.94W	6.77	
Huygens, Mons	19.92N	2.86W	41.97	

Mons/Montes

FEATURE	LAT	LONG	SIZE (km)	NOTES
La Hire, Mons	27.66N	25.51W	21.71	
Maraldi, Mons	20.34N	35.5E	15.9	
Moro, Mons	11.84S	19.84W	13.68	
Penck, Mons	10.0S	21.74E	37.59	
Pico, Mons	45.82N	8.87W	24.42	
Piton, Mons	40.72N	0.92W	22.5	
Rümker, Mons	40.76N	58.38W	73.25	
Usov, Mons	11.91N	63.26E	13.23	
Vinogradov, Mons	22.35N	32.52W	28.73	
Vitruvius, Mons	19.33N	30.74E	44.28	
Wolff, Mons	16.88N	6.8W	32.87	
Blanc, Mont	45.41N	0.44E	21.57	
Agricola, Montes	29.06N	54.07W	159.76	
Alpes, Montes	48.36N	0.58W	334.48	
Apenninus, Montes	19.87N	0.03E	599.67	
Archimedes, Montes	25.39N	5.25W	146.54	
Carpatus, Montes	14.57N	23.62W	333.59	
Caucasus, Montes	37.52N	9.93E	443.51	
Cordillera, Montes	19.44S	94.93W	963.5	
Haemus, Montes	17.11N	12.03E	384.66	

Mons/Montes

FEATURE	LAT	LONG	SIZE (km)	NOTES
Harbinger, Montes	26.89N	41.29W	92.7	
Jura, Montes	47.49N	36.11W	420.8	
Pyrenaeus, Montes	14.05S	41.51E	251.33	
Recti, Montes	48.3N	19.72W	83.24	
Riphaeus, Montes	7.48S	27.6W	190.12	
Rook, Montes	19.49S	94.95W	682.28	
Secchi, Montes	2.72N	43.17E	52.47	
Spitzbergen, Montes	34.47N	5.21W	59.19	
Taurus, Montes	27.32N	40.34E	166.16	
Teneriffe, Montes	47.89N	13.19W	111.98	

Palus

PALUS (Marsh/Swamp)

FEATURE	LAT	LONG	SIZE (km)	NOTES
Epidemiarum, Palus	32.0S	27.54W	300.38	
Putredinis, Palus	27.26N	0.0	180.45	
Somni, Palus	13.69N	44.72E	163.45	

PROMONTORIUM (Cape)

FEATURE	LAT	LONG	SIZE (km)	NOTES
Agarum, Promontorium	13.87N	65.73E	62.46	
Agassiz, Promontorium	42.4N	1.77E	18.84	
Archerusia, Promontorium	16.8N	21.94E	11.21	
Deville, Promontorium	43.31N	1.14E	16.56	
Fresnel, Promontorium	28.63N	4.75E	20.0	
Heraclides, Promontorium	40.6N	34.1W	50.0	
Kelvin, Promontorium	26.95S	33.45W	45.01	
Laplace, Promontorium	46.84N	25.51W	50.0	
Taenarium, Promontorium	18.63S	7.34W	70.0	

Rima/Rimae

RIMA/RIMAE (Rill/Rilles)

FEATURE	LAT	LONG	SIZE (km)	NOTES
Agatharchides, Rima	20.38S	28.56W	54.25	
Agricola, Rima	29.25N	53.42W	125.08	
Archytas, Rima	53.63N	3.0E	90.18	
Ariadaeus, Rima	6.48N	13.44E	247.45	
Artsimovich, Rima	26.66N	38.65W	68.06	
Billy, Rima	14.74S	48.04W	69.82	
Birt, Rima	21.4S	9.28W	54.18	
Bradley, Rima	24.17N	0.6W	133.76	
Brayley, Rima	22.3N	36.35W	327.26	
Calippus, Rima	37.03N	12.66E	40.0	
Cardanus, Rima	11.32N	71.14W	221.93	
Carmen, Rima	19.95N	29.3E	15.02	
Cauchy, Rima	10.42N	38.07E	167.0	

Rima/Rimae

FEATURE	LAT	LONG	SIZE (km)	NOTES
Cleomedes, Rima	27.98N	56.51E	45.54	
Cleopatra, Rima	30.03N	53.8W	14.66	
Conon, Rima	18.69N	1.85E	37.32	
Dawes, Rima	17.58N	26.63E	15.0	
Delisle, Rima	30.87N	32.35W	57.6	
Diophantus, Rima	28.7N	33.67W	201.5	
Draper, Rima	17.37N	25.37W	244.16	
Euler, Rima	21.08N	30.31W	104.97	
Flammarion, Rima	2.38S	4.67W	49.75	
Furnerius, Rima	35.3S	61.17E	65.85	
Galilaei, Rima	12.91N	59.2W	185.88	
Gärtner, Rima	58.84N	35.77E	42.73	
Gay-Lussac, Rima	13.18N	22.33W	40.04	
G. Bond, Rima	32.86N	35.25E	166.85	
Hadley, Rima	25.72N	3.15E	116.09	
Hansteen, Rima	12.09S	52.99W	30.89	
Hesiodus, Rima	30.54S	21.85W	251.46	
Hyginus, Rima	7.62N	6.77E	203.96	
Jansen, Rima	14.5N	29.51E	45.12	
Krieger, Rima	29.29N	46.26W	22.62	

Rima/Rimae

FEATURE	LAT	LONG	SIZE (km)	NOTES
Mairan, Rima	38.28N	46.83W	120.51	
Marcello, Rima	18.59N	27.74E	3.95	
Marius, Rima	16.37N	49.54W	283.54	
Messier, Rima	0.76S	44.55E	74.22	
Milichius, Rima	8.03N	32.87W	140.72	
Oppolzer, Rima	1.53S	1.28E	94.2	
Réaumur, Rima	2.84S	2.47E	30.66	
Reiko, Rima	18.55N	27.71E	4.29	
Rudolf, Rima	19.71N	29.62E	8.29	
Schröter, Rima	1.28N	6.25W	27.25	
Sharp, Rima	46.02N	50.36W	276.67	
Sheepshanks, Rima	58.28N	23.69E	157.49	
Suess, Rima	6.62N	47.14W	156.39	
Sung-Mei, Rima	24.59N	11.28E	3.88	
T. Mayer, Rima	13.26N	31.37W	67.81	
Vladimir, Rima	25.2N	0.75W	10.5	
Wan-Yu, Rima	19.98N	31.43W	13.72	
Yangel', Rima	16.62N	4.79E	30.39	
Zahia, Rima	25.02N	30.46W	15.24	
Alphonsus, Rimae	13.4S	1.94W	87	

Rima/Rimae

FEATURE	LAT	LONG	SIZE (km)	NOTES
Apollonius, Rimae	4.39N	54.33E	89.64	
Archimedes, Rimae	26.34N	4.53W	215	
Aristarchus, Rimae	27.52N	47.25W	175	
Arzachel, Rimae	18.31S	1.38W	57	
Atlas, Rimae	46.82N	44.42E	46.8	
Bode, Rimae	9.54N	3.22W	233	
Boscovich, Rimae	9.87N	11.27E	32	
Bürg, Rimae	44.7N	25.27E	98	
Chacornac, Rimae	29.01N	31.24E	100	
Daniell, Rimae	37.53N	24.33E	140.25	
Darwin, Rimae	19.84S	66.66W	170	
de Gasparis, Rimae	24.99S	50.3W	46.65	
Doppelmayer, Rimae	26.23S	44.53W	7.8	
Focas, Rimae	27.68S	97.54W	61	
Fresnel, Rimae	28.11N	3.73E	75	
Gassendi, Rimae	17.47S	39.87W	70	
Gerard, Rimae	45.54N	84.36W	110	
Goclenius, Rimae	7.84S	42.88E	190	
Grimaldi, Rimae	6.18S	63.9W	162	
Gutenberg, Rimae	4.42S	36.42E	223	

Rima/Rimae

FEATURE	LAT	LONG	SIZE (km)	NOTES
Hase, Rimae	34.71S	67.78E	257.24	
Herigonius, Rimae	13.92S	36.75W	180	
Hevelius, Rimae	0.81N	66.38W	180	
Hippalus, Rimae	25.6S	29.36W	266	
Hypatia, Rimae	0.34S	22.78E	200	
Janssen, Rimae	45.8S	39.26E	120	
Kopff, Rimae	14.68S	88.1W	250	
Littrow, Rimae	22.47N	30.47E	165	
Maclear, Rimae	12.23N	19.9E	94	
Maestlin, Rimae	2.88N	40.48W	71	
Maupertuis, Rimae	51.24N	22.82W	50	
Menelaus, Rimae	17.1N	17.77E	87	
Mersenius, Rimae	20.69S	46.53W	240	
Opelt, Rimae	13.64S	18.14W	55	
Palmieri, Rimae	27.83S	47.17W	27.13	
Parry, Rimae	8.07S	16.52W	210	
Petavius, Rimae	25.23S	60.48E	110	
Pettit, Rimae	25.22S	93.63W	320	
Pitatus, Rimae	29.84S	13.62W	90	
Plato, Rimae	50.88N	3.02W	180	

Rima/Rimae

FEATURE	LAT	LONG	SIZE (km)	NOTES
Plinius, Rimae	17.05N	23.14E	100	
Posidonius, Rimae	32.03N	29.61E	78	
Prinz, Rimae	27.05N	43.51W	10.95	
Ramsden, Rimae	32.93S	31.32W	100	
Repsold, Rimae	50.74N	80.46W	152.05	
Riccioli, Rimae	1.52S	73.07W	250	
Ritter, Rimae	3.5N	17.97E	75	
Römer, Rimae	26.98N	34.86E	112	
Secchi, Rimae	0.99N	44.08E	35	
Sirsalis, Rimae	15.01S	61.36W	405	
Sosigenes, Rimae	8.08N	18.72E	130	
Sulpicius Gallus, Rimae	20.65N	9.99E	80	
Taruntius, Rimae	5.83N	46.83E	35	
Theaetetus, Rimae	33.04N	5.87E	53	
Triesnecker, Rimae	5.1N	4.83E	200	
Vasco da Gama, Rimae	11.55N	84.03W	10.39	
Zupus, Rimae	15.46S	53.76W	130	

Rupes

RUPES (Scarp)

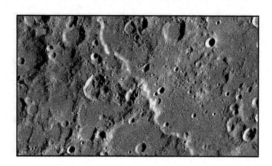

FEATURE	LAT	LONG	SIZE (km)	NOTES
Altai, Rupes	24.32S	23.12E	545.19	
Boris, Rupes	30.67N	33.6W	8.58	
Cauchy, Rupes	9.31N	37.08E	169.85	
Kelvin, Rupes	28.03S	33.17W	85.92	
Liebig, Rupes	25.14S	45.92W	144.78	
Mercator, Rupes	30.21S	22.84W	132.4	
Recta, Rupes	21.67S	7.7W	115.95	
Toscanelli, Rupes	26.97N	47.53W	50.14	

Sinus

SINUS (Bay)

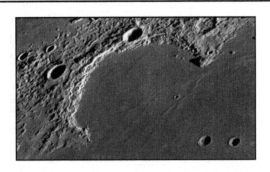

FEATURE	LAT	LONG	SIZE (km)	NOTES
Aestuum, Sinus	12.1N	8.34W	316.5	
Amoris, Sinus	19.92N	37.29E	189.1	
Asperitatis, Sinus	5.41S	27.49E	219.14	
Concordiae, Sinus	10.98N	42.47E	159.03	
Fidei, Sinus	17.99N	2.04E	70.7	
Honoris, Sinus	11.72N	17.87E	111.61	
Iridum, Sinus	45.01N	31.67W	249.29	
Lunicus, Sinus	32.36N	1.85W	119.18	
Medii, Sinus	1.63N	1.03E	286.67	
Roris, Sinus	50.26N	50.86W	195.04	
Successus, Sinus	1.12N	58.52E	126.65	

VALLIS (Valley)

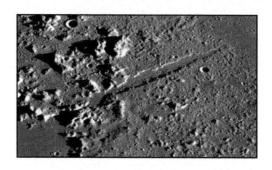

FEATURE	LAT	LONG	SIZE (km)	NOTES
Alpes, Vallis	49.21N	3.63E	155.42	
Baade, Vallis	45.55S	77.23W	206.79	
Bohr, Vallis	10.25N	88.86W	95.32	
Bouvard, Vallis	38.45S	82.32W	287.92	
Capella, Vallis	7.39S	35.04E	106.28	
Christel, Vallis	24.54N	11.08E	2.1	
Inghirami, Vallis	43.95S	72.59W	145.08	
Krishna, Vallis	24.5N	11.26E	2.9	
Palitzsch, Vallis	26.16S	64.64E	110.5	
Rheita, Vallis	42.51S	51.65E	509.07	
Schröteri, Vallis	26.16N	51.58W	185.32	
Snellius, Vallis	30.93S	57.84E	640	

Index

Feature	Page
Abbot	41
Acosta	356
Aepinus	289
Al-Bakri	548, 606
Aldrin	354, 436, 555
Al-Marrakushi	356
ALPO	54
Altai Scarp	215
Ammonius	519
Amontons	429
Ancient Newton	502
Anguis, Mare	199
Anorthosite	35
Anville	608
Apennine Bench	44
Apennine Mountains	43, 152, 411
Apollo 11	10, 436, 555
Apollo 11 - Astronaut craters	359, 436
Apollo 12	12, 357
Apollo 14	12, 225, 357, 478
Apollo 15	10, 56, 152
Apollo 16	14, 38, 161, 185
Apollo 17	10, 148
Apollo 17 (alternative location)	235
Arago - Alpha, Beta Domes	42
Aristarchus Plateau	47, 304, 525, 667
Armacolite	354
Armstrong	354, 436
Artsimovich	178
Asada	608
Auwers	424
Avery	267
Balmer-Kapteyn Basin	66, 297, 358, 389
Bancroft	44
Banding	54, 205
Banting	80
Barker's Quadrangle	493
Barnard	305
Beer	44
Bellot	155, 250
Bliss	502
Bombeli	41
Boulder - Group, Tracks, Trails, Rolls (in Craters)	28, 104, 157, 342, 384, 470, 588, 626
Bowen	407, 601
Caldera	306, 360
Cameron	608
Carillo	267
Carmichael	399, 516
Carrel	313
Cartan	41, 165
Cat's Paw	555
Cassini's Spot	187
Catena, Abulfeda	20, 605
Catena, Davy	171, 473
Catena, Humboldt	278
Catena, Krafft	128, 337
Catena, Littrow	148
Catena, Sylvester	603
Catena, Tauruntius	608
Cayley	168
Cayley Formation	28, 31, 181, 218, 542
Chain of Craters	472, 473, 482

Index

Feature	Page
Chamberlin	486
Chandrayaan 1 (MIP)	12
Chang'E-1	14, 379
Chang'E-3	8, 359
Chappe	497
Chladni	475
Cobra Head	47
Collins	354, 436, 555
Columbia Shuttle	111
Cometry Impact	473
Concentric Crater - Hesiodus A	292
Concentric Crater - Marth	415
Concentric Craters	155, 236, 292, 305, 363, 415
Concentric Craters (information about)	363
Condon	41, 652
Coordinates on the Moon	7
Crater Impact	3
Cryptomare	66, 297
Curtis	493
Curtiss Cross	225
Daly	41
Dark Crater Effect	313
Dark Halo Craters	33, 146, 306
Dark Layer Deposits	460
Daubree	424
Descartes Formation	31, 185, 542
Descartes Highlands	161
Dolland	20, 38, 185
Domes	432
Dome, Kies Pi	327
Dome, Valentine	382
Domes, Arago	42
Domes, Gruithuisen	258
Domes, Hortensius	302
Domes, Milichius	432
Dorsa, Aldrovandi	148
Dorsa, Barlow	640
Dorsa, Cato	608
Dorsa, Geikie	379, 429
Dorsa, Mawson	379
Dorsa, Smirnov	514
Dorsa, Stille	352
Dorsa, Tetyaev	199
Dorsa, Whiston	304
Dorsa, Zirchel	352
Dorsum, Beaumont	75
Dorsum, Buckland	601
Dorsum, Cayeux	608
Dorsum, Gast	43
Dorsum, Guettard	97
Dorsum, Niggli	304
Dorsum, Oppel	481, 602, 674
Dorsum, Termier	493, 580
Dove	386
Downrange	516, 622
Downthrown (slump)	172
Draper	523
Dry Debris Flows	408, 524, 548, 575
Dunthorne	415
Eastern Limb	16
Erlanger	119, 480
Esclangon	399

Index

Feature	Page
Farside	16
Faustini	462
Feuillée	44
Fibiger	119
Floor Fracture Craters	64
Fluidization	113
Formation, Cayley	28, 181, 218, 542
Formation, Descartes	31, 185, 542
Formation, Fra Mauro	181
Formation, Hevelius	169, 279, 293
Formation, Janssen	211
Foucault	581
Fra Mauro (Eta, Upsilon, Zeta)	225
Fra Mauro Formation	181
Fractures	499
Franck	107, 546
Fredholm	399, 516
Fresh Impact Crater	471, 478
Galen	43
Gargantuan Basin	450
Gaseous Features - Meniscus	589
Gerard Kuiper - Egg in a Nest	32
Gerard Q (Inner)	114, 244
Gerard Q (Outer)	244
Gibbs	278
Glaisher	516
Gore	220, 289
Graben	268
Grail Impacts	8, 489
Greaves	376, 493, 674
Grignard	289
Grove	417
Gruithuisen Domes	258
Hale	455
Hargreaves	652
Hausen	63, 497
Heinrich	619
Helmert	328, 659
Herringbone Signatures	239, 373, 523
Hesiodus A - Concentric Crater	292
Hevelius Formation	169, 279, 293
Hill	399
Hinshelwood	220, 480
Hiten	14, 228
Hi-Ti Basaltic Lavas	219
Hortensius Domes	302
Humboldtianum Basin	426
IAU (International Astronomical Union)	359
Ibn Battuta	250, 379
Ibn Yunus	308
Ibn-Rushd	161, 322
Imbrium Sculpture	25, 261, 477
Ina Structure	152
Iron (FeO) Abundances	66, 90
Ivan	515
Janssen Formation	211
Jeans	395
Jehan	499
Joy	43
Kaguya	14
Kao	328, 659
Kies Pi Dome	327

Index

Feature	Page
KREEP	56
Kreiken	328
Laccoliths	166
Lacus, Autumni	565
Lacus, Doloris	104, 407, 601
Lacus, Gaudii	407
Lacus, Mortis	116, 501
Lacus, Odii	407, 601
Lacus, Perseverantiae	57, 217
Lacus, Somniorum	167, 417, 501, 662
Lacus, Spei	130, 428, 570
Lacus, Temporis	130, 203, 426
LADEE	8, 200
Law of Superposition	667
Lawrence	608
Layering	80, 523, 625
LCROSS	12, 121
Light Ray Effect	321, 648
Lineations	294
Lobate Scarp	483
Lomonosov	318
Louville	581
Lovelace	603
LPD (Large Pyroclastic Deposits)	568
LRO (Lunar Reconnaissance Orbiter)	121
Luna 2	8
Luna 5	12, 357
Luna 7	413
Luna 8	8
Luna 9	8
Luna 13	8
Luna 15	10, 481
Luna 16	14
Luna 17	8
Luna 18	10
Luna 20	10
Luna 21	10, 368, 514
Luna 23	10, 481
Luna 24	10, 481
Lunar Orbiter 4	8
Lunar Orbiter 5	12, 565
Lunar Prospector	14
Lunar Reconnaissance Orbiter (LRO)	121
Lunar X	89
Lunokhod 1	8
Lunokhod 2	10, 368
Maestlin	202, 326
Magma Uplift	114
Magnetic Anomalies	185, 308
Mare Anguis	199
Mare Australe	14
Mare Crisium	10, 57, 146, 199, 376, 481, 493, 580, 602, 674
Mare Cognitum	12, 207, 225, 470
Mare Crisium	10, 57, 146, 199, 376, 481, 493, 580, 602, 674
Mare Fecunditatis	10, 14, 155, 250, 260, 376, 379, 574
Mare Frigoris	8, 10, 234, 502, 518, 590
Mare Humboldtianum	10, 426
Mare Humorum	12, 22, 235, 377, 476
Mare Imbrium	8, 44, 132, 178, 352, 359, 492, 502
Mare Insularum	206, 210, 239, 302, 359, 432, 449, 492, 523, 619
Mare Nectaris	14, 75, 236, 400
Mare Nubium	12, 32, 87, 171, 292, 360, 499, 614

Index

Feature	Page
Mare Orientale	12, 198, 275, 458, 487, 565, 579
Mare Serenitatis	10, 43, 80, 148, 167, 172, 514
Mare Smythii	267, 328, 552, 659
Mare Spumans	652
Mare Tranquillitatis	10, 42, 172, 307, 313, 394, 416, 436, 543, 548, 566, 589, 640
Mare Vaporum	10, 411, 673
Marth Concentric Crater	415, 526
Mary Blagg and Karl Muller	359
Melt Fractures	627
Meniscus Features	589
Milichius Domes	432
Mons, Delisle	178
Mons, Gruithuisen (Delta, Gamma)	258
Mons, Hansteen	85, 272
Mons, Maraldi	410
Mons, Penck	161, 322, 679
Mons, Piton	492
Mons, Vinogradov	210, 449
Mons, Vitruvius	640
Montes, Agricola	304, 461, 525
Montes, Apenninus	43, 152, 411
Montes, Carpatus	239, 523
Montes, Caucasus	122
Montes, Cordillera	198, 275, 565, 579
Montes, Haemus	104, 407, 606
Montes, Harbinger	515
Montes, Recti	359, 502
Montes, Riphaeus	207
Montes, Rook	198, 458, 487
Montes, Secchi	574
Montes, Taurus	456
Montes, Teneriffe	502
Morley	652
Mounds (in Saunder)	559
Naming Craters	359
Naumann	376
Nielsen	304, 667
Nomenclature	359
Norman	207
Northeast Quadrant	10
North Pole	17
Northwest Quadrant	8
Oceanus Procellarum	8, 12, 47, 562
Old Crater Near Sacrobosco	511
O'Neill's Bridge	516
Orientale Basin	198, 275, 458, 487, 565, 579
Palus, Epidemiarum	127, 142, 370, 415, 425, 526
Palus, Somni	394
Patrick Moore	529
Peek	604
Peters	50, 453
Petit	41, 652
Pit-like Craters	592
Polygonal Features	656
Posidonius Rille	514
Promontorium, Agassiz	132, 492
Promontorium, Archerusia	606
Promontorium, Deville	132, 492
Promontorium, Laplace	359
Promontorium, Taenarium	360
Pyroclastic Deposits	47, 306, 568
Pyroclastic Vents	250

Index

Feature	Page
Quadrant, Northeast	10
Quadrant, Northwest	8
Quadrant, Southeast	14
Quadrant, Southwest	12
Ranger 6	10
Ranger 7	12
Ranger 8	10, 354
Ranger 9	12, 33
Reiner, Gamma	533
Rille Junction (de Gasparis)	174
Rima, Archytas	518
Rima, Ariadaeus	23, 306
Rima, Billy	85
Rima, Bode	90, 475
Rima, Calippus	122
Rima, Cardanus	128, 469
Rima, Cleomedes	146
Rima, Conon	152
Rima, Dawes	172
Rima, Euler	449
Rima, Flammarion	218, 441
Rima, Furnerius	228
Rima, Galilaei	229
Rima, Gärtner	234
Rima, G. Bond	95, 268
Rima, Gay-Lussac	239
Rima, Hansteen	272
Rima, Hesiodus	142, 292, 327, 415, 425, 656
Rima, Hyginus	23, 306
Rima, Krafft	128, 337
Rima, Krieger	338
Rima, Marius	413
Rima, Oppolzer	218, 471, 528, 537, 575
Rima, Réaumur	528
Rima, Schröter	568
Rima, Sheepshanks	230, 582
Rima, Yangel'	673
Rimae, Arzachel	51
Rimae, Bode	90, 475
Rimae, Boscovich	100
Rimae, Bürg	116
Rimae, Charcornac	138
Rimae, Daniell	167
Rimae, Darwin	156, 169
Rimae, de Gasparis	377, 476
Rimae, Goclenius	260
Rimae, Grimaldi	166
Rimae, Gutenberg	260, 363
Rimae, Hippalus	22, 124, 295
Rimae, Hypatia	307, 436, 543, 555, 566
Rimae, Janssen	315
Rimae, Littrow	148
Rimae, Maclear	397, 548
Rimae, Menelaus	424, 606
Rimae, Mersenius	427
Rimae, Opelt	470
Rimae, Palmieri	476
Rimae, Parry	97, 225, 478
Rimae, Petavius	484
Rimae, Pettit	487
Rimae, Plinius	504
Rimae, Posidonius	514

Index

Feature	Page
Rimae, Prinz	515
Rimae, Ramsden	201, 526
Rimae, Repsold	231, 535
Rimae, Riccioli	539
Rimae, Ritter	543
Rimae, Römer	546
Rimae, Secchi	574, 608
Rimae, Sirsalis	156, 188, 221, 586
Rimae, Sosigenes	589
Rimae, Sulpicius Gallus	43, 601
Rimae, Triesnecker	537, 625
Rimae, Vasco da Gama	634
Rocco	338, 632
Roman Numerals	625
Rozhdestvenskiy	369
Running Man and Flashlight (Euclides)	207
Rupes, Altai	215, 510
Rupes, Liebig	377
Rupes, Recta (Straight Wall)	87, 614
Rupes, Toscanelli	623
Ruth	338, 632
Scarp	468
Scarp - Lobate	483
Schiller (lookalike - Piazzi Smyth V)	492
Schiller-Zucchius Basin	73, 331, 490, 576, 680
Schiller-Zucchius Basin Outline	550
Shoemaker	462
Simultaneous Impacts	46, 165
Sinus Aestuum	8, 90, 206
Sinus, Amoris	107
Sinus, Asperitatis	14, 307, 322, 400, 617, 679
Sinus, Concordiae	170, 394
Sinus, Honoris	397, 548
Sinus, Iridum	8
Sinus, Medii	10, 181, 218, 471, 475, 528, 575, 625
Sinus, Roris	59, 414, 590
Sinus, Successus	41, 652
Slocum	552, 650
Slumping (causes of)	252
SMART-1	12
Smithson	608
Somerville	68, 356
Sommering	441
Southeast Quadrant	14
South Pole	17
South Pole Aitken Basin (SPA)	366, 486
Southwest Quadrant	12
State of Craters (and other features)	495
Statistics of the Moon	7
Stewart	195, 508
Straight Wall	614
Strike-Slip Faults	268
Surveyor 1	12, 219
Surveyor 2	12
Surveyor 3	12, 357
Surveyors 4 and 6	8, 218, 471, 475
Surveyor 5	10, 354
Surveyor 7	12
Svedberg	183, 573
Swift	481, 602
Swirl, Airy	24
Swirl, Ibn-Yunus	308

Index

Feature	Page
Swirl, Reiner	533
Talbot	267, 552, 650
Ti-Rich Deposits	354, 601
Titanium (rich and poor deposits)	354, 601
TLP	33, 47
Tolansky	97, 478
Toscanelli	667
Townley	41, 652
Transient Lunar Phenomena	33
Trapezium	493
Tucker	650
Väisälä	47
Valentine Dome	382
Vallis, Alpes	518, 626
Vallis, Capella	126
Vallis, Palitzsch	474
Vallis, Rheita	532, 675
Vallis, Schröteri	290, 525
Vallis, Snellius	587
Vera	515
Von Baeyer	183, 573
Wallach	416
Wapowski	462, 573
Werner Rays	657
Western Limb	16
Whewell	168, 611
Whipple	480
Wildt	67
Winthrop	371
Wurzelbauer Rille	670
Yakovkin	498
Yutu (Chinese Lunar Rover)	8, 359
Zähringer	608
Zone of Avoidance	516

CPSIA information can be obtained
at www.ICGtesting.com
Printed in the USA
LVOW03s2127130816
500271LV00018B/461/P